PLANTS AND PEOPLE OF **POHNPEI**

Plants and People of Micronesia

PLANTS AND PEOPLE OF POHNPEI

Mark Merlin, Danko Taboroši, James Juvik

iREi

Pohnpei 2026

USP Library Cataloguing-in-Publication Data

Merlin, Mark.

Plants and people of Pohnpei / by Mark Merlin, Danko Taboroši and James

Juvik. -- Palikir, Pohnpei : Island Research & Education Initiative, 2026.

472 p. : col. ill. ; 28 cm.

(Plants and people of Micronesia)

Includes bibliographical references and index.

ISBN 978-982-9170-41-5

1. Ethnobotany--Micronesia (Federated States)--Pohnpei Island.

2. Botany--Micronesia (Federated States)--Pohnpei Island.

3. Plants--Micronesia (Federated States)--Pohnpei Island.

4. Traditional ecological knowledge--Micronesia (Federated States)--Pohnpei Island.

5. Pohnpei Island (Micronesia)--Social life and customs.

I. Taboroši, Danko. II. Juvik, James. III. Title.

GN476.73.M47 2026

581.9966--dc23

Published in 2026 by

Island Research & Education Initiative (iREi)

PO Box PS 303, Palikir, Pohnpei, FM 96941

Federated States of Micronesia

www.islandresearch.org

We dedicate this book

to the youth of Pohnpei State today

and to the future generations yet to come.

It is our hope that they will safeguard and build

upon the traditional resource knowledge passed down

through centuries of sustainable island living.

Above all, we express enduring gratitude to the ancestors

of the Micronesian peoples, who adapted with wisdom

and resilience to life on small islands over thousands of years.

May we continue to learn from their remarkable example,

and may the people of these islands thrive—

drawing upon their deep and shared knowledge

to sustain both themselves and their environment,

now and for generations to come.

TABLE OF CONTENTS

PREFACE

Due to the widespread socioeconomic changes, the influence of modern technology, and increasing global interconnectedness, indigenous people around the world are losing much of their traditional ecological knowledge. In Micronesia and specifically in Pohnpei State, increasing affluence, urbanization, and reliance on imported foods, along with the migration to the urban centers in the region and to foreign countries, especially the United States, have resulted in much of this loss of traditional resource and conservation management practices and enforcement. This precious knowledge is the extraordinary legacy of countless generations whose environmental understanding has been the basis of their resilience and sustainable adaptation to limited natural resources in geographical isolation. In recent decades, a lot of work has been done to reapply this traditional knowledge in modern resource management, with a focus on reviving indigenous concepts of conservation—often community-specific rules about the harvesting of resources on Pohnpei and its outlying atolls. This presents opportunities to develop and build upon a revival and preservation of traditional knowledge, and at the same time, introduce relevant and culturally acceptable modern management concepts to the communities in the islands of Pohnpei State.

Therefore, an urgent need exists for more enthusiastic and well-trained indigenous scientists and resource managers to study, conserve, and restore tropical ecosystems in Pohnpei, Micronesia, and beyond. In the Pacific Islands in general, too few local students are getting "turned on" by science and electing to pursue education leading to careers in applied natural sciences—i.e., agriculture, forestry, fish and wildlife management, protected area management, and soil and water conservation. There is substantial agreement that part of the problem is the lack of published resources that educators and students can turn to for reference and inspiration. Textbooks used by teachers in local schools are often unsuitable because they are usually based on continental, temperate-zone examples and materials imported from other countries, rather than tropical, place-based island materials that are more locally relevant to Pacific island students.

"Plants and People of Pohnpei" is the result of the authors' efforts to produce a useful and inspiring educational introduction to the ethnobotany and ecology of Micronesian islands. It derives from over three decades of research and collaborative work of many people, whose dedication and contributions are recognized in the Acknowledgments section on pp. 12–13. The key purpose of this book is to encourage the young people of the Federated States of Micronesia (FSM) to pursue careers in environmental science and conservation, with the further goal of sustainable natural resource use and preservation of biodiversity in the Pacific. The text aims to make these topics relevant by focusing on the relationship between the people of Pohnpei with the island environment and plant life, using local examples to illustrate scientific concepts. Providing numerous examples of how plants are used by the people of Pohnpei State, we hope to make clear the universal truth of human relationship with our planet:

Cultural diversity goes hand in hand with biological diversity

We have also included proverbs, legends, and stories alongside rich scientific information, aiming to show that careful observation of the environment and a desire to understand it are integral elements of island culture. All of that has led to the book you hold in your hands being relatively large, but keep in mind that it is in fact small in comparison to the richness of Pohnpei's environments and the sophistication of Pohnpeian knowledge of the organisms that inhabit them.

We hope this book will serve as an indispensable and accessible reference for students and teachers in Pohnpei State, supporting not only the public school system but also resource managers, government administrators, and anyone interested in science and life in Pohnpei, the wider Micronesian region, and the Pacific. It is our aspiration that the information shared here—about the deep and dynamic relationships between people and plants in Micronesia—will benefit all Pohnpeians, residents of outlying atolls, visitors to Pohnpei State, and other interested readers. We envision a future where Pohnpeian and other Micronesian scholars expand upon this work, producing increasingly complete accounts of local plants and environments, in tandem with a progressive educational program that empowers students to gain and apply traditional knowledge in sustainable and informed ways. Above all, we hope that the rich environments of Pohnpei, Micronesia, and the Pacific—and the equally rich cultural wisdom developed over centuries, used sustainably, and now being passed on to future generations—will continue to be explored, preserved, and celebrated, and will forever sustain and inspire the people of these islands.

ACKNOWLEDGEMENTS

The inspiration for this book—and much of the foundational information it contains—traces back to the early 1990s, when the John D. and Catherine T. MacArthur Foundation awarded a grant to the Environment and Policy Institute (EAPI) of the East-West Center in Honolulu, Hawai'i. The goal was to support an initiative promoting the protection and sustainable use of tropical island ecosystems. As part of that effort, an educational module focused on the plants and ecosystems of Pohnpei State was created by a team of authors from Hawai'i and Pohnpei, working closely with a group of respected Pohnpeian elders organized by the Pohnpei State Department of Education. These local experts provided invaluable insights and guidance rooted in traditional environmental knowledge: Ioanis Paulino, legends and oral literature specialist from Kitti; Ihper Dison, specialist in chiefly matters of Sokehs; Ioakim David, traditional medicine specialist from U; John Kalio, deer hunter and plant specialist from Madolenihmw; and our project adviser, Dageo Jano, cultural education specialist. Their enthusiasm and expertise were instrumental in bringing the project to life. Additional support came from Teddy John, FSM Office of Historic Preservation; Spencin James, Biology Instructor, College of Micronesia; and Herson Anson, Chief Forester, Pohnpei State Department of Forestry. As the material evolved into a book, important editorial contributions were made by Dr. Len Newell of the US Forest Service; Dr. Kenneth Rehg of the University of Hawai'i; and Alice Kawakami and Rodrigo Mauricio. Coordinated by the late Lawrence S. Hamilton of EAPI and with Dr. Mark Merlin as lead author, the book was published under the title "*Tuhke en Pohnpei*—Plants of Pohnpei" in 1992. It was one of six textbooks produced for a region-wide initiative in environmental education and resource management across Micronesia.

Photo taken in Kolonia, Pohnpei, in 1992 of the core team who created the original "*Tuhke en Pohnpei*—Plants of Pohnpei" book. From left to right: Mark Merlin (lead author), Tom Keene (co-author), and the local experts Ioanis Paulino, Ihper Dison, Dageo Jano, Ioakim David, and John Kalio (see text above for their respective roles in our collaborative research).

Though well-received by educators, the book was never widely adopted and gradually became a rarity—now treasured in libraries across the region.

Building upon that pioneering work, we created the new, greatly expanded and updated book you now hold. It was published in 2026 under the title "Plants and People of Pohnpei" with a modern layout, including numerous color photographs and original illustrations created specifically for this project. Dr. Mark Merlin, faculty member for more than 50 years at the University of Hawai'i at Mānoa, was again the lead author, this time joined by Dr. Danko Taboroši, director of Island Research & Education Initiative (iREi), and Dr. James Juvik, Professor emeritus at the University of Hawai'i at Hilo. Many Pohnpeians and residents of Pohnpei State generously contributed their traditional environmental knowledge, and we are deeply grateful to all who participated—everyone who shared their cultural and intellectual heritage, as well as the photographers, illustrators, designers, and others who helped bring this ambitious project to completion.

More than just a new edition, this book introduces significant updates and original research—most notably, a comprehensive study of local plant names that merged ethnobotany with linguistics. Led by Dr. Taboroši, this work was made possible through years of collaboration with local experts who contributed countless hours to explaining the intricacies of languages spoken on different islands and atolls comprising Pohnpei State. We are especially indebted to Damian Sohl, Robert Andreas, and Rodrigo Mauricio for Pohnpei; Taylor Sehpin and Kenio Frank for Sapwuahfik; Salich Albert and Kalio Lebehn for Mwoakilloa; Weldis Welley and Leilani Ilai for Pingelap; Betty Amon and Benjamin Ludwig for Nukuoro; and Kalio Dikepa and Edgar Lickaneth for Kapingamarangi. Although some of these knowledge-holders are no longer with us, their love of language and commitment to sharing knowledge continue to inspire.

We are also deeply thankful to everyone who worked with iREi during interviews and consultations to help determine and refine the use of specific plant names on different islands throughout Pohnpei State: Miller Benjamin, Tony Sarapio, Nelsin Iriarte, Adelino Lorens, Ryan Ladore, Tony Santos, Francisca Obispo, Bejay Obispo, Epikenio Eperiam, Eugene Eperiam, Bersin Moya, Primo Eperiam, and Lorenzo Gilimete for Pohnpei; Kapino Neior and Albihno Sallel for Sapwuahfik; Ohner Jim, Gibson Mason, Ketchem Harry, Dahi Oliver, Emihner Johnson, and Elna Julius for Mwoakilloa; Ipraim Capelle, Mesaro Solomon, Kensily Boas, Kinchiro Boas, Snyther Biza, Leilani Welley-Biza, Kohsak Keller, Abinor Robert, Ehlis Ernest, and Mwohden Boas for Pingelap; Itaia Fred, Tenny Leopold, Sigfrit Gideon, Johnny, Rudolph, Curtis Charley, Andelise Fred, Serah Charley, and Blumina Leopold for Nukuoro; and Billy Mateak, Stiwan Duiai, Peter Lick, Jano Siproy, Seckson Potek, Skiller Duiai, Simel Lucky, Sheila Duiai, and Mersihda Duiai for Kapingamarangi.

Many of these local contributors collaborated with Mandy Flowers, volunteer assigned to iREi by Australian Volunteers International. Over two years in the FSM, she traveled to and conducted fieldwork on nearly every island and atoll in Pohnpei State and throughout the nation, playing a vital role in the gathering of plant names and cultural insights. We also thank Euclene Joseph, iREi's office manager in Pohnpei, who helped clarify many additional plant names through interviews and correspondence.

Special appreciation

Finally, we express our sincere gratitude to Pressler Martin, Acting Director of the Pohnpei State Department of Education at the time this project began. He recognized the value of this work and imagined it supporting students and teachers across Pohnpei State.

This book honors the lifelong dedication of William "Bill" Raynor and his family for their enduring efforts to protect the Pohnpeian environment and preserve traditional ecological knowledge. We also acknowledge Dr. Michael Balick and his many local and international collaborators for their decades-long commitment to documenting the rich ethnobotanical traditions of the Pohnpeian people. Special recognition is extended to Dana Lee Ling for his longstanding contributions to education and for planting awareness and appreciation of plants and their uses in the hearts and minds of so many Micronesian students.

INTRODUCTION

The islands of Pohnpei State and the wider Micronesian region are home to a wealth of plant species and traditional ecological wisdom about them. The knowledge of how to care for and use plants was passed down through generations and enabled island communities to live sustainably and in harmony with their environment. Today, however, this invaluable knowledge is under threat as life in the islands changes. With many young people learning fewer traditional skills, and even moving far away from their communities, much of the understanding that once guided resource use, conservation, and daily survival is at risk of being lost. Amidst these changes, the need to preserve and revitalize indigenous knowledge systems—especially those related to plants, ecosystems, and cultural practices—has never been more urgent.

To help address this urgent challenge, *Plants and People of Pohnpei* was created as a resource for education, inspiration, and cultural preservation. Drawing on decades of research and collaboration, this book introduces the ethnobotany and ecology of Pohnpei in ways that are meaningful to local students and educators. It celebrates the profound connection between islanders and their natural surroundings, illustrating scientific concepts through local examples and traditional wisdom. By blending cultural stories, practical plant uses, and ecological insights, the book seeks to inspire a new generation of Micronesian environmental stewards—young people who will cherish their heritage and apply both traditional and modern knowledge to care for their islands and shape a sustainable future.

PLANTS AND ORIGINS OF POHNPEI

Pohnpeian legends and ancestral stories emphasize the deep connection between people, land, and plants. Early origin myths highlight the role of plants in protecting land from being washed away (see Merlin et al. 1992: 1).

> "When the first canoe came to what is now Pohnpei, the voyagers worked to make land. They called together stones, and made a level place. Some say this level place was an altar or that the land was sacred. The waves splashed against the land, and began to wash or erode it away. The people called **Katengenihk**, the mangroves, to strengthen the shore and protect it from the sea. But this was not successful as the open sea was too close to the edge of the land. Then they called **Katengenioar**, the barrier reef, to come and surround the land. Together, the reef and the mangroves act as a belt around the island holding it together."

The **Aiau** tree (*Ficus prolixa*) made the land stronger. This large, sprawling tree, known for its many hanging prop roots, typically grows on mountain ridges. It is said to have been one of the first plants to appear on Pohnpei, starting from Kitti:

> "The trunk grew taller and taller and the branches fell over and apart toward what is now Sokehs and Kitti, and that is why that area is now fruitful with food (fruits coming from branches). The roots of the **Aiau** spread toward Madolenihmw and sprouted. It thus held the land together from north to south."

The woody vine, or liana, **Kehoaroahr** (*Derris trifoliata*), also helped strengthen the island. This slender climber can grow very long and, under certain conditions, blanket the ground and cover surrounding vegetation. Found near Pohnpei's shores, it is said to help hold the island together.

These Pohnpeian stories highlight the vital role of native plants in protecting the land. They also stress the importance of safeguarding these plants, which are essential for various purposes. The native plants of Pohnpei today are descendants of those that arrived on the island without human assistance. However, to ensure their survival and the continued safety of Pohnpei, they may now need our help.

NOTES ABOUT LANGUAGE

Although this book is written in English, it honors the indigenous languages of Pohnpei State through the extensive use of local vocabulary. Pohnpei State is a multilingual region, with the primary language being Pohnpeian—spoken as a first language by nearly 25,000 people. Native to the island of Pohnpei, this language exists in two main dialects: the Kitti dialect, spoken in the southern part of the island, and the Northern dialect, spoken in other areas.

The languages of the outlying atolls differ from Pohnpeian to varying degrees. The closest is the language of Sapwuahfik Atoll, considered a dialect of Pohnpeian with distinct features. The languages of Mwoakilloa and Pingelap are also related to Pohnpeian but exhibit greater differences. In contrast, the languages spoken on the atolls of Nukuoro and Kapingamarangi are very distinct. As members of the Polynesian language group, they stand apart from all other languages in the FSM and are considered linguistic outliers within Micronesia.

When not clear from context, non-English words in this book may be followed by an abbreviation or color code indicating their language of origin, as shown below:

P = Pohnpei
Mw = Mwoakilloa
Pi = Pingelap
Nu = Nukuoro
Ka = Kapingamarangi

Irrespective of the language, all non-English vocabulary in the book is indicated through a combination of *italics* and **boldface**. Specifically, **Plant names** are in bold with an initial capital letter, and ***Cultivar names*** are also italicized. All other *local terms* are italicized only. However, when quoting other sources, non-standard or outdated spellings are not set in bold and are placed within single quotation marks. In this way, we indicate that a quoted 'Plant name', '*Cultivar name*', or other '*local term*' does not follow orthographic conventions.

NOTES ABOUT PLACE NAMES

In this book, place names are spelled according to the actual pronunciations and preferences of local people. Over the past several decades, Micronesian languages have transitioned away from colonial-era spellings—often imposed by outsiders—toward modern orthographies developed by, or in collaboration with, indigenous language experts. As a result, the way many words are written has changed, including the names of places. For instance, we now use the modern spelling 'Pohnpei' instead of the outdated 'Ponape', and 'Mwoakilloa' in place of 'Mokil'.

Throughout this book, we adopt the most current and phonetically accurate spellings for place names. However, in cases where local governments use official or widely established spellings that differ from actual pronunciation, we follow those standards. For example, we write 'U' instead of more accurate 'Uh'; 'Nukuoro' rather than uncommon but linguistically correct 'Nuguolo'; and 'Kapingamarangi' instead of the local rendition of 'Kilinid'.

Another important consideration is that some place names have multiple meanings. For instance, before the local government officially adopted 'Sapwuahfik' as the name of the atoll, the entire atoll was commonly known as 'Ngatik'. Today, however, 'Ngatik' refers only to the atoll's main islet. Even the name 'Pohnpei' can be ambiguous—while it should refer only to the island itself, it is sometimes used to describe the entire state of Pohnpei, which includes both the main island and its outlying atolls. In this book, we make that distinction clear: except in the title, which we deliberately kept simple, we use 'Pohnpei State' to refer to the entire state and reserve 'Pohnpei' for the main island. The outlying atolls are identified by name, or collectively referred to as the "outer islands," which is the common term used in everyday speech across Micronesia.

NOTES ABOUT PLANT NAMES

Indigenous languages of Pohnpei State, including Pohnpeian and languages of the outer islands, contain thousands of names for different plant species and their varieties. Despite numerous efforts by scholars and community members to document this over many decades, we faced significant difficulties in pinpointing and selecting plant names to use in this book.

Part of the problem is that many early records of plant names were made by individuals focused on their own fields—perhaps linguists unfamiliar with local flora, or botanists unfamiliar with local languages. That has led to imprecise or incorrect identifications, or at least errors in associating species with their corresponding local names. For example, an author who wrote several books about the Marshall Islands inadvertently switched the Marshallese names of different species in at least three of his books. Over the years, erroneous names that made their way into literature tended to be perpetuated by new researchers as they cite older sources. That has created a situation in which some plants have numerous and often inaccurate names proliferate in literature: multiple versions of the same name, oversimplifications (when more familiar generic names are reported, even though less common and more specific names exist), and even some descriptive terms being recorded when an informant was unsure ("looks like...", "a kind of..."), but the researcher unfamiliar with the language noted and assigned it to a particular plant as its real name.

Even when botanists did record the correct name, they tended to spell it arbitrarily, leading to a confusing range of spellings found in literature today. For example, St. John (1948) and Glassman (1953) both noted local names of plant species and cultivars in Pingelap. They used impromptu spellings, writing Pingelapese words "as they sounded to an American" (St. John 1948: 99) and "spelled exactly as they sounded to the writer" (Glassman 1953: 292). Working just five years apart, these authors faced significant issues when attempting to spell Pingelapese words using English-like rules. For example, Glassman built on St. John's work and expanded his list of local pandanus cultivars by three types, which we can now see are just different subjective spellings of the same names: St. John's '*Nanagaisal*', '*Sonumei*', and '*Aisesewil*' are Glassman's '*Nanagaisak*', '*Sonomuneyah*' and '*Eisesieu-el*'. These renditions are linguistically inaccurate and obsolete now that Pingelapese, like all other indigenous languages in Pohnpei State, has a defined orthography and phonetic spelling system (see, for example, Hattori 2012: 26–31).

In this book, we use modern spellings for all plant names (and other vocabulary) in all languages, except in quoted text or when we could not recognize a word to replace it with a preferred and orthographically correct version. Students and researchers working with names of species and cultivars are advised to ignore the largely arbitrary spellings in older sources (or new sources that copy from them) and strive to determine and record the true names as actually pronounced in different islands today. Ideally, these actual names pronounced by native speakers should be written according to the current spelling rules of a given language, as used in dictionaries and taught in local schools.

While writing this book, we undertook extensive research of plant names, drawing on dictionaries, archival sources, and interviews with local experts (see p. 13). After carefully evaluating the information, we selected the most widely recognized names and presented them using the standardized orthographies of the respective languages. Determining the best names to use for different species in each of the five languages indigenous to Pohnpei State and using them consistently has been a major challenge. To tackle it, we first examined all recorded names we could locate in literature and cross-checked them against dictionaries and other linguistically sound sources. Comparing the published information with data we collected, for all local plant species and across languages, allowed us to clarify doubtful items, find gaps, record missing names, and draw conclusions regarding the names to use in this book. To give you a better idea of how we went about the whole process, we list here the specific workflow and rules we followed:

- Compiled lists of all local names available in published sources.
- Created a photo-book with 300+ plant species images and spent hundreds of hours interviewing local informants to identify illustrated species.
- Made audio recordings of the names pronounced by local experts and transcribed them.
- Built a database comparing plant names from published references with those obtained from all our informants, and then evaluated the reliability of each name based on how often it is corroborated by different sources.
- Cross-referenced plant names and compared them across languages to identify possible errors, keeping in mind that similar names obtained independently from linguistically related languages increase the likelihood of names being correct, whereas very disparate names are considered suspect until confirmed.
- Identified descriptive phrases mistakenly recorded as formal names and removed them (e.g., *Timonius ponapensis*, whose Pohnpeian name, **Kehn**, was incorrectly documented in some sources as '*tuhke duwehte kamal*'—literally, "plant similar to *Pittosporum ferrugineum*").
- Noted that many introduced plants lack indigenous names, though some publications have recorded descriptive phrases in their place (e.g., the unwieldy '*tuhke wah pwonopwon duwehte tehn sele*', meaning "plant resembling a round pepper," used for *Physalis angulata*).

Physalis angulata

- Rendered names of foreign origin, typically loanwords from English, so that their spelling matches local pronunciations and orthography (e.g., *Gossypium barbadense* as **Koatun** rather than "cotton").

Gossypium barbadense

- Merged names when closely related species are named in the literature with slightly different spellings copied from different sources and suggesting that they are distinct names, when in fact they appear to be one and the same word applied to closely related plants (e.g., **Delahrik** sometimes assigned to *Ludwigia hyssopifolia* and **Deleurik** to *L. octovalvis*).
- Identified generic names or names applied to many different species and replacing them with more specific names whenever possible.
- Opted for the most unique and least-shared names when multiple names are available for a particular species (e.g., a plant has three recorded names, but two are shared with other plants).
- Worked to fill data gaps identified from known local names (in dictionaries and other sources) without matching scientific names and vice versa.
- noted that different species can share the same Pohnpeian name—not only when they are closely related, such as members of the same genus, but

even when they are quite distinct and grow in very different habitats. For example, the name **Pwoaki** refers to two sedge species with similarly long, pandanus-like leaves: *Thoracostachyum pandanophyllum* (formerly *Mapania pandanophylla*), which grows in the wet upland forests at high elevations, and *Scirpodendron ghaeri*, which forms dense thickets in lowland freshwater swamps.

Thoracostachyum pandanophyllum
Pwoaki (P)

Scirpodendron ghaeri
Pwoaki (P)

- When identifying gaps, the existence of a Pohnpeian name did not imply that a name must exist in any language from the outer islands language and vice versa (since flora greatly differs between Pohnpei and atolls); however, when a name exists in at least one language from the outer islands, we intensively worked with local informants from other atolls to pursue local names that may be in use there (since flora is often similar between atolls).

- Paid less attention to plants that exist in Pohnpei State but for which no name could be found in Pohnpeian or any language in the outer islands, under the assumption that such plants are not deeply significant to local people or are very recent introductions.

- Eliminated or corrected name versions that violate spelling rules of a particular language (e.g., a name in Pohnpeian cannot have an 'h' after a consonant, except for the special 'h' sound found only in the Sapwuahfik dialect; a name in Nukuoro cannot contain letter 'c', 'f', 'w', or any other not found in their alphabet, etc.).

- Corrected spellings of names to correspond with versions available in dictionaries; if not available, but pronunciation is known or inferred, spelled out the names according to the standard orthography of a given language.

- Included names in Kitti dialect when the difference from the Northern dialect cannot be inferred based on the predictable sound changes (specifically, 'e' to 'oa' in Kitti, and 'oa' to 'o' and 'r' to 'rh' in Sapwuahfik).

- Harmonized spellings for pairs of introduced and native plants that share a base name due to perceived similarities—a pattern in Pohnpeian where foreign plants typically take the native plant's name with the suffix "*en wai*" (meaning "foreign"). For example, **Seir** (*Fagraea berteroana*) and **Seiren wai** (*Cananga odorata*) form such a pair, both recognized for their fragrant scent. In cases where introduced plants become particularly common, the native species may instead adopt the suffix "*en Pohnpei*" to emphasize its indigenous status.

- Similarly reconciled spellings among groups of native species that share a base name due to perceived similarities, but are distinguished by specific traits or habitats. For example, **Mahrek** refers to the ferns *Cyclosorus maemonensis* and *Cyclosorus heterocarpus*, while **Mahrek en nahna** designates the endemic

Cyclosorus maemonensis
Mahrek (P)

Cyclosorus ponapeanus
Mahrek en nahna (P)

high-elevation species *Cyclosorus ponapeanus*, and **Mahrek en leng** applies to *Blechnum orientale*. These names are occasionally extended to other ferns, including as descriptive references when no dedicated name exists, such as *Lindsaea repens* being recorded as '*tuhke duwehte mahrek*' ("plant same as **Mahrek**") in some publications.

Clearly, our efforts to draw conclusions regarding plant names to use in this book would have been impossible without the contributions of a great many people, whose contributions we gratefully acknowledge (see p. 13). In addition, we heavily relied on all available dictionaries for Pohnpeian and other languages of Pohnpei State. These have been incredibly useful references throughout the project and include especially Rehg and Sohl (1983) and Sohl, Rehg, and Andreas (2025) for Pohnpei and Sapwuahfik; Harrison and Albert (1977) for Mwoakilloa; Good and Welley (1989) and Hattori (2012) for Pingelap; Carroll and Soulik (1973) and Drummond et al. (2016) for Nukuoro; and Lieber and Dikepa (1974) for Kapingamarangi. Besides the dictionaries, we also consulted a number of other published sources, such as the checklist of Pohnpei plants (Herrera et al. 2010), ethnobotany of Pohnpei (Balick et al. 2009), and the compendium of Micronesian plant names (Falanruw et al. 1990). We also consulted the few available plant checklists for the outer islands (such as St. John 1948, Glassman 1953, Niering 1956, Stone 1966a). These early records from the atolls were useful because they preserved some names that are nearly forgotten today. While the spellings in those documents are highly approximate and not up to standards, they did help our informants recall the less commonly used names and allowed us to obtain correct pronunciations and spellings. As mentioned previously, when obsolete renditions of plant names are cited in this book, they are in single quotation marks. For example, 'Morishish' of Glassman (1953) corresponds to **Moarjij**, contemporary spelling of the local name on Mwoakilloa Atoll for the plant whose scientific name is *Wollastonia biflora*.

Despite our best efforts, the local plant names used in this book are not definitive and perhaps could never be. The richness of languages and dialects in Micronesia means that plant names vary a great deal between different places, not only between islands, but even between villages on the same island. A name listed in Pohnpeian, for example, may be familiar to some but not all speakers, as people in different parts of the island may know only their local variants. Thus, all local names in this book are offered as resources and references for further exploration, not as the only correct forms. Readers of this book are advised to treat them as such, keeping in mind that the book may also contain erroneous names, though we tried very hard to ensure accuracy.

For more on the origins and significance of local plant names, as well as for more information on scientific nomenclature, please refer to Chapter 5.

1

THE PHYSICAL GEOGRAPHY AND POPULATION OF POHNPEI STATE

CHAPTER 1

THE PHYSICAL GEOGRAPHY AND POPULATION OF POHNPEI STATE

This chapter describes the physical geography of Pohnpei, with summaries of its geographical location, geological history, climate, and soils. We also explore the remarkable diversity of resource areas on Pohnpei and look into the island's population and its changes over time.

1.1. GEOGRAPHY AND GEOLOGY

Pohnpei State is one of the four states that make up the Federated States of Micronesia (FSM), an island nation in the tropical Western Pacific. The FSM also includes three other states—Yap, Chuuk, and Kosrae—each subdivided into smaller political units with their own local governments. The FSM's numerous atolls and high islands are spread out over 2,500 km (1,500 mi) from east to west, spanning a latitude range from 1°N to 12°30'N and a longitude range from 135°E to 165°E. Table 1.1 below provides a summary of the size and population of each state.

Pohnpei State is named after its main island, Pohnpei. It is a large volcanic island with about 345 km^2 (133 mi^2) of land area. Located approximately 760 km (470 mi) north of the equator, it sits around 7°N and 158°E. The island is roughly circular in shape, with coves and jutting peninsulas. Distances across the island average about 24 km (15 mi). Pohnpei's mountainous interior contains rugged ridges and deep valleys. Approximately half of the island is composed of steep slopes. Eleven peaks reach elevations above 600 m (1,970 ft). The highest are Nahnalaud at 782 m (2,566 ft) and Ngihneni at 772 m (2,533 ft). They are also the highest in all of Micronesia.

Surrounding Pohnpei (except on the southeast side) is a large lagoon with a total surface area of about 179 km^2 (70 mi^2). It reaches about 100 m (328 ft) in depth. The lagoon is enclosed by an outer barrier reef, lying on average 3.2 km (2 mi) from the coast. More than 25 smaller islands of volcanic, coral, and mangrove origin are found in the lagoon.

Pohnpei State also encompasses nine coral atolls scattered in the ocean around Pohnpei. Listed in order of increasing distance from Pohnpei, which ranges from only a few kilometers to over 700 km (435 mi), these atolls are And, Pakin, Sapwuahfik, Mwakilloa, Pingelap, Oroluk, Minto, Nukuoro, and Kapingamarangi. They have lagoons of varying sizes. The smallest

Table 1.1. Land area, population, and population density in the four states of the FSM.

State	Political units	Land area	Population (2021)	Population density (persons/km^2)
Yap	19	119 km^2 (46 mi^2)	11597	97
Chuuk	40	127 km^2 (49 mi^2)	49595	391
Pohnpei	11	345 km^2 (133 mi^2)	36896	107
Kosrae	4	110 km^2 (42 mi^2)	6744	61

Map of Pohnpei State. Its location is shown in the globe inset.

Map of Pohnpei.

lagoon is on Pingelap. It measures approximately 1.5 km² (0.57 mi²). The largest lagoon by far is on Oroluk; it covers well over 400 km² (155 mi²).

The number of islets and the land area available on the atolls also varies greatly. With a land area of zero, Minto is the only atoll in the FSM which has no land at all. Huge Oroluk has a single tiny islet, whereas Nukuoro has about 45 islets lined up like beads on a string. With its three islets around the small lagoon, Pingelap is the only atoll in the FSM whose land area exceeds its lagoon area. In total, the combined land area of all the atolls in Pohnpei State reaches only about 8 km² (3 mi²).

Geological research has previously indicated that Pohnpei originated from a volcanic hot spot (melting anomaly), similar to the high islands in the chain of Hawai'i. A more recent study suggests that Pohnpei's volcanic history is more complex and may be the result of both hot-spot activity and fracture-induced subduction-related tectonics. In either case, Pohnpei is the top part of a large shield volcano located mostly under water and about 8.5 million years old. This large volcano was originally produced by thousands of lava flows building up on the ocean floor many thousands of meters below the ocean surface. The upper, comparatively tiny section of this volcano, today stands above sea level as the island of Pohnpei. Because of both subsidence (the sinking of the island due to the massive weight of the volcano) and the sea level rise during the last several thousand years, there are few coastal plains and no natural beaches on Pohnpei. In contrast, the coastal areas support strand vegetation and well-developed mangrove swamps.

Coastal mangroves are endangered ecosystems highly vulnerable to sea level rise, yet they serve as Pohnpei's strongest natural defense against its impacts. To date, the island has experienced minimal coastal erosion, likely due to mangroves acting as vital buffers that trap sediment and dissipate wave energy. Fringing mangrove ecosystems could potentially keep pace with projected sea level rise if they have space to expand inland and continue to trap ample sediment supplies. Safeguarding Pohnpei's future requires a thorough assessment of mangroves' vulnerability to climate change and proactive measures to ensure their survival.

Currently, there is a long-term trend of rising sea levels in the tropical Pacific Ocean caused by global warming and the melting of polar ice caps and glaciers. The rate of sea level rise in recent years is reported to be 3.4 mm/yr (0.13 in/yr). At this rate, the sea level could be expected to rise by about 270 mm (11 in) by the end of the 21st century, thus adding another destructive component to tidal surges and storm wave inundations in populated coastal areas. Sea level rise is of great concern, particularly for the low-lying atolls.

The abundance of rainfall on Pohnpei, acting as an erosional force over millions of years, has greatly weathered the volcanic rocks. This has created deep soils in many places, and a generally rugged, dissected upland landscape. More than 40 streams and rivers on the island can reach flood levels after torrential rainfall events. Although some rainfall seeps into the ground through the permeable soil, a remarkably large percentage (65–70%) of the surface water runs off to the sea, providing a source of the sediment and nutrients for the mangroves.

High islands are volcanic islands with mountains, ridges, and valleys. They reach significant elevation above sea level.

Low islands form from coral sand accumulated on atolls or other reefs. Flat and low-lying, they reach very little above the ocean surface.

Pohnpei's volcanic origins and the forces of erosion have combined to produce the island's most spectacular geomorphological feature, Paipalap, also known as Sokehs Rock. According to one geological interpretation, Paipalap was part of a huge, thick, late-stage lava flow that filled up an ancient island valley that no longer exists. Much of the lava flow that entered that former valley produced a slow-cooling pond of molten material. This semifluid lava cooled and eventually solidified to become very dense, erosion-resistant rock trapped between the walls of this old valley. Later, over a long period of time, the surrounding valley walls eroded away faster than the dense lava they held between them. Today, only that core remains as one of Pohnpei's most striking physical features.

Paipalap, as seen from Pohnpei lagoon.

In addition to Paipalap, the evidence for slow-cooling lava can be seen in the form of jointed, columnar, six-sided basalt columns at Pwisehn Malek, a prominent landmark, as well as several other locations around Pohnpei. It is these natural columns and slabs of basalt that were quarried and transported long distances to provide construction material for building the remarkable structures at Nan Madol.

Columnar basalt

How Nan Madol—the ancient political and ceremonial center of Pohnpei—was constructed from massive blocks of columnar basalt remains a mystery.

1.2. TROPICAL MARITIME CLIMATE

Climate can be defined as the average conditions of the lower atmosphere that prevail in an area in general, or over an extended period of time, such as a year. In contrast, weather refers to atmospheric conditions at any particular moment in time.

The climate of Pohnpei can be characterized as predictable and consistent, with warm tropical temperatures and very high rainfall throughout the year.

1.2.1. Temperature and humidity

Surrounded by the warm tropical Pacific Ocean, Pohnpei experiences little variation in temperature throughout the year. The average daily range of temperature (from minimum to maximum) is quite low, about 6.5°C (12°F). The average annual temperature is 27°C (81°F). Relative humidity (the amount of water vapor in the air in relation to the total that could be held at a particular temperature) is also typically high, averaging between 80 and 90%.

1.2.2. Rainfall

Pohnpei is one of the wettest places in the world, with average rainfall of about 4,820 mm (190 in) per year. This rainfall is generally well-distributed throughout the year, although slightly less falls during January and February (see Fig. 1.1 below), when solar intensity in the northern hemisphere is reduced, which also reduces the heat needed for rising air and convectional rainfall.

To better understand island-wide weather and rainfall patterns, the Water and Environmental Research Institute (WERI) installed a rain gauge network throughout the island in 2003. The recorded data have revealed major differences in rainfall amounts in different locations on Pohnpei. For example, annual (year-long) rainfall in the driest coastal areas averages about 3,175 mm (125 in), while the recorded mean for the mountain summit of Nahnalaud is 7,620 mm (300 in).

In addition, the eastern or windward side of the island (*lepdahla* or *pohnangi*), which faces toward the prevailing trade winds (*enginpar*), tends to receive more rainfall than the western or leeward side (*lepdihla* or *palipah*). Most storms also approach Pohnpei from an easterly direction (*peidak*), as opposed to the leeward, western direction (*peidi*). June is typically a month for strong thunderstorms, which have been known to drop as much as 25 mm (1 in) of rain in 15 minutes! These thunderstorms can produce devastating effects, such as highly accelerated soil erosion, which can be further intensified if upland forest cover on steep slopes is removed or disturbed by human activity.

Figure 1.1. Monthly average rainfall (vertical bars) and temperature (red line for average monthly maximum and blue line for average monthly minimum). The data were recorded at Kolonia, from 1953 to 2001.

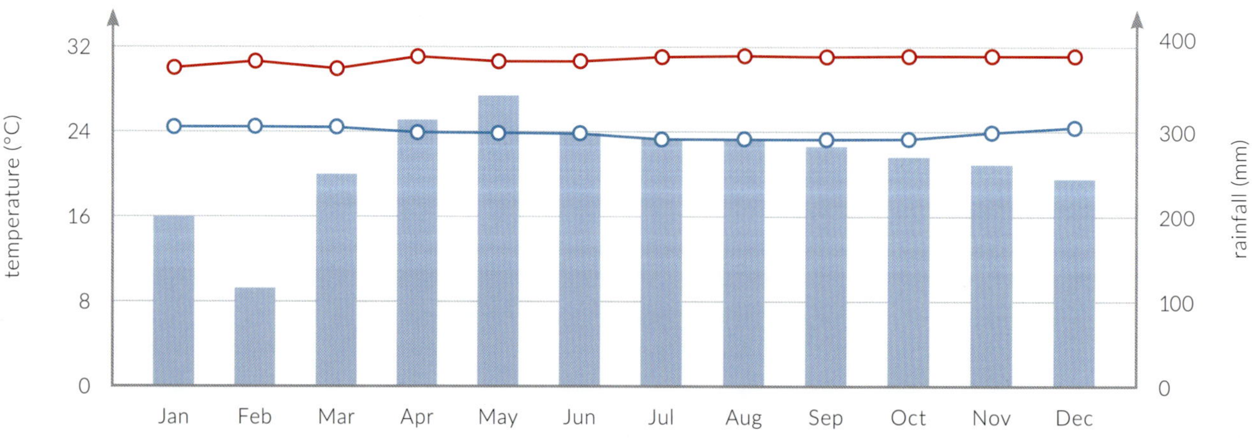

1.2.3. Severe storms

Typhoons are powerful tropical storms that typically affect areas some distance from the equator, including Pohnpei and its outlying atolls. Fortunately, Pohnpei State is impacted only occasionally, with a far lower frequency compared to islands further west. This is due to its location to the east of the most storm-prone area, resulting in fewer than one major storm per year (see Fig. 1.2).

The most severe typhoon ever recorded to impact Pohnpei struck on April 20, 1905 (see the photo below). This huge storm destroyed nearly all the food crops and domestic animals on the island. Over 200 people lost their lives or were seriously injured. Since that disaster, several less destructive yet still significant typhoons have impacted Pohnpei. See Table 1.2 for a list of the most notable events.

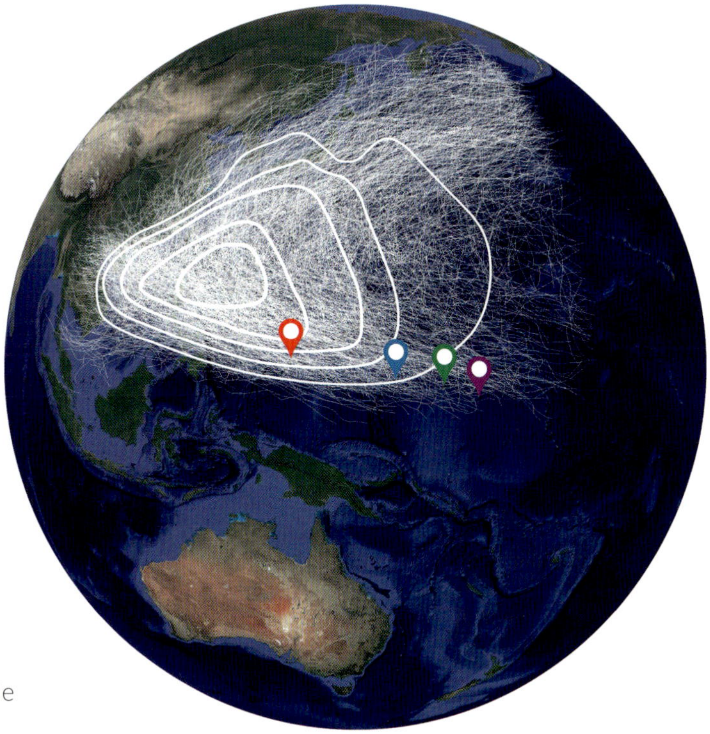

Figure 1.2. Tropical storms and typhoons in the Western Pacific. Thick lines enclose zones with the same storm frequency, one (outermost line) to five (innermost line) per year (based on Landers and Khosrowpanah 2004). Thin white lines trace storms tracks, from 1951 to 2024 (based on data from agora.ex.nii.ac.jp). Place markers show the FSM's main islands: Pohnpei (green), Chuuk (blue), Yap (orange), and Kosrae (purple).

Mission Station on Pohnpei after the typhoon of April 20, 1905. (Reproduced from Müller 1908).

Table 1.2. Timeline of major storms and typhoons in Pohnpei in the 20th century and in the 21st century to date.

Year	Storm/Typhoon name	Effects/Notes
1905	[no name]	The most destructive typhoon to strike Pohnpei in recorded history.
1957	Lola	The first major typhoon to impact Pohnpei in over 50 years.
1958	Ophelia	Significant damage, including roof torn off the US Weather Bureau office in Pohnpei.
1972	Lola	Houses destroyed, especially on Pingelap and Mwoakilloa atolls.
1986	Lola	Third typhoon with the name Lola to strike Pohnpei, and most destructive in nearly 30 years. Although no fatalities were recorded, infrastructure greatly suffered, with many homes damaged and trees and crops destroyed.
1991	Yuri	Millions of dollars in damages, including destroyed radio tower and other infrastructure.
1992	Axel	Power outages, flooding, landslides, and loss of crops on Pohnpei.
1997	Isa and Jimmy	Two storms a week apart trigger over 30 landslides and debris flows on Pohnpei. Many homes destroyed and 20 people killed.
1998	Jelawat	Weather system that later developed into Typhoon Jelawat brought heavy rainfall to Pohnpei, leading to flooding and landslides.
2002	Chata'an	Brought heavy rain, flooding, and ocean swells that damaged low-lying areas, especially in Nukuoro and Sapwuafik atolls.
2015	Dolphin	Heavy winds cause loss of one life and damage to over 200 homes.
2018	Jelawat	Massive landslides and coastal erosion from heavy downpour and high waves.
2019	Wutip	Heavy winds and rain damage homes, clinics, and schools on Pohnpei and cause flooding and destruction of crops on Nukuoro Atoll.
2023	[no name]	Powerful winds from an unnamed storm toppled trees and power lines on Pohnpei.

1.2.4. ENSO

Pohnpei is exposed to a range of weather-related threats that create significant vulnerabilities for all islands within the state. The atmospheric phenomena associated with what scientists refer to as the El Niño/Southern Oscillation (ENSO) determine where typhoon activity forms; they also influence rainfall distribution and sea level.

Although most tropical storm and typhoon tracks in the Western Pacific miss Pohnpei and pass north or west of it (see Fig. 1.2), that can change. Research suggests that during El Niño periods, Pohnpei State experiences a higher risk of typhoons because the storm formation region shifts eastward from its typical position to a position more or less just east of Pohnpei Island. On the other hand, in the year following an El Niño event, as well as in the years scientists refer to as La Niña, the risk of typhoons decreases in Pohnpei State because the storm formation region moves back to the west.

Aftermath of one of over ten landslides that occurred in Pohnpei after extreme downpour in March 2018.

1.2.5. Other hazards

Landslides or mudslides often occur as a result of storm events, whether it is a typhoon or severe thunderstorm. The worst landslide event on record in Pohnpei was triggered by tropical storm Jimmy, which hit the island on April 12–13, 1997. It destroyed homes and caused 19 fatalities.

Significant drought occurrence is very rare on Pohnpei, but droughts do happen occasionally and may become more frequent in the future as a consequence of climate change. The most severe droughts of the 20th century occurred in Pohnpei in 1982–83 and 1997–98. Rainfall between December 1997 and February 1998 amounted to only 16% of the long-term Pohnpei average. More recently, drought conditions occurred in 2014–16, damaging crops, triggering wildfires, and causing an estimated $11 million in economic damage. These extreme dry years coincided with Pacific-wide changes that were connected to the El Niño/Southern Oscillation (ENSO) events and exacerbated by ongoing climate change (global warming).

Although Pohnpei is one of the wettest places on Earth, there are common water resource concerns. According to available data from the 2010 census, 57% of the drinking water throughout Pohnpei is provided by the public/community systems based on surface water sources, especially the reservoir behind Nanpil Dam. Upland forest watershed protection is essential for long-term sustainability of this valuable water resource. In the outer islands, the primary source of water is rainwater catchment systems, and the secondary source is the groundwater (Taboroši and Allen 2015). During droughts, when rainwater catchment tanks are low or even empty, groundwater is used as the backup source, but its amount is also reduced. When rainfall is absent for a long time, the water table (level of water in the ground) is significantly lowered, as can be observed in shallow wells that become dry. When that happens, people can face extreme challenges in meeting their daily needs for water and may even require emergency supplies to be brought in by ship.

In addition to the quantity of water available in the islands, water quality is also a serious problem. In a 2016 survey of 40 streams and rivers, the water quality in every single one indicated it was unsafe to drink. Well over half of the sites investigated were deemed unsafe for recreation. Their contamination ratings were based on significant levels of the *Escherichia coli* bacteria. In the outer islands, quality of water in rainwater catchments and groundwater wells varies greatly based on the amount of care and maintenance by the owners at any given site, but it can rarely be described as clean water completely safe to drink.

1.2.6. Global Climate Change

Today, the great majority of people living in Pohnpei State reside in the lowlands, separated from the open ocean only by coastal vegetation, mangrove swamps, and the surrounding lagoon and reefs. They are threatened by wave surges associated with tropical storms, including typhoons, and this threat may become especially serious due to the global climate change. Recent research predicts more frequent and stronger storm winds and associated waves in some parts of the Pacific, which would lead to increased coastal damage and erosion. Also predicted is an increase in precipitation, which would lead to more runoff, damage to wastewater systems, sewage pollution in nearshore waters, and the reduced health of nearshore ecosystems (e.g., see pi-casc.soest.hawaii.edu/research/research-projects/pohnpei_water_quality/#).

Of greatest concern, however, is the fact that sea level rise associated with global climate change is widely considered an existential threat to human societies as well as ecosystems as a whole. This dangerous risk is especially relevant to islands and particularly so to atoll communities, of which there are many in Micronesia. Sea levels worldwide have risen globally by 10–20 cm (4–8 in) over the past century, with tide gauge data for Pohnpei indicating that the sea level rose by about 15 cm (6 in) between 1974 and 2014. Future climate projections for Pohnpei, and the variety of potential associated environmental impacts for this island, can be found online (e.g., see reefresilience.org/wp-content/uploads/ClimateSummary_Pohnpei.pdf).

Over 20 years ago, Falanruw (2002) summed up the vulnerability and need to protect the special biodiversity of islands, specifically referring to the FSM. Her comments remain extremely relevant as communities today continue to contend with climate change and loss of biodiversity:

"With the increasing rate of global change, the biodiversity of islands represents some of the most vulnerable resources on the planet. This makes islands' conservation efforts especially important. Islands are sentinels of what is to come, as what happens to islands will eventually befall the rest of the world. To stand up among the nations of the world, a country must have something of value. The FSM has something of value in its healthy reefs and island ecosystems, their biodiversity, and a heritage of living with these resources. Our efforts to care for this biodiversity could help guide the way to a sustainable future and become our role in the global community and our gift to the world."

Coastal erosion on Pohnpei, in an area called Keimwin Kiti, from which mangroves have been completely cleared by people.

1.3. SOILS OF POHNPEI

Soils form when rock or coral fragments are gradually broken down by physical forces—such as water, waves, and wind—and by chemical processes, and then mix with organic material, including fallen leaves and the remains of dead plants, insects, and microorganisms. Because the soil layer provides the medium and nutrients necessary for crop and forest growth, differences in soils from place to place may determine which plants can grow there. Soils can differ in many ways including: size and composition of the soil particles; amount of organic material present; ability to hold or drain water; available nutrients; salt content; and depth and layering. These are some of the differences used to classify and characterize soils into distinct groups.

Soil scientists have identified about 18 different kinds of soil on Pohnpei. These can be grouped into three basic categories. The fourth category typifies soils found on atolls (see Table 1.3).

Because of the volcanic nature of the rocks from which the soil originally develops, as well as extreme weathering under high rainfall conditions, the soils on Pohnpei are typically moderate to highly acidic (i.e., have a low pH). High acidity limits the kinds of plants that can grow in such soil. These soils also contain only moderate amounts of organic matter and have comparatively low fertility. In the upland mountains of Pohnpei, soils are well drained and normally very stony. However, use of these upland areas for agriculture is usually limited by steep, often unstable slopes with significant erosion hazards and overly cloudy and wet conditions.

People often think that because dense forests grow well on these mountain slopes, crop plants would also prosper. This is not generally true. Native species that grow well in the low-fertility upland soils of Pohnpei have been naturally selected over countless generations for their adaptation to these soil conditions. Attempts to grow agricultural plants in such highly acidic, low-fertility soils can be unrewarding for the farmer and disastrous in the long run for vegetation on steep, naturally forested areas. Soils in bottomland or swampy areas are also not ideal for farming. They normally have poor drainage and can only be used for cultivation of specifically adapted crops, especially swamp taro, **Mwahng** (*Cyrtosperma merkusii*).

Table 1.3. Soil types in Pohnpei State.

Soil types	Description
Oxisols	Old, highly weathered, and acidic soils predominantly found in tropical regions. Oxisols are rich in iron and aluminum oxides and exhibit a distinctive reddish hue due to the oxidation of iron minerals derived from underlying volcanic rock. In Pohnpei, they dominate the lowland areas.
Inceptisols	Young, moderately fertile soils formed from layers of volcanic ash. Inceptisols are in the early stages of weathering (both chemical and physical) and have not yet fully developed. In Pohnpei, they are commonly found on steep mountain slopes.
Entisols	Young, poorly developed soils with high organic content. Entisols typically form in coastal wetlands and swampy environments. In Pohnpei, they are primarily found in river floodplains and mangrove zones.
Atoll soils	Young, shallow, and alkaline soils, developing on coral atolls and similar low-lying islands. Atoll soils are composed mostly of weathered coral fragments and are sandy and nutrient-poor. Due to their high porosity, they drain rapidly and do not retain a lot of water. In Pohnpei State, they occur on the outer islands and small coral islets surrounding Pohnpei's lagoon.

1.4. TRADITIONAL RESOURCE AREAS OF POHNPEI

The rugged landscapes and extremely humid climate of Pohnpei have contributed to the preservation of much of its native plant and animal life. Traditional cultural values emphasizing sustainability have also played a significant role in conserving the island's native wildlife. One key factor in preserving the biological diversity of both native and cultivated species is the cultural recognition of distinct environmental zones, each supporting specific traditional resource use. For generations, the people of Pohnpei have identified these diverse marine and terrestrial environments as areas where organic and inorganic materials could or could not be collected, hunted, or cultivated. Many of these traditional resource areas are detailed in Table 1.4.

Key traditional resource areas as seen from Dekehtik Causeway in Kolonia: shallow lagoon, coastal mangroves, cultivated vegetation in the lowlands, and wild forest in the uplands.

Table 1.4. Traditional resource areas of Pohnpei.

Pohnpei name	Description
deke	coral or reef island, islet, atoll
kahp	clan resource areas in forest for gathering and cultivation
nampwel	seagrass beds
naniak	mangrove swamps lagoon holes
nanlehpwel	swampy areas where taro root crops can be cultivated
nanmad	shallow water areas in the lagoon
nanmadau	the open ocean
nanmal	grassland/fernland in *nansapw* or *nanwel*
nanmweli	stony areas dangerous to walk across due to instability
nanmwoakot	areas of good soil in *nansapw* or *nanwel*
nannahna	dwarf forest "on top of mountains"
nannamw	deep water areas in the lagoon
nanpahina	the barrier reef
nanpidir	impenetrable forest, jungle of secondary vegetation
nanpwoaki	non-cultivated swamp areas difficult to move through
nansapw	lands where humans settle, reside, and cultivate
nansed	the lagoon
nanwahu	valleys that are difficult to reach, mostly in *nanwel*
nanwel	wild forest lands or lands outside human authority
ni oaroahr	the shore (between *naniak* and *nansapw* on Pohnpei)
nierehr	steep areas or cliffs
pahniak	seaward edge of the mangrove swamp
paip	uncultivated forest
pidiriniak	impenetrable areas of mangrove
pohnnahna	tops of the mountains where dwarf forest can be found
pweleng	mountain ridges
sampei	hill or volcanic islet (from *sapw* = land; *pei* = stone pile)

Note: Similar to Pohnpei, the people living on the outer islands also identify distinct land, reef, lagoon, and ocean areas suitable for cultivating or obtaining specific resources. They have developed unique and highly specialized classifications and vocabularies within their own languages.

On Pohnpei, the most important distinction in land use is between *nansapw*, the cultivated lands shaped by human care and daily life...

... and *nanwel*, the wild forests and mountains, left untamed and imbued with spiritual power.

1.5. FORESTED AREAS OF POHNPEI

If we compare the plants and vegetation on Pohnpei with those on the other high islands of Micronesia and Polynesia, we can see that Pohnpei still benefits from a relative abundance of largely undisturbed or recently disturbed upland forests and coastal mangroves. These native forests contain a rich diversity of native plants and animals. A scientific survey undertaken in 2006 indicated that approximately 95% of Pohnpei was covered by some sort of forest: native forest (*nanwel*), agroforest (*nansapw*), or secondary vegetation (*nanpidir*). A significant portion of the secondary plant growth on Pohnpei consists of vines, shrubs, and small trees, which are part of the vegetation undergoing ecological succession in areas recently disturbed by natural events or human activities, such as the expansion of upland cultivation of the **Sakau** plant (*Piper methysticum*), more commonly known in English as kava. Over time, if disturbance is reduced or eliminated for a long enough period of time, the secondary vegetation is normally replaced by mature native forest through the natural process of ecological succession. In an extensive vegetation survey of Pohnpei, Donnegan et al. (2011: 6) summarized the plant life patterns on Pohnpei and a major threat to the stability of forest and soil resources on this island:

> "The native lowland vegetation has been heavily modified over thousands of years of human activity. Tree gardening and agroforestry has been practiced for millennia by using unique systems of guided vegetation succession and spatial rotation (Raynor and Fownes 1991), resulting in remnant agroforestry species occurring throughout the extensive, formerly settled landscape. Throughout the Pacific Islands, plants used for medicine and food are found in now-remote areas, likely owing to those areas formerly supporting more dense human populations. Most of these agroforest species are found on gently sloping or flat lands. A recent and more alarming trend is to cultivate species of commercial value on steeper slopes at higher elevation (e.g., **Sakau**), where forests are more intact and soils are much more likely to erode."

Figure 1.3. Relative areas of different land classes on Pohnpei in 2006 (based on Donnegan et al. 2011). The label "forest" includes all forest types except *nansapw*, which is separately labeled as "agroforest."

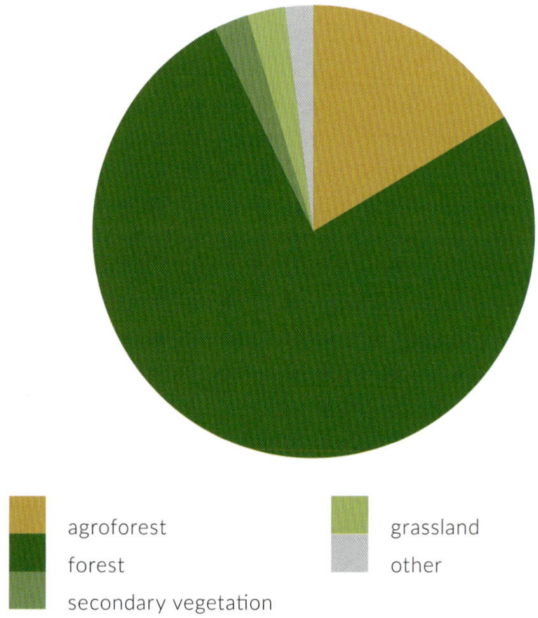

Figure 1.4. Approximate area of different forest types as a percentage of total forest areas on Pohnpei in 2006 (based on Donnegan et al. 2011). The label "other forest" includes freshwater swamps, atoll forest on lagoon islets, palm forest, coconut plantations, etc.

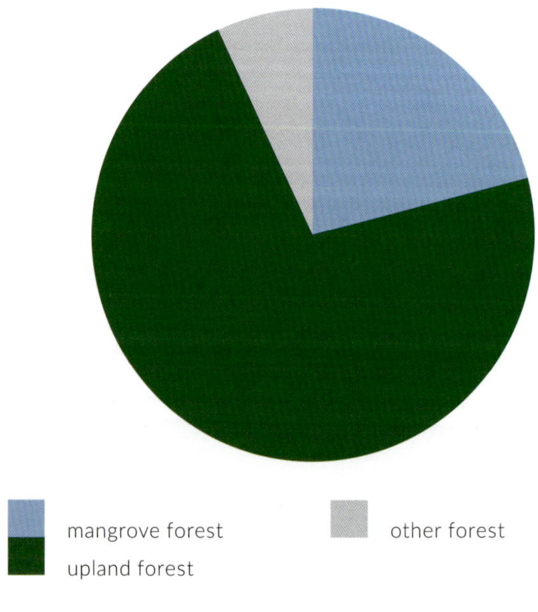

Lush, humid landscape in the *nanwel* forest on Pohnpei.

Cultivated trees grow together with wild plants in the *nansapw* landscape on Pohnpei.

Indeed, the rise in the popularity of the traditional beverage made from **Sakau** in Pohnpei during the latter part of the 20th century prompted its rapidly expanding cultivation in areas hidden from view. That resulted collectively in a widespread decrease in native mature forest, locally cut down to plant **Sakau**). We must remember that native mature forest is crucial for watershed protection in the uplands as it allows for fresh water recharge and limits the extent and destruction of erosion, especially during heavy rainfall events. Moreover, the native mature forest contains much of Pohnpei's biological diversity and vulnerable endemic species of plants, animals, and other organisms.

The broad-based transition of local consumption of beverage made from the **Sakau** plant from a mainly traditional ritual use to broader commercial consumption in Pohnpei began with the opening of *sakau* bars in Kolonia in the early 1970s. It further accelerated after the unprecedented El-Niño-driven drought of 1982–83 (referred to on p. 31) that severely depressed all crop production in coastal lowland areas. Farmers noted that the cloud- or mist-shrouded upland forests remained relatively moist, and **Sakau** planting, for both local and export production, began moving progressively to steeper slopes at higher elevation on Pohnpei (Merlin and Raynor 2004).

Inspecting sakau planted in a clearing in the upland forest, where its cultivation is ecologically harmful.

More recently, **Sakau** has become a major export crop from Pohnpei to other areas of the Pacific and beyond, generating significant cash income for local farmers. This ongoing trend has led to destruction, or significant disturbance, of more than 70% of the remaining mature upland forest watershed which is so important, as noted above, for sustainable water

Figure 1.5. Primarily due to population growth, the need for space to cultivate food crops, and the clandestine (hidden) cultivation of **Sakau**, Pohnpei's remaining native upland forest has dwindled over the past decades. These maps illustrate the decline in native upland forest (*nanwel*—dark green) between 1975 and 1995.

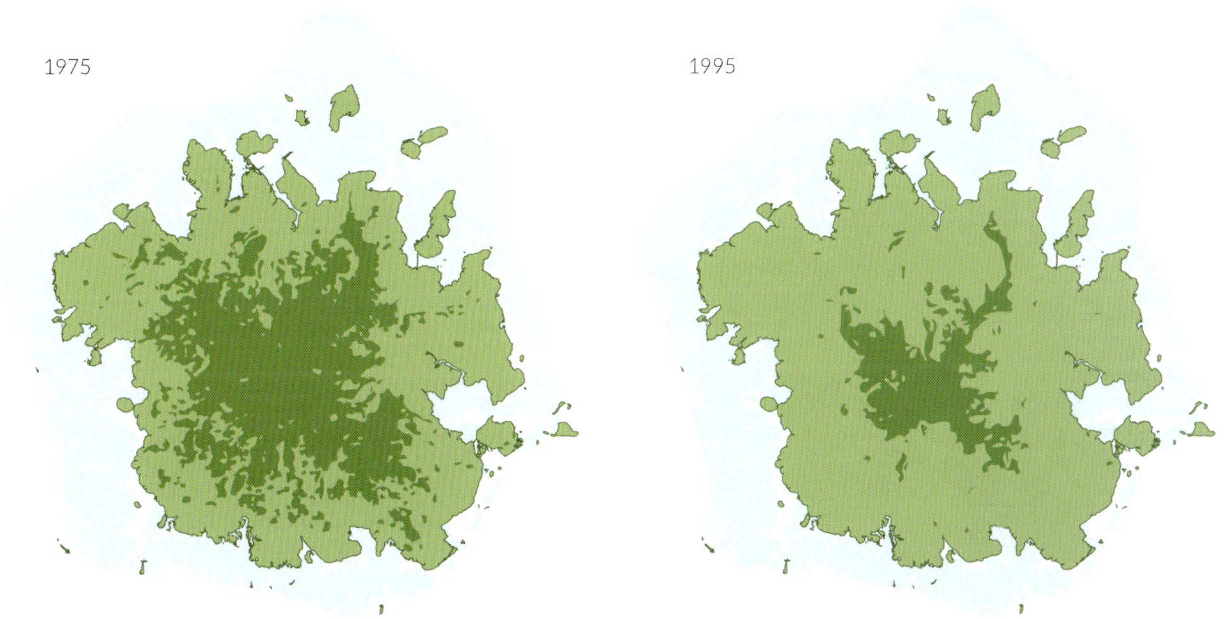

quality, erosion control, and the availability of useful traditional forest resources including culturally important products such as medicinal plants (Merlin and Raynor 2005, Balick et al. 2009). In response to a growing human population and continued upland expansion of commercial **Sakau** production over the last few decades, the Pohnpeian government and traditional chiefs, with the technical help of long-time resident conservationist William Raynor (1957–2015), initially created a **Pohnpei Watershed Forest Reserve** (1987).

Shortly before his untimely death, Bill Raynor spoke to the *Mwoahl en Wahu Ileile en Wein Pohnpei* (The Paramount Chiefs Council of Pohnpei Island), at which time he thanked them for 35 years of hospitality and respect, during which they made him the first American village chief (*Pelien Dahl*). Raynor urged the council to revitalize action involving the Pohnpei Forest Watershed and the island marine protected area conservation activities around the state, and to continue the partnerships of 30 years to improve the lives of all of those living on Pohnpei, especially the poorest people who depend on Pohnpei's natural resources for their livelihoods and survival. Despite the success of the "Grow Low" campaign (Fig. 1.6), which had begun many years ago to encourage cultivation of **Sakau** in the lowlands and not in the vulnerable uplands of Pohnpei, consistent vigilance and adaptive management must continue to protect the island's natural resources and ecosystem health.

Figure 1.6. Logo for the "Grow Low" Campaign: *Sakau en Nansapw* = "kava in the traditionally cultivated lowland regions" (e.g., see Merlin and Raylor 2005).

Ongoing global challenges, especially the climate change, growing commercialization, and loss of traditional knowledge, are increasingly affecting Pohnpei and the other islands in Micronesia. In his meeting with the Paramount Chiefs in January of 2015, Bill Raynor called for renewed commitment to the **Pohnpei Watershed Forest Reserve**:

> "So here is my request—many people think that the Pohnpei watershed is about trees, fish, birds, seafood, but no, our work is to protect lands in Pohnpei from the shore to the top of the mountains. It's about the very survival of the people, because our people survive on the fruit of the land, the mountains, the rivers and the ocean—not just the rich people, but the poor people that have nothing else to survive on. If our people can't find anything in the forest to eat; if the water comes down from the forest dirty and polluted; if they can't eat or drink anything from the river and forest, what's going to happen? If people can't shoot birds for their food, what are they going to do for food? If there are no more medicine plants in the forest because we abuse our forest for fast cash, what are people to use to get their health back? So our watershed work is not about animals or plants, it's about our people!"

Although protection has been in place for many years, in the face of continued upland forest disturbance, the **Pohnpei Paramount Chiefs Council** has been promoting a renewed watershed protection initiative and protection campaign since 2015 under the slogan: *Ahi Mour, Ahi Pwukoah* (My Life, My Community, My Reef, My Responsibility).

The fact that relatively large areas of biologically diverse native forest remain on Pohnpei today can be explained by at least two main cultural factors important to the people of Pohnpei. First, Pohnpeians have long recognized the importance of distinct traditional resource areas such as those listed in Table 1.4, on p. 35. Second, according to Pohnpeian traditional knowledge, each resource area is a living system, each requiring specific kinds of human behavior which limit overuse of its natural products and help ensure its continued survival.

Ultimately, significant native forest remains on Pohnpei because plants in general and the traditional uses of many forest products remain very important in Pohnpeian life. For many hundreds of years, the people of Pohnpei have used plants for food, fiber, shelter, transportation, medicine, and other common things in daily life as well as in ceremony and ritual. The value of plants to the people of Pohnpei and the traditional limitations placed upon the use of many of these plants have helped preserve the diversity of plant life still found on the island. The biodiversity of plant life on Pohnpei is described in some detail in Chapter 2, and the remarkable variety of uses and other types of relationships between the people of Pohnpei State and the plants in these islands are described in much detail in Chapter 3.

Will this diversity of plant life continue on Pohnpei—especially the diversity of plant life in the native forests or *nanwel*? The answer to this question depends to a large extent on: (1) the number of people who need to use these plant resources; (2) how well they are informed about the traditional uses and values associated with this diversity of plant life; and (3) a widespread, public understanding of the principles of ecology and sustainability.

River flowing over polished basalt rocks, making its way through Pohnpei's verdant and diverse native forest.

1.6. THE IMPORTANCE OF BIODIVERSITY

Biodiversity describes the sum of all the variation among living things. This biological variation can be considered at many different levels. One can measure the genetic variation within a population, the number of different species in a region, or the number of biological evolutionary lineages that exist. Regardless of what level one decides to measure, a high amount of biodiversity is extremely beneficial for humans and for the health of an ecosystem. First of all, this variation provides for a greater likelihood of resistance to environmental changes, making the ecosystem more resilient. Secondly, humans benefit from the many goods produced by a diverse ecosystem. Humans already receive countless natural resources, crops, and medicines, and there are still more potentially useful species to be identified. Biodiversity also produces valuable ecosystem services, such as nitrogen fixation and the regulation of hydrologic and carbon cycles. Lastly, biodiversity is important because it has an intrinsic value to humans. People generally have an inherent appreciation of nature and utilize it for anything from spiritual to recreational activities.

Being a large, complex, and geologically old island, Pohnpei boasts relatively high biodiversity. It has many significant terrestrial communities, ranging from mangroves at the coast to the upland forest in the mountains. Unfortunately, much of the original upland forest is being converted into agroforest (Wortel 2010). This reduces the area of native ecosystems but provides resources for the people. Since agroforests still maintain a higher level of biodiversity compared to monoculture plantations, they help Pohnpei retain a wide array of ecosystems, offering diverse ecological niches for various species.

While endemism in Pohnpei State is not as high as on some other more isolated high volcanic islands in the Pacific region, Pohnpei is still home to many unique species. Over 100 plant species are considered endemic, accounting for about 40% of all forest plant species. Many animals in Pohnpei's native forests are also endemic, and include insects, at least 21 species of terrestrial snails (Pelep and Hadfield 2011), several freshwater fishes, two reptiles (Buden and Taboroši 2016), about eight bird species, and one species of bat (Waldien et al. 2000).

The remaining forests in Pohnpei have a relatively high level of biodiversity and are important ecosystems to conserve. They provide significant amounts of goods and services, which the people of Pohnpei depend on for subsistence. If the ecosystem is overexploited, resources will be depleted and a great deal of biodiversity will be lost.

1.7. HUMAN POPULATION GROWTH ON POHNPEI

Based on archaeological research, scientists believe that Pohnpei, and at least some of the atolls now politically part of Pohnpei State (FSM), have been occupied by people for about 2,000 years and potentially earlier. Some of this research has produced evidence for early slash-and-burn agriculture on Pohnpei, which progressed towards more permanent areas of agroforest cultivation relatively soon following the island's initial settlement.

This 1883 drawing of Nan Madol, from *Südsee-Erinnerungen* by Franz Hernsheim, shows that Pohnpei's symbol of growing population and centralized power had already been abandoned when the first Westerners arrived.

As the human population expanded over time, Pohnpeians moved towards "a more hierarchical political system." The production of food and other crops increased as it provided yields from the "prestige economy" that began to evolve.

Scholars studying the history of Pohnpei suggest that about 15,000 people lived on the island when European ships first arrived there in the early 19th century.

Following this early foreign contact, large numbers of Pohnpeians died because of the disastrous effects of unintentionally introduced diseases brought in by the early visitors. By 1880, the human population on Pohnpei was very low; perhaps only a few hundred had survived. The population then began to increase gradually during the late 19th and early 20th centuries (Fig. 1.7 below).

The human population has been growing rapidly on Pohnpei since the Second World War due to increased medical facilities and other factors. Between 1940 and 1960, the population doubled, and between 1960 and 1980 it doubled again. As of 2025, approximately 40,000 people lived on Pohnpei.

Traditional agricultural practices still take place in mixed managed agroforests and cultivated swampy areas. Among the most important cultivated plant species are breadfruit (*Artocarpus* spp.), yams (*Dioscorea* spp.), banana (*Musa* spp.), kava (*Piper methysticum*), coconut (*Cocos nucifera*), and taro (e.g., *Cyrtosperma merkusii*, *Colocasia esculenta*, and others).

Presently, most people reside along or near the coastline, although some are living in settlements inland. Today, the great majority of people living on Pohnpei reside in the lowlands near the surrounding lagoon. In the outer islands, communities concentrate in leeward lagoon-facing areas. Therefore, wave surges associated with tropical storms, including typhoons, threaten the human populations both on the island of Pohnpei as well as in the outer islands. In addition, new research indicates that global climate change is predicted to produce more frequent and stronger storm winds and waves that could increase coastal damage.

In the future, with potential growth of the human population, more land will probably be converted from *nanwel* (wild forest area) to *nansapw* (cultivation areas) to produce the food. In addition, more imported food may be necessary to satisfy the needs of the growing population. Rapid human population growth is not unique to Pohnpei. It is occurring in many areas of the world. Many of these other areas are also under pressure to convert more of their remaining native forests to cultivated land.

Figure 1.7. Human population growth on Pohnpei since 1840.

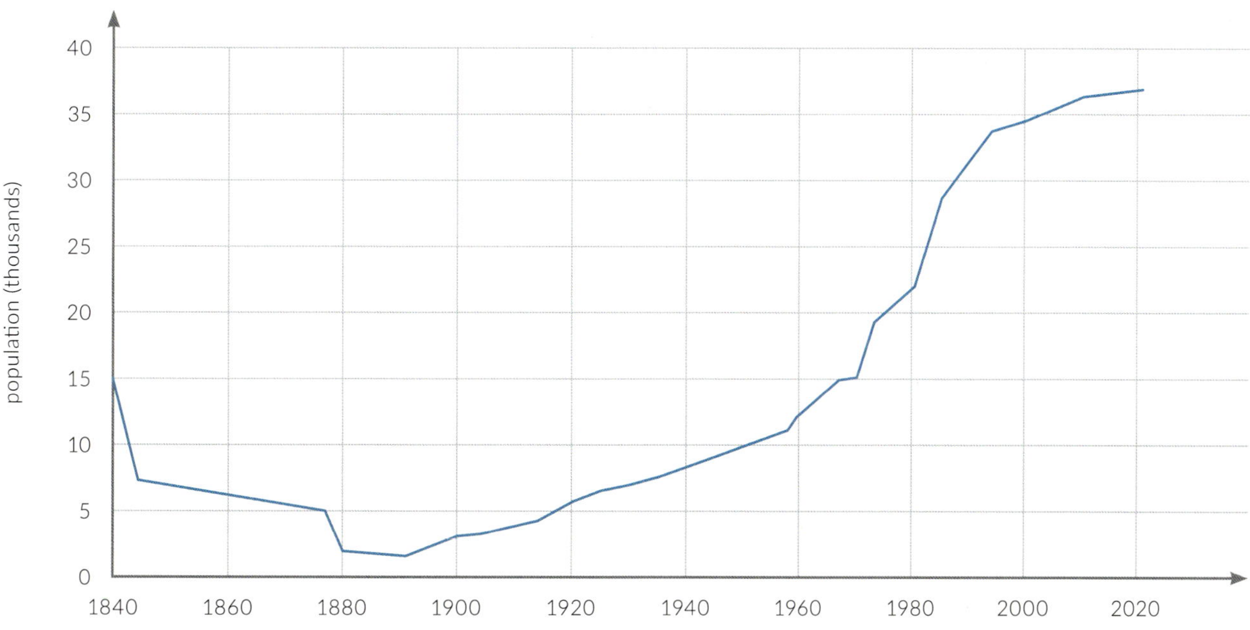

2

VEGETATION OF POHNPEI STATE

CHAPTER 2

VEGETATION OF POHNPEI STATE

Vegetation is a collective term referring to the structure and organization of plant communities. It encompasses all types of plants, including herbs, shrubs, trees, vines, and epiphytes—plants that grow on other plants and obtain moisture from the air. Vegetation is generally classified into broad categories such as rainforests, grasslands, and other ecosystem types.

In contrast to vegetation, which is a general term for all plants in a given area, a plant community is the group of plants that live together in a specific environment. Plant communities have distinctive structural properties, shaped by the physical forms of different plant species. These communities vary based on environmental factors such as latitude, elevation, soil composition, animal interactions, human impact, and other influences. Botanists generally agree that plant communities with similar form and function can be categorized as distinct vegetation types. Each vegetation type is typically adapted to a specific environment, or habitat, such as sandy coastal areas, saltwater swamps, steep mountain terrain, and many other kinds of places. Additionally, some vegetation types, such as cultivated tree gardens (agroforests), taro patches, and open grassy or fernland areas, have developed as a result of intentional or accidental human activity over time. As we will see in the following pages, human actions have significantly altered the vegetation of Pohnpei State, both on Pohnpei and the outer islands.

2.1. ORIGINS OF THE SPECIES IN POHNPEI AND THE OUTLYING ATOLLS

Many plants arrived in the islands of Pohnpei State without human assistance, with most establishing themselves long before the first people arrived. However, numerous other plants were introduced by humans. Today, many of the plants found in these islands serve various purposes, including food, building materials, medicine, shade, and decoration.

Scientists refer to plants that reached the islands naturally, without human intervention, as **native species**. These plants arrived millions of years ago, with their seeds or spores carried by air currents, ocean currents, or migrating birds. Native plants thrive in different environments without requiring human cultivation. Over time, as they adapt to local conditions, some of these plants may evolve into unique new species found nowhere else.

Among native species, those that occur naturally in Pohnpei but are also found in other regions are called **indigenous species**. For example, **Dohng** (*Campnosperma brevipetiolata*), a common rainforest tree, is indigenous to Pohnpei but also grows in the rainforests of other high volcanic islands in the Federated States of Micronesia. By comparison, **endemic species** are found only in a specific location. **Koahi** (*Claoxylon carolinianum*), a small understory tree, is endemic to Pohnpei, meaning it exists nowhere else (see Costion and Lorence 2012, for a checklist of endemic species in Micronesia, including Pohnpei). Endemic species are typically evolved from indigenous species that arrived in a new area naturally but, due to isolation and unique environmental conditions, developed distinct characteristics over time and became new species distinct from their ancestral populations. Many endemic species also tend to be classified as **endangered species** if their survival is threatened by habitat destruction or other disturbances (see Conservation Society of Pohnpei 2013).

In this book, unless otherwise specified as **endemic**, all native plant species described here are considered to be **indigenous** to Pohnpei. This also applies to coastal and lowland plants found in the outer islands of Pohnpei State. On the other hand, any species brought to the islands by humans, either intentionally or accidentally, is referred to as an **alien species** or **introduced species**.

Many species introduced by humans to the Pacific Islands are highly significant. Early Pacific voyagers deliberately introduced valuable plants to Pohnpei and other Micronesian islands to provide food, fiber, medicine, and other essential resources. In contrast, many unwanted plants have also taken root, growing near houses or in disturbed areas. These plants are commonly referred to as **weeds** (*dihpw* in Pohnpeian)—defined as plants growing where they are not wanted. Since the arrival of Europeans and other foreigners, many additional plant species have been introduced, both intentionally and accidentally. While many of these introduced species are valuable in Pohnpeian and outer-island cultures, some have escaped cultivation, spreading into natural habitats where they have become common weeds.

Pohnpei's luxuriant, ecologically diverse vegetation, made up of both native and introduced species, is sustained by high rainfall and streams cascading from the uplands toward the coast.

2.2. ENVIRONMENTAL ALTERATION BY PEOPLE IN POHNPEI

People have been altering the terrestrial environments of Pohnpei and the outer atolls since their first arrival. However, environmental changes caused by ancient societies likely occurred at a slower pace than those observed in modern times. Early Pohnpeians and atoll-dwelling peoples created agriculturally productive landscapes, but in doing so, they also transformed native vegetation. On Pohnpei, villages replaced cleared areas, and native lowland forests were cut down to make way for cultivated lands filled with introduced trees and edible plants.

On the atolls, people similarly converted portions of lowland forests into agroforests of useful species, such as coconut palms (*Cocos nucifera*) and breadfruit (*Artocarpus* spp.). In some locations, they excavated pits to reach the freshwater table beneath the larger islets, filling these depressions with decomposing plant material (mulch) to create fertile marshes. These artificial or expanded natural wetlands were then cultivated with wetland food crops such as swamp taro (*Cyrtosperma merkusii*) and true taro (*Colocasia esculenta*).

A defining feature of lowland coastal vegetation throughout Pohnpei State and particularly on atolls is the coconut palm (*Cocos nucifera*), known as **Nih** in Pohnpeian. Some lowland areas still contain uniform plantation groves of coconut palms, originally established in the late 19th century to support commercial export—an industry that has since declined. The widespread planting of coconuts replaced much of the original vegetation, including traditional cultivated tree gardens. One of the most significant species that were largely replaced by this "mono-cropping" is breadfruit (*Artocarpus altilis*), an essential subsistence crop known as **Mahi** on Pohnpei and by other names on outlying atolls.

In recent times, external influences have significantly reshaped parts of Micronesia. People from Europe, North America, and Asia have altered or replaced vast areas of vegetation for military, commercial, and government purposes. In some regions—particularly the Marshall Islands—modern warfare and weapons testing have severely damaged ecosystems, devastating native habitats and indigenous plant and animal life. As a result, these landscapes are now dominated by a limited number of alien species. Fortunately, unlike other areas in the Micronesian region, Pohnpei and its outer islands were largely spared from the widespread impacts of World War II and later military activities.

The following discussion focuses on the natural and human-made vegetation types of Pohnpei, highlighting the typical plant species associated with these different ecological units.

Dredging and construction have drastically altered Kolonia Bay, disrupting natural water circulation and degrading water quality and ecological health.

2.3. TYPES OF VEGETATION IN POHNPEI STATE

This section offers an overview of the rich diversity of vegetation types found in Pohnpei State. It includes descriptions of each type, along with lists of plant species commonly associated with them. Before proceeding, it is important to note that vegetation is made up of individual plants. These plants are adapted to particular vegetation types or plant communities and grow in a variety of distinct forms.

Figure 2.1. Plant growth forms:

a) **Trees.** Tall, woody plants with a single main trunk.

b) **Shrubs.** Multi-stemmed woody plants of relatively low height.

c) **Palms and palm-like plants.** Plants with a single, upright stem and a cluster of large leaves at the top; or similar-looking plants, such as pandanus.

d) **Herbs.** Small, non-woody plants.

e) **Grasses**. A category of herbs with long, narrow leaves and jointed stems.

f) **Climbers**. Vines and plants with long stems that grow upwards by attaching themselves to or twining around supporting surfaces.

g) **Creepers.** Vines and plants with long stems that grow horizontally along the ground, rooting at nodes as they spread.

h) **Epiphytes.** "Air plants" that grow on other plants, including various ferns and other small herbaceous species.

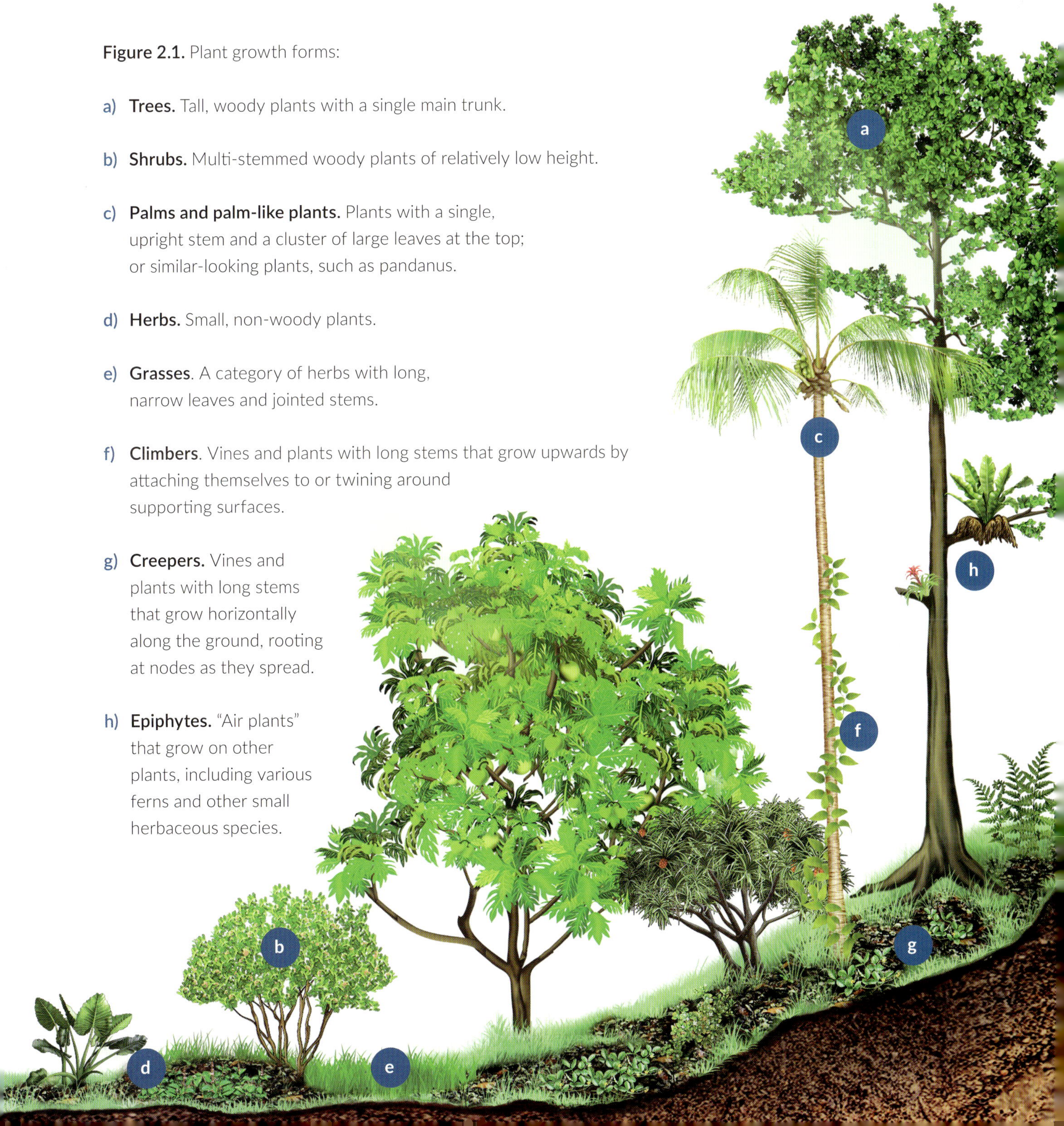

The plant life of Pohnpei and its outlying atolls can be described as distinct vegetation types, which are groups of plants growing together in particular environments. These vegetation types are summarized in the table below.

Table 2.1. Vegetation types in Pohnpei State.

Saltwater aquatic vegetation	Seagrass communities that include relatively few species adapted to shallow saltwater environments.
Coastal strand and lowland vegetation	Forest or shrub communities in the coastal lowlands of Pohnpei and in the outer islands of Pohnpei State; plants in this community type are found on sandy and rocky shores of the high and low islands.
Mangrove vegetation	Saltwater forest communities found along sheltered coastal areas and lower streams on Pohnpei, including other plants found in the landward edge of mangroves.
Swamp forest and marsh vegetation	Freshwater wetland communities containing wild and cultivated plants adapted to standing water that is present most of the year.
Agroforests	Cultivated plant communities, especially with tree crops, but often interspersed with taro patches and intermittent mixed gardens.
Secondary vegetation	Areas where ecological succession is active, normally in places where agriculture is in the fallow stage or where natural disturbance has occurred.
Grasslands or savanna	Open or disturbed areas covered largely by herbaceous or non-woody plants such as grasses, sedges, and ferns.
Upland and valley forests	Rainforest communities found in steeper inland areas containing a wide variety of plants, including many native species.
Summit dwarf forests	Rainforest in the highest parts of Pohnpei, on mountain summits and ridges, consists of stunted trees coated with ferns and mosses, all nearly permanently shrouded in mist.
Atoll forests	Mixed broadleaf forest that naturally develops in the interior of coral atoll islands. In Pohnpei State, such habitats were once common in the outlying atolls, but now nearly all have been converted to agroforests.

Note: for Pohnpeian names for these as well as other types of places on Pohnpei, please see Table 1.4 on p. 35.

Saltwater aquatic vegetation.

Coastal strand and lowland vegetation.

Mangrove vegetation.

Swamp forest and marsh vegetation.

Agroforests.

Grasslands or savanna.

Upland and valley forests.

Summit dwarf forests.

2.3.1. Saltwater aquatic vegetation

This type of marine wetland vegetation is, more or less, continuously flooded by saltwater. It consists of seagrass beds or meadows (known in Pohnpeian as *nampwel*) that often cover large areas of near-shore ocean habitat. These marine habitats typically include only a few seagrass species, which are flowering marine plants. Seagrass beds are located in shallow seawater up to about 7 m (23 ft) deep. Below this approximate depth, there is not enough sunlight for the seagrasses to survive. Although many species of marine algae are found in seagrass beds, the number of seagrass species in this marine vegetation is limited. In fact, an increasingly smaller number of seagrass species are distributed naturally as you proceed geographically from west to east among the many Caroline Islands of Micronesia (see Lobban and Tsuda 2003). For example, marine botanists have documented 10 species of seagrasses in Palau, 7 species in Yap, 5 species in Chuuk, and only 3 species in Pohnpei (also see Tsuda and Sukhraj 2016).

The three seagrasses known to be found in the saltwater vegetation in Pohnpei include *Thalassia hemprichii*, *Enhalus acoroides*, and *Cymodocea rotundata*. Like other seagrass species, these three that are found in near-shore marine habitats of Pohnpei are rarely found with flowers or seeds; thus it is usually difficult to identify them scientifically.

The seagrass *T. hemprichii*, known in Pohnpeian as **Oaloahd**, is probably the most common type of seagrass found in the saltwater vegetation of Pohnpei. This species is a dioecious plant, which means it has some plants that produce only male flowers, while others, which have longer leaves, produce only female flowers. Key characteristics of **Oaloahd** are hooked or curved

Seagrass thrives in shallow waters, rooted in soft sandy or muddy bottoms.

leaves that are strap-like and have red flecks; the leaves are also rounded with a slightly serrated tip. **Oaloahd** grows submerged in shallow coastal bays with its roots embedded in sand, mud, or in pools in tidal flats.

The second most common seagrass in the saltwater vegetation of Pohnpei is *E. acoroides*, known in Pohnpeian as **Oaloahdilap**. Similar to *T. hemprichii*, *E. acoroides* is also a dioecious plant, and thus has only male or female flowers on an individual plant. The leaves of this seagrass are produced from a solid creeping rhizome (root-like plant stem) that has tough black bristles. These bristles at the base of the plant are the fibrous remains of old leaves. In some areas of Micronesia, its strong leaves and base fibers are traditionally used to make nets for catching reef fish (see Chapter 3).

A third species of seagrass, *Cymodocea rotundata*, is also found in the shallow ocean area around Pohnpei. It is a smaller species than the other two described above. *C. rotundata* is identifiable because of its thin, ribbon-like leaves, smooth herbaceous rhizomes, and rounded leaf tips. Although this species was not documented as occurring in the marine environments of Pohnpei until the research done by McDermid and Edward (1999), these scientists observed *C. rotundata* as the dominant seagrass in the shallow near-shore waters (5–45 m from shore) in some areas, such as near Nahpali Island in Pohnpei. Nevertheless, because *C. rotundata* is apparently not very common in the seagrass beds of Pohnpei, no local name has yet been documented (e.g., see Herrera et al. 2010: 44).

Cymodocea rotundata

Seagrass beds in Madolenihmw.

Note: There is a common misinterpretation of the taxonomic relationship of seagrasses and species in the "true" grass family (Poaceae). Seagrasses actually have no relation to true grasses; this is supported by systematic and phylogenetic evidence. Nevertheless, the term "seagrass" suitably defines a group of flowering plants that are specially adapted to grow in estuaries and marine ecosystems. Overall, seagrasses are a group of marine plants that includes 13 genera and approximately 72 species belonging in the following families: Zosteraceae, Potamogetonaceae, Posidoniaceae, Cymodoceaceae, Ruppiaceae, and Hydrocharitaceae. The two major seagrass species on Pohnpei, **Oaloahd** and **Oaloahdilap**, belong in the family Hydrocharitaceae. Interestingly, the freshwater aquatic plant **Oaloahdenpil** (*Blyxa aubertii*) is also a species in the family Hydrocharitaceae (see freshwater wetland vegetation, p. 72).

The importance of seagrasses

Seagrass meadows occur widely along the coast of Pohnpei, but until only recently the ecological role of these seagrasses has remained relatively undocumented and unfortunately unappreciated. The saltwater aquatic vegetation of seagrass beds forms a very productive ecosystem, but it can be damaged, sometimes severely, if surrounding land or sea areas are disturbed, either by natural or by human causes. For example, excessive soil run-off due to increased erosion on land may muddy the ocean water and thereby limit the sunlight available to the algae and seagrasses; this can stunt the growth of the plants, or even kill them. In turn, reduced algae and seagrass growth may limit the growth and development of marine animals that are dependent upon these saltwater organisms.

Like the mangrove forests (see p. 62), seagrass beds serve as traps for silt and sediments washed into the sea from land areas. Seagrass beds, known in Pohnpeian as *nampwel*, provide food and shelter for a number of animal species, including sea cucumbers, clams, some fish—such as **Palapal** (the gold-spotted rabbitfish, *Siganus punctatus*) and **Umwule** (the forktail rabbitfish, *Siganus argenteus*)—as well as sea turtles, all of which are commonly referred to as **Wehi** in Pohnpeian. Sea turtles generally feed on seagrasses, and therefore these marine flowering plants are referred to as "turtle grass." Two species of sea turtles that live in marine areas of Pohnpei are **Kalahp**, the green sea turtle (*Chelonia mydas*), and **Sapwake**, the hawksbill sea turtle (*Eretmochelys imbricata*). **Sapwake** is a totem (a sacred symbol) for the Soun Kawad clan, so they do not eat this turtle.

Chelonia mydas
Kalahp (P, Mw, Pi), **Honu** (Nu), **Hono** (Ka)

Eretmochelys imbricata
Sapwake (P), **Japwikoa** (Mw), **Saepwake** (Pi), **Gea** (Nu), **Gee** (Ka)

As noted above, seagrasses are important, underappreciated plants that inhabit coral reef flats and other marine habitats. Extensive, naturally occurring seagrass beds or meadows provide key ecological and economic resources for many coastal ecosystems worldwide. Consequently, an increase in scientific studies, effective resource management, and a major rise in public awareness of the environmental importance of seagrasses as coastal resources and vital habitats are more important than ever.

A growing concern in Micronesia has developed "for the conservation of sea turtles, for the management of coastal fish populations, and for the protection of coral reefs," and there is an evident and pressing need for increased "understanding of the role played by seagrasses in grazing and detritus food webs, nutrient recycling, sediment retention, and shoreline stabilization" (McDermid and Edward 1999). Major environmental threats face several Micronesian island estuaries, including Pohnpei and its seagrass beds, as described briefly by Raynor (2006: 109):

> "The sources of degradation of estuaries are varied and numerous, the most serious being sedimentation and pollution from land-based activities. Dredging and construction of causeways with insufficient openings for current and tidal flow have also further degraded a number of estuaries in the FSM. Sea grass beds are threatened by sedimentation and pollution, but have been spared some of the ravages of estuaries especially where healthy mangrove forests still line the coast and can absorb land-based impacts. To a lesser extent, boat traffic (propeller damage) has had local negative impacts on sea grass beds, especially along regular transportation routes in lagoons. Storms and wave action particularly those resulting from typhoons occasionally impact sea grass beds, as does increased freshwater..."

Seagrass beds often stretch in shallow waters just seaward of mangroves, forming a green belt in front of the trees.

2.3.2. Coastal strand and lowland vegetation

The plants that grow in the lowland areas of Pohnpei and the outer islands are able to adapt to the special habitat requirements of the often harsh conditions of the land at the ocean's edge. Coastal strand vegetation is found in a narrow zone, near the ocean's edge on open, sandy, and rocky shores. The main island of Pohnpei lacks natural, sandy beaches, has few rocky shores, and is mainly surrounded by tidal flats and mangrove swamps. However, there are sandy shores on some of the small islands located on the barrier reef that encircles the Pohnpei Lagoon. Atoll islets in the Outer Islands also have extensive sandy and rocky shores.

Plants with different life forms, from trees to small herbs, are found in the strand vegetation of Pohnpei State. Although they can differ greatly, in some ways, the strand plants share several things in common. Many produce seeds which float and are able to survive for long periods of time immersed in saltwater. The seeds of these plants are often dispersed over long distances by ocean currents to other suitable beach habitats.

The coastal strand plants have to be able to survive in very sunny, often windy, salty conditions along the seashore. Near the edge of the sandy and rocky shores, the plants are usually non-woody (herbs) which commonly have fleshy leaves and salty sap. Behind the beaches and rocky shores, a variety of shrubs, trees, and other non-woody plants may be found.

Much of the coastal, non-wetland vegetation (e.g., non-mangrove areas) of Pohnpei State is found in a narrow zone near the edge of the ocean on open, sandy, and rocky shores. This vegetation is referred to scientifically as littoral vegetation (*litoris* means shore in Latin). This vegetation can be divided into at least four subtypes of plant communities:

The outer islands are dominated by coastal strand vegetation, yet Pohnpei's coastline is primarily mangroves, only interspersed with small coastal strand patches.

1. **Herbaceous strand:** This non-woody vegetation grows in a narrow zone above the high tide mark on both sandy and rocky shores. The typical native plants in this zone are creeping vines, especially *Ipomoea littoralis*, *Ipomoea pes-caprae*, and *Vigna marina*. See Table 2.2 for the local names of these and other species in Pohnpeian and other languages of Pohnpei State.

Vines creep across sandy beaches on the outer islands.

2. **Littoral shrubland:** This shrubland is found on windy coastal ridges and slopes, or on the seaward edges of coastal forests. Only a few plants are common in this coastal zone. These include *Scaevola taccada*, *Heliotropium arboreum*, and *Guettarda speciosa*.

Scaevola taccada scrub lining the edge of an islet on Mwoakilloa Atoll.

3. **Pandanus scrub:** This shrubland occurs on rocky, often exposed and windswept shores. As indicated by its name, this community is dominated by relatively small or stunted pandanus trees, *Pandanus tectorius*, known as **Kipar** (P) (Mw) (Pi) and in Pohnpei's Polynesian-speaking atolls as **Hala** (Nu) and **Binu** (Ka). *Pandanus tectorius* is also known in English as the "screwpine" because of the twisted form of the long leaves as they develop on the top portion of the trunk, and the shape of its compound fruits that look somewhat like pineapple fruits (though pineapples grown on herbaceous plants close to the ground, not on trees like pandanus).

A grove of pandanus trees on Pingelap Atoll.

4. **Littoral forest:** This type of vegetation is common on the outer islands but much less abundant on Pohnpei. Typical native trees in this community include: *Barringtonia asiatica*, *Hernandia nymphaeifolia*, *Terminalia samoensis*, *Calophyllum inophyllum*, *Cordia subcordata*, and others.

Mixed forest dominates the shorelines of all atoll islets in Pohnpei State.

Almost all the native plants of the coastal vegetation on Pohnpei are indigenous. However, since the majority of people on Pohnpei live relatively close to the shore and utilize the coastal zone for many purposes, the coastal vegetation has been modified significantly by human activity. Over the centuries, a great many introduced trees have been planted in the coastal environments of Pohnpei, including many coconut palms, *Cocos nucifera*,

A sprawling **Koatoa** (*Sonneratia alba*) tree extends its branches over the tranquil waters near Temwen Island.

breadfruit trees, both “seedless” *Artocarpus altilis* and “seeded” *Artocarpus mariannensis*, and a number of other woody species. Several of these plants are (or were) valuable sources of food, fiber, medicine, and other uses (see Chapter 3). Many of these same plants are also found in places other than the low coral or rocky coast, such as along the landward edge of the mangrove forests or in the cultivated lowlands of Pohnpei.

Many of the common coastal and lowland plants in Pohnpei State are familiar and useful species. Some are (or were) sources of food products; others are important as sources of fiber for weaving; and many are still used in traditional medicine. Some, especially the coconut palm, are true multi-purpose species with a very wide range of uses. See Chapter 3 for detailed descriptions of traditional uses of plants in Pohnpei State.

Table 2.2. Some plants found in the coastal and lowland areas of Pohnpei State.

Scientific name	Pohnpei name	Mwoakilloa name	Pingelap name	Nukuoro name	Kapingamarangi name
Artocarpus altilis	Mahi	Moai	Maei	Gulu	Gulu
Artocarpus mariannensis	Meipa / Meikole	Moai koaloa	Maei koalae	Gulu de henua	Gulu golee
Asplenium nidus	Tehnlik	Kardoap	Saehlik	Lau gadaha	Logo hoo
Barringtonia asiatica	Wihnmoar / Wihnsed	Wi	Wi	Gavaausu haahine	Laagau iha / Laagau hagadali eidu
Calophyllum inophyllum	Isou	Ijoau	Saepaeng	Hedau	Heedau
Cassytha filiformis	Likotokotasahu	Kosoakos	Kosokos	Hiisoa	Uihoo
Clerodendrum inerme	Ilau	Ilau	Ilau	Manu mmala	Hia
Cocos nucifera	Nih	Ni	Ni	Nui	Niu
Cordia subcordata	Ikoik	Kanau	Ikoik	Ganava	Laagau mmee
Dalbergia candenatensis	Selen kaikes	–	–	–	–
Guettarda speciosa	Ihd	Id	Eles	Bua laamodu	Bua
Hernandia nymphaeifolia	Pingiping	Pingping	Pingiping	Bingibingi	Bingibingi
Ipomoea littoralis	Emp / Omp	Oahp	Oip	Hue lauligi	Huuwe
Ipomoea pes-caprae	Sonsol	–	–	Hue haangai biigi	Huuwe
Morinda citrifolia	Weipwul / Wempwul	Wɛhnpwul	Uhpwul	Buga liaa	Nonu
Pemphis acidula	Ngih	Kaingi	Kaingae	Nngie	Laagau hai goo
Pandanus tectorius	Kipar	Kipar	Kipar	Hala	Binu
Scaevola taccada	Remek	Roamoak	Raemaek	Manu gaba aasanga	Nau
Tacca leontopetaloides	Mwekimwek	Mwoakmwoak	Mwaekaemwaek	Bie	Mogumogu
Terminalia catappa	Wahn dipwoapw	Wehn doapwoapw	Wahn dipwoapw	Hua laagau	Gegebinu
Terminalia samoensis	Dipwoapw	Win	Win	Kini	Laagau daliha
Thespesia populnea	Pone	Pɛnɛ	Paenae	Milo	Mila
Heliotropium arboreum	Titin	Sisin	Sesen	Manu gili maau	Dogodogongo
Vigna marina	Wahpaili / Oangoahng	Doaudol	Saudul	Gaulimago	Dua gimoo*
Wollastonia biflora	Moaresed / Ngkahu	Moarjij	Kisiwael	Aigi mea	Dilidai

* The same name is also used for *Canavalia cathartica*; to refer specifically to *Vigna marina*, people may say **Dua gimoo lobuge dono akai**.

Barringtonia asiatica

Calophyllum inophyllum

Clerodendrum inerme

Guettarda speciosa

Pemphis acidula

Scaevola taccada

Terminalia catappa

Thespesia populnea

2.3.3. Mangrove vegetation

In general, what are mangroves?

"Mangroves form a unique and dominant ecosystem comprised of intertidal marine plants, mostly trees, predominantly bordering margins of tropical coast-lines around the world. These halophytic (salt tolerant) plants thrive in saline conditions and daily inundation between mean sea level and highest astronomical tides, and they provide vital structure as habitat and food for similarly adapted resident and transient fauna. Mangrove plants exchange gases from exposed roots using special lenticels [raised pores in the stem of a woody plant that allows gas exchange between the atmosphere and the internal tissues], while flooding tides allow uptake of river-borne nutrients and frequent dispersal by their buoyant propagules. The ecological limits defined by the diurnal tidal range explain the setting and why just 70 species around the world are considered to be mangroves (Tomlinson 1986, Duke et al. 1998), compared with adjacent rainforests that may have hundreds of tree species per hectare. Specialized morphological and physiological characteristics largely define and characterize mangrove plants, such as buttress trunks and roots providing support in soft sediments, aboveground roots allowing vital gas exchange in anaerobic sediments, and physiological adaptations for excluding or expelling salt. Fewer than 22 plant families have developed such essential attributes, representing independent instances of co-evolution over millions of years to form today's mangrove habitats." (Duke 2006: 2–3)

Much of human knowledge and culture is learned from observing nature. Proverbs and legends often refer to the value of plants, the many ways they have been used, and express the wish to explain nature and people's place in it.

Low tide reveals the twisted, knotted roots of **Sohmw** mangroves in Kitti.

Mengihngihn oapwoat ahk
The core of the mangrove log

The phrase above refers to the hard, strong core of some mangrove plants, and to the strength and durability of old men who have been "purified of the laziness, fickleness, and ignorance of youth" (see Riesenberg and Fischer 1955).

Mangroves grow in saltwater-influenced swamps that surround much of Pohnpei. These forests are composed of trees adapted to areas affected by ocean tides. Mangrove forest communities occur in many subtropical and tropical areas around the world.

A survey in the latter part of the 20th century estimated that mangrove forests covered 55 km^2 (21 mi^2) along Pohnpei's fringing coastline (Petteys et al. 1986). A more recent survey in 2006 confirmed approximately the same area of mangroves.

Protecting mangrove forests is important for a number of reasons. The plants of the mangrove forests of Pohnpei are generally located along the muddy reef flats and are essential for environmental protection. They stabilize coastal areas by trapping and holding sediment washed down from inland areas. Rainfall run-off carries soil eroded from the interior of Pohnpei down to the mangrove swamp areas where the mass of tangled roots of the woody plants prevent much of this mud from washing out and smothering the coral reefs. Mangrove plants also offer some coastal protection from normal wave action and strong storm waves that surge over the outer barrier reef.

Mangrove rooting systems

As "producers" in the natural environment, the mangrove plants convert energy from the Sun through the process of photosynthesis (also see Chapter 4). This process produces useful proteins and carbohydrates in

Perched safely above high tide mark, moss, ferns, and other plants thrive atop **Koatoa** mangroves in Madolenihmw.

their leaves and other parts of the plants. Eventually, these plant parts drop off into the water and are carried out with the tides into nearshore waters, where they form part of a rich "nutrient soup." This "soup" serves as food for a variety of reef animals. In Pohnpei, mangroves provide an important source of food for the animals living within the mangrove swamps. These swamps also serve as nurseries for many kinds (species) of young fish, mollusks, crabs, other marine organisms, and birds.

All trees that live in the mangrove swamp must have a means of obtaining oxygen when the tide is in and water covers the roots. Some woody mangrove plants have roots above the ground. These aerial prop roots have special air canals that allow the trees and shrubs to absorb oxygen, which is difficult to get from the thick mud where the mangroves grow.

Their specially adapted roots are a common way of identifying different species of mangroves. **Akapah** (*Rhizophora apiculata*) and **Akelel** (*R. mucronata*), often lumped together as **Ahk**, are easily recognized by their hanging, aerial prop roots.

Other mangroves, **Koatoa** (*Sonneratia alba*) and **Weingal** (*Lumnitzera littorea*) can be recognized by exposed, cone- or pencil-shaped projections (pneumatophores) that rise from the extensive, cable-like roots and stick out from the muddy soil. These exposed roots serve as breathing organs in the swampy areas. Because mangrove trees need oxygen, even a thin coating of oil can block the breathing pores and kill the trees. It is very important to protect mangroves from oil spills.

Still other mangrove trees, such as **Pwulok** (*Xylocarpus granatum*) can be identified by their buttressed (or arched) roots. The aerial, cable, and buttressed roots spread out, allowing the mangrove trees to grow in the loose soil or thick mud. As noted above, these wide-spreading roots effectively trap sediment washed down from the uplands and thus stabilize the soil.

Elegant, arching roots stretch outward to strengthen anchoring of **Ahk** mangroves in Sokehs.

Figure 2.2. Major types of mangrove rooting systems:

a) **Akapah** (*Rhizophora apiculata*) and **Akelel** (*Rhizophora mucronata*) have hanging, aerial prop roots that anchor the trees into the mud;

b) **Koatoa** (*Sonneratia alba*) root system includes upward-sticking cone- or pencil-shaped roots;

c) **Weingal** (*Lumnitzera littorea*) possesses cable-like roots partly visible above ground like twisting and winding snakes;

d) **Sohmw** (*Bruguiera gymnorrhiza*) uses knobby, knee-like breathing organs (pneumatophores) that stick up out of the mud.

Along their most landward edges, mangroves occur alongside trees associated with freshwater swamps and coastal forests.

Inland areas, sheltered yet influenced by tidal changes, support the greatest diversity of mangrove species.

Species diversity declines seaward as fewer species can tolerate harsher conditions.

Mangrove vegetation zones

The mangrove vegetation can be divided into zones from the edge of the island out into the lagoon.

1. ***Ni oaroahr:*** Going from inland areas toward the sea, we first encounter the intertidal border zone where lowland vegetation and *nansapw* meet the mangroves. The actual shoreline is a narrow zone along the high-tide line with little or no vegetation. This zone usually has shallow mud deposits with patches of shrubs and trees typical of coastal vegetation (see p. 56). Some muddy areas in this zone emerge at low tide. This inner margin of the mangrove forest is usually quiet and protected. In places where it can be reached from the sea by channels through the mangroves, it makes good landing places for canoes or boats. It is also an important type of environment for collecting **Elimoang**, the mangrove crab (*Scylla serrata*). This valuable resource is a popular food.

2. ***Weliniak:*** The mangrove forest vegetation at the edge of the intertidal, border zone next to the shore often has more species of mangrove plants than other mangrove zones. This comparatively diverse mangrove forest is often the best site to find and harvest relatively large, valuable mangrove trees. In places further away from the open sea, in water that is less salty, we can find **Soumw** (or **Sohmw**), the oriental mangrove (*Bruguiera gymnorrhiza*) with its looped aerial roots. Another plant associated with the mangrove forest is **Parem**, the nipa palm (*Nypa fruticans*), which has many uses.

3. ***Naniak:*** Further from the shore and closer to the sea, we get deep inside the mangrove swamp, where there are fewer tree species. Much of the mangrove areas around Pohnpei falls within this zone, where the only thriving plants are those that can handle increased salinity and almost permanently flooded conditions. The typical species here are **Akelel** (*Rhizophora mucronata*) and its close relative **Akapah** (*R. apiculata*). These mangrove trees develop interlocking prop roots that grow outward from high positions on the trunk.

4. ***Pidiriniak:*** **Akelel** and **Akapah** often create dense tangles of trees, branches, and prop roots. This is especially so in areas dominated by shorter, dwarfed forms of the same species, which are smaller than normal because the soil is much firmer due to less water circulation. These smaller growth forms are known as ***Ahktik*** and make truly impenetrable parts of the mangrove swamp. Only *Rhizophora* trees typically grow here and animal life is also less abundant in such areas.

5. ***Pahniak:*** The seaward parts of the mangrove swamp differ from the rest by their unique ecological conditions. This is where the influence of the sea is the strongest and the only trees are *Rhizophora* species and *Sonneratia alba*. Parts of this area are flooded even at low tide, and numerous marine organisms live among the roots.

6. ***Likiniak:*** The very fringe of the mangrove swamp, the outermost edge where the vegetation stops and the lagoon begins, is known in Pohnpeian as *likiniak* (*likin* = outside, *iak* = mangrove). Here, we often see an overhang of branches above the water. The overhanging branches have two basic forms, depending upon the dominant mangrove tree species fronting the reef flat or lagoon waters. The more common form of overhanging branches is dominated by **Akelel** and **Akapah**. These large trees (5–10 m or 15–30 ft tall), with a green bushy canopy and reddish brown aerial prop roots, often form a dense wall of vegetation. Less common is the form of overhanging branches dominated by groves of **Koatoa** (*Sonneratia alba*), the mangrove tree with a single thick trunk, open understory, and conical breathing roots (pneumatophores) sticking out of the mud.

Winding through the different zones of mangroves in many places are natural or human-made channels. They enhance water circulation through the swamps, especially during changing tides, and include broad endings of river valleys, narrow tunnel-like passages overhanging with tree crowns, small tidal channels called *dewerik*, and human-made canals.

Interiors of some mangrove forests are nearly impenetrable due to dense tangles of intertwined roots.

Near the seaward edges, mangroves face extreme conditions where only the most salt-tolerant species can survive.

The outermost edge of mangroves forms a wall of roots and foliage facing the quiet waters of the lagoon.

There are several plant species that comprise the mangrove forests on Pohnpei and a number of additional plants that grow in the landward edge of the mangroves. Together, these specially adapted plants form a vegetation zone fringing much of the coast of Pohnpei and surrounding the smaller coral and volcanic islets in the lagoon. Mangroves are also found alongside the lower river channels of Pohnpei. The most common woody plants found in the mangrove forests of Pohnpei are listed in Table 2.3, and typical plants of the mangroves' landward fringe are listed in Table 2.4 below.

Table 2.3. Common trees found in the mangrove forests of Pohnpei.

Scientific name	Pohnpei name	Mwoakilloa name	Pingelap name	Nukuoro name	Kapingamarangi name
Bruguiera gymnorrhiza	Soumw / Sohmw	Johmw	Soal	Donga	–
Lumnitzera littorea	Weingal	Waingal	–	–	–
Nypa fruticans	Parem	Param	–	–	–
Rhizophora apiculata	Akapah	Ak	Ak	–	–
Rhizophora mucronata	Akelel	Ak	Ak	–	–
Sonneratia alba	Koatoa	Koasoa	Koasa	–	–
Xylocarpus granatum	Pwulok	Pwulok	Pwuloak	–	–

A number of other species live on the outskirts of mangrove swamps or on the mangrove trees themselves. A feature of mangrove areas in Pohnpei, which generally experience high rainfall, is the presence of many epiphyte species. Epiphytic plants, by definition, grow upon other plants (e.g., a fern growing upon a tree trunk). Epiphytes are also known as "air plants."

Table 2.4. Common plants growing as epiphytes or in landward edge areas of the mangrove forest of Pohnpei.

Scientific name	Pohnpei name	Mwoakilloa name	Pingelap name	Nukuoro name	Kapingamarangi name
Asplenium nidus	Tehnlik	Kardoap	Saehlik	Lau gadaha	Logo hoo
Barringtonia racemosa	Wihnmoar pwetepwet / Wihn Pohnpei	–	–	Gavaausu daane	–
Derris elliptica	Peinuhp / Uhp	–	–	–	–
Derris trifoliata	Kehoaroahr	–	Kainipil	–	–
Heritiera littoralis	Mworopwinsed / Merepwinsed	–	–	–	–
Hibiscus tiliaceus	Keleu / Kolou	Kɛlɛu	Kaelaeu	Goehau	Hau
Phymatosorus grossus	Kideu	Kamkam / Kino	Kideu	Maaile	Gideu
Pandanus tectorius	Kipar	Kipar	Kipar	Hala	Binu

Bruguiera gymnorrhiza

Derris trifoliata

Heritiera littoralis

Lumnitzera littorea

Nypa fruticans

Rhizophora sp.

Sonneratia alba

Xylocarpus granatum

The importance of mangroves

Mangrove forests provide raw materials for many human uses, including building materials, wood fuel, craft wood, charcoal, resins, and bark products. Unfortunately, in many parts of the tropical world, mangrove forests are being destroyed by overharvesting, dredging for land reclamation projects, and by upslope forest clearing. We should remember that mangrove forests play important roles as natural stabilizing and cleansing barriers along the reef. They also drop leaf and other natural, nutrient-rich litter into the water.

Probably the greatest overall value of these mangroves involves their role in supporting marine food chains, protecting coastal areas, and improving water quality. Among the environment advantages of protecting and maintaining the native mangroves (or even introduced mangroves in some cases) include the following: stabilization of the soil, as well as provision of natural windbreaks, woodlot reserves, wildlife habitat for indigenous species, and food for native animals including birds, forage for bee colonies, and environmental refuges for fish and other biota in the marine food webs (see Duke 2006, 2016). In the case of soil stabilization, *Rhizophora* red mangrove forests hold wet soils in place with their durable, overlapping prop roots that reduce water movement and promote sediment build-up in areas that normally would be eroded. Preservation of indigenous *Rhizophora* mangrove vegetation acts generally as a windbreak along the margins of the coastal areas, while also offering wildlife places of refuge during strong storms such as typhoons.

Different parts of the mangrove environment provide essential habitat for many animal species that depend upon the mangrove plants for food and shelter. Some of them are shown on the colorful illustration on these pages (based on Knox 1978).

Points to remember about the mangrove forest:

1. The destruction of *weliniak* or mangrove forest reduces the breeding and nursery grounds for many animal species.

2. Since the roots of mangrove forest trees stabilize the coastal areas by trapping sediment, excessive removal of these trees allows damaging sedimentation (smothering) of the reefs located near the mangrove forests.

3. This reduces the amount of fish and mollusks that live on and can be harvested from the reefs lying beyond the mangrove forests.

2.3.4. Freshwater wetland vegetation

Freshwater wetlands are regularly flooded or saturated by ground or surface water long enough to kill plants that cannot tolerate waterlogged soils. Freshwater wetland habitats consist of swamps, marshes, bogs, and similar areas with saturated soils. Freshwater wetland vegetation can be divided into five types: freshwater aquatic; freshwater marsh; freshwater swamp; cultivated wetlands; and ruderal wetlands.

1. **Freshwater aquatic vegetation:** This type of vegetation develops in permanently flooded areas. Freshwater aquatic vegetation is found in slow-moving streams, reservoirs, irrigation ditches, natural and artificial ponds, and open bodies of water surrounded by freshwater marshes. The plants in this aquatic vegetation grow submerged, partially submerged, or floating. They lack well-developed structural support provided by the rigid trunks, stems, and branches of plants growing in soil on land. Most of the species in this type of vegetation are annuals (plants which complete their whole life cycle—germinate, grow to maturity, develop flowers, set seed, and die—in a single year or less). A few species found in the freshwater aquatic vegetation are perennial plants (those which live longer than one year). Some alien species, if introduced into the freshwater aquatic vegetation on Pohnpei, could become troublesome weeds that might multiply rapidly and choke waterways. Two examples of aquatic plants that can produce this kind of problem are "waterlily" (*Nymphaea* spp.) and "water hyacinth" (*Pontederia crassipes*). Both of these plants are alien species introduced into Pohnpei during the 20th century as ornamentals (e.g., see Herrera et al. 2010).

River pool at Pilen Seleur in Kitti.

As noted previously in the section on seagrasses, there is a native aquatic herb found in the freshwater vegetation of Pohnpei that scientists

Figure 2.3. Generalized diagram of wetland vegetation types on Pohnpei and other Micronesian islands (based on Stemmermann 1981).

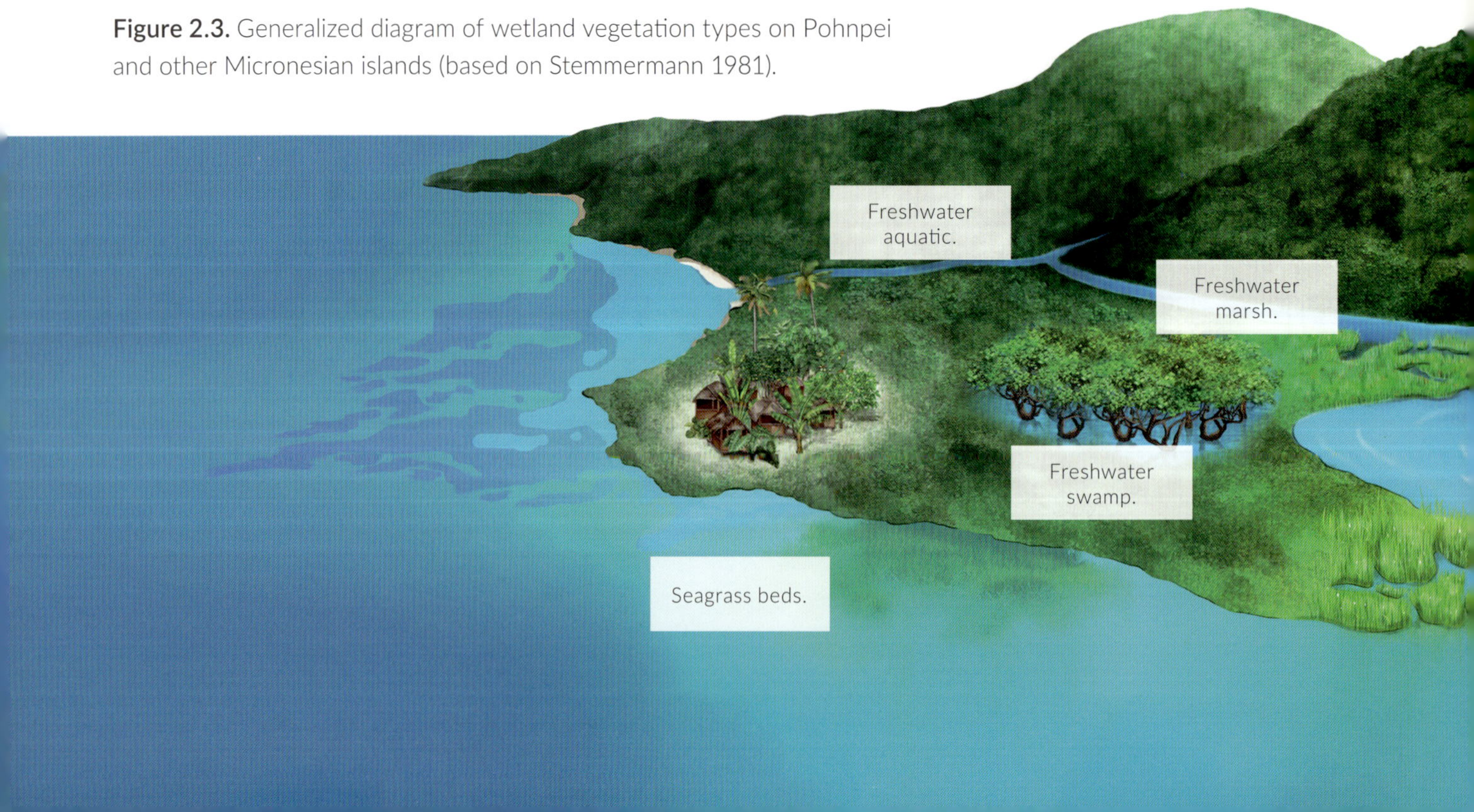

recognize as *Blyxa aubertii*, which is known in Pohnpeian as **Oaloahdenpil**. Reportedly used for medicinal purposes, its Pohnpeian name shows its association with two common native saltwater herbs (seagrasses) in Pohnpei; these are *Thalassia hemprichii*, known in Pohnpeian as **Oaloahd**, and *Enhalus acoroides*, known in Pohnpeian as **Oaloahdilap**. Botanists believe that all three of these species belong to the same family: Hydrocharitaceae. Indeed, *Blyxa aubertii* is a true freshwater aquatic plant species that is normally found in a wide variety of habitats such as streams and rice paddies. Its adaptation to being submerged in water, and its similarity in appearance to **Oaloahd** (*T. hemprichii*) inspired its Pohnpeian name, **Oaloahdenpil** ("**Oaloahd** of fresh water").

2. **Freshwater marsh:** The freshwater marsh type of vegetation on Pohnpei, known as *nanpwoaki*, contains mostly non-woody species, especially plants that have roots in the waterlogged soil such as large grasses (Poaceae), and sedges (Cyperaceae), with scattered pandanus trees. In addition to many other botanical differences between sedges and grasses, if you compare the stems of most sedges and grasses, you will find that "sedges have edges and grasses [stems] are round."

Marshy area in Palikir.

A few of the herbaceous species found in the freshwater marshes of Pohnpei include: *Phragmites karka* (**Lirau**), *Scirpodendron ghaeri* (**Pwoaki**), *Cyperus javanicus* (**Sapasap** or **Use**), and *Ludwigia octovalvis* (**Delahrik** or **Deleurik**). Crops can be cultivated in many marshy areas if they are drained. In the past on Pohnpei, many marsh areas were drained and cleared of native vegetation for farming purposes. This occurred especially in places with rich muddy soil where true taro, **Sawa** (*Colocasia esculenta*), and swamp taro, **Mwahng** (*Cyrtosperma merkusii*), could be cultivated.

3. **Freshwater swamp:** Surveys of the total land area covered by forest of different types on Pohnpei were carried out in 1976 (see MacLean et al. 1986) and more recently by Donnegan et al. (2011). According to the data recorded in these surveys,

the total area covered by freshwater swamp forests alone declined from 421 ha in 1976 to 99 ha in 2006. Swamp forests still exist on Pohnpei but are rare today; they are found where soils are flooded with fresh or slightly salty water. Although some smaller plants such as the "beaksedge" *Rhynchospora corymbosa* (**Matamat**) occur in the forests of freshwater swamps, these forests are mainly dominated by woody plants, such as *Terminalia carolinensis* (**Kehmah**), *Campnosperma brevipetiolata* (**Dohng**), and *Barringtonia racemosa* (**Wihnmoar pwetepwet** or **Wihn Pohnpei**); these species can grow on soils that are often, if not always, waterlogged. The swamp forests are found in depressions (areas sunk below the surrounding terrain) or other poorly drained places, usually behind the beach strand or mangrove areas. In the past, many freshwater swamps (as well as marsh areas) were cleared or severely altered for farming purposes, especially in places with rich muddy soil where true taro, **Sawa** (*C. esculenta*) and swamp taro, **Mwahng** (*C. merkusii*), could be cultivated. Many of these former freshwater swamps are presently covered with secondary vegetation. Although several plant species of the original swamp forests of Pohnpei still can be found, in those wetlands where considerable disturbance has taken place and no root crops are being cultivated, successional plants such as *Hibiscus tiliaceus* (**Keleu** or **Kolou**), and *Phragmites karka* (**Lirau**) are commonly found.

Swampy area in Nandaku.

4. **Cultivated wetland vegetation:** The cultivated wetland vegetation of Pohnpei exists in areas where traditional "root crops" are grown. As noted above, cultivated plants have replaced natural freshwater marsh and swamp plants in many wetland areas of Pohnpei, often those near the coast fringing

Rather than growing deep into waterlogged soil, swamp trees develop wide, sprawling roots that provide stability and access to oxygen.

mangrove swamps. True taro, **Sawa** (*C. esculenta*), and swamp taro, **Mwahng** (*C. merkusii*), are the two most important cultivated plants in this largely artificial or human-made vegetation. Although **Sawa** is generally preferred to **Mwahng**, it is cultivated to a lesser degree than the larger traditional aroid root crop. **Mwahng**, which has much larger corms than **Sawa**, is normally planted in very muddy, deep soils, while **Sawa** is often grown at the drier margins of these cultivated wetlands. It should be noted that many so-called root crops, such as taro, are grown for the edible corm (the swollen underground portion of the stem, similar to a plant bulb, and not a true root).

5. **Ruderal wetland vegetation:** Ruderal wetlands develop in human-influenced areas or abandoned "waste" lands, such as roadside ditches, drainage canals, excavated places, bomb sites, rubbish dumps, neglected taro patches, and other areas with poor drainage. They typically support hardy, fast-growing plant species that thrive in disturbed or nutrient-rich environments. Such plants are basically weeds, which in Pohnpei are called *dihpw* and generally refer to undesirable, uncultivated plants. While not naturally occurring, ruderal wetlands can still have useful ecological functions, such as flood control, water filtration, and habitat for aquatic invertebrates, insects, and other small animals.

Taro patch at Pohnrakied in Kolonia.

Roadside pools of rainwater in Sokehs.

Unaffected by tides, water levels in freshwater swamps rise and fall with the weather, surging during heavy rainfall.

Nypa fruticans

Barringtonia racemosa

Blyxa aubertii

Cynometra ramiflora

Hibiscus tiliaceus

Table 2.5. Common plants in freshwater wetlands on Pohnpei.

Scientific name	Pohnpei name	English name or description
Barringtonia racemosa	**Wihn Pohnpei**	powder-puff tree
Blyxa aubertii	**Oaloahdenpil**	–
Campnosperma brevipetiolata	**Dohng**	–
Colocasia esculenta	**Sawa**	taro
Cynometra ramiflora	**Kemeu** / **Koamoau**	–
Cyperus javanicus	**Sapasap** / **Use**	marsh sedge
Cyrtosperma merkusii	**Mwahng**	swamp taro
Heritiera littoralis	**Mworopwinsed** / **Merepwinsed**	looking glass tree
Hibiscus tiliaceus	**Keleu** / **Kolou**	coastal hibiscus
Ludwigia octovalvis	**Delahrik** / **Deleurik**	swamp primrose
Nypa fruticans	**Parem**	nipa palm
Phragmites karka	**Lirau**	troopical reed
Rhynchospora corymbosa	**Matamat**	golden beaksedge
Scirpodendron ghaeri	**Pwoaki**	–
Terminalia carolinensis	**Kehmah**	Micronesian terminalia

Phragmites karka

Rhynchospora corymbosa

Scirpodendron ghaeri

Terminalia carolinensis

2.3.5. Vegetation of cultivated areas

Plants and vegetation in the coastal plains and the lower, more gently sloping mountain areas of Pohnpei have been significantly altered by human activity. These profound changes were brought about by the labor of many generations of Pohnpeians, driven by their need to access and utilize agriculturally fertile lands for food production.

Over the centuries, these areas became the *nansapw*, where people traditionally reside and cultivate crops. These lands offer rich soil, ideal for growing useful plants, and are typically located closer to the lagoon, where fish and other marine resources can be caught and collected. As a result, most of the people of Pohnpei live in these areas of coastal lowlands and lower mountain slopes.

Before the first people arrived in Pohnpei, the coastal and lowland plains were probably covered by thick forests. Most of these forests have been cut down and replaced with breadfruit (**Mahi**) and other cultivated trees, as well as yams (**Kehp**), bananas (**Uht**), taro (**Sawa** and **Mwahng**), kava (**Sakau**), and additional crops. Moreover, areas of lowland native vegetation have been replaced by homesteads, paths, roads, and other structures built by people.

Replacing the original, wild forests, these new areas of cultivated trees (and some wild plants) can be considered artificial (human-made) forests. Scientists refer to them as agroforests. This type of traditional agroecosystem includes trees with associated understories of shrubs and herbs, and combines native and introduced species (see Fig. 2.4 below).

Figure 2.4. Diagram showing general view and different sizes of some of the plants cultivated or growing wild in the *nansapw* agroforests of Pohnpei (iREi, based on Raynor 1989).

A tall breadfruit tree stands beneath the majestic Pahntakai waterfall.

Kolonia town nestled within Pohnpei's cultivated lowlands.

Paved road winding through *nansapw*.

A house clearing surrounded by useful plants.

Traditionally, all lands on Pohnpei are classified broadly according to the type of vegetation and its use. In Pohnpeian culture, the two basic, contrasting classifications are *nanwel* and *nansapw*.

Nanwel refers to "wild" lands, or lands not under human management:

- the mountainous areas which are typically covered with dense, wild forest, and
- land that has been abandoned long enough to have reverted to secondary forest.

Nansapw, on the other hand, refers mainly to the "humanized" lands where people settle, reside, and cultivate crops:

- flat areas and narrow plains between the mountains and sea, and
- cleared lands that are cultivated or used in some other way.

Nansapw is further broken down into different types of land based on current use, including:

- *pahnwel* (indigenous agroforest),
- *mwetiwel* (forest clearing for farming),
- *lehpwel* (cultivated or uncultivated swamps),
- *dihpw* (open areas overgrown with grass and weeds, i.e., "weedy"),
- *wasahn kousoan* (house clearings), and
- *mwaht* (gardens and plantations).

Although Pohnpeians make use of various foods and other products extracted from the *nanwel*, most of the everyday subsistence and prestige foods and other products are cultivated in or extracted from the *nansapw* areas.

The two areas of *nansapw* and *nanwel* further contrast in that, through diligence and hard work, *nanwel* areas can be transformed into *nansapw*, and conversely, through abandonment, *nansapw* can eventually revert back to *nanwel*. The process of farming is known as *sapwasapw* (literally "making land"), and newly cleared land in the secondary or mature forest is known as *sapwkapw*.

Pahnwel.

Mwetwel.

Lehpwel.

Dihpw.

A lovingly maintained ornamental garden.

A commercial plantation of black pepper.

What is agroforestry?

Agroforestry, especially in the form of the indigenous agroforestry system, is the main agricultural production system in Pohnpei. Therefore, agroforestry is of greatest importance to the people of Pohnpei in the role it plays in providing the fertile environments where the important crops and domesticated animals are produced.

Technically, agroforestry has been defined as any sustainable land-use system that maintains or increases total yields by combining food (annual) crops with tree (perennial) crops and/or livestock on the same unit of land, either alternately or at the same time, using management practices that suit the social and cultural characteristics of the local people and the economic and ecological conditions of the area.

Agroforestry is the major agricultural land use in Pohnpei, as it has been for centuries. In fact, nearly all production of subsistence, local market, and export crops is done by family units using local agroforestry technology.

> "The most important agricultural products are breadfruit (*Artocarpus altilis*), bananas (*Musa* sp.), yam (*Dioscorea* spp.), kava (*Piper methysticum*), coconut (*Cocos nucifera*) and taro (*Colocasia esculenta, Cyrtosperma merkusii, Xanthosoma sagittifolium, Alocasia macrorrhizos*). Food production on the island is used to support both day-to-day subsistence and a "prestige economy" that reinforces social status through redistributive feasts (Bascom 1948, 1965). There is [archaeological and oral history] evidence that this prestige economy has been present for a large portion of the island's occupation." (Levin 2016b: 7; also see Haun 1984 and Ayres and Haun 1985)

In Pohnpei, agroforestry is a complex, sustainable system of land use.

Figure 2.5. A comparison of Pohnpei natural ecosystems (such as *nanwel, nanlehpwel,* and *naniak*), essentially containing only native species without human disturbance...

Agroforestry does not generally exhaust soil fertility and thus continues to provide a stable source of much food and other resources, such as timber, fiber, and medicine for the people of the island. Nevertheless, although the traditional cropping system of agroforestry has significant value in its provision of nutrition, cultural preservation, and rural economic stability, this long-lasting local farming system is changing. That is happening because of an inadequate understanding among some people of the important role that the traditional cropping system and practices play in food production, local economics, and ecological balance.

Before we discuss the problems associated with the current changes and decline of traditional agroforestry and the relationships of this decline to human population growth and commercial development, let us briefly review the historical development of Pohnpei agroforestry and some key cultural factors involved.

When early Micronesian settlers first arrived in Pohnpei about 2,000 years ago, they found an island covered with forest, rich with plant and animal species, including many unique to the island. The early settlers learned to make use of the native species in their newfound home. Settlement and adaptation most likely started on the coasts and progressed inland. Plants brought in by these early settlers, especially breadfruit and yams, probably became major components in their agricultural system soon after their arrival.

Scientists suggest that the early settlers adapted to the island which they had discovered and modified the vegetation to fit their own survival needs. Over a long period of time, numerous crop plants and agricultural techniques were introduced through continued waves of human migration and cultural exchange. Therefore, local agroforestry on Pohnpei is the result of many hundreds of years of development and refinement.

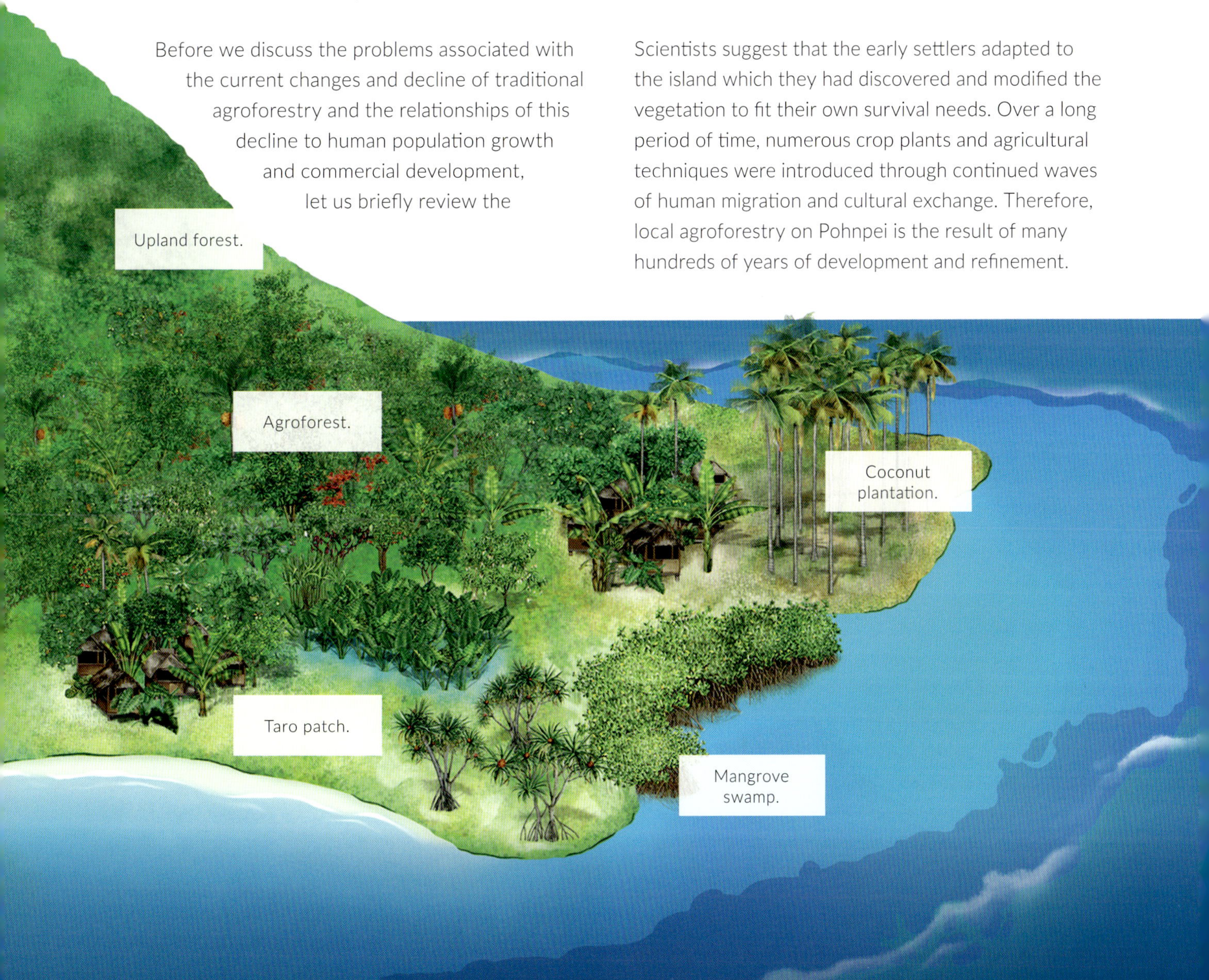

... and traditional agroecosystems (*nansapw*), consisting of cultivated forests, open areas, taro patches, intermittent gardens, and other land used to cultivate plants.

Outline of the history of agroforestry development on Pohnpei

1. The first farmers of Pohnpei probably practiced shifting cultivation in which fields are cleared, the cuttings burned if possible, and crops are planted for two or three years; these fields were then allowed to grow wild (through ecological succession) for several years before starting over, while other areas are brought into cultivation through the same process. Although shifting cultivation predominated during the early settlement phase, Pohnpeians had developed an elaborate agroforestry system with permanent tree crops and a multi-layer canopy of vegetation at least several hundred years ago. This system of permanent tree crops was most likely an ecologically induced (a natural environmental) response of a growing human population to the very wet, cloudy environment and rocky, steep topography.

2. Unlike on many other islands in Micronesia, Pohnpeian men, rather than women, did most of the farming, as well as the deepwater fishing.

3. Excess production of both food and non-food crops, such as **Sakau** (*Piper methysticum*), giant yams **Kehp** (*Dioscorea* spp.), and pit breadfruit, were and are an important feature of an elaborate prestige economy that was well-developed by the time the first Europeans arrived.

4. Most of the crops that are important in both subsistence and prestige agriculture today, such as **Sakau** (kava), **Kehp** (yams), and **Mahi** (breadfruit), are not native, but alien species brought to Pohnpei by the early Micronesian settlers before the first Europeans arrived. However, many new cultivated varieties of these crop species have been introduced since Western contact began in 1826. While Pohnpeians have accepted many new crops (introduced cultivated plants) and even a few new methods of growing these plants, they have eventually integrated all of these new crops and cultivation techniques into their traditional agriculture system.

A typical view of cultivated vegetation in Pohnpei, including herbs, vines, shrubs, tree-like herbs (papaya, banana), and trees.

Table 2.6. Some cultivated trees in the *nansapw*.

Scientific name	Pohnpei name	Mwoakilloa name	Pingelap name	Nukuoro name	Kapingamarangi name
Annona muricata	Sei / Soai	Joi	Saei	Ssai	Tatap
Areca catechu	Pwuh	Pwu	Pwu	Buu	Buu
Artocarpus altilis	Mahi	Moai	Maei	Gulu	Gulu
Ceiba pentandra	Koatun	–	–	–	–
Citrus aurantifolia	Karer / Laman / Laim	Lɛmɛn / Laim	Karaer	Laim	Beleni mmala
Cocos nucifera	Nih	Ni	Ni	Nui	Niu
Fagraea berteroana	Seir / Pwuhr	Seir / Pwur	–	–	–
Falcataria moluccana	Tuhke karisihn	Koarsihn	Karesihn	–	–
Inocarpus fagifer	Mworopw / Mwoaroapw	Mwoaroapw	Mwoaroapw	Ihi	–
Mangifera indica	Kehngid	Koahngid	Kaehngid	Gaangidi	Geengidi
Plumeria obtusa / P. rubra	Pwohmaria	Pwohmaria	Pwohmaria	Bua maliaa	Laagau hai akai
Psidium guajava	Kuahpa	Kuahpa	Kuahpa	Guaafaa	Kuabaa
Swietenia macrophylla	Mahokani	Moakini	–	–	–
Syzygium jambos	Apel en wai	–	–	–	–
Syzygium malaccense	Apel en Pohnpei	Apɛl	Apol	Aabele	Aabele lauwehe

Table 2.7. Some cultivated herbs and shrubs in the *nansapw*.

Scientific name	Pohnpei name	Mwoakilloa name	Pingelap name	Nukuoro name	Kapingamarangi name
Bambusa vulgaris	Pehri	Pehri	Pehri	Madila	Madila
Capsicum frutescens	Sele	Jeli / Pɛɛpɛr	Saele / Paepaer	Silibeeba	Telee
Carica papaya	Memiap / Moamiap	Kaniahpo	Kaeiniap	Mami	Memeabu
Colocasia esculenta	Sawa	Jawa	Sawa	Dalo	Dala
Cordyline fruticosa	Dihng	–	–	Manu lauligi	–
Dioscorea alata	Kehp	Koahp	Kaehp	Iaama	Eeme
Musa × paradisiaca	Uht	Wus	Wis	Gahudi	Hudi
Piper methysticum	Sakau	Jakau	Saekau	Sagau	Tagaao
Saccharum officinarum	Sehu	Doa	Seu	Dolo	Mee ngau

Artocarpus sp.

Capsicum frutescens

Colocasia esculenta

Mangifera indica

Manihot esculenta

Musa sp.

Plumeria sp.

Syzygium sp.

In addition to the cultivated trees, shrubs, and herbs in the *nansapw*, there are also many woody and non-woody plants that grow wild. Several types of plants (different species) are not cut down and removed from the cultivated areas. There are a variety of reasons for leaving certain wild plants to grow among the cultivated plants of the *nansapw*, including the resource values (various uses) these wild plants may have. The Table 2.8 lists some of the plant species commonly seen growing wild within *nansapw* areas of Pohnpei.

Table 2.8. Some wild plants in the *nansapw*.

Scientific name	Pohnpei name	Mwoakilloa name	Pingelap name	Nukuoro name	Kapingamarangi name
Aglaia mariannensis	Marasau / Maroasoa	Marjau	–	–	–
Alocasia macrorrhizos	Ohd / Wehd	Wod	Woad	Ngaungau	Ngaungau
Atuna racemosa	Ais	Aij	–	Adidi	Adidi
Campnosperma brevipetiolata	Dohng	Dohng	Doahng	–	–
Cananga odorata	Seiren wai / Pwuren wai	Jeir in wai	–	–	–
Claoxylon carolinianum	Koahi	–	–	–	–
Elaeocarpus carolinensis	Sadak	Jadak	–	–	–
Ficus prolixa	Aiau	Au	Aiau	Aoa	Iaiau
Ficus tinctoria	Nihn	Kuwɛhn	Nin*	–	–
Glochidion ramiflorum	Mwehk	–	–	–	–
Ixora casei	Ketieu	Koassoau	Kaelaeseu	Gadiou	–
Metroxylon amicarum	Oahs	Oaj	–	Gala mau	Oodi
Myristica hypargyraea	Karara	Karara	–	–	–
Pandanus conoideus	Mwatal	Mwasal	–	–	–
Parkia korom	Kurumw	–	–	–	–
Pittosporum ferrugineum	Kamal	–	–	–	–
Sida rhombifolia	Koiohnleng	–	–	–	–
Sphaeropteris nigricans	Katar	Kasar	Kaesar	Manu haihale	–
Syzygium carolinensis	Kehnpap	–	–	–	–
Syzygium stelechanthum	Kirekinwel	–	–	–	–

* Another name used on Pingelap is similar to that from Mwoakilloa and was recorded as "Kawain" by St. John (1948).

Aglaia mariannensis

Alocasia macrorrhizos

Campnosperma brevipetiolata

Cananga odorata

Ficus tinctoria

Glochidion ramiflorum

Ixora casei

Myristica hypargyraea

Agroforestry and taro patches on the outlying atolls of Pohnpei State

Agroforestry, including wetland taro cultivation, remains the major agricultural land use on the outlying atolls of Pohnpei State, as it has been for centuries. Nearly all production of crops is done by family units using local agroforestry technology. Commercial monocropping involving coconut plantations was developed in historic times, but has been steadily declining in recent decades.

Similar to the customary lowland farming on Pohnpei, agroforestry areas in the outer islands consist of cultivated trees mixed with some wild trees and associated understories of shrubs and herbs (see Table 2.9). This type of human-created forest, or traditional agroecosystem, combines native and alien species and is a complex, sustainable system of land use. Unlike monoculture, this system is sustainable and has supported human life on Micronesian atolls for about 2,000 years. Unfortunately, this time-honored system is changing with new socioeconomic realities as well as due to climate change and sea level rise. For example, the use of centuries-old taro pits to cultivate Swamp taro, a key staple food of Micronesian atolls, is declining on many islands. This is due to importation of now preferred foreign foods, such as rice and flour, as well as saltwater intrusion overland during storms or into groundwater and the root zone during droughts.

Coconut grove on Pingelap Atoll.

Taro patch on Mwoakilloa Atoll.

Table 2.9. Some cultivated plants in the outer islands.

Scientific name	Pohnpei name	Mwoakilloa name	Pingelap name	Nukuoro name	Kapingamarangi name
Annona muricata	Sei / Soai	Joi	Saei	Ssai	Tatap
Artocarpus altilis	Mahi	Moai	Maei	Gulu	Gulu
Carica papaya	Memiap / Moamiap	Kaniahpo	Kaeiniap	Mami	Memeabu
Citrus aurantifolia	Karer / Laman / Laim	Lɛmɛn / Laim	Karaer	Laim	Beleni mmala
Cocos nucifera	Nih	Ni	Ni	Nui	Niu
Colocasia esculenta	Sawa	Jawa	Sawa	Dalo	Dala
Cyrtosperma merkusii	Mwahng	Mwɛhng	Mwaeiang	Daogoli	Bulaga
Inocarpus fagifer	Mworopw / Mwoaroapw	Mwoaroapw	Mwoaroapw	Ihi	–
Mangifera indica	Kehngid	Koahngid	Kaehngid	Gaangidi	Geengidi
Manihot esculenta	Kehp tuhke / Dapiohka	Dapiohka	Dapioka / Maenioak	Dabiogaa	Tabiogaa
Musa × paradisiaca	Uht	Wus	Wis	Gahudi	Hudi
Pandanus tectorius	Kipar	Kipar	Kipar	Hala	Binu
Plumeria obtusa / P. rubra	Pwohmaria	Pwohmaria	Pwohmaria	Bua maliaa	Laagau hai akai
Saccharum officinarum	Sehu	Doa	Seu	Dolo	Mee ngau
Tacca leontopetaloides	Mwekimwek	Mwoakmwoak	Mwaekaemwaek	Bie	Mogumogu
Terminalia catappa	Wahn dipwoapw	Wehn doapwoapw	Wahn dipwoapw	Hua laagau	Gegebinu

Carica papaya

Cocos nucifera

Crinum asiaticum

Cytrosperma merkusii

Musa × paradisiaca

Pandanus tectorius

Tacca leontopetaloides

Terminalia catappa

2.3.6. Agroforestry and Pohnpeian culture

Pohnpei, like other islands of Micronesia, has a modern administrative system of government. In addition, there is the traditional political system which continues to function in Pohnpei, with its authority still dominating the social and agricultural life of the island, especially in the rural areas.

Each of the five politically semi-autonomous municipalities (*wehi*) still has its well-established leadership headed by the paramount chiefs. This leadership is in the form of diarchy, with two complementary offices titled the *Nahnmwarki* and the *Nahnken*. These two chiefships are inherited through a political or cultural succession based on the female line of the family (matrilineality). These leaders exert their political influence through the authority of Pohnpeian cultural tradition and by controlling a stock of titles known as *mwahr*. Although there are no distinct classes, the common people in each municipality are organized into levels of ranked status or titles.

Tiahk en wahu

The chiefs also serve as the center of the traditional prestige economy, which is based on complex rituals of redistribution of agricultural production. This Pohnpeian system is governed by the concept of *tiahk*, the customary code of the island. This traditional code calls for *wahu*, roughly translated as respect, and is clearly expressed through respectful or valued behavior—*tiahk en wahu*.

Traditional social activities, including those involving the cultivation, exchange, and use of plants, are generally referred to as *doadoahk en wahu*, or work of respect (for more detailed discussion, see Shimizu 1982). The German colonial administration introduced "Western" or European land ownership concepts in 1907, giving individual title to a large percentage of the land on Pohnpei. Nevertheless, because of the ongoing influence of Pohnpeian *tiahk*, the high chiefs still control the land in their respective municipalities, at least spiritually. The paramount chiefs continue to "lend" land to their subjects for their use.

Pohnpeian *tiahk* requires that this chiefly generosity of land-lending be compensated by tribute offerings mainly consisting of agricultural products such as pigs, yams (**Kehp**), kava (**Sakau**), and breadfruit (**Mahi**). The chiefs redistribute these offerings, and this encourages an ongoing cycle of mutual exchange (reciprocity).

Those people who continue to accumulate and increase their tributes are recognized by their leaders, who give them higher titles. This, in turn, encourages continued participation in this traditional *tiahk en wahu* system among the entire Pohnpeian population; therefore, even today the agroforestry system of the island remains closely linked with the traditional cultural system.

A display of numerous **Sakau** plants, prepared for a significant cultural occasion.

A vehicle in Pohnpei's Cultural Day parade, beautifully adorned with local plants.

Doadoahk en wahu

Work of respect, *doadoahk en wahu*, is divided into two basic types of contributions to the chiefs:

- *tou*, voluntary contributions, and
- *pwukoah*, required contributions.

The obligatory *pwukoah* contributions take two principal forms:

- *nohpwei*, a tribute of "first fruits," and
- *kamadipw*, which is the practice of large-scale feast ceremonies.

Nohpwei

Nohpwei is the required tribute of *mwowe* (the first one) of specific crops to the chiefs. The meaning of the *mwowe* differs from plant to plant, and therefore results in many different types of *nohpwei* tributes: For example, for seasonal crops, such as breadfruit (**Mahi**), pandanus (**Kipar**), and to a lesser extent, yam (**Kehp**), the *mwowe* refers to the first fruits of the new season, while for non-seasonal crops, such as kava (**Sakau**), sugar cane (**Sehu**), banana (**Uht**), and swamp taro (**Mwahng**), *mwowe* is defined as the first harvest for a particular field or planting.

Table 2.10. *Nohpwei* system of tributes on Pohnpei.

Seasonal *nohpwei*	**Mahi** (breadfruit)
	Kehp (yam)
	Kipar (pandanus)
	Kehngid (mango)
Instrumental *nohpwei*	**Sakau** (kava)
	Sehu (sugar cane)
	Uht (banana)
	Mwahng (swamp taro)
Occasional *nohpwei*	**Sakau** (kava)

Adapted from Shimizu 1982.

Tribute of **Sakau** delivered to a respected traditional leader.

Seasonal *nohpwei*

The seasonal *nohpwei* tributes are still practiced widely, especially for breadfruit and yams. The *nohpwei* for these two important crops is defined by the various stages of their growth. Seasonal *nohpwei* tributes are performed by each *kousapw* (subsection of the respective Pohnpei municipalities) in the form of a presentation to the paramount chiefs. Later, members of *kousapw* may also perform the *nohpwei* tribute to their own *kousapw* chief, the *soumas*.

Instrumental *nohpwei*

Instrumental *nohpwei* differs from the seasonal *nohpwei* in that the *mwowe* (or "first offerings") of these plants is defined in terms of the specific fields in which they are cultivated. When a person clears land or enlarges an existing planting and then successfully harvests the crop, he brings the first crop from that endeavor as tribute to the chiefs. This type of *nohpwei* is only rarely practiced today.

Occasional *nohpwei*

Occasional *nohpwei* involves only the **Sakau** (*Piper methysticum*) plant. That is because **Sakau** transcends all other crops in its *wahu* value; thus, it is the key component of nearly all traditional activities. **Sakau** is presented and prepared during meetings, when asking for favors or asking forgiveness for particularly serious transgressions, and during marriage ceremonies and various other occasions when a commoner needs to meet with higher titled Pohnpeians or paramount chiefs.

Table 2.11. Breadfruit *nohpwei*.

kehmei	Presentation of stone oven (*uhmw*) cooked **Meikole** (fruit fermented for two days). After this, people can eat **Meikole** in its intact form, not cut or sliced.
kerismei	The presentation of newly formed young fruit, indicating that main breadfruit season (*rahk*) is coming.
lihli	Tribute of pounded *kanu* (fruit fermented for a day), roasted over the *uhmw*. After this, people are permitted to eat *lihli* and breadfruit roasted, seasoned with coconut, and/or sliced.
pahini	Tribute of cooked **Meikole** breadfruit in a huge *kiamoroh* (coconut frond) basket, indicating that the full season for breadfruit has arrived.
sakalap	Tribute of large amount of breadfruit cooked in an *uhmw*. Notification that abundant breadfruit is available.
songmar	Presentation of fermented breadfruit (*mahr*) cooked in an *uhmw*, after which people are permitted to eat *mahr*.
umwen luhwen mei	Tribute of *uhmw*-cooked **Meinsahrek** (rough-skinned breadfruit type), which tend to ripen late in the season. Notification that breadfruit have been exhausted and that the yam season is to start.

Table 2.12. Yam *nohpwei*.

idihd	Tribute of cooked yams which are grated and seasoned with coconut milk. After this, eating of grated yam (*idihd*) is permitted.
kamadipw en wahu	A feast in which kehi (entire yam cluster attached to carrying pole) are presented. After this, people are permitted to use *kehi* in their feasts.
keidisol	A feast which terminates the use of *kehi*. After this, people can only use single tubers at feasts.
kotokehp	Tribute of *uhmw*-cooked tubers of yam with freshly cut vine attached. After this, people are permitted to eat whole yam tubers.
umwen kehp pwon	Presentation of *uhmw*-cooked whole yam tubers. Notification that people have been eating yam only in the unsliced form.
umwen pelien kehp	Tribute of cooked divided tubers. Notification that yams have grown enough to be divided and cooked. After this, people are permitted to eat sliced yam tubers.

Kerismei, *pahini* and *sakalap* occur rarely today; *songmar* and *umwen luhwen mei* now often occur with other *nohpwei* (Raynor 1991, after Shimizu 1982).

Kamadipw

Kamadipw are large-scale feasts generally held to recognize crucial events in peoples' lives, for example the peak of the yam harvest, the coronation of chiefs, acceptance of titles, deaths, etc. *Kamadipw* are performed with a single intricate ritual, regardless of the event or size of the gathering. Paramount chiefs, or high-titled family members at smaller family feasts, preside over the *kamadipw* as honored guests. The central aspect of all *kamadipw*, regardless of purpose, is the tribute of the five important agroforestry prestige items: yams, breadfruit, **Sakau**, and pigs (or rarely nowadays dogs). These items are required for all mature male participants, and the sheer size and amount of these items are aptly described by the English translation of the word *kamadipw*, which literally means "to beat the bush." As a result, much of Pohnpei's agroforestry system, beyond meeting basic food, medicine, and material needs, focuses on prestige tribute commodities. The production and consumption of these prized crops are closely linked to traditional social customs and the two main agricultural seasons recognized on Pohnpei: *rahk* and *isol*.

Rahk

Rahk season is based primarily on **Mahi**, breadfruit, the most important traditional staple food of Pohnpei. *Rahk* generally lasts between April and August and is considered the "time of plenty."

Rahk is defined by the harvesting of breadfruit. In addition, many other crops bear their seasonal fruits during the *rahk* season. These include ylang ylang (**Seiren wai**, *Cananga odorata*), mango (**Kehngid**, *Mangifera indica*), Polynesian chestnut (**Mworopw** or **Mwoaroapw**, *Inocarpus fagifer*), local apples (**Apel en wai**, *Syzygium jambos*, and **Apel en Pohnpei**, *Syzygium malaccense*), pandanus (**Kipar**, *Pandanus tectorius*), and others.

The *rahk* season is also known as the "lazy time," because all a farmer has to do is go out and harvest fruit. Production is not considered to be under human influence during this season; instead, it is believed to be a function of nature (although a bad *rahk* season is often traditionally blamed on human divergence from the accepted ways of *tiahk*).

Figure 2.6. Yam and breadfruit seasonality on Pohnpei.

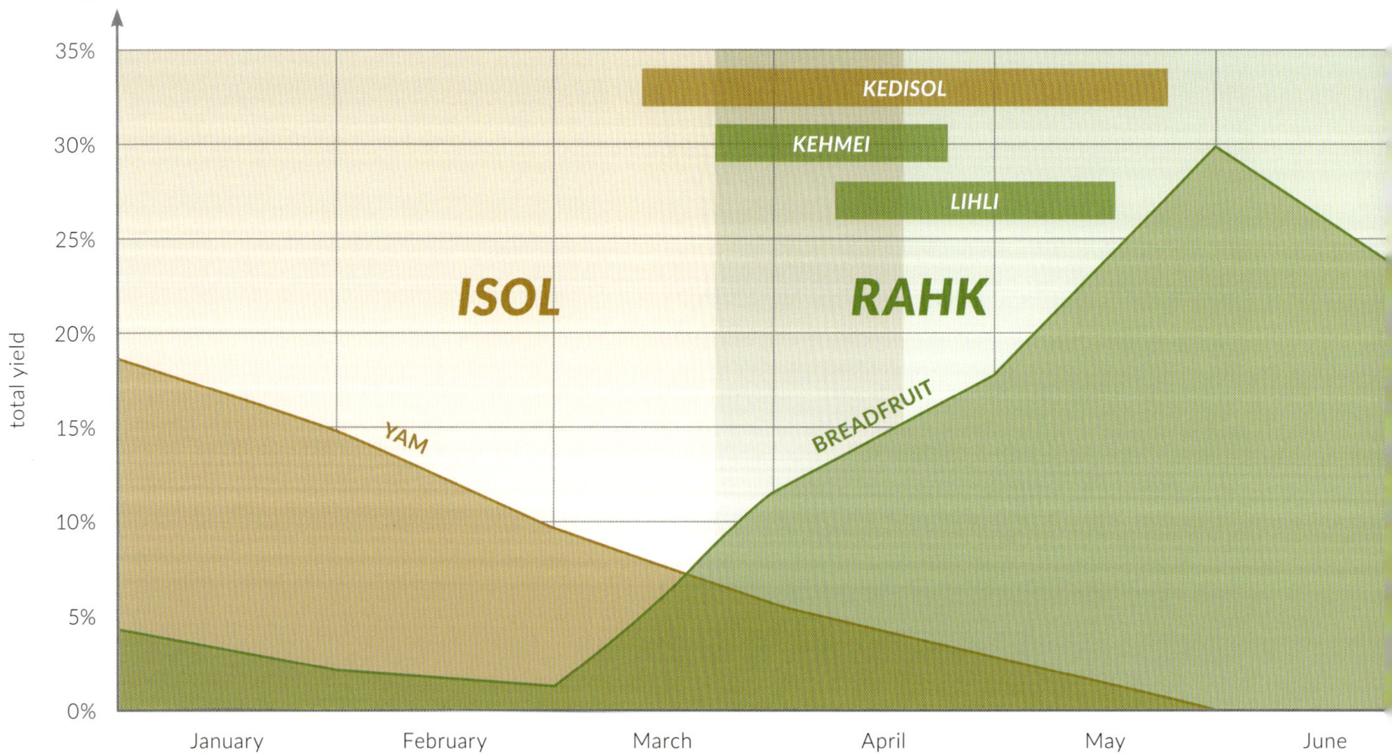

The beginning of the breadfruit season on Pohnpei is known as *menpwukoadoak*, and the end of the breadfruit season is referred to as *menpwukoaloak*. In this context, the main period of the breadfruit season is known as *rekenleng*, which derives its name from *pwukemenleng*, literally "responsibility for the things of the air."

Isol

Isol season is based primarily on yams, **Kehp**, the most important prestige crop on Pohnpei. *Isol* generally lasts between September and April and is considered the "time of scarcity."

Isol differs from the *rahk* in time of occurrence and cultural perception of the food resource availability. Compared to *rahk*, the season of *isol* is considered a time of reduced food availability.

Besides the dominant yam crop of the *isol* season, other common foods are harvested during this period, including true taro (**Sawa**, *Colocasia esculenta*), swamp taro (**Mwahng**, *Cyrtosperma merkusii*), New World taro (**Sawahn Awai**, *Xanthosoma sagittifolium*), and cassava (**Kehp tuhke**, *Manihot esculenta*).

The *isol* season is considered a "time of hard work," because the crops being harvested require human labor for their production, as well as for the harvest itself (digging up the crops).

The traditional agroforestry of Pohnpei is therefore divided into two distinct and differing seasons, the *rahk*, which is controlled by nature, and *isol*, which is influenced much more by human farming activities. The chiefs, using the cultural system of Pohnpei, especially the timing of *nohpwei*, first fruit tributes (see pp. 96–97), determine the duration of each season. Depending on the breadfruit production during the *rahk*, the high chiefs can delay or hasten the digging of the first yams (*kotokehp*), which then starts the *isol*. The chiefs then end the yam season when the following year's breadfruit season begins, with the *kedisol* ("end of *Isol*") yam *nohpwei*.

The result is one of the most striking features of Pohnpei traditional agroforestry: the complimentary seasonality of the major food crops so that food is available at all times of the year. This is most clearly shown in Fig. 2.6 below, which is based on 1988 records of yam and breadfruit production (Raynor 1989).

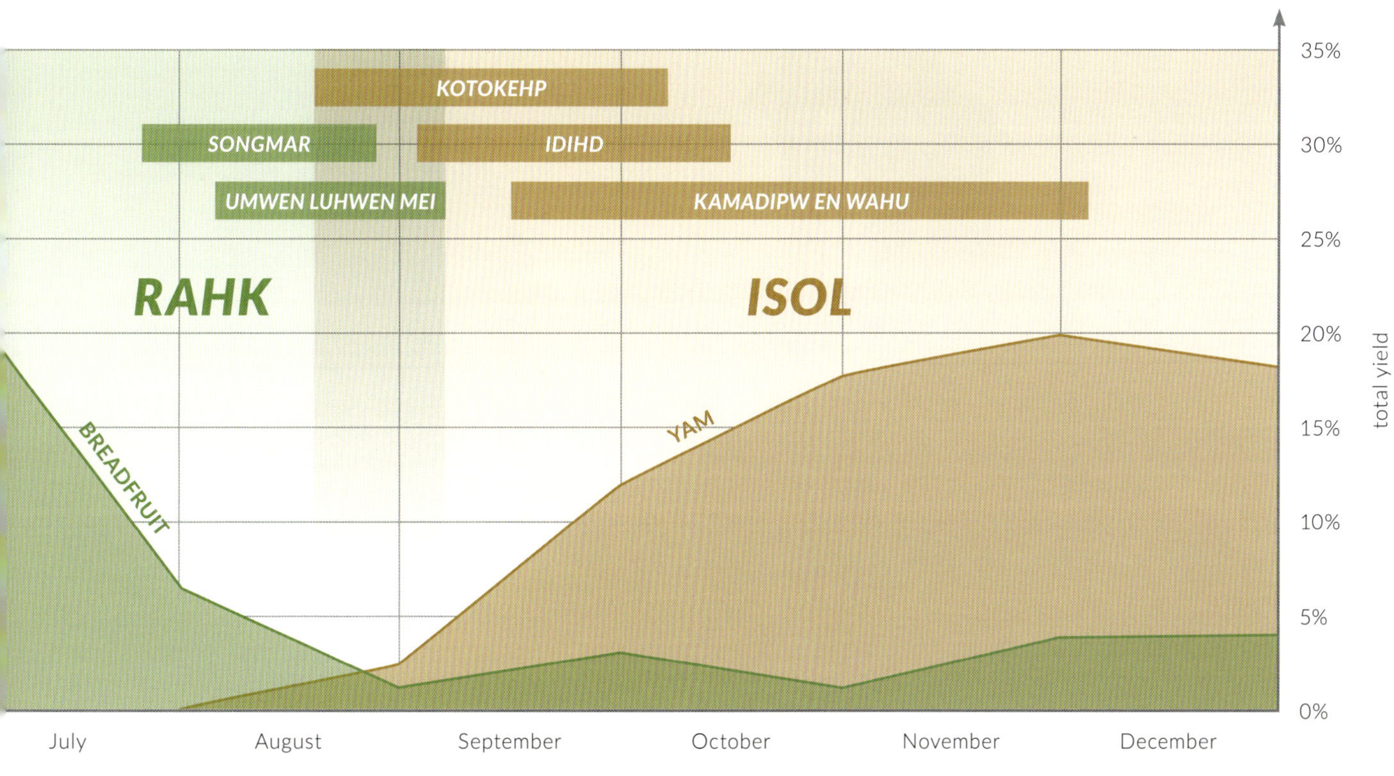

2.3.7. Forest clearing and secondary vegetation

Over many generations, much of the original, native forests on the lower slopes of the mountains of Pohnpei has been repeatedly cleared for cultivation of crops and other purposes. Cutting tools and occasionally fire are still used to clear more of the native forest. After the trees and shrubs have been cut and dried, "bush" fires are usually started as a way of clearing off the wild vegetation and making open space available for cultivation of crop plants.

Sometimes, the clearing fires get out of control and destroy sizable sections of the forest. After the trees and other plants are gone, the soil may be exposed to erosion and washed away, and it is especially likely to happen after heavy rainfall, which is common in Pohnpei. If the crops and their protective covering of leaves and other decaying plant materials (mulch) are not put into place soon after the clearing of the forest, accelerated soil erosion will take place, especially on the steeper slopes.

When farming plots are abandoned and left uncultivated, secondary forest develops naturally. During this process, the land progresses through ecological succession. That is, certain kinds of plants "succeed," or follow others as the soil is enriched by accumulating organic material, from grassland to scrubland to fairly dense secondary forest vegetation.

Secondary vegetation consists of fast-growing small trees, shrubs, and vines. This secondary vegetation acts as a natural "bandage," protecting disturbed soils from the heavy rains of Pohnpei. This bandage of secondary vegetation slows down erosion and allows the seedlings of the larger mature forest trees to grow in their shade. If left to grow up to maturity, the larger trees will eventually shade out the smaller trees and shrubs of the secondary vegetation (see Chapter 4).

Some plants in the *nansapw*, such as those described earlier, are cultivated, while others appear and grow without human assistance. Among the "wild" plants that do not need human help to grow in the *nansapw*, many are indigenous, or in some cases endemic, to Pohnpei; others were alien (introduced by people) to the island environments and now reproduce and establish themselves without human help. The wild plants in the *nansapw* areas which are indigenous or endemic can be called native; the plants brought into Pohnpei by people, which now grow on their own can, be referred to as alien species in the farmlands of Pohnpei.

A neglected taro patch and cultivated forest, now overrun by aggressive, fast-growing vines.

Secondary forest growing on the outskirts of Kolonia town.

Ecological succession in the cultivated areas

When farmers abandon a cultivated area for some time, because the soil is temporarily exhausted or relatively low in the nutrients necessary for adequate crop growth, many plant species invade the old cultivated plots. The first plants to arrive on their own are called pioneer plants (in the *nansapw* these are commonly referred to as "weeds"). These are usually herbaceous (non-woody) plants such as grasses and small ferns. Following this, the seeds of larger herbs and small woody shrubs germinate and grow under the shade in the moister conditions provided by the grasses and small ferns. Eventually, the seeds of small trees and larger ferns germinate and grow up to form a canopy covering the smaller plants. After sufficient time has passed to allow the natural, ecological processes to re-establish the soil fertility, a farmer may decide to burn or cut back (girdle) this secondary vegetation to plant his crop plants once again. Some of the plants listed in Table 2.13 below are left growing in cultivated areas (agroforests) of Pohnpei because of their ecological and/or economic value. They also sprout and grow wild without human help in the secondary vegetation that develops in abandoned areas of former cultivation.

Table 2.13. Some common plants in the secondary vegetation on Pohnpei.

Scientific name	Pohnpei name	English name or description
Adenanthera pavonina	**Kaikes**	false sandalwood
Aleurites moluccana	**Sakan** / **Soakan**	candlenut tree
Commersonia bartramia	**Kahil**	successional shrub
Falcataria moluccana	**Tuhke karisihn**	white Albizia
Glochidion ramiflorum	**Mwehk**	glochidion
Hibiscus tiliaceus	**Keleu**	wild hibiscus
Ixora casei	**Ketieu**	native ixora
Macaranga carolinensis	**Apwid**	native macaranga
Melastoma malabathricum	**Kisetikmei** / **Pisetikmei**	white-flowered shrub
Merremia peltata	**Iol** / **Salomp**	white-flowered vine

Note: *Adenanthera pavonina*, *Falcataria moluccana*, and probably *Aleurites moluccana* are alien species introduced after European contact and now naturalized in Pohnpei.

Adenanthera pavonina

Aleurites moluccana

Clerodendrum paniculatum

Commersonia bartramia

Costus speciosus

Falcataria moluccana

Macaranga carolinensis

Merremia peltata

2.3.8. Savanna grassland vegetation

Several low-lying areas or gently rolling slopes of Pohnpei are now covered with a vegetation of scattered shrubs, occasional trees, many ferns, and several species of grasses. In other parts of the world, areas with few trees among many grasses and/or ferns in more-or-less open environments are called savanna in English, and on Pohnpei they are referred to as *nanmal*. Some savanna areas in Pohnpei may be natural because of nutrient-poor soil conditions. Most of these open *nanmal* areas with partially or severely degraded soils are more likely the result of repeated human cutting and burning over centuries.

Sometimes, human-caused erosion can be severe, removing great quantities of valuable top soil. Without the precious, fertile top soil, it is difficult for most native forest trees and other plants to grow. In many cases, crop plants have been cultivated in these open areas or on the now-grassy fernland slopes (sometimes for a number of generations). In others cases, such as with pineapple, **Pweinaper** (*Ananas comosus*), cultivation in these open, burned-over areas of Pohnpei has only taken place during the 21st century.

Continued burning changes the vegetation and substrate (soil or rock underneath the vegetation). When the vegetation is removed by fire or other means, the rich topsoil may be washed away after heavy rains, especially on steeply sloping areas. Eventually, eroded slopes or canyons with hard, nutrient-poor soils are created, and only a few, fire-resistant plants are able to survive on these degraded areas. These eroded, scarred lands are then of little or no use for farming, and most of the native organisms cannot recolonize the land for long periods of time!

Hillslopes with degraded soil are naturally colonized by resilient ferns and hardy grasses.

Table 2.14. Some plants of the grasslands and fernlands on Pohnpei.

Scientific name	Pohnpei name	English name or description
Dicranopteris lineris	**Dipwinmal / Mwedilinmal**	false staghorn fern
Ischaemum polystachyum	**Rehpadil**	paddle grass
Lycopodiella cernua	**Kidi en mal**	club moss
Miscanthus floridulus	**Sapeleng**	sword grass
Nephrolepis acutifolia	**Rehdil**	sword fern
Saccharum spontaneum	**Ahlek**	wild cane

Dicranopteris linearis

Hedychium coronarium

Melastoma malabathricum

Morinda citrifolia

Saccharum spontaneum

Sphagneticola trilobata

2.3.9. Vegetation of upland forest

The *nanwel* or native forested areas of Pohnpei contain mostly indigenous and endemic plants. *Nanwel* is mainly found on the interior, upland, weathered volcanic slopes of the island. Much of this upland native forest at the lower—up to about 100 m (300 ft)—elevations, has been altered through centuries of cultivation and other human activities (and is now known as part of the *nansapw*). Over time, the native forest on some higher slopes has been transformed as well because of increased cultivation. In areas where rainforest has been cleared, landslides and other forms of erosion can create scars of varying size, visible from the lowlands. In recent times, unauthorized cultivation of **Sakau** (kava, *Piper methysticum*) in the uplands of Pohnpei has accelerated the human disturbance of the native *nanwel* vegetation, which has played a crucial role in watershed preservation and erosion control (for more details see the **Sakau** section in Chapter 3).

Nevertheless, the remaining upland native forest areas are good examples of tropical oceanic island rainforest vegetation. In these remaining forest areas, some of the trees are huge. They form a dense canopy and are hung with thick, woody vines. The trunks of the great buttressed trees are covered with carpets of mosses, and the branches of the trees have trailing curtains of ferns and orchids.

The major plants found on the many steep ridges (*pweleng*) of the *nanwel* forests include sweet-smelling **Seir** or **Pwuhr** (*Fagraea berteroana*), the large fig trees of **Aiau** (*Ficus prolixa*), the liana or **Rahra** (*Freycinetia ponapensis*), and tree ferns such as **Katar** (*Sphaeropteris nigricans*).

Beyond the indigenous native plants of Pohnpei that are shared across tropical Micronesia (including Kosrae, Chuuk, Yap, and Palau), the island's upland forests harbor numerous endemic species. These plants—uniquely evolved on Pohnpei and found nowhere else in the world—often bear the island's name in their scientific nomenclature (e.g., *Discocalyx ponapensis* and *Melicope ponapensis*). As biological treasures exclusive to Pohnpei, these endemic species are unique parts of the island environment and should be conservation priorities.

Discocalyx ponapensis

Melicope ponapensis

There is little doubt that the native forest has been reduced greatly in its range on Pohnpei due to human impact over many generations. Furthermore, large sections of this forest have been altered in extent and structure. Nonetheless, the native plants found there, and forest vegetation as a community, still offer a variety of ecological, economic, social, and ritual resources to the Pohnpeian people and should continue to do so. In 2005–2006, a forest survey, including a quantitative inventory of plant species in FSM, revealed that the different types of forest vegetation covering the islands of Chuuk, Kosrae, Pohnpei, and Yap amounted to about 665 km^2 (~253 mi^2), which equaled about 94% of the total land area of FSM (Donnegan et al. 2011: 11). More specifically, this study indicated that forests still cover most of the land in Pohnpei, with more than 50% currently covered by mature forest in upland, mangrove, and swampy areas. This survey also estimated that about 5% of Pohnpei was covered by secondary (not original) vegetation, and another 33% is covered by agroforest. Our remaining forest resources provide many ecological services and have great natural and culture value. They need to be appreciated and protected. These considerations and others are discussed in Chapter 4.

The upland forest of Pohnpei covers the slopes and tops of the mountains. Originally, it was an uninterrupted dense forest composed only of indigenous and endemic plants, but it people have changed it over the centuries. It now contains scattered introduced and even invasive species, patches of cultivated trees, and clearings invaded by ferns.

With Pohnpei being one of the rainiest places in the world, the interior of the upland forest is very wet. The landscape is green and lush, prolific with plants competing for space and smaller ones growing atop others, and traversed by many permanent freshwater streams fed by rain and soaked soil.

A grove of native palms and ferns in the upland forest, with diversity of life so dense that plants grow on top of each other. Tree trunks are covered with multitudes of mosses, vines, and herbs taking root in every spot available.

Table 2.15. Native trees in the *nanwel* forest.

Scientific name	Pohnpei name	English name or description
Angiopteris evecta	**Paiwed pwetepwet**	giant native fern
Atuna racemosa	**Ais**	native tree
Campnosperma brevipetiolata	**Dohng**	native tree
Cinnamomum carolinense	**Madeu**	Caroline cinnamon
Clinostigma ponapense	**Kotop** / **Koatop**	endemic palm
Elaeocarpus carolinensis	**Sadak**	native tree
Fagraea berteroana	**Seir** / **Pwuhr**	native tree
Freycinetia ponapensis	**Rahra**	climbing pandanus
Garcinia ponapensis	**Kehnpwil** / **Koahnpwil** / **Nikenpiri**	endemic tree
Glochidion marianum	**Luwekidinloal**	native tree
Pandanus patina	**Piht**	endemic pandanus
Ponapea spp.	**Kedei**	endemic palms
Sphaeropteris nigricans	**Katar**	endemic tree fern

Endemic *Timonius ledermanni* and other rare species emerge through the low canopy and drifting clouds near Nahnalaud.

Atuna racemosa

Campnosperma brevipetiolata

Cinnamomum carolinense

Elaeocarpus carolinensis

Fagraea berteroana

Glochidion marianum

Sphaeropteris nigricans

Syzygium stelechanthum

2.3.10. Vegetation of summit dwarf forest

The *nannahna* refers to the unusual and interesting dwarf, cloud forest on the wet summit ridges and peaks (*pohnnahna*) of Pohnpei, where heavy rainfall and frequent cloud immersion (fog-drip) provide continuous abundant moisture for plant growth. The vegetation on these exceptionally humid mountain summits—typically found along crests and ridges above 615 m (2,000 ft)—is often called the "moss forest" due to the luxuriant growth of mosses and epiphytes that cloak every surface. These moisture-loving plants create a lush, otherworldly landscape where mist-shrouded vegetation plays visual tricks on observers. A striking example is **Didimwerek** (*Hymenophyllum pallidum*), a small filmy fern whose bluish-green, iridescent leaves are said to appear fluorescent in the dark. Intriguingly, the same name also describes the shimmer of certain fish at night, as well as **Saleng en eni** (*Filoboletus manipularis*)—eerily luminescent mushrooms whose name means "ears of the ghosts" and which emit a greenish glow in the dark forest. Many other rare mosses, ferns, and non-woody epiphytic plants, such as the indigenous orchid **Lamwahk** (*Pseudovanilla ponapensis*) and the endemic fern **Mahrek en nahna** (*Cyclosorus ponapeanus*) thrive here, draping the twisted trunks and branches of the local stunted, bent, or slanting trees and shrubs.

Because of special climatic and soil conditions in the wet, often windswept summits of Pohnpei, the plants in the *nannahna* usually are less than 6 m (20 ft) tall. The most common larger plants in the dwarf or moss forests include **Sadak** (*Elaeocarpus carolinensis*), **Duhduhnwoal** (*Astronidium ponapense*), **Katar** (*Sphaeropteris nigricans*), and, to a lesser extent, **Dohng** (*Campnosperma brevipetiolata*), **Madeu** (*Cinnamomum carolinense*), **Kedei** (*Ponapea* spp.), and **Akenwel** (*Gynotroches axillaris*).

In some exposed crest and ridge areas of the upland summits, the vegetation is dominated by thick patches of a fern species: **Dipwinmal** or **Mwedilinmal** (*Dicranopteris linearis*). This is the same aggressive, spreading fern of disturbed areas, which is also commonly found in the lowland grasslands and fernlands of Pohnpei.

Hymenophyllum pallidum
Didimwerek (P)

Pseudovanilla ponapensis
Lamwahk (P)

Some of the plant species listed in table 2.16 as typical of dwarf forest areas of Pohnpei (*nannahna*) are also referred to in the previous sections of this chapter because they also grow in other areas (both *nanwel* and *nansapw*).

Note: There are many additional species of plants found in the various vegetation types of Pohnpei which have not been listed in this chapter. For further information regarding the plants of Pohnpei, see Chapter 3 and references in the bibliography.

Mist slipping through a maze of prop roots in a *Pandanus patina* grove near Nahnalaud.

Table 2.16. Some native plants in the dwarf *nannahna* forest on Pohnpei.

Scientific name	Pohnpei name	English name or description
Astronidium ponapense	**Duhduhnwoal**	endemic shrub
Campnosperma brevipetiolata	**Dohng**	native tree
Cinnamomum carolinense	**Madeu**	Caroline cinnamon
Elaeocarpus carolinensis	**Sadak**	native tree
Gynotroches axillaris	**Akenwel**	native tree
Hymenophyllum pallidum	**Didimwerek**	native filmy fern
Lycopodiella cernua	**Kidi en mal**	native club moss
Thoracostachyum pandanophyllum	**Pwoaki**	native large sedge
Pandanus patina	**Piht**	endemic pandanus
Ponapea spp.	**Kedei**	endemic palms
Pseudovanilla ponapensis	**Lamwahk**	endemic orchid
Sphaeropteris nigricans	**Katar**	endemic tree fern

Dwarf forest canopy bathed in clouds.

Elaeocarpus carolinensis

Gynotroches axiliaris

Hymenophyllum pallidum

Lycopodiella cernua

Pandanus patina

Ponapea sp.

Pseudovanila ponapensis

Thoracostachyum pandanophyllum

2.3.11. Atoll vegetation

From the dwarf forests perched on the highest peaks of Pohnpei, shrouded in clouds and mist, we now descend to sea level and journey to the farthest reaches of Pohnpei State to explore the unique vegetation of its coral atolls.

The native plant species of Mwoakilloa, Pingelap, Nukuoro, Kapingamarangi, and other atolls in Pohnpei State are remarkably similar and well-adapted to the harsh environmental conditions of the low-lying islets of these atolls. In stark contrast to the mountainous, wet, and lush landscapes of Pohnpei, coral atolls are characterized by low elevation, limited freshwater resources, and nutrient-poor soils—conditions that few plants can endure. Despite these constraints, atoll vegetation is both resilient and thriving, sustaining not only the natural ecosystem but also the human communities that depend on it. This distinctive vegetation is dominated by hardy plants that are tolerant of frequent salt spray, able to thrive in sandy alkaline soils, and capable of dispersing across vast ocean distances, with seeds that float on ocean currents or are carried by seabirds. Together, these indigenous plant species form the biological foundations of the unique terrestrial atoll habitats, particularly the atoll forests that once covered those coral islets that received adequate yearly rainfall.

A typical atoll islet forest is a mixed, fully developed ecosystem with low to medium-height trees and a closed canopy, leaving little or no open space between the uppermost layers of leaves, branches, and twigs. These forests are composed of relatively few tree, shrub, and herb species. On relatively humid atolls, some ferns and other epiphytes (plants that grow on other plants) are also common. One of these, for example, the large bird's nest fern, *Asplenium nidus*, thrives on Pohnpei State atolls but is absent from drier atolls farther northeast in the northern Marshall Islands. Many of the herbaceous and woody plants found in these mixed atoll forests also grow in the strand vegetation along rocky or sandy shores. In fact, the boundary between strand vegetation and inland forests on coral atolls is often gradual or unclear. Nevertheless, a typical zonation pattern exists,

Beach on Mwoakilloa Atoll with its typical mix of species adapted to life with lots of salt, sun, and sand.

based on individual species' adaptations to specific ecological conditions. Strand species like *Scaevola taccada*, *Heliotropium arboreum*, and *Guettarda speciosa* tolerate the harsh conditions of windward, ocean-facing shores. Trees like *Terminalia samoensis*, *Intsia bijuga*, and *Ficus tinctoria* prefer more sheltered areas inland, while *Calophyllum inophyllum*, *Cordia subcordata*, and *Thespesia populnea* thrive along lagoon-facing sandy beaches. Over time, atoll forests may become dominated by one or a few tree species, forming pure stands of large trees like *Pisonia grandis* and *Ochrosia oppositifolia*. These are sometimes referred to as mature forests: stable ecosystems that have reached their maximum development. Such ancient atoll forests can still be found on a few remote or uninhabited atolls in Micronesia, though in Pohnpei State they have largely been altered by human activity.

Over many generations, human settlement has significantly transformed atoll vegetation. By cultivating food crops and other useful plants, people have gradually converted atoll forests into agroforests human-managed ecosystems that provide food, medicine, fiber, and construction materials. Remnants of original atoll forest only persist today in small interior patches on larger, less disturbed islets, particularly as crescent-shaped strips along windward shores, just behind dense thickets of strand vegetation. People have deliberately left these areas intact to serve as natural windbreaks which help protect agroforests, taro patches, and coconut plantations from wind and salt spray. This traditional practice offers multiple benefits: it maintains natural habitats, preserves biodiversity, and supports food and copra (dried coconut meat) production in the challenging atoll environment.

As a result of human alteration, much of the vegetation on atolls now consists of plants introduced by people. Coconut palms (*Cocos nucifera*) are perhaps the most iconic and economically important local plants. They provide nourishment, shelter, and materials for the local communities, who rely on them for food, hydration,

An islet on Ahnd Atoll, still dominated by coconut palms descended from a historical copra plantation that is no longer maintained.

crafting tools, building homes, and producing copra for trade. The palms also play a crucial ecological role by stabilizing the sandy soils and reducing erosion caused by wind and waves. In addition to coconut palms, most important atoll trees are breadfruit (*Artocarpus* spp.) and pandanus (*Pandanus tectorius*). Breadfruit trees supply a starchy staple food and wood for canoe building and other construction purposes, while pandanus yields nutritious fruits and leaves valued for weaving mats, baskets, and other items. These multipurpose trees exemplify the ingenuity of atoll communities in utilizing different parts of plants for diverse needs.

The understory of atoll vegetation typically includes a mix of native shrubs, grasses, and herbs, alongside introduced species cultivated for food, ornamental value, or other uses. Paramount among them is swamp taro (*Cyrtosperma merkusii*), which requires swampy conditions and is grown in depressions first excavated centuries ago and carefully managed for generations so that taro can thrive close to the water table. Another important understory plant is island arrowroot (*Tacca leontopetaloides*), which grows semi-wild and was historically a crucial source of starchy flour, preserved for emergencies when storms or droughts disrupted other crops. These agricultural practices also highlight the resourcefulness of atoll communities in adapting to their environment and their ability to modify the environment to suit their needs.

Despite their resilience, atoll ecosystems are fragile and highly vulnerable to environmental changes. Rising sea levels, increased storm intensity and frequency, and prolonged droughts—all linked to climate change—pose significant threats to both plant life and the human communities that depend on atoll ecosystems. Conservation efforts, including the preservation of traditional knowledge and the promotion of sustainable land management practices, are critical to safeguarding the vegetation, biodiversity, and livelihoods of those living on coral atolls in Pohnpei State and across the wider Pacific.

Asplenium nidus ferns are shown here covering much of the forest floor in a patch of atoll vegetation on Pingelap Atoll.

Table 2.17. Some native trees and shrubs representative of atoll vegetation in Pohnpei State.

Scientific name	Pohnpei name	Mwoakilloa name	Pingelap name	Nukuoro name	Kapingamarangi name
Calophyllum inophyllum	Isou	Ijoau	Saepaeng	Hedau	Heedau
Cordia subcordata	Ikoik	Kanau	Ikoik	Ganava	Laagau mmee
Ficus prolixa	Aiau / Oaioa	Au	Aiau	Aoa	Iaiau
Guettarda speciosa	Ihd	Id	Eles	Bua laamodu	Bua
Heliotropium arboreum	Titin	Sisin	Sesen	Manu gili maau	Dogodogongo
Morinda citrifolia	Weipwul / Wempwul	Wɛhnpwul	Uhpwul	Bugaliaa	Nonu
Ochrosia oppositifolia	Kidi	Kajpwar	Kaesaepwar	Gahao	Ganniu
Pemphis acidula	Ngih	Kaingi	Kaingae	Nngie	Laagau hai goo
Pisonia grandis	Muk	Mɛhs	Mahs	Buga	Buge
Scaevola taccada	Remek	Roamoak	Raemaek	Manu gaba aasanga	Nau
Soulamea amara	Marat	–	–	Hakapini	–
Terminalia samoensis	Dipwoapw	Win	Win	Hua laagau	Laagau daliha
Thespesia populnea	Pone	Pɛnɛ	Paenae	Milo	Mila

Children going to school along a historical path through the agroforest on Mwoakilloa Atoll.

A boy posing with a huge **Aiau** tree on Sapwuahfik Atoll.

SHARKS

3

TRADITIONAL USE OF PLANTS IN POHNPEI STATE

CHAPTER 3

TRADITIONAL USE OF PLANTS IN POHNPEI STATE

Plants have been an essential resource for people ever since their early arrival in Pohnpei and atolls that are now included in Pohnpei State. Several plant species were carried on the long-distance voyaging canoes that brought the first peoples to these islands. These useful introduced plants, which have provided food, fiber, medicine, and other natural products for many generations, are described and discussed in detail in this chapter.

The islands of Pohnpei State contain highly diversified environments. These ecological habitats include mountaintops shrouded in mist, valleys with spring-fed rivers and freshwater aquifers, tropical freshwater swamps and magnificent mangrove stands, and bountiful and life-sustaining coral reefs and lagoons. Humans arrived in Pohnpei and its outlying atolls about 2,000 years ago and have been present continuously since then. Most of the economically important plants on Pohnpei and the Outer Islands are introduced cultigens (highly selected farm crops). The majority of these were brought into Pohnpei long ago, but other cultivated plants have been introduced more recently (e.g., Athens and Stevenson 2012; Levin 2015a; Levin et al. 2019; Merlin et al. 1992).

Introduced plants, such as bananas, yams, breadfruit, taros, coconut, kava, and others, have long been important to the people of Pohnpei and other islands. This traditional crop diversity was crucial for their wellbeing and survival:

> "In the Pacific, food security was maintained by a high diversity of subsistence resources that were mobilized using extensive environmental knowledge." (Bayliss-Smith and Hviding 2015, also see Clarke and Thaman 1993; Kennedy and Clarke 2004; Balick et al. 2009)

In addition to the plants brought in by early Pacific Islanders, it is also essential to focus on the native biodiversity of subsistence resources (both plants and animals) available for people on Pohnpei and the Outer Islands. These indigenous and endemic organisms were useful because of the traditional knowledge people accumulated and passed on from generation to generation about the natural habitats, life histories, and human usage of these plants and animals. This chapter focuses mainly on the traditional or contemporary utilization of both introduced and native plants of Pohnpei and the Outer Islands.

We must acknowledge not only the generations of people who have flourished on Micronesian islands because of their intimate relationship with nature, but also the extensive impact of the modern world on these local environments and traditional cultural practices. In modern times, the natural abundance of Micronesia is threatened by a series of human-induced and natural forces on both local and global scales. These social, cultural, economic, and environmental forces, affecting sustainability in Micronesia and beyond, require immediate and long-term action to control them.

> "Traditionally, complex resource management practices maintained these resources while providing subsistence and prestige goods for the island's inhabitants. However, in the last 50 years, population growth, unsustainable resource use, and the destruction of forest and marine habitat now threaten these unique environments with

serious consequences for the island's human inhabitants." (Raynor and Kostka 2003: 55)

The human impact on Pohnpei and the Outer Islands is discussed at other points in this book, but this chapter also addresses the manmade environmental impacts affecting the individual plants, the ecosystems in which they occur, and the traditional biocultural knowledge which should be maintained for present and future generations of Pacific Islanders (for example, see The Federated States of Micronesia National Biodiversity Strategy and Action Plan 2018–2023).

The early human colonizers learned how to utilize the native, wild-growing species they found in the various environments of these islands. Over many centuries, a large body of knowledge about these plants has been gathered and passed on from generation to generation. This traditional knowledge about both the native plants and the introduced plants includes their importance as food during periods of famine, and their religious significance, cultural history, economic value, medicinal use, and other resource applications:

> "The role of traditional knowledge is increasingly recognized in playing an essential part in the conservation of biodiversity. Most management strategies and plans now consider the importance of traditional knowledge. The four states have Historical Preservation Offices and there have been numerous studies to record traditional knowledge. The national and state Constitutions also respect the importance of traditional knowledge and culture." (SPREP 2019: 151—FSM State of the Environment 2018, underlined for emphasis).

Scientists who study traditional knowledge and use of plants in different societies are known as **ethnobotanists**. The most comprehensive study of this knowledge on Pohnpei Island involved a decade-long research effort by many local experts, experienced practitioners, and many scientists and scholars; see The Ethnobotany of Pohnpei by Michael Balick and his many collaborators (2009), University of Hawai'i Press (also see Merlin et al. 1992 and other published sources in the bibliography at the end of this book).

It is important to consider here the ethics of ethnobotanical research. The declaration below is adapted from the statement articulated by Dana Lee Ling, Instructor, College of Micronesia (FSM):

> "Ethnobotanists record and study traditional uses of plants by different cultures. In the globalized world of modern technology and communication, and sometimes theft or misuse of traditional cultural knowledge, it is expected that an ethnobotanist should obtain permits and permission from all applicable national, state, local, and traditional authorities prior to engaging in their work. However, this may vary if the ethnobotanist is a member of the indigenous people that the ethnobotanist is studying. In any case, when working with traditional indigenous peoples, the ethnobotanist should be careful to ensure that their intellectual property rights are protected." (See the following website for further advice and guidance regarding intellectual property rights and ethical behavior as an ethnobotanist: www.comfsm.fm/~dleeling/ethnobotany/text/ethnobotany.html).

With that in mind, we can proceed with this chapter and explore different plants and what they mean to the people of Pohnpei State.

Table 3.1 shows some of the numerous uses and functions plants can have in the Pacific Islands. This information is based on research conducted in the high islands and atolls (e.g., see Thaman 1990) and demonstrates the great diversity of resources that plants provide.

Table 3.1. Uses and functions of plants on tropical Pacific islands (Thaman 1990).

Ecological uses and functions of plants	
Shade	Temperature control
Erosion control	Animal habitats
Wind protection	Plant habitats
Soil improvement	Flood/runoff control
Wild animal food	Weed/disease control

Cultural and economic uses of plants	
Timber	Hats
Tools	Plaited ware
Fuel	Baskets
Weapons	Fertility control
Hunting	Mats
Containers	Scents/perfumes
Torches/lamps	Ritual exchange
Sails	Toilet paper
Fibers/fabrics	Recreation
Dyes	Fire making
Glues/adhesives	Dancing grounds
Toys	Cultivar vine support
Switches	Staple food
Brushes	Supplementary food
Aphrodisiacs	Snack/wild/emergency food
Abortifacient	Parcelisation/wrapping
Magico-religious	Fishing gear
Cages/roosts	Boat/canoe building
Poison	Boundary markers
Insect repellent	Teas
Toothbrush	Body ornamentation
Totems	Meat tenderizer
Brooms	Preservatives
Decoration	Medicines
Abrasive	Musical instruments
Insulation	Metaphors/symbols
Cordage/lashing	Commercial products
Floats	Subjects of mythology
Caulking	Deodorant
Spices/seasonings	Embalming corpses
Masticants	Religious images
Drugs	Secret meeting grounds

3.1. SOME LOCAL USES OF PLANT MATERIALS IN POHNPEI STATE

As an introduction, we present a brief overview of some of the general uses of plants by the people in Pohnpei and the Outer Islands. Many of these uses are traditional more or less throughout Micronesia. Following this overview, we focus on a majority of the more important plants used by people in the region and examine them species by species. As you explore this chapter, please keep in mind that there is much more to document and describe about the traditional knowledge of plants and environments in Pohnpei State and Micronesia. Hopefully, our presentation will encourage further research by students, teachers, administrators, resource managers, and interested people both within and without the region. Documenting and preserving traditional knowledge is crucial for cultural and ecological survival and sustainability.

3.1.1. Staple foods

Local food sources from the land and the sea are varied and have developed over many, many generations through experimentation, refinement, and communication about known food sources for human consumption and long-term health. We can divide the plant foods into two main categories: the major and the minor groups of food plants. The "major" group includes the traditional staple food plants which we might refer to as "keystone" cultural plants. They have long been among the most important cultivated food plants in Pohnpei State; they include species of taro plants—the giant swamp taro (**Mwahng**) and sweet or true taro (**Sawa**)—along with one or two species of breadfruit trees (**Mahi**), yams (**Kehp**), bananas (**Uht**), and coconut palms (**Nih**). Other traditional crops, such as arrowroot (**Mwekimwek**), were formerly important. Cultivated yams are especially valued on Pohnpei for food and ceremonial purposes. All these significant staple food plants long used in Pohnpei and the Outer Islands are assumed to have originated in Southeast Asia or Remote Oceania, from where they were purposely transported by humans in prehistoric times to various Pacific islands, including Micronesia. Some additional and now important staple plants, such as cassava (**Kehp tuhke** or **Dapiohka**), have been introduced after European contact.

3.1.2. Supplementary foods

Among the "minor," or supplementary, food plants in Pohnpei and the Outer Islands are several native and introduced plants, which are also described in some detail in this chapter. These supplementary edible resources sometimes serve as food for snacks and for survival during famines. However, we need to recognize that there has been a profound "Evolution of Pacific Islander Food Culture" over the past few centuries; this "evolution" or change "From Yam to Spam" in food sources and human diets should be identified and evaluated (see Abraham et al. 2012). The preservation of traditional knowledge about food plants, meals, long-term health, and island sustainability in the globalized world is at stake:

> "The introduction of Western concepts into Pacific Island cultures greatly affected their way of life, whether through the adoption of Western clothing, religion, and style of government. But perhaps the one lasting effect Western society had on Pacific Islanders was the introduction of Western food and food preparation. This shifting resulted not only in a greater emphasis of importing food from other places but created detrimental effects on the health of Pacific Islanders throughout Oceania. The evolution of Pacific Islander food culture has created a unique identity for Pacific Islanders, one that infuses both Western and traditional foods into the local islander diet, embracing not only the family values found in traditional food styles but also accepting and celebrating Western food culture." (Abraham et al. 2012; also see Sievert et al. 2019; Durand et al. 2022)

Native peoples of the Pacific region, including those living in the islands of Pohnpei State, have indeed experienced key changes in eating patterns because of foreign influence and colonization. These influences

This skillfully mixed garden provides its owners with a variety of local foods and other resources.

Simon's Market in Kolonia offers a diverse supply of healthy local produce.

have introduced an assortment of foods (including novel food and spice plant species). In recent times, foods that are considered to be traditional or foreign in origin are consumed to various extents. However, the classifications of traditional, recently introduced, and assimilated ("acculturated") foods can be ambiguous as many foreign foods have been incorporated into the diets of Pacific cultures. Recently, the first study to identify and classify traditional, acculturated, and locally grown foods from the perspective of the native people living in the islands of the US Affiliated Pacific Islands, including Pohnpei State, has been conducted:

> "Understanding these similarities and differences in how food is classified and identified, through the lens of those from the Pacific, is crucial for nutrition education, and understanding what foods are locally grown is important for future sustainability." (Dela Cruz et al. 2022)

This chapter discusses the relative importance of the diversity of uses of the different plant species for food. Often, this discussion of comparative significance focuses on differences in long-term traditional patterns of food consumption versus more modern integration into the diets of people inhabiting the islands of Pohnpei State.

3.1.3. Medicine

Medicinal plant use began in prehistoric times, when ancient people discovered that ingestion or application of certain plant materials from one or a combination of species was effective in treating some of the ailments that afflicted them. Use of plant medicine is a fundamental part of all cultures, including the Micronesian islands of Pohnpei and the Outer Islands. Most traditional Pohnpeian medicines contain materials from one or more plant species. Some of these Pohnpeian medicines and their treatments are common knowledge. Other medicines and treatments are less well known.

Even today, herbal medicine is used at one time or another by a large percentage of the Pohnpei State population, especially during infancy and childhood. While plants used for food, shelter, dyes, and many others aspects of the material culture of Pohnpei

On Pohnpei, yams are the most prestigious crop, and growing them demands sustained effort and skill.

Once dug up, yams can be rinsed in a stream, with fresh fern leaves used to gently brush away the soil.

Coconuts are an essential crop and husking them is a core traditional skill across Micronesia.

Breadfruit is Pohnpei's primary local staple, enjoyed baked and prepared in many other ways.

Although introduced to Pohnpei, **Likehdou** has local therapeutic uses, particularly for pregnancy and childbirth.

The bark of the native **Madeu** tree can be boiled for medicinal and preventative purposes.

Sakau is Pohnpei's most significant psychoactive plant, central to numerous cultural practices and values.

Fermented coconut flower sap is consumed across Micronesia as a mildly alcoholic drink.

are easy to see, study, and understand, the use of plants for medicines is more obscure—and, in some cases, secretive. Some basic medicinal procedures, including plant ingredients, methods of preparation, and applications, appear in this book. General references to the medicinal use of plants in Pohnpei came from knowledgeable Pohnpeians or Outer Islanders. However, since some knowledge is private, only general references to specific plant uses are made (except in the context of quotes from select published sources). Some simple descriptions of remedies, however, may be more specific and are derived from published accounts involving ethnomedicine; in these cases, source citations are included. In any case, this book is not meant to be used as a practical guide for taking or administering herbal medicine, since the information was collected from the healers on the understanding that it was not for this purpose.

3.1.4. Psychoactive plants

In most of the world, some parts of plants are used for psychoactive (mind-altering) purposes; tobacco and marijuana are two examples of plants used to produce an altered state of consciousness. In Micronesia, two common plants have long been used for their psychoactive effects: the fruit of the betel nut palm, *Areca catechu*, which is technically classified as a stimulant; and the rootstock of the kava shrub, *Piper methysticum*, which is known as **Sakau** in Pohnpeian and **Seka** in Kosraean and technically classified as a narcotic as it can induce relaxation and potentially sleep (i.e., narcosis). The traditional and more recent recreational uses of these two keystone cultural species are described in some detail in this chapter.

Some psychoactive plants, and the mind-altering substances made from them, are known as intoxicants. A true intoxicant is a substance which is toxic to the human body, especially in relatively large dosage. Alcoholic beverages, made from the partial digestion (glycolysis) of plant sugars by yeast, can be extremely intoxicating when large amounts are consumed. In the middle of the 20th century, Glassman (1952: 15) referred to several types of beverages that were being consumed on Pohnpei, including some that were fermented into alcoholic beverages and others that were fortified with larger amounts of alcohol per volume (usually illegal at the time).

3.1.5. Spices, extracts, dyes, and food coloring

A wide range of plants are utilized as sources of dye, condiments, and food coloring. Examples of traditional knowledge involved in extracting plant dye and spice materials from a variety of native and introduced species, as well as the methods of preparing and use of these products, are also described in this chapter.

3.1.6. Perfume

The people of Pohnpei State make and use a great variety of perfumes, including those used exclusively on the body and others especially for use on clothing or ceremonial objects. Plants play an important traditional role in providing the raw materials used in preparation of fragrant oils and other substances used to enhance one's appearance, for example in cosmetics. Some plants provide materials from which body oils or coloration are produced for cultural ornamentation or special events.

3.1.7. Body ornamentation

Various species found in the islands of Pohnpei State provide plant materials to be displayed on people, including parts of ferns and flowering plants. Some of these plants are native, and others have been introduced.

3.1.8. Garlands

The most popular form of body ornamentation is the use of flower garlands for the neck, head, and other parts of the body. Garlands are used to adorn persons during important occasions and in everyday life. People wishing to show affection or respect may present this kind of flower gift. Many fragrant and attractive flowers and other parts of various plants,

In a *sakau* ceremony, *tehn wehd* (giant taro leaves) are placed around the pounding stone. Known as *pwoaikoar*, they catch pieces of *sakau* that fall, and pounders are expected to place their feet on them.

Traditional dancers from Nukuoro and Kapingamarangi perform adorned with beautiful, fragrant flowers.

A dancer from Mwoakilloa wears a necklace and bracelets made of green fronds and an imposing crown woven from pandanus leaves.

Pohnpeian dancers line up for their performance, wearing vibrant, skillfully braided flower garlands.

Constructed with wooden posts and thatched roofs, Pohnpeian *nahs* is a traditional meeting house, central to the cultural and social life of the communit

Every element of a *nahs*—each beam, post, or piece—has a distinct name, purpose, and cultural significance in a time-honored construction tradition.

such as ferns, are used to make these garlands, known throughout Micronesia as *mwaramwar* and referred to in English as *lei* (a Polynesian word).

3.1.9. Tools and carvings

Many different species of plants are utilized as woody resources to make tools and other carvings in Pohnpei State. This chapter presents examples of plants that provide wood for housing and canoe construction (see below), canoe paddles, adze handles, tattooing tools, planting sticks, coconut husking stakes, fish hooks, spears, and other useful and symbolic traditional wooden objects.

3.1.10. House construction

Traditionally, all houses on Pohnpei have been made of plant materials available on the island. That includes residential homes, supplementary structures such as cookhouses, and, most importantly, the Pohnpeian feast house, the *nahs*, which is by far the most significant type of traditional structure on the island. Building it requires wood from various local plants. In fact, it has been reported that "there are at least 333 lumbers that make up a standard *nahs*, the *koupaleng*" (Liwy 2010: 7, quoting Rodrigo Mauricio). Roofs for houses and other structures in Pohnpei are traditionally made of thatch, which is known as *oahs* in Pohnpeian. The word **Oahs** is also the plant name in Pohnpeian for the large palm tree, the Caroline ivory nut palm (*Metroxylon amicarum*), indicating that the fronds of this palm tree species have long been a significant thatching material on the island. In addition, the Pohnpeian word *oahs* refers to "roof." Furthermore, wood in the trunk of an **Oahs** palm is also traditionally used to make floorboards for local huts, as well as for support of a hut's roof. This is in contrast with the outlying atolls, where pandanus and coconut leaves are the main materials used for thatching. This thatching difference between Pohnpei and the outlying atolls can probably be explained by the inability of **Oahs** palms to grow well on the low-lying coralline islands.

The stylized coconut-grating stools from Nukuoro are famed for both their functionality and artistic value.

A piece of wood transforms into art under the hands of a skilled Kapingamarangi carver.

Construction of a traditional meeting house on Pakin Atoll, built entirely from local plant materials.

Across most of Micronesia, thatch comes from pandanus or coconut leaves, but on Pohnpei, leaves from the **Oahs** palm are preferred.

The hull of a small canoe is being carved from a breadfruit log on Pingelap Atoll.

A Pohnpeian man crosses a short distance over water on a bamboo-pole raft.

Bundles of chopped firewood are prepared for sale in Pohnpei.

Some non-traditional plants, notably black pepper, have become economically significant on Pohnpei and are cultivated on plantations.

Today, however, many Pohnpeians no longer use attractive and readily available local plant species for sources of lumber, thatch, and other materials to construct a *nahs*, which are now built with the use of imported materials such as tin. Modern use of non-traditional materials in *nahs* structures is often preferred because of their durability. For example, the use of concrete will strengthen a *nahs*, allowing it to "withstand typhoons and other natural disasters that may happen," as well as reducing "the hardship of going out and actually cutting and delivering the local products that are always available in Pohnpei" (Liwy 2010: 2). Nevertheless, even with the use of imported materials in construction, "the powerful meanings of the *nahs* remain strong," still reminding its owners of their place in society and their obligations to society (Liwy 2010: 6).

3.1.11. Canoe building

Plants from both the *nansapw* and the *nanwel* were used in the traditional construction of canoes. Canoes have been very important to Pohnpeians for many centuries. They brought the first people to the island. They were used for transportation between areas within the Pohnpei lagoon and up the navigable streams. They were also used for inter-island voyaging. They are constructed from many of the same plants now as they were in the past. For example, **Sadak** (*Elaeocarpus carolinensis*), **Dohng** (*Campnosperma brevipetiolata*), and **Mahi** (*Artocarpus altilis*) are good for the hull; **Kedei** (*Ponapea ledermannianum*) for the gunnel; **Keleu** (*Hibiscus tiliaceus*) for the booms, float, spar(s), and stringers; and **Ais** (*Atuna racemosa*) for the oil extracted from its large fruit, which makes a useful canoe paint.

3.1.12. Cordage

Cordage of various types (rope, twine, string, etc.) has been essential to the traditional way of life in Pohnpei State: for creating canoes (to tie the wooden parts together and to secure the canoe sails), building houses, making fishing gear, and other purposes. The best known cordage has long been fabricated from natural plant fiber extracted from coconut husks. Nowadays, however,

Long, narrow Pohnpeian canoes, made from various woods including **Mahi**, **Ketieu**, **Karara**, **Pone**, **Mworopw**, **Pingiping**, **Kurumw**, **Sadak**, and **Titin**, are perfectly suited to glide through calm lagoon waters.

A young couple in Kitti with a unique cart made from local wood and salvaged shock absorbers.

Performers of the Pingelapese traditional stick dance wear plant-fiber skirts and headbands, their oiled skin adorned with scattered green leaves.

In Pohnpeian culture, hanging the *lapwkopwou* basket symbolizes generosity and the offering of food to guests.

coconut sennit cordage is rare and has been largely replaced by nylon and plastic ropes. Nevertheless, Pohnpeian farmers still rely on local cordage made from the fibrous bark and wiry, strong stems of other woody plants, especially *Hibiscus tiliaceus*.

Koahl, skirts traditionally made from the inner bark of the **Keleu** tree, are commonly worn by men at ceremonies.

3.1.13. Fabrics and clothing

> "Men wore a loincloth made from the outer layers of the banana trunk [*Musa textilis*]; later, they would weave a *koahl* 'grass skirt' from strips of [*Hibiscus tiliaceus*] bark. Women wrapped themselves in pieces of cloth shaped from the inner bark [*likou meimei*] of the breadfruit tree [*Artocarpus altilis*]." (Hanlon 1988: 6; also see Bernart 1977: 16)

A section of a Kolonia supermarket showcases local products: wooden and plant-fiber handicrafts, along with coconut-oil-based cosmetics.

Traditional weaving of textile materials for clothing and other cultural products in the Caroline Islands, including Pohnpei, involved the use of a horizontal tension backstrap loom. The main fiber used in weaving on this backstrap loom was high-quality, soft, silky, and fine. It was traditionally extracted from the special kind of banana plant, *Musa textilis*. Known in English as "Manila hemp" and in Pohnpeian as **Utinwel** or **Tikahp**, it is cultivated specifically as a fiber resource. Its fruits are not eaten. It spread together with the knowledge of the backstrap loom and weaving teachniques from Southeast Asia into northern Melanesia and parts of Micronesia, including Pohnpei.

The extremely rare and highly prized *lakiot*, a cloth made from banana leaf and hibiscus fibers, is now only found in special collections and museums.

In Pohnpeian tradition, backstrap loom weaving is believed to have been provided to people by divine beings, and as was the case in many other Micronesian islands, the art of loom weaving was highly developed among the women, who have used basic backstrap looms to create fabrics of excellent artistic intricacy (Liwy 2010). Examples of the most complex and colorful Micronesian weavings include the *dohr*, a large sash formerly produced on Pohnpei. These colorful traditional Pohnpeian woven belts were worn by high-ranking men. They were made by women and were "the most valuable loom woven textiles [used] to decorate or *kapwata* the bodies of high status or *soupeidi* men, as signifiers of rank and authority" (Kihleng 2015: 89). An example this type of special clothing, dated to the late

No longer seen on Pohnpei, *dohr* are intricately woven belts, such as this example made nearly 200 years ago and now held in the British Museum.

A row of *kiam* baskets filled with fresh mangoes.

A basket woven from a single coconut frond split lengthwise.

A simple yet versatile round food basket woven from coconut leaves.

A contemporary purse-style basket sold as a local handicraft.

19th–early 20th century, has been on display in the Metropolitan Museum of Art in New York City:

> "Reserved exclusively for high-ranking men, tor were ceremonial regalia, worn as belts over grass skirts during dances and feasts. The sashes were woven from fiber obtained from the tough, stringy trunks of banana trees [*Musa textilis*], either left in its natural light tan color or tinted red, black, or light purple with vegetable or mineral dyes. Adorned with intricate geometric motifs, the ornamentation of the sashes varied greatly and the right to use some patterns and colors belonged exclusively to particular families. In the late nineteenth and early twentieth [centuries], weavers frequently, as here, embellished the indigenous banana fiber by interweaving it with red [cotton] yarn obtained from European traders." (www.metmuseum.org/art/collection/search/314527; also Kihleng 1996: 60)

Additional fibers used in traditional weaving included those extracted from the bark of **Keleu** (*Hibiscus tiliaceus*). Everyday clothing in Pohnpei was also produced without weaving, through extraction and processing of local plant resources. Today, cotton and modern synthetic fibers have been introduced and are used widely in those areas where weaving is still practiced in Micronesia (see collections.tepapa.govt.nz/object/205813).

3.1.14. Mats

Throughout Micronesia and the other inhabited tropical Pacific islands, useful large and small mats have been traditionally made from local plants. They are used to sit on, sleep on and under, cover parts of houses, cover canoes, and can even be worn as clothing for certain occasions. The plant particularly suitable for weaving mats is pandanus, thanks to its long, light, malleable, and durable leaves. Many varieties of *Pandanus tectorius* are most commonly used in the islands of Pohnpei State. People extract, dry, and further process selected leaves from these plants to make their mats. The best mats, which are highly labor-intensive, prestigious, and used by high-ranking persons, are made from *Pandanus patina*, a species endemic to Pohnpei.

3.2. DESCRIPTIONS, DISTRIBUTIONS, AND USES OF PLANTS IN POHNPEI STATE

"Biodiversity and traditional culture in the FSM cannot be considered in isolation from each other. Many native and endemic species are utilized in various aspects of daily living, including customary practices and traditional medicine, amongst many others. Historically, traditional knowledge, practices and modes of resource management have protected and conserved the FSM's biodiversity. This sustainable management has undergone a huge shift as the FSM's societies have changed over the past decades. The role of traditional knowledge is, however, increasingly recognized as playing an essential part in the conservation of biodiversity." (Hall 2018: 11)

The plants described and discussed in this chapter represent the great majority of those species which have been most important over time to the people living in Pohnpei State. The information given for each plant includes the following:

1. Pohnpeian and the available Outer Islands names. These names in different languages are identified by the following abbreviations given in parentheses:

    - **P** = Pohnpei
    - **Mw** = Mwoakilloa
    - **Pi** = Pingelap
    - **Nu** = Nukuoro
    - **Ka** = Kapingamarangi

2. Scientific name of the plant. Although listed after the local names, the scientific names are the basis for alphabetical order of plants in this chapter (as this approach ensures that related species are presented next to each other). To find the botanical family of each species, see the Index (pp. 461–465).

3. Botanical description.

4. Geographical distribution and habitat in which the plant is found. This often includes reference to the species being native (indigenous or endemic) versus introduced (see pp. 46–47 and 389–390).

5. General and regional uses of the plant in Pohnpei and the Outer Islands. Knowledge regarding the traditional and more recent uses for the plants has been obtained from a variety of published and oral sources (see pp. 12–13 for acknowledgments and pp. 439–459 for bibliography).

The special focus on the traditional usage of these plants has been presented with the hope that students and others will make appropriate verifications, additions, and changes where needed in this text. This will help facilitate preservation and perpetuation of the vast biocultural knowledge that these people have brought together and utilized effectively and sustainably for many centuries in their island homelands.

Close-up of woven pandanus leaves, a practical and culturally significant material widely used across Micronesia.

▸ **Osen likendinikep** (P), **Suhkoa dokdok** (Mw), **Geinga duduia** (Ka)
Achyranthes aspera

Description: *Achyranthes aspera* is a coarse, woody herb or small shrub up to about 4 m (13 ft) tall, but normally much shorter. In English, it is known by some as "chaff-flower" or the "devil's whip." The stems, especially young individuals, have small hairs. The leaves are produced in opposing pairs, up to 12 cm (4.8 in) long. The greenish, prickly flowers are about 6 mm (0.25 in) long and clustered in erect terminal spikes up to 20 cm (8 in) long. Its fruit are orange to reddish purple or brown. Overall, it can be identified by its opposite leaves, branched stem, and spiny bracts that readily adhere to animals and clothing.

Distribution: *A. aspera* is common in various areas of the tropics. It is either a native species or was introduced to Pohnpei during the prehistoric period. It is sometimes found in open coastal areas of atolls and high islands, frequently becoming weedy in lowland plantations along roadsides. Its spikelet-like fruits easily become attached to human clothing or to fur or feathers, and this is the plant's major dispersal mechanism.

Uses in Pohnpei: *A. aspera* is known in Pohnpeian as **Osen likendinikep**. The smoke produced by burning its root is reputedly used to treat the sting of a stonefish: "[A]s many roots as needed to fill half a coconut shell (*Cocos nucifera*) are pounded and placed in the shell, covered with a hot stone, and then covered with the other half of the shell, which has a small hole on the end, and the steam and smoke that come out are directed to the place where the sting is..." (Balick et al. 2009: 320). Another closely related plant species, *Achyranthes bidentata*, is also found on Pohnpei and shares the local name of **Osen likendinikep**, but as of yet without any documented use by people (e.g., see Balick et al. 2009: 320; Herrera et al. 2010: 80).

Uses in Mwoakilloa: Glassman (1953: 293) recorded this species as a common "weed in coconut grove," but did not note any local uses for it.

▸ **Kaikes** (P), **Kaikɛs** (Mw), **Gaigese** (Nu), **Gaigeti** (Ka)
Adenanthera pavonina

Description: *Adenanthera pavonina* is a large tree that can reach heights of 25–30 m (up to about 100 ft). The trunk of this introduced tree may appear twisted. It is medium-sized and deciduous with leaves that each have many small pairs of leaflets, typical for many species in the bean family (Fabaceae), to which this tree belongs. Each leaflet is about 2–5 cm (~1–2 in) long, thin and rather pale green. The flowers are produced in slender, elongated racemes that are produced in the leaf axils (the angle between the upper side of a leaf and the stem from which it grows). The flowers are yellowish and sweetly fragrant with 8–10 stamens. The calyx is 4–5-toothed, and its pods are slender, linear, curved or contorted, dehiscent, thin, flattened, brown, smooth, and shiny inside. The dark or bright red seeds (10–12 per pod) are hard and about 8 mm (~0.3 in) wide.

Distribution: *A. pavonina* is reportedly dispersed by birds, with a wide native range in tropical southeastern Asia, including tropical areas in Australia, Bangladesh, much of Southeast Asia, and Near Oceania (e.g., New Guinea and the Solomon Islands). It is an alien (brought in by people) species in Micronesia. It may have been an early human introduction to parts of Melanesia and western Polynesia, but was most likely introduced only recently to other areas in Remote Oceania, including Micronesia. On some islands of Micronesia, *A. pavonina* is found in abandoned agricultural (agroforest) areas, and it is also sometimes cultivated as a protected tree in the garden areas of high volcanic islands. On Pohnpei, *A. pavonina* is naturalized (growing on its own) near roads, in dry forest, and sometimes in dense forest from near sea level to about 600 m (2000 ft) in elevation. It is usually not found on atolls.

Uses in Pohnpei: *A. pavonina* is known in Pohnpeian as **Kaikes**. Glassman (1952: 74) observed a "planted" specimen "20 foot tall" on Pohnpei in 1949. Today, **Kaikes** is a relatively common tree on Pohnpei found typically in the *nansapw* areas where agricultural cultivation and human settlements are located (e.g., see Merlin et al. 1992; Herrera et al. 2010). Although **Kaikes** has many known uses, it also has a reputation as a potential threat as an invasive species in those islands where it has been introduced in Micronesia. Moreover, the timber of this introduced tree splits easily and therefore is not useful for lumber. Nevertheless, its green wood burns, producing a very hot, useful fire; indeed, the wood of **Kaikes** is reportedly used on Pohnpei mainly to heat the rocks in the traditional Pohnpeian stone oven (*uhmw*). Balick et al. (2009: 379) tell us that the seeds of **Kaikes** "are eaten as food by young children [and its] wood is used as construction material for small structures, such as *imwen kuk* (cookhouses)" on Pohnpei.

▸ **Marasau** or **Maroasoa** (P), **Marjau** (Mw) *Aglaia mariannensis*

Description: *Aglaia mariannensis* is a small, native (indigenous) understory tree 2–6 m (6–18 ft) in height when mature, with rusty-brown, hair-like scales that cover the branches, leaves, and inflorescence. It produces alternate compound leaves that have leaflets arranged opposite along both sides of an elongated axis 60–80 cm (24–32 in) long, with an unequal number of leaflets (9–13). New leaves are pale green, almost white, but darken as they get older. Its large leaflets are oblong, 15–25 cm (6–10 in) long, and 6–8 cm (2.5–3.5 in) wide, acuminate at the apex, sometimes rounded toward the base, with 12–16 lateral veins in pairs. The young leaves appear like about 2.5 cm (1 in) long and rusty or orange in color (see Kitalong et al. 2008; Raulerson and Rinehart 1991).

Distribution: *Aglaia* is a genus of over 100 woody species mostly found in the moist tropics, mainly in Southeast Asia and tropical Northern Australia. *A. mariannensis* is generally found in a number of the tropical islands of Micronesia. In these islands, *A. mariannensis* occurs as "small understory trees with new growth that has a conspicuous reddish brown pubescence," with the population in the Mariana Islands restricted to limestone forest communities, while the populations of *A. mariannensis* on Palau and in the FSM are also found on volcanic terrain (Costion and Lorence 2012: 80–81). Note: *A. mariannensis* was found to occur in relatively large numbers in non-limestone areas of the volcanically active Pagan Island (Mariana Islands), where it is abundant in association with the tree snail, *Partula gibba*, an endemic invertebrate species that is rare elsewhere in these islands (see Hadfield 2010). In Palau, and probably elsewhere such as Pohnpei, forest birds feed on the fruits of *A. mariannensis* and most likely disperse the seeds (Kitalong et al. 2008). In Pohnpei, *A. mariannensis* appears to be a woody component in the understory of native lowland forest areas.

Uses in Pohnpei: This tree is known in Pohnpeian as **Marasau** or **Maroasoa**. Kanehira (1935: 30) referred to it as "very common in thickets and forests from low altitudes up to the highest elevations on the island." In the past, the stems of this tree were used to catch feral chickens. It has also been used as firewood, which burns rather slowly because of its relatively abundant resin, and as a deterrent to keep rats out of crop areas on Pohnpei. It has even been used as a kind of magic charm to protect people from spirits, as well as to ward off thieves trying to steal **Sakau**, yams, or other crops. In addition, **Marasau** has medicinal uses for treating amoebic dysentery and inducing vomiting (Balick et al. 2009: 438–440).

▸ **Kehnmand** (P)
Aidia racemosa

Description: *Aidia racemosa* was formerly classified in the genus *Randia*. It is a native shrub or relatively small tree in the coffee family, Rubiaceae, and can grow to heights of about 15 m (45 ft) as an attractive woody plant with a layered growth habit. Like other members of the family Rubiaceae, the leaves of *A. racemosa* are oppositely arranged along the stems, and stipule scars frequently can be seen extending around the stems. The flowers are cream colored, strongly scented, and borne in small clusters along the branches. Its fruits are oval, about 1 cm (~0.5 in) long and bright red when mature. The fruits and flowers of *A. racemosa* attract birds and butterflies.

Distribution: *A. racemosa* is found in various areas of the Indo-Pacific region, including parts of Remote Oceania, and reaches as far as the southern Marshall Islands in Micronesia, where it is reported only on Arno and Jaluit atolls. *A. racemosa* is sometimes common in somewhat undisturbed coastal or atoll scrub forests of Micronesia or in coconut plantations, such as those in Palau, Yap, Fais, and Guam. Kitalong et al. (2008: 31) referred to *A. racemosa* as "an attractive plant which produces fragrant white flowers and would thus make an excellent landscaping species on limestone soils" and could "also be a good species for restoration of vegetation."

Uses in Pohnpei: *A. racemosa* is known in Pohnpeian as **Kehnmand**, a local name that reportedly also refers to two other plant species found in Pohnpei: *Geniostoma repestre* (J.R. Forst. and G. Forst) and *Cyclophyllum barbatum* (G. Forst and N. Hallé). Both of these plants also have traditional uses in Pohnpei, e.g., see Herrera et al. (2010: 169). The leaves of *A. racemosa* are reportedly used to treat an unspecified skin disease that afflicts children, in which "several leaves are pounded, wrapped in cloth, squeezed into a cup of water, and consumed daily for four or eight days..." (e.g., see Balick et al. 2009: 478).

▸ **Sakan** or **Soakan** (P)
Aleurites moluccana

Description: *Aleurites moluccana* is a small to medium-sized tree that grows up to 15–25 m (~50–80 ft) in height, with branches that are wide-spreading or pendulous and often many-angled or crooked, and a canopy that can stretch out to about the same length as its height. The bark is gray-brown to blackish and the leaves and inflorescence have a mealy-scurfy whitish to light brown covering of fine hairs. The leaves are simple and ovate, three-lobed or rarely five-lobed, and acute at the apex; the leaf blades are generally about 10–20 cm (4–8 in) long on an equally long slender petiole. Its abundant, small, whitish flowers are produced in many-flowered panicles 10–15 cm (4–6 in) long. The roundish fruits have a slightly fuzzy, olive-green rind, are about 6 cm (~2.5 in) in length, and contain 1–2 oily seeds (endosperms or nuts). The seeds are held in a bony endocarp about 4–6 cm (1–1.5 in) in diameter.

Distribution: *A. moluccana* is an early secondary successional or large-gap woody species, or in many human-impacted areas, it may occur as a long-lived pioneer species. *A. moluccana* can be found in places from near sea level to higher upland areas in some of the high islands of Micronesia. This useful tree is native to the Indo-Malaysia region and today is widespread throughout the tropics. It was originally introduced to areas of Oceania as a "canoe plant" and was transported to many of the remote tropical Pacific islands in pre-European times; today, it is still a common tree of some areas of this broad region, especially in much of Polynesia. However, on some islands where *A. moluccana* trees are no longer protected, replanted/reseeded, or otherwise maintained, its distribution may shrink or it may even become extirpated, for example in areas in the higher Cook Islands in eastern Polynesia or Kosrae in the Federated States of Micronesia.

General uses: *A. moluccana* has long had multiple uses on many tropical Pacific islands for construction, illumination, tattoo ink, medicine, relish food, skin care, manure, and mulch. Indeed, this species is one of the world's more impressive multipurpose plants with many traditional and modern uses of its wood, leaves, flowers, seeds, and seed oil. The endosperm or "nut" of *A. moluccana* is often cooked and used in Indonesian and Malaysian foods; for example, on Java Island the oily nut is used to make a thick sauce that is eaten with vegetables and rice.

Uses in Pohnpei: Known in Pohnpeian as **Sakan** or **Soakan**, this is a relatively common alien tree in the secondary vegetation of the cultivated agroforests (*nansapw*) of the Pohnpei lowlands (e.g., Raynor and Fownes 1991; Merlin et al. 1992; Balick et al. 2009: 10). According to Glassman (1952: 69), *A. moluccana* was known in Pohnpeian as "Sakan in the Nett District" and "Shakan in Kiti District" and the oil extracted from its "seeds is used as an ointment, and it is burned for illumination." Thus, its oily seeds were still being used as a useful body oil and being burned to provide light at night in the middle of the 20th century. Indeed, the common English name of this plant is the "candlenut tree," referring to its widespread traditional use for light by many Pacific Islanders, including Pohnpeians. In old Pohnpei, the ash from the burned nutshells (endocarp) was apparently used as a dye source for tattooing, which was very important traditionally before the arrival of the Europeans, after which missionaries discouraged the Pohnpeians from tattooing their bodies (e.g., see Furey 2017). For example, the tattoo made with **Sakan** dye, illustrated here, shows the "tattooed right hand of Ounpohn of Tomwara from the clan of Dipwinmen Toantoal."

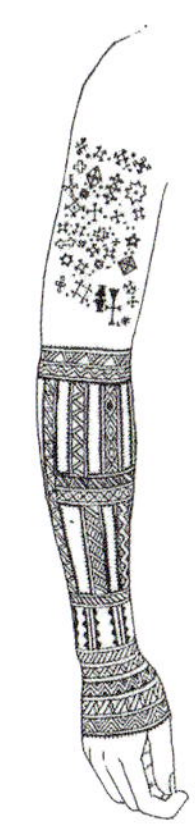

▸ **Sekesek** (P), **Kidahk** (Mw) (Pi), **Dala halu** (Nu), **Daga halu** (Ka)
Allophylus timorensis

Description: *Allophylus timorensis* is a native scandent shrub or small tree with a narrow trunk in the soapberry family (Sapindaceae). It can grow to heights of 4 m (~13 ft) and produces trifoliate leaves (three leaflets per leaf) that are arranged alternately along the stems. The inflorescences of tiny white flowers are produced on stalks (racemes). The round or oval-shaped fruits are up to about 1–4 cm (~0.5–1.5 in) in diameter, turning from green to red as they mature.

Distribution: This woody species is widespread, generally found in the inner coastal strand forests of some areas of tropical Asia and across the Pacific to Micronesia and eastern Polynesia. Its distribution stretches eastward from Timor in Southeast Asia to the Philippines and to Tonga, Niue, Samoa, the Marquesas, and the Society Islands, as well as to Micronesia. Although many "endemic" species have been named as *Allophylus* trees in different areas of this geographical range, they all probably belong to the single species listed here as *A. timorensis*. On the atolls of Micronesia, this shrub or small tree is one of several woody species still found in the broadleaf "windbreaks" left on the windward sides of islets to protect crops inland. Fruits of *A. timorensis* are reportedly eaten by Pacific pigeons (*Ducula pacifica*), at least on some islands such as those in forested areas of the Tonga (McConkey et al. 2004), and these pigeons have probably played a major role in long-distance dispersal of the seeds to many tropical islands of Remote Oceania, including those in Micronesia.

Uses in Pohnpei: *A. timorensis* is widely used by different people across its wide natural distribution. It is noted as a source of medicine and other applications, although its wood apparently varies in quality. Known on Pohnpei as **Sekesek**, it has a number of traditional uses. For example, it is used to make sticks for traditional stick dancing, and its wood has also been used to make baseball bats, as firewood, and for fishing poles. Additionally, parts of this species are used as a charm to cast away evil spells and for medicine, such as *soumwahu en naniak* (mangrove sickness), diarrhea, painful urination, back, muscle, and joint pains, and to treat bruises or sprained ankles, wrists, or knees. Furthermore, it is planted as an ornament and used for decoration in Pohnpei (e.g., see Balick et al. 2009). Lee at al. (2010: 106) referred to the consumption of **Sekesek** on Pohnpei as an anti-stress medicine: "To reduce stress, four round **Sekesek** fruits are collected before sunrise, chewed and swallowed every morning for four or eight days."

Uses in Mwoakilloa and Pingelap: *A. timorensis* is known as **Kidahk** on both Mwoakilloa and Pingelap. In the middle of the 20th century, Glassman (1953: 304) observed it as a frequent tree in the coastal strand vegetation, averaging about 15 ft (~5 m) tall. This species generally yields good-quality timber, which is durable, hard, and can be used for various purposes like construction of buildings, bows, fishing equipment, and as fuel. For example, the saplings on Pingelap Atoll are used traditionally for making handles of nets for flying fish and are useful for fabricating fish traps and fishing poles (e.g., see Damas 1994: 237; Merlin 2002).

▸ **Ohd**, **Sepwikin** or **Wehd** (P), **Wod** (Mw), **Woad** (Pi), **Ngaungau** (Nu) (Ka)
Alocasia macrorrhizos

Description: *Alocasia macrorrhizos* is known in English as "giant taro" among other names. It is a large, spreading, starchy aroid species in the arum family Araceae, to which its relatives "true taro" (*Colocasia esculenta*) and "giant swamp taro" (*Cyrtosperma merkusii*) also belong. *A. macrorrhizos* can reach up to 3 m (10 ft) or more in height. It develops a thick, erect, and ringed stem up to 20 cm (8 in) wide and 1–2 m (3–6 ft) long, which grows out of the underground part of the stem known as the corm. The heart-shaped shiny green leaves are huge, with 4–7 primary lateral veins per side; these big leaves are up to about 1.5 m (5 ft) long and tend to point upward. The multitude of tiny flowers are borne on a fleshy spadix surrounded by a whitish to yellowish (or greenish) oblong spathe. The berry fruits are ovoid, 8–10 mm long, and about 5–8 mm in diameter.

Distribution: The giant taro is most likely native to tropical south and southeast Asia, where it commonly occurs in humid, low to medium elevation valleys. It is naturalized (grows wild) along stream banks, mainly in secondary or disturbed forests and in old clearings. Although giant taro is found throughout much of the tropics today, it is probably a pre-European introduction into many tropical Pacific islands, including high islands and numerous atolls. For many generations, giant taro has served on tropical Pacific islands as a famine crop, needing approximately a year to mature. Its above-ground starchy stems are rich in thiamine and energy-rich starch, but like the other cultivated aroids, the potentially edible parts of this species must be cooked for a relatively long time to become safely edible, that time being largely dependent on the variety of giant taro being cooked.

General uses: Mueller-Dombois and Fosberg (1998: 293) described the special food usage and general plant associations of giant taro (*A. macrorrhizos*) in some of the islands of the FSM, where it is found growing near villages and scattered dwellings among cultivated "bananas, cassava, small garden plants and numerous ornamentals." As a large spreading taro with very large, upright, pointed leaf blades, giant taro is a common component in the "coconut woodland":

> *Alocasia* [giant taro], which grows well on drained soil, is extensively used as food in the central and western Caroline atolls, though it is generally treated as a famine food elsewhere."

When used as either a major or minor component of the diet in tropical Pacific islands, giant taro is mainly cultivated for its above-ground, starchy upright stems; however, these stems are only edible after lengthy roasting, baking, or boiling for most varieties (e.g., see Murai et al. 1958; Manner 2011a). Apparently, the stems of this taro can only be harvested after one year of growth and can remain in the ground for several years without spoilage (e.g., see Pollock 1992). Englberger et al. (2009: 138–139) referred to the benefits of cultivating and observing the growth of *A. macrorrhizos*:

> There is a belief in Pohnpei that if *Alocasia* is growing on a piece of land, this shows that the land is good for growing all kinds of crops."

Uses in Pohnpei: *A. macrorrhizos* is known as **Ohd**, **Sepwikin**, or **Wehd** in Pohnpeian. It was introduced into Pohnpei by early Micronesian settlers and formerly served as an important root crop on the island (see Glassman 1952 and Mahoney 1960); this plant's former high regard is indicated by the large number of known *A. macrorrhizos* cultivars in Pohnpei (see p. 421). However, today *A. macrorrhizos* is very underused on Pohnpei:

> The neglect of *Alocasia* has been so drastic that this food, documented in "The Book of Luelen" (Bernart 1977) as the second oldest food in Pohnpei, is no longer recognized as a food by many young Pohnpeians" (see Englberger et al. 2009: 160; note: "The Book of Luelen" is a comprehensive, oral history of Pohnpei written by a much respected local author).

Indeed, in the middle of the 20th century, Bascom (1949) indicated that *A. macrorrhizos* was the most preferred aroid-based food on Pohnpei: "[T]he corm is baked or boiled and is the most common substitute for fresh breadfruit and yams" (Lessa 1977: 142). Furthermore, it has been reported that all families on Pohnpei formerly cultivated *A. macrorrhizos*, "with some families having more than a hundred plants of one or more varieties" (Englberger et al. 2009: 133). *Alocasia* taro was grown "on dry land to avoid acridity," something for which all taro species, especially *A. macrorrhizos*, are notorious because serious skin, mouth, or throat irritations can occur if the potentially edible parts of these plants are not prepared with special care.

Today, "true taro" (*Colocasia esculenta*, known in Pohnpei as **Sawa**) and the huge "swamp taro" (*Cyrtosperma merkusii*, known in Pohnpei as **Mwahng**) are more commonly cultivated in Pohnpei than *A. macrorrhizos*. The latter is still found on Pohnpei around some houses, and even at fairly high elevations in the mountains within farmed land (*nansapw*); in addition, it is also sometimes found growing wild in uncultivated areas (*nanwel*). Raynor and Fownes (1991) listed *A. macrorrhizos* as a perennial root crop occurring commonly in the lower canopy of the tree gardens or agroforests of Pohnpei (also see Balick et al. 2009: 10). Although *A. macrorrhizos* is currently not considered to be as palatable a food source as true taro or giant swamp taro on Pohnpei, it can serve as a source of food during periods of famine, as well as a source of medicine, such as in the healing treatment of large wounds (e.g., see Glassman 1952: 108). Lee et al. (2010: 93) provide us with some aspects of the traditional therapeutic uses of *A. macrorrhizos* on Pohnpei, for cleaning wounds and for treatment of back pain:

> For cleaning wounds, the rotten stem of **Ohd** [*A. macrorrhizos*] is wrapped in a leaf of **Mahi** [*Artocarpus altilis*], and dried over a fire. **Ohd** is removed from the **Mahi** leaf and the juice is squeezed onto the wound. This is said to prevent infection and should be applied once daily for four days. For the treatment of back pain, the roots of **Ohd** are gathered, cleaned and combined with the meat of one whole coconut fruit, and then wrapped in a piece of cloth with a hot stone. The medicine is massaged onto the painful area." (Lee et al. 2010).

On Pohnpei today, perhaps the best-known uses for *A. macrorrhizos* are that its leaves are used to cover the stone oven (*uhmw*), as well as "to line the floor around the *sakau* [pounding] stone" (see Balick et al. 2009: 255), and sometimes when it is raining the large leaves can serve as umbrellas (e.g., see Ashby 1993).

Cultivars in Pohnpei: According to Bascom (1946), in the mid-20th century there were nine cultivars of *A. macrorrhizos*, including one poisonous and two considered to have been introduced since European contact (see Glassman 1952: 108; also see Bascom 1949: 119). Raynor (1991) lists ten cultivars. See p. 421 for the list we created by cross-checking these sources with the entries in the Pohnpeian–English dictionary.

Uses in Mwokilloa and Pingelap: This taro is called **Wod** in Mwoakilloa and **Woad** in Pingelap. Glassman (1953) observed it growing on both atolls around the edges of the *C. merkusii* wetland. Glassman (1953: 297) noted that giant taro (*A. macrorrhizos*) is "not a major source of food." However, researching food security in the FSM, Soto (2014: 30) emphasized the importance of this taro as a famine food and a somewhat "forgotten" wild edible crop:

> " On small islands and atolls, imported foods and medicines may not arrive for more than a month if a typhoon or an oil crisis occurs. Therefore, it is very important to re-discover plants already naturalized on each island, for food security. From this perspective, *Alocasia macrorrhizos* and *Tacca leontopetaloides* are very important starchy crops for emergency [situations]. However, the younger generation [on Pingelap and other islands] has never eaten them and/or does not know how to cook [them]. It is necessary to document how to eat these plants for the next generation." (Underlined for emphasis).

Cultivars in Mwokilloa and Pingelap: According to Glassman, a couple of varieties exist: one said to be edible and another regarded as poisonous on Mwoakilloa (1953: 297) as well as on Pingelap (1952: 301) (see pp. 421–422).

Uses in Kapingamarangi: It is noteworthy that *A. macrorrhizos* was eaten more often in the past on many Pacific atolls, and mostly as an emergency or famine food only. That is because its large stems are acrid with crystals of calcium oxalate and difficult to process before becoming edible. Kapingamarangi has been an exception in that *A. macrorrhizos* reportedly was not introduced until the early 20th century (Emory 1965). It is now known as **Ngaungau** on this atoll and is said to be relatively rare, cooked and eaten only occasionally (Bayliss-Smith 2012: 109–113).

▸ **Iouiou** (P) (Mw)
Alpinia carolinensis

Description: *Alpinia carolinensis* is a giant herbaceous species, one of the largest in the ginger family, Zingiberaceae. Like other gingers, *A. carolinensis* plants lack true stems, producing what are referred to as false or pseudostems. These false stems are composed of overlapping leaf sheaths and can reach remarkable heights of up to about 8 m (25 ft) tall. The leaves of *A. carolinensis* are lanceolate to oblong, and the rounded, dry, or fleshy capsule-shaped fruits are produced in large clusters, with individual fruits turning from green to bright red as they mature.

Distribution: *Alpinia* is the largest and most widespread genus in the ginger family. It consists of approximately 230 species distributed in the Asia–Pacific region. *A. carolinensis* is a native ginger species on the high islands of Micronesia, where it is a common component of the forest understory (Costion and Lorence 2012: 70). The ripe fruits of this large native ginger are eaten by birds and possibly bats on Pohnpei, and presumably they are thus dispersed by these flying animals, especially the Pohnpei lory (*Trichoglossus rubiginosus*), the state bird of Pohnpei, known in Pohnpeian as **Serehd**, and maybe fruit bats (e.g., *Pteropus molossinus*), known in Pohnpeian as **Pwehk**.

General uses: Many herbaceous *Alpinia* species are popular as ornamentals, including two well-known ornamental ginger relatives of *A. carolinensis*: *A. purpurata* (red flowering ginger) and *A. zerumbet* (shell flowering ginger), which are commonly cultivated in tropical environments for their attractive flowers and foliage. *A. carolinensis*, Micronesia's "giant ginger," is endemic to some high islands in Micronesia but has, for the most part, not been exploited as an ornamental cultivated species, though many plant enthusiasts express unreserved astonishment at its size (Costion and Lorence 2012: 70). *A. carolinensis* plants are aromatic because of their essential oils.

Uses in Pohnpei: This large native ginger species is called **Iouiou** in Pohnpei and grows wild. The banana-like leaves of this huge herb are utilized as plates by people to cover the earth oven (*uhmw*) as well as for lining pits used to produce fermented breadfruit (*mahr*). Therapeutically, **Iouiou** is also reported to have been used on Pohnpei for treating arthritis and painful swollen joints, physical ailments said to be brought on by spells associated with a type of ghost sickness known as *soumwahu en sokolei* and caused by "a dwarf leprechaun that lives in the forest" (see Balick et al. 2009: 309–311 for reference and additional medicinal uses of **louiou** on Pohnpei).

▸ **Daureleng** (P)
Amyema artensis

Description: *Amyema artensis* is a small, non-woody plant that grows on other plants and is therefore known as an epiphyte (or air plant). *A. artensis* can grow to heights of 3–4 m (9–12 ft) but often is much smaller. It produces leaves that are opposite, glabrous, and thick, and its scentless flowers are borne in bunches, laterally along the stem. The flowers are bisexual and 3 cm (~1 in) wide, with reddish corollas that are orange near the tips of its five petals.

Distribution: *Amyema* is a genus of semi-parasitic shrubs ("mistletoes") which are found naturally in the islands of Southeas Asia and in Australia. More specifically, *A. artensis* is distributed from New Guinea to several tropical Pacific islands, including the Solomons, Louisiades, and New Caledonia and adjoining islands, as well as in Vanuatu, Samoa, and at least one high volcanic island in Micronesia. In New Guinea, *A. artensis* has been recorded from primary and secondary rainforests, where it is parasitic on various host plant species. Its distribution in Micronesia is apparently restricted to Pohnpei, where this species is indigenous (i.e., it is a species that is native to this island but not unique (endemic) to it).

Uses in Pohnpei: *A. artensis* is known as **Daureleng** in Pohnpeian, and its red fruit are reportedly used in traditional Pohnpeian medicine, as noted in "The Book of Luelen" (see Bernart 1977: 21, where this plant is referred to as '*Tauralan*'; also see Rehg and Sohl 1979 under its Pohnpeian name and under "vine" in English).

▸ **Paiwed pwetepwet** (P)

Description: *Angiopteris evecta* is an enormous perennial fern with a massive rhizome (or caudex) at its base, which is a somewhat spherical, trunk-like unit that can grow up to about 120 cm (~3.25 ft) tall and 100 cm (~3 ft) in diameter. This giant rhizome structure produces two flat, rounded, dark brown, leathery, stipule-like growths about 10–15 cm (4–6 in) long. These growths are found on either side of the petiole or stalk that attaches the leaf or frond blade to the stem and can bear many buds that may develop into additional plants when they break off. *A. evecta* also produces huge leaf blades from its base near the ground. These leaf blades are often taller than a person and can be up to 8 m (25 ft) long. The sporangia are clustered in short, double rows of three to seven with no indusium (thin membranous covers serving as a shield covering a sorus—a cluster of spore-producing receptacles on the underside of a fern frond). The *A. evecta* fern is apparently very long-lived.

Distribution: *A. evecta* is reportedly indigenous to parts of northern and eastern Australia, as well as to various areas of the South Pacific, including Polynesia, Melanesia, Micronesia, and other tropical areas. Unfortunately, this huge herbaceous species has established invasive alien populations in Hawai'i, Costa Rica, and Jamaica (e.g., see Funk 1987), and thus its widespread cultivation as a decorative plant should be discouraged. On the other hand, *A. evecta* is now a rare plant in part of its putative native range in eastern and northern Australia, due largely to habitat alteration. *A. evecta* occurs in both tropical and subtropical forests, generally in moist areas and rainforest at low to middle elevations, but especially in ruderal or disturbed areas, and in wetlands. In Micronesia, it is reported from Guam in the Mariana Islands as well as on Palau, Yap, Pohnpei, and Kosrae.

Uses in Pohnpei: The large *A. evecta* fern is known as **Paiwed pwetepwet** in Pohnpeian. Its huge fronds are said to be used in building temporary huts in forests, as well as for bedding in these short-term refuges. Medicinally, this giant fern is reportedly the source of therapeutic remedies for a variety of sicknesses caused by evil spirits or sorcerers and has been used to treat different forms of dizziness, such as vertigo and seasickness (*mehn madau*); In addition, **Paiwed pwetepwet** has also been planted as a shade provider near **Sakau** plants (*Piper methysticum*) (Balick et al. 2009: 240–242). However, because these very large ferns are known to alter the natural environment, which may reduce local biodiversity: "it is advisable to restrict the trade and spread of the species and to discourage its cultivation as an ornamental" (Christenhusz and Toivonen 2008: 1215).

▸ **Sei** or **Soai** (P), **Joi** (Mw), **Saei** (Pi), **Ssai** (Nu), **Tatap** (Ka)
Annona muricata

Description: Known in English as the "soursop" tree, *Annona muricata* is a cultivated woody species which is usually shorter than 8 m (25 ft). This small, many-branched evergreen tree produces shiny, dark green, oblong or oblong-elliptic leaves which are pointed at the apex and rounded at the base. Its large, irregularly heart-shaped, soft-spined, yellowish green fruits are more or less oval or heart-shaped. The soursop fruits can reach about 10–30 cm (4–12 in) in length and are covered by a leathery skin with regularly spaced, short, slightly curved, fleshy spines. The edible pulp is cream colored, granular, and separates easily from the mass of white, juicy segments surrounding the central core. The pulp of the soursop fruit has a somewhat pineapple-like aroma, but its musky, sub-acid to acid flavor is distinctive. The seeds are tough and dark brown to black (Stone 1970, Morton 1987, also see Falanruw 2015: 34).

Distribution: *A. muricata* is native to tropical America, probably originating "in the Antilles in the Caribbean, Central America and South America [and now is] widely distributed in Asia from south India to northern Australia and Polynesia, [as well as] in Africa in the hot lowlands in the west and east" (Lim 2012: 191). In addition, the soursop tree was probably introduced for its food value to some islands of Micronesia, such as Yap, before World War II. It is often cultivated in gardens, especially near homes on the high islands, but it can be found elsewhere, typically in lowland areas below 1200 m elevation. It is a hardy fruit tree which thrives in many tropical regions today because it can tolerate relatively poor soils (PIER 2002).

General uses: Although *A. muricata* (soursop or graviola) of the Annonaceae family is best known as a tasty fruit pulp resource, it has also been used widely in medicinal remedies for many years; for example, it reportedly has a broad range of therapeutic effects, including "antibacterial, antiviral, antifungal, antitumor, antihelmin, analgesic, hypotensive, anti-inflammatory, and immune enhancing effects" (Dongre 2014; also see Moghadamtousi et al. 2015).

Uses in Pohnpei: The stems of the introduced soursop tree, known in Pohnpeian as **Sei** or **Soai**, have been used in Pohnpei to make tool handles. In addition, in recent times, the juice of ripe soursop fruits is sometimes extracted and allowed to ferment for about a week and then consumed as an alcoholic beverage. Medicinally, parts of this tree have been used in Pohnpei to treat seizures (*lusulus*), asthma, headaches, upset uterus, and to inhibit vomiting. It has also been reported that the bark is scraped, pounded, and wrapped in a piece of cloth to use as a charm to cast away an angry spirit of an ancestor (*winahnihn kouseni*, see Balick et al. 2009: 325). Lee et al. (2010: 122) provide a brief description of how parts of **Sei** are mixed with parts of seven additional plants to treat "shaking sickness due to high fever" on Pohnpei:

> "The stem of **Sei** (*A. muricata*), **Topwuk** (*Premna serratifolia*), **Weipwul** (*Morinda citrifolia*), and **Laman** (*Citrus limon*) are scraped and added to four young shoots of **Ilau** (*Clerodendrum inerme*), four leaves of **Oahs** (*Metroxylon amaricum*), and four young roots of **Nih** (*Cocos nucifera*). These plants are pounded together and wrapped in

a piece of cloth. Four drops of the juice are squeezed into a cup of hot water and one to two cups per day are reported to stop the shaking and reduce fever."

Uses in Mwoakilloa: Murphy (1950: 73) noted the presence of "a few soursop trees (*Annona muricata*)" on this atoll. Glassman (1953: 293) indicated that it was cultivated and probably introduced to Mwoakilloa from Pohnpei.

▸ **Pwuh** (P), **Pwu** (Mw) (Pi), **Buu** (Nu) (Ka) *Areca catechu*

Description: The betel nut palm (*Areca catechu*), also known as the *Areca* palm, is an attractive, well-known tree that grows to heights of about 20 m (~65 ft) or more. The trunk is dark gray ringed, straight, and slender, with a diameter of about 15 cm (6 in). The trunk bears a crown of pinnate fronds (leaves); these are normally more than 1 m (3 ft) long, with the lowest frond usually drooping. The leaf sheaths form a distinct cylinder at the top of the trunk, as in other palm species. Where the trunk and sheaths join, a branched flower cluster develops. Each betel nut palm produces flowers of both sexes, with the small male flowers borne on the upper part of the flowering stalk, and the larger female flowers at the base. Fruits are produced in bunches and are 1.5–5 cm (0.5–2 in) in diameter, turning from green to yellow or orange when fully mature. Each fruit holds a single "nut" (endosperm or seed). The nut, which is grayish brown and 0.5–4 cm (1–1.5 in) wide, is enclosed in a fibrous husk with no edible flesh. Under ideal growing conditions, the betel nut palm begins fruiting when it is about 4–5 years old, but more often starts when it is 7–8 years old. Maximum fruit production is reached at about 10–15 years or more. Fruits need 6–8 months from bloom to maturity. Betel nut trees in some tropical areas reportedly produce fruit even after 60 years of age, and others are said to live and bear fruit even when they are beyond 100 years of age.

Distribution: Today, the betel nut palm is cultivated widely throughout India, Sri Lanka, Indonesia, the Philippines, and Malaysia (where it may be native). It is also grown and used extensively in many islands of the tropical Western Pacific, including New Guinea, the Solomon Islands, Palau, the Mariana Islands, and Yap. The betel nut palm prefers lowland tropical habitats on well-watered fertile soils. In open grass or fern lands, the betel nut palm is rarely found due to dry soil conditions. Archaeological evidence indicates that this useful palm species was introduced to western Micronesia in prehistoric times (e.g., see Fitzpatrick et al. 2003; Zumbroich 2007–2008; Paulino et al. 2011). The betel nut palm has apparently spread on its own on some tropical islands following its introduction in the Pacific, and some have argued that this increased distribution has been useful in reforestation as betel nut palms establish themselves spontaneously in secondary growth regions, usually not too far from cultivated areas. More recently, betel nut "chewing" has spread beyond its traditional areas of use in western Micronesia to some islands in eastern Micronesia, such as Pohnpei and the Marshall Islands.

General uses and health problems: In modern times, the betel nut is utilized for mind-altering and other physiological purposes by hundreds of millions of people, mainly in Asia. It is generally used as a mild to strong stimulant by both men and women, and among various age groups and social classes. With its very long history, such use of the betel nut is deeply ingrained in many sociocultural and religious practices. A "betel quid" is normally prepared by wrapping slivers of the betel nut, along with slaked lime (calcium hydroxide), in the leaf of a betel vine (*Piper betle*). Chewing or sucking on a "betel quid" has been a cultural tradition for thousands of years throughout much of tropical Asia and the Western Pacific. The betel nut is reported to be the fourth most consumed psychoactive substance in the world, preceded only by alcohol, nicotine, and caffeine (e.g., Boucher and Mannan 2002; Warnakulasuriya et al. 2002). Betel nut use today includes approximately 600 million people globally, who mostly live in low to moderate income countries in the Asia–Pacific region (Gupta and Warnakulasuriya 2002). The betel nut fruit (technically a berry) is commercially obtainable today in dried, cured, and fresh states, depending on cultural and personal preferences. The husk of the fresh fruit is soft enough to be cut through with a sharp knife; however, as the fruit ripens and dries, the exterior of the fruit husk becomes yellow, orange, or reddish, while the interior hardens considerably and thus can only be sliced using a special scissors-like cutter. Although the chewing or sucking on the betel quid is an ancient and still a very widespread culture trait, the consumption of the betel quid for psychoactive purposes has been called into question with regard to both short- and long-term health issues, especially when more modern additives, such as tobacco, are mixed into the betel quid.

Garg et al. (2014: 3) reviewed the medical literature regarding the safety of "chewing" betel over time:

> "[The] [a]reca nut [betel nut] is widely consumed by all age groups in many parts of the world, especially south-east Asia. The objective of this review is to systematically review and collate all the published data that are related to the systemic effects of *Areca* nut."

Based on their review, they concluded:

> "There is substantial evidence for carcinogenicity of Areca nut in cancers of the mouth and esophagus. *Areca* nut affects almost all organs of the human body, including the brain, heart, lungs, gastrointestinal tract and reproductive organs. It causes or aggravates pre-existing conditions such as neuronal injury, myocardial infarction, cardiac arrhythmias, hepatotoxicity, asthma, central obesity, type II diabetes, hyperlipidemia, metabolic syndrome, etc. *Areca* nuts affect the endocrine system, leading to hypothyroidism, prostate hyperplasia, and infertility. It affects the immune system leading to suppression of T-cell activity and decreased release of cytokines. It has harmful effects on the fetus when used during pregnancy." (Underlined for emphasis).

These conclusions strongly suggest that the betel nut is not a harmless substance. Consequently, there is an urgent need to recognize that long-term use of the betel nut alone, or especially when it is combined with tobacco, can be a harmful substance. As suggested by Garg et al. (2014): "Strict laws may be necessary to regulate the use of betel *Areca* nut" (also see Warnakulasuriya et al. 2002).

Another review of the literature that has been published over the past 50 years focused on betel nut use as a recognized but often neglected public health issue in the Asia–Pacific region (Joo et al. 2020). This review characterized "the current situation on betel nut use at a global level regarding the epidemiology, pharmacology and toxicology, health effects, treatment options, policies and strategies." The extensive literature these researchers surveyed indicated "that there is a lack of up to date statistics on betel nut use," and there has only been "limited research on the effect of policies and strategies to control betel nut use." Given these results of their widespread survey, the authors concluded that in countries where this habit is widespread, citizens and their governments "should focus on understanding which populations chew betel nut and implement appropriate policies [along with] educational and cessation programs to help control betel nut use" (also see Buente et al. 2020).

Uses in Micronesia: The consumption of the masticated mix of human saliva and the chewed betel quid varies in method of ingestion and the diverse reactions that occur in the human body, depending upon the psychoactive and physiological qualities of the individual components of the betel quid (seed/nut, lime, and leaf). One method of consumption involves consuming a quantity by swallowing the liquid mix but not the quid itself. Another ingestion pathway involves allowing some of the liquid mix to be absorbed through the mucous membrane that lines the inside of a person's mouth and periodically spitting out most if not all of the remaining liquid mix. Furthermore, there are cultural factors affecting motivations to indulge, as well as the psychoactive effects brought on by chewing betel quids:

> "Betel chewing releases a complex set of biologically active components into the blood stream which result in diverse physiological and psychosomatic responses. Betel chewers experience a sense of wellbeing, heightened alertness, a warm body sensation, improved digestion and increased stamina. This wide range of effects has been validated by an understanding of some of the underlying [physiological] mechanisms." (Zumbroich 2007–2008: 90)

Today, many Micronesians chew on a combination of the soft immature fruit of the betel palm after it has been split open, partially covered with lime (hydrated calcium hydroxide), then wrapped with a leaf of the *Piper betle* vine. This "betel quid" is placed in the side of the mouth, like a plug of tobacco, then chewed (i.e., masticated and sucked on). Betel users commonly have to spit out the excessive, usually reddish, saliva generated by chewing the quid. Too much of the powder lime can burn the mouth, which over time may lead to serious health consequences. More recently, some Micronesians have also begun including tobacco in their betel quid, and others soak the tobacco in alcoholic vodka before putting it into the quid. Anecdotal reports indicate that small doses of the nut, lime, leaf and tobacco mix produce a sense of euphoria and increased energy, while large doses often result in sedation. Although all these quid ingredients may contribute to these effects, most experts attribute the majority of the psychoactivity to the alkaloids found in betel nuts themselves (e.g., arecoline).

In addition, to the use of betel "nuts" for their stimulating effects, the trunk of the betel nut palm tree has long been utilized widely for traditional building purposes. For example, its long trunk is hollow in inside but hard on the outside, providing raw material for several traditional uses in parts of Micronesia. These include construction use for building houses, benches, pig pens, and other items. The leaves of the betel palm also have a variety of traditional uses. Beyond the multiple uses of the betel palm and the individual biological responses to ingestion of betel nut, the betel quid carries deeply symbolic connotations and has long played a role in the social fabric of many Asian and Pacific island cultures. The significance of the betel nut quid in various contexts of private and public life centers around the theme of facilitating and structuring relationships. As noted above, the betel palm, and the associated betel vine (*Piper betle*), have long been cultivated and used extensively in many island areas of the tropical Western Pacific. However, it has only been during the latter part of the 20th century and the early 21st century that betel nut use has been spreading from Yap and elsewhere to other areas of the FSM, such as Pohnpei, and to the Marshall Islands. Note: On Mwoakilloa Atoll, the betel nut palm (*A. catechu*) is reportedly cultivated "in sands heavily mulched with organic matter" (Staples and Bevacqua 2006: 74).

Uses in Pohnpei: The betel nut palm is known in Pohnpeian as **Pwuh**. It was introduced to Pohnpei in historic times by Saipanese and Yapese laborers during the Japanese Mandate (1914–1945). However, chewing betel nut quids has only become widespread throughout Pohnpei since the end of the Second World War, primarily among young men in the early stages of its spreading use on this island. Furthermore, some Pohnpeian betel nut users add chewing tobacco to their quids, as is occurring elsewhere in Micronesia today. A more recent change is that the betel nut, previously not grown in Pohnpei, has now become an important cash crop on the island, with locally grown nuts partly replacing imports from Yap.

Betel nut use among students on Pohnpei: A small but comprehensive health-oriented research survey (Milgrom et al. 2013) revealed a high rate of betel nut use among "adolescents" (seventh- and eighth-grade students) in Yap and Pohnpei. These students were using betel nut at least once per week during the period of one month before the survey. A key health concern has been the relative lack of knowledge about whether betel nut use, with and without tobacco, leads to addiction among young people: "Limited research on adults suggests that chronic users of *Areca* nut [betel nut] alone develop dependence." A more recent study (Milgrom et al. 2016) described "dependence symptoms among adolescents using betel alone or with tobacco." Part of this study focused on 269 students in the ninth grade in Pohnpei schools and revealed that more than a third (38%) had used betel nut at least once, and most (67%) had ingested betel nut within the month before the survey. Among those who had used betel in the previous month, 90% reported using betel with tobacco! Given the long-term negative health aspects of tobacco use, this trend is disturbing, and it has been suggested that betel nut users who include tobacco in their betel quid "may benefit from tobacco cessation strategies" (i.e., anti-addiction counseling). Another recent and more expanded study (Abraham et al. 2018) involving student use of drugs in Pohnpei focused on "non-communicable disease (NCD) related behaviors among adolescents," which have been increasing worldwide and in the Pacific region. This study was "conducted among more than 80% of the thousands of secondary school students on Pohnpei." Among the students who responded, "21.7% of students self-reported smoking tobacco in the past 30 days, 30.3% self-reported drinking alcohol in the past 30 days, 40.9% self-reported chewing betel nut in the past 30 days, and 21.2% self-reported chewing tobacco with or without betel nut in the last 30 days." (Underlined for emphasis).

These studies of betel nut use alone or in combination with tobacco provide evidence that there is reason to be concerned about long-term health consequences. The youthful use of betel nut and tobacco should be evaluated seriously and without bias by students, parents, teachers, and administrators. Unfortunately, because of the growing popularity, relative scarcity, and high cost of betel nut, theft of the betel fruits from palms on Pohnpei has become more common and is becoming an increasing social problem. (For more information regarding the effects of betel nut, see www.comfsm.fm/~dleeling/ethnobotany/text/ethnobotany.html).

A freshly harvested cluster of betel nuts.

Preparing betel nuts for sale.

Betel nuts packaged for sale. They are typically chewed with **Kapwoi** leaves and powdered lime usually included in each package.

A more positive aspect of betel nut palm cultivation on Pohnpei is shown by a recent endeavor to utilize betel nut palm leaf sheaths to fabricate "eco-plates." The project started in 2018 and its objective has been to help reduce the number of single-use plastic and Styrofoam plates imported into Pohnpei and discarded on island. Plastic products are difficult to recycle and can readily pollute land and sea environments locally as well as far beyond the island. This betel nut eco-plate project was developed by the Awak Farmers Association on Pohnpei and represents "a modern evolution in ethnobotanical material culture." These environmentally friendly and fully biodegradable "eco-plates" are made on Pohnpei from the flexible but sturdy leaf sheaf portion of the petiole of a betel nut palm:

> "Betelnut palm is abundant and a natural resource in the community and currently just often discarded. The betelnut eco-plates are hypo-allergenic, food and human safe, and easily decompose after use. The project not only aims to solve an environmental issue, but also provides an alternate income source for the villagers. With such a promising project, the Awak Farmers Association is a recent grant recipient of the Direct Aid Program through the Australian Embassy in Pohnpei. The grant funds will go towards securing four press machines for production of the betelnut palm eco-plates." (see "Kaselehlie Press" from October 18, 2018).

There are arguably both positive and negative aspects of betel nut use in Pohnpei, especially when we consider whether betel nut is "chewed" with or without tobacco or other psychoactive substances. More recent studies provide additional perspective on betel nut use. For example, a review of the literature over the past 50 years (Joo et al. 2020) focused on betel nut use as a recognized, but often neglected, public health issue in the Asia–Pacific region. This research review described "the current situation on betel nut use at a global level regarding the epidemiology, pharmacology and toxicology, health effects, treatment options, policies and strategies." The extensive literature these researchers surveyed indicated "that there is a lack of up-to-date statistics on betel nut use," and there has only been "limited research on the effect of policies and strategies to control betel nut use." Given these results of their widespread survey, the authors concluded that in countries where this habit is widespread, citizens and their governments "should focus on understanding which populations chew betel nut and implement appropriate policies [along with] educational and cessation programs to help control betel nut use" (also see Buente et al. 2020 for an "Instagram-Based Study to Understand Betel Nut Use Culture in Micronesia").

▸ BREADFRUIT SPECIES

The common breadfruit species, *Artocarpus altilis*, is one of the most important plants for people living in Micronesia and much of Polynesia. Here we provide information on the botanical, geographical, and ethnobotanical aspects of this multipurpose cultivated tree. There is some confusion relating to the taxonomy of the common "seedless" form, *Artocarpus altilis*, and the closely related breadfruit species, *Artocarpus mariannensis*, which produces seeds and is native to islands in western Micronesia. Both are described and discussed in this book. Today, what is referred to as "breadfruit" generally includes *A. altilis* and botanical hybrids resulting from crosses of *A. altilis* and *A. mariannensis* (e.g., see Fosberg 1960; Ragone 1997; Zerega et al. 2005). The closest wild relatives of common breadfruit, *A. altilis*, are *A. camansi* and *A. mariannensis* (Lincoln et al. 2018; also see Zerega et al. 2006 for a thorough description of each of these *Artocarpus* species and a botanical key for identification purposes).

The common breadfruit, *A. altilis*, is a tree introduced to Micronesia in ancient times by early Pacific islanders as part of their "transported landscape," which included important cultivated plants and domesticated animals. Breadfruit trees are most important as a staple food source for peoples on tropical Pacific islands, but they also have many other uses, which are discussed below.

William Dampier (1651–1715) was an English explorer and buccaneer who explored Australia and the Western Pacific and was the first person to sail around the world three times. He visited some tropical islands in the Western Pacific, including Guam in the Marianas, and in 1697, he was the first European to describe breadfruit (most likely to be a fruit from a cultivar of *A. altilis* because of its lack of seeds):

> "The fruit grows on the boughs like apples; it is as big as a penny loaf, when wheat is at five shillings the bushel, it is of a round shape, and hath a thick, tough rind. When the fruit is ripe, it is yellow and soft, and tastes sweet and pleasant. The natives of Guam use it for bread. They gather it when full grown, while it is green and hard, then they bake it in an oven, which scorcheth the rind and maketh it black; but… the inside is soft, tender and white, like the crumb of a penny loaf. There is neither seed nor stone in the inside, but all of a pure substance, like bread. It must be eaten new, for [if] it be kept above twenty-four hours, it grows harsh and choky, but it is very pleasant before it is too stale. This fruit lasts in season eight months in the year, during which the natives eat no other sort of bread-kind." (quoted by Sacks 1998: 213; also see Preston and Preston 2004: 143–144)

Vaughn MacCaughey (1917) eloquently lauded the significance of the *A. altilis* breadfruit in comparison to all its other widespread cultivated species:

> "The breadfruit is by far the most important member of the genus [*Artocarpus*], outranking in economic value and in geographic range of all the others combined. It is one of the world's great trees, and for its beauty, unique features, and human interest, has a place beside the orange tree and the coco palm."

▸ **Mahi** (P), **Moai** (Mw), **Maei** (Pi), **Gulu** (Nu) (Ka)
Artocarpus altilis

Description: Common breadfruit trees (*A. altilis*) are most often seedless, and can grow up to about 25 m (80 ft) in height. Breadfruit trees release a very sticky, milky latex when cut. Their leaves are large, up to about 60 cm (2 ft) in length, glossy green, and often have toothed to deeply lobed margins. The trunk can grow to be as wide as 2 m (~6 ft) in diameter. Both male and female flowers are produced by an individual breadfruit tree, with the male inflorescence appearing first. The male flowers are 5 cm (2 in) in diameter and up to 45 cm (18 in) long. Female inflorescences consist of 1500–2000 reduced (small) flowers attached to a spongy core. The multitude of tiny flowers fuse together and form the fleshy edible fruit. The large, oblong fruits develop at the branch tips and vary in size, color (yellow to green) and texture (smooth to bumpy). There is great morphological variability in these breadfruit trees throughout the tropical Pacific due to domestication (artificial selection) by humans over thousands of years. This long-term artificial selection by people has created many differing forms and functions of common breadfruit trees, as Wilder (1928: 18) recognized long ago:

> "The breadfruit of Oceania has been the object of human care for many hundreds of years. It is therefore natural that selection has improved the standard of its excellence. By close observation, Man has suited varieties to their environment; some thriving better in very moist localities, others growing well under dry conditions. The trained eye and discriminating palate of those whose chief article of diet is the breadfruit are quick to notice slight variations in form or quality. Many surprises must have appeared in the way of 'sports'. Apparently new varieties were thus created and duly named."

Distribution: Domesticated in island Southeast Asia and/or Near Oceania, breadfruit is a member of the mulberry family, Moraceae. The term "breadfruit" usually refers to the species *A. altilis*, although, as noted above, it is occasionally used to refer to other members of the genus, especially *A. mariannensis* and *A. altilis* × *A. mariannensis* hybrids. Genetically, the closest relatives of the common breadfruit, *A. altilis*, are the truly wild breadfruit tree species, *A. camansi* and *A. mariannensis*. Recent DNA fingerprinting data (amplified fragment length polymorphisms = AFLP) from over 200 breadfruit cultivars belonging to *A. altilis*, along with DNA data from 30 trees of *A. camansi* and 24 of *A. mariannensis*, were used to investigate the relationships among these species. The results indicate that the wild, seeded, ancestral form of most breadfruit varieties in Melanesia and Polynesia is *A. camansi*, known as "breadnut," which is native to New Guinea and possibly Indonesia and the Philippines. "In contrast, most Micronesian breadfruit cultivars appear to be the result of hybridization between *A. camansi*-derived breadfruit [i.e., *A. altilis*] and *A. mariannensis*" (Zerega et al. 2004; also see Ragone 1997).

Gradually, the efforts of the English captain Bligh and some French voyagers resulted in the introduction of a limited series of seedless breadfruit varieties into Caribbean in the late 1700s. These and other breadfruit varieties were then spread around the tropics and are now grown in many countries.

According to the National Tropical Botanical Garden on Kaua'i, Hawai'i, breadfruit generally requires a temperature range of about ~16–38°C (60–100°F), an annual rainfall of 200–250 cm (80–100 in), and a relative humidity of about 70–80%. According to many reports, the breadfruit tree requires deep, fertile, well-drained soil; however, other authorities on South Pacific plants point out that the seedless breadfruit does well on sandy coral soils, and seeded types grow naturally on relatively wet "coralline limestone" islands in Micronesia. Over the past two centuries, breadfruit varieties have been introduced to other tropical regions of the world, and more recently they have been promoted as a potentially important staple food globally (Lucas and Ragone 2012).

Historical uses in Pohnpei: Breadfruit trees are most important as a staple food source for peoples on tropical Pacific islands and have many other uses and an immense cultural significance:

> "The Pohnpeian words *engin aramas* ("the scent of human beings") are spoken when referring to the planting of breadfruit trees in the proximity of human dwellings in the belief that the trees will bear more fruit." (Rehg and Sohl 1979)

Here we focus on some historical aspects of breadfruit use on this island, including some limited archaeological evidence, as well as reference to legendary explanations about the origins and long-term traditional use of breadfruit on the eastern high volcanic islands of both Pohnpei and Kosrae. Cultivation and consumption of breadfruit have long played a fundamental role in the subsistence systems of these high islands as well as of the low-lying coralline outer islands. For example, according to Levin (2015b: 70–71), "In eastern Micronesia [Pohnpei and Kosrae], breadfruit is an important staple crop and has probably been part of the diet since initial settlement." Direct evidence of breadfruit pollen and charcoal recovered from archaeological excavations on Kosrae has confirmed the use of breadfruit on that island for almost 2000 years (see Athens et al. 1996).

On Pohnpei, archaeological research by Ayres and Haun (1985, 1990) revealed ancient breadfruit fermentation pits at Awak (in U District) and at Wene (in Kiti District), dating back 1,500 years. These archaeologists also documented an increasing scale of cultivation which they associated with agricultural intensification over time and its role in the rise of chiefdoms on Pohnpei (also see Ayres 1990; Haun 1984). More recently, Levin (2015a, 2017) used ancient botanical evidence to identify what appear to have been breadfruit fermentation pits as part of a "historical feature" on Temwen Island, Pohnpei, the location of the prehistoric city of Nan Madol, now an archaeological site part of UNESCO World Heritage. Indeed, recent remote sensing technology used to survey all of Temwen Island "revealed a vast archaeological landscape." On Temwen, Levin recovered prehistoric charred breadfruit exocarp (fruit skin) material, indicating that breadfruit was being roasted in ancient times on this island. Levin suggested that when early agriculture on Pohnpei transitioned from slash and burn shifting cultivation to "permanent agroforestry, breadfruit played a major role" (also see Levin 2018 for more details).

Modern uses in Pohnpei: Today, cultivation, cooking, and fermentation of breadfruit continue to play an important role in the Pohnpeian subsistence system as it is a major traditional staple and ceremonial food. The main breadfruit species on Pohnpei, in terms of human use, is the "seedless" *A. altilis*, which is known in Pohnpeian as **Mahi**. Mature breadfruit can be eaten after baking, boiling, or frying. Overripe fruits can be fed to pigs. Breadfruit can also be fermented for a few days, or in special leaf-lined pits for several months or even years, and then used as a significant emergency food in times when food is scarce (for example, following disastrous storms or other natural calamities). More details about fermented breadfruit are described below.

According to Englberger et al. (2003c: 8), breadfruit generally is the most consumed traditional staple food in the FSM. Breadfruit has been important in traditional diets because of its high dietary fiber and potassium, as well as vitamin C, calcium, and folate, combined with low levels of fat and sodium. Essentially, breadfruit is a very nutritious carbohydrate food, and when seeds are available in the fruit, they can provide protein and thiamin. In addition, although breadfruit trees are very sensitive to dry periods, they can recover relatively fast, and if they are growing near a stable source of underground water, they can become quite large and live for long periods of time (for more detail, see Lawrence et al. 1964; Merlin et al. 1992; Ragone 1997, 2002; Petersen 2006; Ragone and Raynor 2009).

- **Breadfruit seasons:** In much of the tropical Pacific, breadfruit trees bear fruits more or less continuously year-round; however, there are two or three main fruiting periods. In the islands of the FSM and the Gilbert Islands, the main ripening season is May–July or September, and in the Society Islands and New Hebrides, the main ripening season is November–April, with a secondary crop that becomes ripe in July–August. Captain James Cook, in the latter part of the 18th Century, wrote approvingly about the resource-richness and seasonality of the breadfruit tree:

> "If a man plants ten of them in his life time, he will as completely fulfill his duty to his own and future generations, as the native of our less temperate climate can do by plowing in the cold of winter and reaping in the summer heat, as often as these seasons return." (Quoted by Gerrit Parmile Wilder, in his 1928 manuscript on "The Breadfruit of Tahiti").

- Like elsewhere in the tropical Pacific, in Pohnpei and Kosrae, breadfruit trees bear fruits during most of the year. However, fruit is most abundant during the rainy season from May through August, which is the main breadfruit season on Pohnpei, known as the *rahk*. According to Lawrence et al. (1964: 46–47), there are four additional, minor breadfruit seasons on Pohnpei. Spelled according to the Pohnpeian–English dictionary, these seasons are *edihsou* (October–November), *soupar* (during January), *ketingen angak* (in March), and *dilin* or *dilinrek* (during March or April). However, these breadfruit seasons may be shifting due to the related impacts of global warming and associated climate change on the soil environment. The 5th National Report by the FSM to the Convention on Biological Diversity (2014: 15) stated the following:

> "In Pohnpei, stakeholders reported that changing weather patterns are affecting the soil and reducing crop yields. Changing soil conditions may also be behind recent outbreaks of fungus affecting more than 100 varieties of yams in Pohnpei and **Sakau** plants in Pohnpei, and Kosrae farmers reported that changing weather patterns have already resulted in different harvesting patterns than previously known. For example, <u>species of breadfruit are ripening at different times during the year</u>, which is impacting traditional and cultural practices that revolve around harvest seasons. These observations are in line with scientific findings about <u>changes in soil and weather patterns</u>." (Federated States of Micronesia 2012, underlined for emphasis).

- **Preparation of breadfruit food on Pohnpei:** As elsewhere in Micronesia, the fruit of the breadfruit is a major traditional staple and ceremonial food of the Pohnpeian people. Edible breadfruit can be prepared in various ways. Mature fruit can be eaten after baking, boiling, or frying.

Today, the most common everyday cooking practice involves roasting breadfruits in an *uhmw* (stone or earth oven). Stones are placed on the ground and heated with firewood. Breadfruits and other foods are then placed on top of the stones and covered with leaves to cook. This frequently takes place in cookhouses. Although such cooking structures are commonly used in many areas of the tropical Pacific, such as the pit cooking structures used in Polynesia, the Pohnpeian earth ovens were constructed on level ground, making them less readily identified by archaeologists interested in early food cooking techniques:

> "Nevertheless, some archaeological cook houses have been previously identified in the Awak area, north-east Pohnpei Island (Ayres et al.1981). Thus, while these features may not present as a pit, they have the potential to contain a great deal of botanical evidence about cooking practices, including charred firewood, breadfruit and other edibles, and phytoliths from leaves and wood." (Levin 2015b).

- **Fermentation and storage of breadfruit on Pohnpei:** Preserved breadfruit, known as *mahr* in Pohnpeian, is produced through fermentation of **Mahi**. On Pohnpei, these fruits have been fermented traditionally using two different methods. The main method involves collection of at least 100 ripe breadfruits, peeling of the skin using "a scraper made from the back of a tiger cowry shell by removing the two ends of the shell and sharpening one side of the 'bridge' remaining across the middle section of the shell's back." Following the peeling, the fruit flesh is removed from the core and prepared for placement in the fermenting pits. "In some instance[s] the flesh is first washed in fresh water or (if a particularly good form of preserve is desired) in ocean water" (Lawrence et al. 1964: 54).

The Pohnpeian communal fermentation pits are traditionally approximately 1–20 m (3–60 ft) across and 0.5–2 m (1.5–6 ft) in depth and can be circular or oval shaped. Individual households produce smaller pits for private use. Commonly, the larger fermentation pits have a stone lining (Hunter-Anderson 1991), which is similar to those identified by archaeologists as ancient large pit features on Pohnpei (Haun 1984). These pits served as both fermentation chambers and storage areas. Dug into clay soil to prevent water seeping into them, these pits are lined not only with specially chosen stones, but also with leaves of banana plants (*Musa × paradisiaca*, **Uht**) and native turmeric (*Curcuma australasica*, **Auleng**) to prevent soil from mixing with the breadfruit paste. Hundreds of fruits are peeled, the cores are removed and soaked, and the flesh is placed in the pit for fermentation.

If prepared and placed in the pits appropriately, the preserved breadfruit can be carefully stored in these special leaf-lined pits for several months or even years, and then used as an important emergency food following disastrous storms or other natural calamities. In fact, this longtime storage of fermented breadfruit has been important for food security by utilizing excess harvest during the normal *rahk* (breadfruit harvest) season. In addition, well-aged, fermented breadfruit can bestow prestige on the people who prepared and stored it. Bascom (1948, 1965) reported that some of these larger fermentation pits could contain up to 10,000 fruits and that fruit could be stored in them for up to 100 years. In addition, special parcels wrapped in prepared leaves that contain baked fermented paste wrapped in leaves also provide transportable, long-lasting food for sea voyages (e.g., Lawrence et al. 1964; Ragone 2002; Clarke 2008; Balick et al. 2009).

A second method of preparing *mahr* produces *mahrenleng* or *mereleng* ("*mahr* of the air"); this preparation of *mahr* is practiced occasionally by harvesting a ripe breadfruit from near the main trunk of the tree, and then covering it up into a bundle using the same kind of leaves used in the *mahr* pit processing method (see above). The bundle is then tied to a branch of the breadfruit tree and allowed to ferment in place:

> "After a few weeks the fruit turns into a form of *mahr*. There is a certain amount of prestige attached to this form of *mahr* and it is said that a man who prepares a number of fruit for this type of *mahr* may well be looked upon favorably as an intelligent and reliable person by the *Nahnmwarki*." (Lawrence et al. 1964: 56)

Although cultivation, cooking, and fermentation of breadfruit continue to be part of the native Pohnpeian subsistence system, breadfruit fermentation in pits "is a practice that is disappearing, and along with it, the associated cultural knowledge." Therefore "recording breadfruit fermentation practices of the past helps to increase and preserve cultural knowledge in cases where revitalizing this tradition becomes desired or necessary" (e.g., see Levin 2015a for some details regarding how Pohnpeians have traditionally carefully selected appropriate soil conditions for locations of their breadfruit fermentation pits, now and in the past; also see Hunter-Anderson 1991; Merlin et al. 1992; Merlin and Juvik 1996; Ragone 2002; Petersen 2006; Ragone and Raynor 2009).

- **Multiple uses of breadfruit trees on Pohnpei:** The great traditional value of the breadfruit food crop is honored in the *lepin kahs* (proverb): *Wahn pehn eni* ("fruit of the hands of the gods"), which refers to breadfruit trees which grow freely and require no care beyond their initial planting. The breadfruit crop is, therefore, believed to be a "divine gift" with no real prestige value. In contrast to yams, which require considerable cultivation effort (see Reisenberg and Fischer 1955; and for more detail on the prestige values of yams, see the sections focused on the *Dioscorea* species).

Breadfruit produces large leaves all year, which offer ample protection from the intense sunlight of the tropics, and thus it serves as a shade tree. The trunks of breadfruit trees serve as a living support or trellis for yams (**Kehp**, *Dioscorea* spp.), a very important source of food and prestige on Pohnpei. Yams are planted at the base of breadfruit, and then the vines are trained up into the canopy of these trees along ropes made out of the bark fiber of **Keleu** (*Hibiscus tiliaceus*) or the vines of **Iol** (*Merremia peltata*). In addition, the trunks and large branches of the breadfruit trees can be used to make the bodies of canoes, or sometimes for lumber, and fallen branches are used for firewood. Previously, the pounded inner bark fibers of breadfruit trees were used in the fabrication of women's clothing such as wraparound bark cloth (Anson and Raynor 1993: 144). Furthermore, the leaves of breadfruit trees serve as fans and plates and have a number of other uses, such as covers for cooking breadfruit during the food preparation process (Ragone and Raynor 2009). Several parts of breadfruit trees are also used in a variety of traditional Pohnpeian medicines; for example, Lee et al. (2010: 67) provided this remedy:

> "To treat a wound, remove the bark [of breadfruit], scrape off the inner stem just beneath the bark, pound together, and squeeze the juice onto the cut. Apply this once daily for four or eight days or until the wound has healed."

Another valuable product of the common breadfruit tree is its sticky white sap, which was formerly used to seal leaks in canoes, as well as to catch birds: "To make a bird trap, the sap is mixed with ash and put on a tree or on the terminal bud of a banana tree. Birds get stuck on the sap/ash mixture and are caught for eating..." (Balick et al. 2009).

Cultivars in Pohnpei: In the Pohnpeian language, **Mahi** is the generic name for the common breadfruit, *A. altilis*. Legends suggest that many different cultivars were developed from cuttings taken from a few original introduced breadfruit trees. These varieties have been artificially selected in Pohnpei over generations for special purposes. The cultivar referred to as ***Mahi mat*** (also as ***Meimat***) produces a soft, ripe breadfruit that "is favored by older people who have difficulty chewing." In fact, many insights into traditional culture can be gained from examining the names of cultivars:

> "Some of the names are of particular interest because they would seem to suggest specific "best uses" for the respective fruits, such as "Crying Child Breadfruit" (***Meiserihseng***), being a variety which is known to cook to tenderness very quickly and, therefore, is recommended for a quick meal to still the child crying from hunger; "Sardine Breadfruit" (***Meisaip***) which may be especially suitable when served with this small fish; or "Luhk forest" (***Lukuwal***) characterized by a long season and fruit which remains in a state of ripeness for a long period without rotting on the tree, the name itself seems to be associated with a durable cultural hero called Luhk (Lawrence et al. 1964: 45).

In Pohnpei, breadfruit varieties tend to be divided into two major classes: the smooth-skinned cultivars (***Meiniwe***), and the rough-skinned cultivars (***Meinsahrek***). Trees of ***Meiniwe*** group normally bear spherical-shaped fruit that ripen earlier, taste sweeter, and have firmer flesh. They are preferred for making *mahr* (preserved breadfruit). Trees of ***Meinsahrek*** group bear fruits that are somewhat elongated in shape and ripen later, may taste bitter, have looser flesh and occasionally produce seeds. Within those two classes, there are some 130–150 varieties on Pohnpei. The early estimates of 40–50 were given by Christian (1897) and Kanehira (1931), followed by 40–80 by Sasuke (1953), Lawrence et al. (1964), and Soucie (1978). For their names of 65 Pohnpeian cultivars of breadfruit, Lawrence et al. (1964: 60–65) provide known or probable translations into English. Raynor (1991) identified 131 known cultivars, while Clarke (2008: 88) referred to "150 named varieties of breadfruit." We provide on pp. 414–415 a list of 144 varieties compiled from a combination of the above sources.

Uses in Mwoakilloa: Breadfruit is known generally as **Moai** on this atoll, where people have long grown this useful tree crop in their agroforests. Based on archaeological recovery of shell tools and charred plant material, this ancient breadfruit cultivation seems to have begun at least 1,500 years ago when the first inhabitants arrived on the atoll, which is now believed to have been just a few hundred years after the earliest documented settlement of the high islands in the region (i.e., Pohnpei and Kosrae). This was also just about the time that Pingelap Atoll was first settled by people (discussed below; see Levin et al. 2019). Thus, breadfruit had been a very important cultivated tree in the local culture of Mwoakilloa for many centuries when Murphy (1950: 73) described the changing agricultural scene on this atoll in the mid-20th century:

> "Breadfruit appears to have declined in importance in [Mwoakilloa]. It is not a commercial crop and thus has lost out to the coconut. As a food, breadfruit suffers relatively since it is not as reliable as [giant swamp taro, *Cyrtosperma merkusii*]. It is available for only a few months of the year, and the yield varies considerably from year to year. With the atoll's ever-increasing pressure of population, breadfruit's relative decline is likely to continue."

On the other hand, a few years after Murphy's report, Glassman (1953: 295) indicated that breadfruit (*A. altilis*) was "very common" and still one of the most important food crops on this atoll. Glassman also referred to a report by Christian (1899) that the wood of one breadfruit variety on Mwoakilloa, known as in the local language as '*Maimat*', was "highly prized for house building; and the wood of other varieties is used in the construction of canoes." Murphy (1950: 72–73) did report that breadfruit trees were being grown widely throughout the islets of this atoll with "three areas of concentration," determined largely by "better-than-average soil and a forest dense enough to provide shade for the young trees." Murphy referred to the use of a long breadfruit "picking pole" because of the height of the mature trees, and described the ripe breadfruit as being "harvested while still pea green in color" and then "baked, fried in strips, boiled, or used in various combination dishes."

Because of its relatively "short season" with June and July the prime months, Murphy described breadfruit on Mwoakilloa as having limited food importance, but, as elsewhere in much of Micronesia, food use of cooked breadfruit has been prolonged on this atoll by mixing it with coconut milk and storing it in "pits lined with leaves" which were then "covered with coral rock to prevent damage by rats" and also "protected against rain by special roofed sheds without walls." More recently, Nagaoka (2014: 313) described the role of breadfruit trees in the building of a traditional sailing canoe on Mwoakilloa Atoll:

> "Building a canoe begins with the selection of a suitable breadfruit tree, the only kind of tree on the island for building the canoe hull. According to Mwoakilloa tradition, a dead breadfruit tree can be cut down anytime, but it is forbidden to cut down the living breadfruit tree from the time when the trees start to bear fruit to the end of the breadfruit season [known as *roak* on Mwoakilloa]. It is believed that breaking the taboo will bring a bad harvest of breadfruit, the most important crop on Mwoakilloa. It is, however, possible to cut down the living tree if it has been killed through burning during the prohibited period. By custom, the leftover chips from carving cannot be burnt." (Underlined for emphasis).

Although breadfruit trees and their products remain significant on Mwoakilloa in the 21st century, their use was apparently not as important in the past. The following information on traditional and recent breadfruit use on the atoll was provided in 2000 by a student in an ethnobotany course at the College of Micronesia on Pohnpei:

> "We like to use the fruit during important occasions and also usual occasions. The main part of the plant that we use now is the fruit. If we are cooking in an *uhmw* [earth oven], then we also use the leaves for covering up the *uhmw*. How do we use them? During funerals, we just bring fruits as they are and distribute them among the guests after the burial ceremony. However, during birthday and anniversary parties or other parties for that matter, we cook the fruit and bring it in a dish. There are various recipes that can be obtained for this particular fruit, and [people on Mwoakilloa Atoll] love it. [For other uses,] I was told that our ancestors use to make canoes out of the trunk. I also heard that they use the roots for medical purposes. My grandparents passed away already so I could not get information on how exactly the canoes and medicines are

made from this tree. My parents lack that knowledge. These traits are obviously not passed down from our ancestors to us, the later generation." (Underlined for emphasis).

Uses in Pingelap: Breadfruit is known generally as **Maei** on Pingelap, where people apparently have cultivated it for at least 1,500 years. In fact, archaeologists have used excavated animal and plant remains, including charcoal, to help develop a better understanding of how food production has changed over the past one and a half millennia on Pingelap Atoll. Although farming patterns on Pingelap have clearly seen significant changes over the course of human habitation, interviews with local inhabitants of this atoll indicate that their cultivation systems "continue to function under the guidance of experienced and skilled farmers employing largely traditional methods" (Levin 2016; also see Levin et al. 2019). When St. John (1948) visited Pingelap in the mid-20th century, he described breadfruit as "abundant and vigorous [with trees of this species] attaining a height of 20 m (~65 ft) or more."

More recently, Damas (1994: 53–54) referred to breadfruit as "the second major crop of Pingelap," after swamp taro, *C. merkusii*. Damas noted that breadfruit produces ripe fruit on this atoll from late May to the beginning of September, and although Damas also referred to a declining population of breadfruit trees on Pingelap (when he made his visits to this atoll from 1975–1983), he further noted that these trees had traditionally provided "important building material, especially for canoes, in addition to their fruit for human consumption." Following on from an informant's description, Damas briefly noted the method of "replanting" breadfruit by vegetative reproduction (cloning) on Pingelap:

> " One of the roots which spread from the trunk well above ground is cut. From this cut a shoot usually sprouts. When the shoot reaches a height of three or four feet it is planted in a hole fertilized with grasses together with a section of the root to which it is attached. It takes seven to ten years after planting for the trees to bear fruit."

Damas also described in some detail how breadfruit is cooked and fermented in Pingelap, much of which is similar to the same processes on other islands in the tropical Pacific:

> " Fresh breadfruit is cooked by baking it in the earth ovens as well as by frying it as crescent shaped 'chips' in coconut oil. The Pingelapese also preserve by a method known throughout Oceania, that of burying the fruit in the ground, where it ferments. In this process Pingelapese soak the fruit in salt water for at least 24 hours and then it is peeled before being placed in pits about three feet in depth and of about that [in diameter]. The now softened fruit is wrapped in leaves and covered with them as well as with [soil]. In time the fruit assumes a thick, moist consistency, dark orange in [color], and appears at feasts in sheets [about two inches thick], resembling the slabs of young cheese which I have seen in cheesemakers' vats. My informants told me that this mar [fermented breadfruit] is still edible after a year or more in the ground if the leaves are renewed periodically (about once a week). Nowadays plastic garbage bags are used more frequently than leaves to contain the fermenting fruit in the pits."

Cultivars in Mwoakilloa and Pingelap: Breadfruit varieties on these atolls included seeded and unseeded types; see p. 416 for the list we compiled from available sources.

Uses in Kapingamarangi: Breadfruit is an important tree crop on Kapingamarangi, where its local name is Gulu. One use of its non-food uses involves extraction of the bast fiber from the bark to make the molo gulu, a loincloth garment (see Lieber and Dikepa 1974: 795). *Huaagulu dunu* in the Kapingamarangi language refers to breadfruit cooked over coals with its skin on. Cooked paste of ripe breadfruit is also prepared traditionally on remote Kapingamarangi. This paste is pressed into sheets about 1.5 m (5 ft) long and 50 cm (20 in) wide, dried in the sun on coconut leaf mats, then rolled into cylinders, wrapped in pandanus leaves, and then stored for at least three years. This preserved breadfruit paste (*bagu gulu*) from Kapingamarangi was examined by Murai et al. (1958), who provided information on the nutritional values of fresh or cooked food and the pounded paste of breadfruit:

> " The pounded paste: is a concentrated, nutrient-rich food containing only 21% water, compared with the 60–70% water content of fresh or cooked fruits. It contained 68 g of carbohydrates, double the amount that the authors reported for fresh or cooked breadfruit. Protein and fat levels were substantially higher in the paste, with almost five times the average amount of protein and fat found in fresh or cooked breadfruit. Calcium and phosphorus levels were notably higher than that of fresh or cooked breadfruit. Iron, thiamin and riboflavin were the only nutrients with lower levels in paste."

Hiroa (1950: 33) provided insightful descriptions of breadfruit use and observations of consumption of uncooked ripe breadfruit (*gulu bala*) on Kapingamarangi; Hiroa was "surprised at the delicious taste of the food which was as good as any fruit salad" he had ever tasted, and indicated that he had "never seen breadfruit treated in a similar manner in any part of Polynesia."

Cultivars in Kapingamarangi and Nukuoro: Breadfruit diversity in these two Polynesian atolls is lower than in other parts of Pohnpei State and has not been well documented. We were able to find three named varieties from Kapingamarangi and four from Nukuoro (see p. 417). As is the case for other islands discussed above (Pohnpei, Mwoakilloa, Pingelap), these lists of varieties include both *A. altilis* and the seeded *A. mariannensis*, which is examined in more detail below.

▸ **Meipa** or **Meikole** (P), **Moai koaloa** (Mw), **Maei koalae** (Pi), **Gulu de henua** or **Gulu golee** (Nu), **Gulu golee** (Ka)
Artocarpus mariannensis

Description: Breadfruit trees of this seeded species (*Artocarpus mariannensis*) are large, up to about 66 ft (20 m) or more in height, with a trunk stabilized by buttresses developing at the base. A single fertile tree may yield up to about 50 kg (~110 lb) of fruit each year. Leaves are broadly oval to elliptical and are entire or have a few shallow lobes. *A. mariannensis* is a monoecious species, and its inflorescences produce very large numbers of male and female flowers, each attached to a central spongy core. The seeded fruits are generally cylindrical or asymmetrical with dark green skin and deep yellow flesh. Falanruw (2015: 150) provided an identification key to differentiate the Mariana Islands seeded breadfruit (*A. mariannensis*) from the common breadfruit (*A. altilis*) because both produce sizeable fruits on branches: (1) Leaves pinnately lobed, lower surface without hairs, fruit often over 20 cm long, usually seedless = *A. altilis*; (2) Leaves simple, generally entire or with few shallow lobes, lower surface hairy, fruit mostly 15–20 cm long and with seeds = *A. mariannensis*.

According to Ragone and Manner (2006), *A. mariannensis* seeded breadfruit trees start producing fruit after about five years, then continue fruiting in season with a peak period in July–September. These authors also described how *A. altilis* and *A. mariannensis* are perceived locally by Micronesians:

> "[Seeded breadfruit] *Artocarpus mariannensis* is native to the Mariana Islands and Palau and is closely related to [seedless] breadfruit (*A. altilis*), with which it has naturally hybridized. [*A. mariannensis*] is recognized on those islands as being distinct from [*A. altilis*] breadfruit. However, this species and the numerous interspecific hybrids in Micronesia [including Pohnpei and the outer islands] are considered to be "breadfruit," whether they are seeded or seedless."

Distribution: Native Micronesian seeded breadfruit trees (*A. mariannensis*) grow wild in Palau, Guam, the Northern Mariana Islands, and less commonly in Yap and Chuuk; they are also cultivated throughout much of Micronesia, especially on moderately wet to very humid atolls. *A. mariannensis* is adapted to limestone ridges, outcrops, and shallow calcareous soils, but it can thrive in volcanic soils with adequate drainage. It is generally found in areas from near sea level to about 150 m (~500 ft). On some of these islands, the trees serve as an important food for endangered fruit bats (*Pteropus* spp., flying foxes). The truly wild, seeded breadfruit tree (*A. mariannensis*) is probably native to the Mariana Islands (known there as **Dokdok**). It is also most likely a native tree in Palau, but Spaniards or Germans may have introduced it into Yap during the early historic period. Beyond some areas of western Micronesia, *A. mariannensis* breadfruit trees have all been introduced by people. Many breadfruit tree varieties in Micronesia are hybrids between *A. mariannensis* and *A. altilis*. Both species and their cultivated varieties or hybrids have been transported across Micronesia from west to east. Humans, through artificial selection, have produced numerous varieties of the hybrid crosses between *A. mariannensis* and *A. altilis* in Micronesia as well as in other tropical regions of the world. Hybrid varieties of breadfruit trees do well in the shallow sandy soils of coral atolls.

General uses: Self-sown indigenous forest populations of *A. mariannensis* are declining in some areas of its native range. For example, on Guam and the Rota Islands, *A. mariannensis* remains a relatively important part of the cultivated landscape, providing various resources in regions of Micronesia, either as a separate species or as individual hybrid trees. For example, its ripe fruit can be eaten raw, and the cooked seeds are also consumed: "Boiled or roasted [breadfruit] seeds provide a rich, tasty food" (Falanruw 2015: 152). The nutritious seeds are high in protein, low in fat, and a good source of vitamins and minerals. The seeds are cooked in the fruits and eaten, especially on some atolls, where cooked, sprouted seeds are a delicacy (Ragone and Manner 2006: 10). Today, cultivated seeded breadfruit trees (*A. mariannensis*) can be found growing throughout many of the islands of Micronesia. They are also particularly important on some atolls because some of its varieties have a high salt tolerance. In contrast, they are not found in Melanesia or Polynesia, other than on the low coralline islands of Tokelau and Tuvalu. On some tropical Pacific islands, especially in western Micronesia, large trunks of the *A. mariannensis* are used for canoes, and its sticky latex is used for caulking and as a glue in boat building. The wood is also used for handicrafts, and several parts of the tree have medicinal uses in the tropical Pacific:

> "All parts are used medicinally, especially the latex, leaf tips, and inner bark. The latex is massaged into the skin to treat broken bones and sprains and is plastered on the spine to relieve sciatica. It is commonly used to treat skin ailments and fungus diseases such as "thrush," which is also treated with crushed leaves. Diluted latex is taken internally to treat diarrhea, stomachaches, and dysentery. The sap from the crushed petioles (leaf stalks) is used to treat ear infections or sore eyes. The root is astringent and used as a purgative; when macerated, it is used as a poultice for skin ailments. The bark is also used to treat headaches in several locations." (Ragone and Manner 2006: 10)

Uses in Pohnpei: **Mahi** is a general Pohnpeian name used for all breadfruits trees (*A. altilis*, *A. mariannensis*, and hybrids). *A. mariannensis*, specifically, is called **Meikole** in Pohnpei (e.g., Rehg and Sohl 1979). All breadfruit have traditional multipurpose usage in Pohnpei, both as a key food source and also for building canoes, carvings, medicine, glue, and previously for fabricating women's clothing such as wraparounds made of pounded inner bark fibers from breadfruit trees (e.g., see Anson and Raynor 1993: 144). Breadfruit generally is believed to have been introduced by early Pacific islanders who settled on Pohnpei. See the previous discussion about seedless breadfruit (*A. altilis*) for more detail about the origins, distribution, local uses, and breadfruit cultivars in Pohnpei and other islands of Pohnpei State. Below we list just two examples specifically related to seeded breadfruit (*A. mariannensis*).

Uses in Mwoakilloa: In Mwoakilloa, seeded breadfruit is known as **Moai koaloa** and comes in three cultivars (see p. 416). This is similar to Pingelap, where it is known as **Maei koalae**. Working on Mwoakilloa to examine water resources available to the community there, Taboroši and Sánchez Collazo (2011: 95–96) also investigated the consumption and relative importance of different liquids on that atoll in modern times:

> "In general, everyone drinks far more water than coconut juice. In addition, water-based powdered drinks such as instant coffee and Tang® are the most popular, but rarely available. More common is a drink called *limpoak*, also known as the "local coffee" and made of breadfruit seeds."

Uses in Kapingamarangi: In Kapingamarangi, the seeded breadfruit is known as **Gulu golee** (same as in Nukuoro, where another name is **Gulu de henua**). It appears that the wood of this species was historically preferred for making canoes, which were traditionally made only from locally grown wood (though if large, well-preserved drift logs were available, they were preferred because they were already "seasoned"). In Kapingamarangi, this was reportedly a different type of breadfruit tree than most of those found on Pohnpei. Described as having a different leaf shape, fruit with seeds, and wood considered "to be of superior quality," this species was probably *A. mariannensis*. Also, it was this species of breadfruit (or a hybrid of *A. mariannensis* and *A. altilis*) that supplied the reduced, sticky sap (called *daubugu*) used to patch up holes and flaws in the canoes.

Kolonia has grown into a town, yet breadfruit trees still surround many homes, providing food and sustaining traditions as they have in the past.

Not all breadfruit varieties on Pohnpei are truly local, indigenous crops. For example, ***Meiuhpw*** is not considered traditional and is excluded from tribute and other customary uses for that reason.

Throughout Micronesia, breadfruit trees have provided essential wood for carving canoes, such as this one being made on Sapwuahfik Atoll.

As one of Pohnpei's most important traditional staple foods, breadfruit is sold seasonally in the local markets.

Mahi uhmw, traditional local dish of whole breadfruit roasted over hot stones.

Mahr, traditional dish of fermented breadfruit, here wrapped in leaves and baked.

Breadfruit is healthy and highly nutritious, but non-traditional methods—like deep-frying it into chips—greatly reduce its benefits.

Non-traditional foods can be made healthier using local ingredients, such as these pancakes made from breadfruit flour.

▸ **Tehnlik** (P), **Kardoap** (Mw), **Saehlik** (Pi), **Lau gadaha** (Nu), **Logo hoo** (Ka)
Asplenium nidus

Description: *Asplenium nidus* is a large, relatively slow-growing fern in the spleenwort family, Aspleniaceae. *A. nidus* is known widely in English as the "bird's nest fern." In fact, in Latin, *nidus* means nest. *A. nidus* develops as an epiphyte (air plant), perched above ground on other plants, especially trees, but also sometimes on rocky surfaces which have very poor or no soil; however, it can develop and grow directly on the ground in appropriate terrestrial environments. *A. nidus* can reach heights of 1.2 m (4 ft), producing a light- to dark green rosette or cluster of long, overlapping fronds (leaves) that are smooth, leathery, or glossy, and narrow-oblong in shape with undulating, smooth margins. The fronds are usually 0.5–1 m (2–4 ft) or more in length and 5–20 cm (3–8 in) wide, with a prominent, well-rounded, dark midrib underneath. Spreading upward and outward, the fronds form a roundish, bird's-nest-like receptacle into which organic matter, including leaves of surrounding plants, naturally collects. The spore cases (sporangium) have a linear arrangement along the veins on the underside of the fronds.

Distribution: *A. nidus* (bird's nest fern) is indigenous to many tropical rainforest areas of Southeast Asia, Australia, Madagascar, and beyond to parts of eastern Africa, as well as in many islands in Near and Remote Oceania, including the islands of Micronesia. It occurs in tropical rainforests at various elevations, preferring shady locations from sea level lowland rainforest areas to the upland montane rainforest regions. This fern is also used as a decorative house plant in some temperate countries.

General uses: In some areas of Micronesia, especially on atolls, the small, young leaves of the bird's nest fern (*A. nidus*) provide ingredients for medicine and for soup. For example, parts of this fern are reportedly used for contraception in Vanuatu in eastern Melanesia (e.g., see Bourdy et al. 1996). According to Barrau (1961: 57), the young leaves of this fern "are sometimes utilized as food during periods of shortage in Micronesia, the Marshalls, and the Carolines." In addition to its use as a casual and famine food (also see Thaman 1994: 172) and medicine, the large, attractive fronds of this fern are also used ornamentally, as a source of material to line the bottom of preserved breadfruit pits, and as temporary roofing for short-term shelters.

Uses in Pohnpei: The *A. nidus* is known in Pohnpeian as **Tehnlik** or other local names; it is broadly used on this island for decorating special plants, as well as for general decoration or for ornamental design. In addition, parts of its dark green leaves are consumed "as food and contain a medium amount of provitamin A, as well as a high content of lutein and zeaxanthin" (Engleberger et al. 2003). The leaves of the **Tehnlik** are also used to line prepared breadfruit pits to preserve the fermented breadfruit. Epiphytes such as the **Tehnlik** (bird's nest fern) not only help make cultural offerings more attractive, but also present a larger, more impressive gift; for example, the entire plant is used to adorn yams (*Dioscorea* spp.) presented at special occasions. In some Pohnpeian cultural events, such as feasts, birthday celebrations, and burial ceremonies, whole **Tehnlik** ferns are tied on to the yam, signifying that a *nopwei* (first-fruit tribute) has been performed. Leaf tips of this fern are also used as an ingredient in soup, reportedly to signify that the *nopwei* ceremony has been performed. Furthermore, this fern has been used medicinally as a contraceptive as well as to treat boils, diarrhea, a pregnant woman's labor pains, and to speed up and ease the birthing process. In addition, parts of the fern are also used to treat excessive uterine pain and also excessive stomach pain in both men and women, as well as painful and discolored urination in women (Balick et al. 2009: 223–234). Based on her archaeobotanical research on Pohnpei, the archeologist Levin (2015a: 158) noted that "*A. nidus* leaves do not produce sufficient identifiable phytoliths to document their past presence, which in ancient cooking features on Pohnpei might help identify [its] use for decorating yams or for lining of the roasting pits."

Uses in Pingelap: This fern is known in Pingelapese as **Saehlik**; on this atoll, it grows on rotten logs or coral and its leaves are "wrapper[s] around breadfruit baking" (Damas 1994: 231). Balick collected a specimen of this fern in a mangrove swamp "along a sea wall" on Pingelap in 2002.

▸ **Duhduhnwoal** (P)
Astronidium ponapense

A unique Pohnpei plant species: *Astronidium ponapense* is regarded tentatively as an endemic woody shrub in the wet upland forests of Pohnpei; in other words, this is a rare and unique plant species that is reputedly only found in remote, humid, mountain environments of Pohnpei. It produces clusters of drooping reddish-pink flowers and has clearly visible veins running lengthwise down the leaves, which is typical of plants in the melastome family (Melastomataceae) to which *A. ponapense* belongs; for example, see the leaves of **Pisetikmei** (*Melastoma malabathricum*), found more commonly in open areas in the lowlands. Known in Pohnpeian as **Duhduhnwoal**, it is common in some exposed dwarf forest areas above 2,000 ft on Pohnpei. The word *mwoal* refers to the fruit color of the **Duhduhnwoal** plant (which is also the color of a rare edible sea snail). This unique and relatively rare plant and its Pohnpeian name need further study.

▸ **Ais** (P), **Aij** (Mw), **Adidi** (Nu) (Ka)
Atuna racemosa

Description: *Atuna racemosa* is a large canopy tree that grows straight and tall to heights of as much as 25–30 m (80–100 ft), although often less. *A. racemosa* produces large, round, reddish brown fruits 6–8 cm (2–3 in) in length. Its leathery leaves are simple, oblong, smooth on both surfaces and along the margins, and mostly pointed at the tips. The showy, fragrant flowers are up to 2.5 cm (1 in) across, and are produced in short axillary racemes. The brown fruits (nuts), which are spherical in shape and more or less flattened on two sides, develop on the tips of the branches. The fruits, which contain useful oil, drop to the ground upon ripening. The kernel within the fruit is hard and surrounded by a thick wall. The wood of this tree is firm.

Distribution: *A. racemosa* is an indigenous species spreading from insular Southeast Asia, including Indonesia and the Philippines, all the way to Fiji and the FSM (Clarke and Thaman 1993; also see Prance 2004). Although this tree species is found in some cultivated areas in various high islands of Micronesia [such as Yap proper], it may occur in almost pure stands in some forested places, especially in those areas where recent disturbance has occurred. In these seemingly wild, forested areas, this tree may have been consciously introduced for its valuable products. Although this tree is commonly found on the high islands of Palau, Yap, Chuuk (upland areas of Fefan, Tol, and Weno), Pohnpei, and Kosrae, it may not be a native species in one or more of these Micronesian islands since its large fruit probably would not have been brought to the islands by birds, wind, or ocean currents, although it has undoubtedly been moved around or within islands by severe hurricanes/typhoons. In some high islands in Micronesia, it may occur in places recently disturbed by natural or human-induced causes.

Uses in Pohnpei: Known in Pohnpeian as **Ais**, its hard wood is resistant to rot and is therefore very important in house construction, especially where the wood of several important mangrove trees is not available. The fruits are used for medicine and as an insect repellent (see Balick et al. 2009: 349; Merlin et al. 1992). In addition, valuable oil can be produced from the fruit. This oil is very long-lasting and was used to paint canoes. First, a reddish-brown soil cover is applied to the surface of the canoe, and then oil from the fruit of **Ais** is applied to seal in the earth color and protect the canoe. Balick et al. (2009: 349) provided the following description of the production and use of a cement or dye from nuts of **Ais** trees:

> "To make a glue or paint, as many fruits are needed are cut in half, then the nuts are taken out, scraped, and mixed with a type of red soil (that can only be found at Nanpwoarenais, Madolenihmw), then put in a pot over a fire and stirred until it turns into a thick red liquid... The resin used to make paint to coat and protect the outside of canoes... Alternatively, paint is made by putting the seeds in a metal container and roasting them in an *uhmw* for thirty minutes. The seeds are then grated and squeezed to obtain the liquid, which is collected in a container and painted onto canoes, using *inipal* (coconut fiber) as a brush, to protect the wood from rotting... The kernels are mixed with yellow earth as a varnish on canoes." (Also see Brosi et al. 2007).

▸ **Pehri** (P) (Mw) (Pi), **Madila** (Nu) (Ka)
Bambusa vulgaris

Description: All bamboo species belong to the grass family (Poaceae). *Bambusa vulgaris* is a common, large-stemmed bamboo in the Pacific. This huge, woody, segmented, perennial grass grows in clumps and has short, thick rhizomes. Its green to yellowish stems usually grow erect but sometimes arching to heights of approximately 15 m (50 ft) and are about 10 cm (4 in) wide with moderately thick walls, no spines, and slender branches. When this bamboo does flower (which is only very occasionally), it produces large, leafy floral clusters at the stem nodes.

Distribution: *B. vulgaris* is a large, useful bamboo species that is native to tropical Asia and was probably brought to Pohnpei and other tropical Pacific islands soon after the early arrivals of Europeans. It is planted and in some situations has become naturalized (growing and reproducing on its own), often forming large stands, especially in areas with alluvial soils and sufficient available water. Cultivated widely in warm regions, this species probably originated in India or in Indonesia. *Schizostachyum lima* is a smaller species of bamboo that is native in Palau and Yap.

General uses: Bamboo plants have an extraordinary number of uses in the tropical areas where they occur. The bamboo's woody stems are widely used in construction, fencing, handicrafts, making rafts, cages, knives, net-mending needles, fishing poles, floats for fishing nets, fruit-picking poles, irrigation conduits, musical instruments, fencing, containers for food, water, cooking food over an open fire, and as fuel wood (e.g., see Clarke and Thaman 1993). In addition, some parts of bamboo are used medicinally; for example, in areas of the FSM, leaves of the *Bambusa* plants are commonly used to stop bleeding (Wilkinson and Elevitch 1994).

Uses in Pohnpei: The large bamboo species, *B. vulgaris*, is known in Pohnpeian as **Pehri**. For food in Pohnpei, young stems of bamboo are cleaned, cut up, and mixed into soups, and the young shoots are also eaten. The long stems are used as pipes to move water from one place to another. In addition, mature bamboo stems are used to make rafts and fishing poles. The young stems are used as a trellis to support growing yam (e.g., *Dioscorea alata*) plants. "Sharpened tips of the stems were used in traps for enemies during war and small stems were used as the barrel of airguns," and parts of bamboo are used as a contraceptive as well as an ingredient in a concoction to make hair grow thick, shiny, and healthy (see Balick et al. 2009: 295). Another bamboo species now found on Pohnpei is *Bambusa multiplex* ("hedge bamboo"); this species is known in Pohnpeian as **Pehri en Sapahn** and is cultivated in clumps to make hedgerows in settled areas (e.g., see Herrera et al. 2010: 63).

▸ **Wihnmoar** or **Wihnsed** (P), **Wi** (Mw) (Pi), **Gavaausu haahine** (Nu), **Laagau iha** / **Laagau hagadali eidu** (Ka)
Barringtonia asiatica

Description: The large "fish poison tree" (*Barringtonia asiatica*) is a thick-trunked, native coastal tree with prominent, attractive white flowers. It can grow to heights of 20 m (65 ft), often as a mangrove associate that occurs on sandy and rocky shores. Its bark is grayish and smooth, and its large, glossy, mature leaves are simple, narrow obovate (more or less egg-shaped), about 20–40 cm (8–24 in) long and 10–20 cm (4–8 in) wide. The leaves are produced in rosettes at the ends of branches. The young leaves are bronze-colored with pinkish veins, while the older leaves turn yellowish. Flowers appear on a long spike from the center of a leaf group. The flowers are "pompoms" of long white stamens tipped with pink and have four small white petals. The flowers open at night and attract large pollinating moths and nectar feeding bats with their heavy scent. The following morning, the flower stamens are usually found strewn beneath the tree. The fruits are relatively large, somewhat squared, about 10–12 cm (4–5 in) across and lantern-shaped. They are green at first, turning brown when ripe, are one-seeded, and can float in water.

Distribution: *B. asiatica* is a large, tropical woody species found growing on rocky shores along the coasts of East Africa to the Western Pacific, including all of Micronesia. (Note: The reported local names in the Philippines for this tropical coastal tree are closely related to the local names in Palau, Yap, the Mariana Islands, and Chuuk, but not in Pohnpei and further east.) After maturing on the tree, the fibrous fruits drop off and are carried out to sea by the tides. These buoyant fruits then drift along with the currents on the surface of the ocean. They can remain floating for years and cross great distances. Eventually, they wash up many tropical Pacific islands. If they are still viable at that time, they may sprout and grow into large trees on shores very far from their places of origin. Since this shade-providing tree is a useful species, it may have been introduced into some islands in the eastern part of its present tropical Pacific range, and it is possible that it was subsequently introduced by humans to some more remote tropical Pacific islands where it has become naturalized (i.e., growing, reproducing, and establishing new plants on its own). If so, this probably happened before European contact. This may also apply to the coconut palm in some areas of the Pacific (see p. 189).

General uses: The fruits of *B. asiatica* are one of the most common traditional sources of fish poison among Pacific Islanders who have long inhabited islands where this tree grows naturally (or was introduced by people) in times before European contact; this includes islands in Micronesia. All parts of the large *B. asiatica* tree contain saponin, a potentially poisonous substance. The seeds and other parts of the plant are pounded, pulped, or grated to release the poison, which is used to stun fish in stagnated seawater in the intertidal zone as well as in some lower areas of freshwater streams. The floating fruits of *B. asiatica* are sometimes used as fishing floats. As *B. asiatica* is a colorful shady tree, it is commonly planted as a roadside tree in some tropical areas of Southeast Asia. Furthermore, in some tropical regions its wood is used in construction (e.g., for houses and canoes). The species also has traditional medicinal uses. DaSilva et al. (2004: 907) noted the medicinal use of *B. asiatica* in most of the islands in Polynesia and Micronesia and summarized its use as an "anti-fungal agent," and noted that the grated leaves and seeds of this species have been "mixed with coconut cream to treat burns, wounds and ringworm."

Uses in Pohnpei: The large, coastal *B. asiatica* tree is known in Pohnpeian as **Wihnmoar** or **Wihnsed**. As noted above, the seeds of *B. asiatica* have been used to stupefy fish in many places. To stun fish in tidal pools and quiet streams, the firm white seed can be crushed, mixed with water, and thrown into pools where fish are found. This method of stupefying fish does not appear to harm the flesh of the fish as food. The leaves of *B. asiatica* are used to cover an earth oven (*uhmw*) when it is used to cook food. The leaves are also used to cleanse and smooth the skin and for treating diarrhea in children, as well as for treatment of food and drink poisoning when someone has poisoned another person. "To treat skin diseases, a fruit that is floating in the water is cut in half to get to the nut, which is then scraped and wrapped in a piece of cloth and rubbed onto the skin" (Balick et al. 2009). The following traditional Pohnpeian story involves the buoyant fruits of **Wihnsed**:

> "There was a man on Pohnpei who while working on his land cut his finger. Blood poured out from his cut. He took a leaf of **Sawa** [taro—*Colocasia esculenta*] and put his blood in this leaf. He then left the leaf with his blood on the ground and went back to work. When he looked back at the leaf on the ground, he saw it shaking and a pretty girl (his daughter) emerge out of the leaf. The man and his new daughter walked along the shore, where she saw a **Wihnsed** fruit floating nearby. She told her father that she wanted to get on the fruit and drift away because there were no people there. When she stepped onto the fruit the seedling emerged. She held on and floated away. The fruit and the girl eventually arrived in Madolenihmw where they named the fruit *Likin* ("outside"). After the girl and the fruit become established there, she sent a messenger to her father who told him that she wanted to eat a breadfruit. Her father collected a special breadfruit variety and then paddled [to meet his daughter] with this breadfruit. There this breadfruit is called *Temwen mei* and it eventually evolved into *Meinpwuht*, a type of breadfruit now only found in this area of Madolenihmw, a place where many **Wihnsed** trees are also found today. When the huge, thick-trunked, native fish poison coastal tree drops its leaves, certain fish such as the **Pwilak** (unicorn fish), are found in abundance on the reef." (See Merlin et al. 1992).

Uses in Pingelap: According to St. John (1948), *B. asiatica* (known as **Wi** on Pingelap Atoll) was most likely "introduced by the natives because of its value as a fish poison." Damas (1994: 238) also noted that its seed is "mixed with ashes and used as fish poison."

Uses in Nukuoro: The seeds of the fruits of this tree, known locally as **Gavaausu haahine**, are used traditionally to extract "a stupefying drug affecting fish" (Stone 1966a).

▸ **Wihnmoar pwetepwet** or **Wihn Pohnpei** (P), **Gavaausu daane** (Nu) *Barringtonia racemosa*

Description: *Barringtonia racemosa* is a small to medium-sized tree with a straight, unbranched stem that leads to a rounded crown and is usually 4–8 m tall (12–24 ft), but occasionally reaches 15 m (~45ft) or taller. *B. racemosa* is apparently a native species on a number of Micronesian islands. It has leaves 15–30 cm (6–12 in) long and 8–10 cm (3–4 in) across and is wider at the base (obovate). The specific name *racemosa* refers to the long racemes on which the flowers and fruit are produced; indeed, many prominent white to pinkish flowers with four petals hang suspended from branches when the plant is in bloom. Its fruits, which are buoyant and are dispersed by ocean currents, are ellipsoid in form and 4–7 cm (1.5–3 in) long.

Distribution: Botanically, *B. racemosa*, known in Pohnpeian as **Wihnmoar pwetepwet** or **Wihn Pohnpei**, is closely related to *Barringtonia asiatica*, known on Pohnpei as **Wihnmoar** or **Wihnsed**. *B. asiatica* is a large native tree which is usually found in the coastal strand vegetation near the ocean shore on Pohnpei; on the other hand, *B. racemosa* is a small native tree which can be found near the ocean shore, behind mangrove swamps, and along or near streams. Indeed, *B. racemosa* normally occurs close to water, along riverbanks and in freshwater swamps, and occasionally in the less saline areas of mangrove swamps, where it may develop pneumatophores (specialized aerial roots that enable plants to breathe air in habitats that have waterlogged soil). *B. racemosa* is adapted to a variety of wet tropical, moist topical, and wet subtropical regions from East Africa through Indo-Malaysia to parts of Remote Oceania, including islands in Polynesia and Micronesia.

Uses in Pohnpei: *B. racemosa* is a very important source of medicines with many uses on Pohnpei. For example, *B. racemosa* is reportedly used to relieve the pains after childbirth and for relief of earache. A variety of additional uses of *B. racemosa*, including medicinal applications, are referred in the following:

> The timber is used in construction… The fruits can be used as laundry soap… The small fruits are put into the mouth to soften, then used as airgun pellets… [Medicinally, *B. racemosa* (**Wihnmoar pwetepwet**) is used on Pohnpei]: To treat *soumwahu en kau* (sickness caused by sorcery)… To treat *soumwahu en nansapw* (sickness from the cultivated fields)… To reduce the need to go to the bathroom when a person will be attending a big funeral party and there will be too many people around… To stop diarrhea… To treat back pain… To treat headache… To treat poisoning caused by eating food and drink that may have been hexed by sorcerers… To treat *soumwahu en ohl*… To reduce a high fever… To treat or clean the uterus… To treat *salengamat* (ear infection) in babies (symptoms include a bad smell from the painful ear accompanied by copious amounts of white sticky fluid coming from the ear)… To treat babies with cracked skin where the ears join the head (there is not a local name for this illness)… To treat rash similar to chicken pox… To treat a rash on a child… To treat a bad rash given by a sorcerer due to jealousy… To treat skin rash… To treat itchiness… To treat *soumwahu en ohl* (men's sickness)… To treat a fungal infection…" (Balick et al. 2009: 420)

Lee et al. (2010: 58, 154) provided brief descriptions of the traditional treatment of skin rash:

> One fruit is cut in half and the inner flesh scraped out and combined with the meat of one whole coconut. Both plants are wrapped in a piece of cloth, and the juice is squeezed out, and rubbed onto the affected area. This treatment is applied twice daily until symptoms are gone."

According to the same source, Wihnmoar is also used to treat back pain:

> The skin of the [**Wihnmoar**] stem is cut into two pieces, cleaned, pounded, and wrapped in a piece of cloth. The juice is squeezed into 4 L of water and 2 L is taken daily for eight days."

▸ **Mahrek en leng** (P)
Blechnum orientale

Description: *Blechnum orientale* is a large tropical fern with attractive, arching fronds that needs plenty of sunlight. It is often confused with *B. cartilagineum*. The rhizome of *B. orientale* forms into an erect trunk. Sterile and fertile fronds are similar, ranging in length from about 22–130 cm (8–46 in) and about 8–54 cm (3–22 in) wide.

Distribution: The *B. orientale* fern ranges from India, Nepal, and South China throughout Southeast Asia to southern Japan, Australia, and many tropical Pacific islands. This species is adapted to open areas and is thus common along road cuttings in rainforest and extending into drier regions. In fact, *B. orientale* is a primary colonizer of forest areas opened up naturally or by human impact, and it may remain as a dominant plant species in such areas if repeated fires occur.

General uses: In at least some areas of its tropical Asian and Pacific distribution, the tender young unfurled fern fronds ("fiddleheads") of *B. orientale* are eaten as a vegetable. For example, among the "Shompens, a dwindling and critically endangered Mongoloid aboriginal tribe inhabiting the Great Nicobar Island" in the Indian Ocean, the fronds of this fern are "boiled and eaten as [a] vegetable" (Sharief and Rao 2007; also see Piggot 1996: 400–401 for reference to human consumption of the fronds of *B. orientale* by some people in Malaysia). In addition, in some tropical areas, preparation of the root of *B. orientale* for human consumption may involve the rhizome of this fern being washed, roasted, and ground into flour to make a type of unleavened bread (e.g., see Winter and Amoroso 2003). Furthermore, *B. orientale* is used in traditional medicine for "the treatment of various skin diseases, stomach pain, urinary bladder complaints and sterilization of women" (Lai et al. 2010), and it is also said to be customarily used by native peoples in its geographical range of distribution "as a poultice to treat wounds, boils, ulcers, blisters, abscesses, and sores on the skin" (Lai et al. 2016).

Uses in Pohnpei: *B. orientale* is known in Pohnpeian as **Mahrek en leng**, and sometimes as **Limenkasar** (a name also used to refer two other ferns: *Nephrolepis hirsutula* or *Pteris vittata*; e.g., see Rehg and Sohl 1979). According to Balick (2009: 226–227), a beverage including parts of this fern is used as an "energizer" for pregnant women who are said to "drink this tonic to make the baby's head smaller so childbirth will be easier." It is thus used to treat *soumwahu en lih* (women's sickness) or an upset uterus, helping induce or facilitate childbirth. Parts of this fern are also consumed with *sakau* to treat depression; in addition, *B. orientale* is used to treat boils (see Balick et al. 2009 for various traditional medicinal treatments involving this fern species).

▸ **Soumw** or **Sohmw** (P), **Johmw** (Mw), **Soal** (Pi), **Donga** (Nu)
Bruguiera gymnorrhiza

Description: *Bruguiera gymnorrhiza* is a mangrove tree that grows to heights of about 25 m (80 ft) or more. It has a buttressed, fissured trunk and knobby breathing roots (pneumatophores) that spread out over the muddy soil in which the tree grows. These knee-like breathing roots stick out above the mud during low tides and allow the plant to take in oxygen in the swampy soil environments. *B. gymnorrhiza* produces attractive and long-lasting red to yellow flowers about 2.5–4 cm (1–1.5 in) in length. The fruits, which germinate while still attached to the mother plant (vivipary), can grow to be 15–35 cm (6–14 in) long before becoming detached from the trees and falling into the water, mud, or sand below. In fact, mangroves have a special reproductive strategy. Reminiscent of mammals, mangroves are viviparous, which means that they bring forth live young. They do not produce dormant seeds like most flowering plant species. Instead, they grow propagules with varying amounts of embryonic development while the propagules are still attached to the parent tree. The propagules are then released into the water below the parent tree, which depends on the ocean to disperse its young.

Distribution: *B. gymnorrhiza*, the large mangrove tree known in English as the "black mangrove," is found in tropical, swampy coastal areas from East Africa to the Western Pacific, including many waterlogged soil environments of Micronesia. It is a very common tree in the middle to landward parts of mangrove swamps. This tree is often planted by people who have extended its distribution into various natural or artificially excavated swampy areas. For example, it was possibly brought into the Marshall Islands by people. However, it is most likely a native plant in most other areas of Micronesia and so reached most of these islands without human intervention. According to Clarke and Thaman (1993), *B. gymnorrhiza* is said to be "[i]ndigenous to East Africa, the Indian Ocean, and tropical Asia to Tonga and Samoa in Western Pacific and to Nauru and the Marshall and Gilbert Islands in eastern Micronesia," but it may have been introduced by humans to some of the islands (including some atolls) in the eastern regions of Micronesia. St. John (1951) observed this woody mangrove species in the Marshall Islands atolls of Aur, where it was about 6 m (20 ft) tall and rare "at [the] edge of [a] swampy excavation in [the] forest" on Tabal Islet, and on Majuro where it was growing at the border of a "swampy *Cyrtosperma* [large cultivated swamp taro] patch." Obake (1941d: 9) cited S. Kusano as stating in his book "Forests of the South Sea Islands" that the mangrove tree species **Soumw** (*B. gymnorrhiza*) "is not indigenous but was introduced from some other islands with a view to dye fibers for mats etc."

General uses: This mangrove tree has many uses in Micronesia including as rot-resistant timber, firewood, and even as a food source. A comparison of its use for food in the tropical Maldive Islands (atolls) in the Indian Ocean is instructive; in the Maldives, the propagules or green pods (hypocotyls) of the *B. gymnorrhiza* mangrove tree are eaten as a cooked vegetable. This involves special preparation, including peeling and then boiling, with the water being thrown away and replenished four times or more. Thaman (1992: 12) referred to the fruit of the mangrove species *B. gymnorrhiza* being eaten after it is cooked in "Nauru, Palau, and Yap in Micronesia, and throughout most of Melanesia." Merlin, co-author of this book, has personally consumed the propagules of this mangrove in cooked meals served to him by indigenous peoples in rural areas of the Solomon Islands.

Uses in Pohnpei: This mangrove is known in Pohnpeian as **Soumw** or **Sohmw**. As on many other Micronesian islands, its timber is very durable and is used as house posts because it is resistant to rot. Poles of *B. gymnorrhiza* wood can be put in soil that is often moist because they will not deteriorate, whereas the hard wood of other mangrove trees, e.g., **Ahk** (Pohnpeian name for *Rhizophora* species), will decay under those soil conditions. *B. gymnorrhiza* is also used on Pohnpei for firewood. Other uses and reasons for usage of *B. gymnorrhiza* on Pohnpei can be found in Balick et al. (2009: 475), for example:

> "To charm a fishing net, the bark is rubbed on the net, coating it to attract fish... The wood is used for house posts, as it is nice and straight... The wood is used for construction material and handicrafts... To use as a charm to protect a person from eating a poisonous food or drink... To treat diarrhea... To treat dysentery... To cause a person to vomit after food poisoning... To treat numbness..."

Lee et al. (2010: 165) referred to this abundant mangrove species as **Sohmw** and **Kihrinsohmw** and noted that it can be used to treat sting of **Nehu** (stonefish):

> "The roots of the seedlings are cut and cleaned. These are then chewed up and placed on the sting as a poultice. This is applied immediately after injury and is said to remove poison from the wound and speed the healing process."

Although the pain relief that the chewed mangrove seedling root treatment supposedly provides is supported by some pharmacological testing (e.g., see Homhual et al. 2006), Lee et al. (2010: 165) also added this cautionary note:

> "The practice of chewing an herb and spitting on the wound is a traditional method of application, however it is important to note that it is not medically hygienic and can transmit hepatitis in an open wound."

▸ Uhpen Iap or Uhpen Palau (P)
Callicarpa candicans

Description: *Callicarpa candicans* is a relatively small shrub that can grow up to 2–4 m (6–12 ft) in height. Its broadly elliptical leaves vary in size, being about 3.5–18 cm (1.5–7 in) long by 1.5–9 cm (0.5–4 in) wide with about 5–7 lateral veins on either side of the midrib. The small mauve or violet flowers of *C. candicans* are produced in clusters. Its fruits are small and fleshy drupes with a hard inner layer enclosing a single seed.

Distribution: *C. candicans* is found in thickets and secondary forests at low altitudes from southern China westward to eastern India and southward throughout Southeast Asia to tropical parts of Australia and the Pacific, including the Mariana Islands and the FSM. Although *C. candicans* is normally considered to be an indigenous plant on Pohnpei and possibly other Micronesian islands, it may have been introduced by people given its effective pre-modern use in catching fish. Indeed, traditional preparations of *C. candicans* have been considered to be the most potent forms of naturally occurring fish poison on some islands in Micronesia, such as Pohnpei and Kosrae. However, but because it kills both small and large fish, as well as other marine organisms, its modern use for commercial and other non-traditional purposes has been banned in both Pohnpei and Kosrae States (e.g., see Des Rochers 1992: 11–13).

General uses: Traditional use of various parts of plant species in the genus *Callicarpa* includes preparations used as fish poisons, insect deterrents, and traditional medicine; for example, in Malaysia, the tender leaves are boiled and made into a drink for stomach ache. The genus *Callicarpa* consists of 40 or more species, approximately 20 of which have been utilized by people "in ways suggesting that the genus is a rich source of biologically active natural products." Indeed, studies of phytochemical aspects and biological extracts from *Callicarpa* support the probable pharmacological efficacy of the traditional ethnomedicinal uses of these species "and suggest that this genus may offer a rich supply of bioactive secondary metabolites" (Jones and Kinghorn 2008). In Micronesia, the leaves of *C. candicans* have been used, in some cases for many hundreds of years, to stupefy fish.

Uses in Pohnpei: In some areas of Pohnpei, the leaves of *C. candicans* var. *ponapensis* were traditionally pounded and then used or mixed with parts of other plant species as a fish poison (e.g., see Balick et al. 2009: 402):

> "To use as a fish poison, the leaves are pounded on a pounding stone (much like *sakau* is pounded), mixed with sand, and formed into small balls. These are then used in channels or areas where water is contained or blocked off, such as a mangrove channel; the fish eat the sand particles and get docile and stunned, and then people catch them."

Using parts of *C. candicans* or any other plant or chemical material to stun or poison fish is now illegal in Pohnpei, so such fish "poisons" are no longer used and the knowledge about the traditional use of *C. candicans* is being lost or already forgotten in Pohnpei (Balick et al. 2009: 403).

▸ **Isou** (P), **Ijoau** (Mw), **Saepaeng** (Pi), **Hedau** (Nu), **Heedau** (Ka)

Calophyllum inophyllum

Description: *Calophyllum inophyllum* is a medium to large slow-growing tree that can reach heights of 20 m (65 ft) and has very hard wood. Typically, this tree species produces a broad, low-branching, spreading crown. The simple, glossy leaves have many parallel, obliquely lateral veins, and the fragrant flowers are white with yellow stamens (male sex organs) in the center. The spherical fruits are about 2.5 cm (1 in) or more wide and are hard like golf balls. The green fruits turn yellow when mature and have a thin leathery skin covering a hard shell. The kernel of the fruit is somewhat toxic and oily.

Distribution: *C. inophyllum* is a widespread, useful tree believed to be a wide-ranging indigenous tree species "from tropical Africa to eastern Polynesia and Micronesia, but is possibly an aboriginal introduction to some islands" (Clarke and Thaman 1993). Indeed, *C. inophyllum* was probably brought to several of the islands of Remote Oceania, including some in Micronesia, by early Pacific voyagers who discovered and settled these far-flung islands. However, in the western regions of Micronesia, this tree may have been native and dispersed by large bats such as flying foxes (*Pteropus* spp.). For example, the Marianas flying fox (*Pteropus mariannus*, an endangered species primarily due to human consumption) has been a principal disperser of large fruits on those Pacific islands where they currently exist or did in the past. *C. inophyllum* grows equally well in the strand vegetation close to ocean shores or in lower elevation locations inland, developing best in well-drained, sandy soils (e.g., see Friday and Okano 2006a). In addition, *C. inophyllum* is traditionally a very important plant on atolls where it is often planted by seed as a windbreak near both the ocean and lagoon sides of the islets, providing relatively easy access to a valued timber resource.

General uses: *C. inophyllum* is an ethnobotanically important but slow-growing tree. It is common to abundant in coastal forests and is the dominant component of the pre-phosphate-mining vegetation of Nauru, Banaba (Ocean Island), and Makatea (in the Tuamotu Archipelago). It also occurs in lowland forests on atolls and high islands, and occasionally *C. inophyllum* is planted or protected in home and inland gardens, as well as along roadsides. It provides good timber for construction of several things, including canoes (especially on atolls). Furthermore, preparations using latex extracted from its bark are reportedly used in some areas to poison rats and stun fish. Although nowadays *C. inophyllum* is best known in some areas as a large, hardy ornamental plant because of its decorative leaves, fragrant flowers, and spreading canopy crown that provides shade (e.g., see Friday and Okano 2006a), the leaves and the oil extracted from its seed have some medicinal qualities that have been widespread in their applications within traditional Pacific cultures (see Friday and Ogoshi 2011 for a scientific review of the oil extracted from the fruit of *C. inophyllum* and its long-time traditional and cosmetic use in Pacific island cultures).

Uses in Pohnpei: Besides providing excellent timber, the fragrant flowers of *C. inophyllum* (**Isou** in Pohnpean) are used in *mwaramwar* (garlands worn on the head), and its sap is applied as glue. In a Pohnpeian chant for breadfruit, **Mahi**, the fruits of **Isou** are said to resemble small breadfruit. Some believe that this culturally important tree was introduced to Pohnpei by people because the word *sou* means "migrant." As suggested above, this tree, a source of very hard and useful wood, may have been brought to Pohnpei and some other Micronesian islands either by the early humans who discovered and settled on these islands or by early European settlers. On Pohnpei, the sap or latex from a large stem of this tree is sometimes "boiled and used as glue to seal the glass in making diving goggles (the ones with a wooden frame)," and the sap is also "used as a type of caulk" presumably for canoes (e.g., see Balick et al. 2009).

Uses in Mwoakilloa and Pingelap: This tree is known as **Ijoau** on Mwoakilloa. Glassman (1953: 294) spelled it 'Isho' and described it as "a tree 50 to 60 feet high along the seashore." Glassman noted that *C. inophyllum* was a relatively rare tree on Pingelap, where it is called **Saepaeng** and is found on the upper margin of the beach. He also reported that the wood was used in canoe hull construction and that the fruit was used medicinally and also as "a source of oil." Damas (1994) referred to the use of *C. inophyllum* on Pingelap for walking sticks, timber, knife handles, furniture, etc., as well as for squeezing oil to be used by a special person "to give power." However, he contradicted St. John by indicating that *C. inophyllum* is unsuitable for canoe construction because of the "knottiness" of its wood. Another point where the authors disagreed is that St. John (1948) suggested the local name, **Saepaeng**, is a geographic reference to Saipan Island in the Marianas, from where it is assumed to have been introduced to Pingelap, whereas Damas (1994: 237) suggested St. John was wrong about that connection and that *C. inophyllum* is a native tree on Pingelap Atoll. It is possible there is some confusion here with the *Mammea odorata* tree. It is quite similar in general appearance to *C. inophyllum* and also grows on atolls, but their local names and uses would differ. The confusion of these two species is not unlikely, as can be seen from the Pohnpeian name **Luwehs** for *M. odorata*, which is a cognate of the Marshallese **Lukweej** for *C. inophyllum*.

▸ **Dohng** (P) (Mw), **Doahng** (Pi)
Campnosperma brevipetiolata

Description: *Campnosperma brevipetiolata* is a large, native tree that is a common and sometimes dominant species in the upland forests of the Micronesian high islands (except for Chuuk). It has potential use for lumber and reforestation efforts within the FSM (e.g., see Sheely and Meagher 1996). It is a fairly rapid-growing, straight tree, reaching heights of approximately 30 m (~100 ft), with large leaves that are mainly produced as rosettes and are located at the ends of the branches. Leaves of young trees (saplings) are larger than those on mature individuals. The flowers and fruits of the tree are quite small. The fruits are a greenish-yellow color at first, becoming dark-colored when they mature. These small fruits are round, about 0.5 cm (~0.2 in) across and contain relatively large stones or pits. The peak fruiting period is in March and typically *C. brevipetiolata* trees mature in about four years.

Distribution: *C. brevipetiolata* apparently originated in the Indo-Malaya region and is generally assumed to have been naturally dispersed widely throughout its native range. It has been suggested that the early human colonizers of Micronesia brought this species with them on canoe voyages. However, this tree is common and probably native to the high islands of Micronesia (including Pohnpei, but excluding Chuuk). *C. brevipetiolata* is also native in the Solomon and Bismarck Islands, New Guinea, and the Moluccas. Indeed, the fruit of this wet forest tree is food for the Micronesian imperial pigeon, *Ducula oceanica*, and other forest birds which may have dispersed the plant to some or even many islands of Remote Oceania. In Palau, this species is associated with mid-successional to mature forests and is an important component of the forest canopy, especially in redeveloping vegetation. *C. brevipetiolata* naturally regenerates both within the forest matrix and in highly disturbed areas, like roadsides and forest clearing areas (e.g., see Endress 2002, Canfield et al. 1992): "Fruit of [*C. brevipetiolata*] are eaten and transported by birds [and it is] a timber species with some potential as it grows readily in forest gaps and open areas" (Falanruw 2015: 24). Indeed, this relatively large native tree species could serve as a good landscape or reforestation tree in some areas of Micronesia.

Uses in Pohnpei: *C. brevipetiolata* is called **Dohng** in Pohnpei, where it is often one of the largest and most common forest plants. It is stronger but heavier than **Sadak** (*Elaeocarpus carolinensis*), another common native tree in Pohnpei's high island forests. Although **Dohng** trees produce good-quality wood used for many purposes, including logs for canoe bodies, this outstanding woody species is easily damaged by fire. Occasionally, yam vines are trained to grow up into the open canopy of **Dohng**, and the large leaves of **Dohng** are sometimes used for food plates, fans, to cover transplanted seedlings, and to build pig pens. Parts of the **Dohng** trees, such as the bark, are used in traditional medicine; indeed, the medicinal uses of **Dohng** in Pohnpei are apparently diverse. For example, according to Balick and his colleagues (2009: 323–323), uses of parts of **Dohng** trees include:

> "To treat cuts, bark is scraped into a clean cloth and juice is squeezed onto the cut to stop the bleeding and clean and treat the cut... To treat diarrhea, the bark scrapings are wrapped in a piece of cloth and squeezed to extract the liquid, which is then consumed only once... To treat tuberculosis, the bark is pounded, wrapped in a piece of cloth, squeezed into... water, and consumed... The plant is used as a painkiller for toothache... To treat the pain from a broken tooth..."

In addition, Pohnpeians use **Dohng** to trap birds such as **Mwuroi** (Micronesian imperial pigeon, *Ducula oceanica*), which feed on that tree's fruits. Bird hunters traditionally place their traps in the **Dohng** trees, and when the **Mwuroi** come to feed, they get caught inside the traps. Unfortunately, due to increasing human population, clearing of native forest and excessive hunting, often with rifles, the **Mwuroi** has been brought to "Near-Threatened" status on the IUCN Red List.

▸ **Seiren wai** or **Pwuren wai** (P), **Jeir in wai** (Mw) *Cananga odorata*

Description: *Cananga odorata* is a relatively fast-growing tree that can reach heights of about 10–20 m (~33–65 ft) or more, preferring moist soil rich in humus. It produces a crooked trunk, smooth, gray bark, and its drooping, brittle branches with large glossy dark green leaves, up to about 20 cm (8 in) in length, are arranged along a plane, and are alternate, simple, entire, elliptic-oblong in shape and have a drip tip. *C. odorata* trees flower throughout the year in clusters of 4–12 yellowish green inflorescences which consist of long, twisted, wavy, linear-lanceolate, drooping petals in thick bunches at the leaf axis. These are followed by small, greenish fruit which are fleshy, containing many seeds. It is sometimes referred to as a "perfume tree" because its flowers are very fragrant and is widely known as "Ylang Ylang."

Distribution: *C. odorata* is a useful tree native to Southeast Asia, the Philippines, and northern Australia, and it may be truly wild as far east as the Solomon Islands and perhaps some Micronesian islands. However, it was probably introduced to several Pacific islands by people during prehistoric times, and it is a more recent human introduction in others. This relatively fast-growing tree, whose height can increase by more than 2 m (6 ft) per year, was probably brought to all or most of the Micronesian islands after early European contact (e.g., see Manner and Elevitch 2006a). In some cultivated or disturbed tropical Pacific island areas where it is native, or more likely alien, *C. odorata* sprouts from seeds scattered by birds (especially pigeons) and other animals. It is thus common in home gardens and is occasionally planted or protected in other areas.

General uses: In the Pacific, *C. odorata* can be used for many purposes. For instance, the wood is used in the construction of small canoes and furniture, as well as for fuel. It can serve as a trellis support for vine crops such as yams. The fragrant flowers are also used in making garlands and the essential oil extracted from the flowers is used in cosmetic products such as perfumes, soaps, and shampoos, and in foods and aromatherapy (e.g., see Elevitch 2006).

Uses in Pohnpei: *C. odorata* is known in Pohnpeian as **Seiren wai** or **Pwuren wai**. It is used on Pohnpei as an understory trellis tree for yams in agroforest areas, where it is sometimes ring-barked for this purpose (see Manner and Elevitch 2006a).

Women use the fragrant flowers as ornaments, weaving them into *mwaramwar* head garlands. The flowers are collected with the use of a hooked pole made of **Keleu** (*Hibiscus tiliaceus*) or **Ahlek** (*Saccharum spontaneum*). Since this tree will grow back (coppice) from a cut stump, it is chopped down sometimes to harvest the sweet-smelling flowers. *C. odorata* is used medicinally:

> " To treat fungal infection, as many flowers as needed are boiled in as much water as needed; then, after the water has cooled, it is used to bathe the body… To treat sunburn, red spots, and other skin problems, coconut oil (*Cocos nucifera*) is heated and the flowers are added and left to stand for a while, after which the flowers are strained off and the purified oil is obtained, then applied to the skin…" (see Balick et al. 2009)

▸ **Sele** (P), **Jeli** or **Pɛpɛr** (Mw), **Saele** or **Paepaer** (Pi), **Silibeeba** (Nu), **Telee** (Ka)
Capsicum annuum and *Capsicum frutescens*

Description: This section covers two species of chili peppers: *Capsicum annuum* and *Capsicum frutescens*. Distinguishing them can be difficult because they have many diverse cultivars (e.g., see Eshbaugh 2012). Even their status as two distinct species is provisional, and they may be one single, not yet fully resolved taxonomic species encompassing numerous types, including all with names such as pepper, chile, chili, aji, paprika, and capsicum. These plants belong to Solanaceae, the nightshade family, which includes tobaccos, tomatoes, potatoes, eggplants, and many other interesting and often medicinally or economically significant species and hybrids. They are shrubby, relatively short-lasting perennials (where frost does not occur). They generally grow to heights of 1 m (3 ft) or more, with woody glabrous stems that form at the bottom of the plant and branch out near the top. The leaves are ovate to lanceolate with entire margins and petioles. The greenish white flowers grow in the leaf axils alone, or in pairs or threes. The ripe fruit, or pepper, is a many-seeded pod with a leathery exterior in various shades of red or yellow. The fruits also vary according to the degree of their pungent taste.

Distribution: The widely cultivated chili peppers of *Capsicum* are important as spice and vegetable crops over a huge range of distribution. They are among the most diverse series of artificially selected plant species in the world! The genus *Capsicum* appears to have originated somewhere within Central and South America, and it was probably among the first New World cultivated plants, with domestication occurring possibly before the initiation of farming in what today are the tropics of Latin America (e.g., see Pickersgill 1969; Long-Solis 1986; Perry et al. 2007). Eshbaugh (2012: 14) described some the key aspects of chili pepper taxonomy and its history of human use:

> " When Europeans discovered the New World they began what became known as the Colombian exchange, future explorers (exploiters) of the Americas would focus on the material riches of these new lands. However, the real significance of the discovery was an initiation of transfer of previously unknown plants from the west to the east (and vice versa) in the form of new species, many of which later became economically important in the Old World. One was a plant whose unique fruit possessed a "bite" that added zest to the Caribbean diet. Eshbaugh (1993) concluded that few could have imagined the impact of the discovery of a spice so pungent that it rivaled the Old World black pepper. Five centuries later chilli peppers (*Capsicum*) dominate the world hot spice trade and are grown everywhere in the tropics and in warm temperate regions. Not only have hot peppers come to command the world's spice trade but a genetic recessive no-pungent form has become an important "green" vegetable crop especially in temperate regions." (Underlined for emphasis).

The present range of *Capsicum* species is now very widespread in many parts of the world, including numerous Pacific islands:

> " The species is intolerant of shade: it will grow with broken overhead shade and moderate competition from grass

and forbs, but fruits best in full sun. Red peppers can be cultivated in areas that receive from 30 to 430 cm [12–170 in] of annual precipitation at elevations from near sea level to more than 2,000 m [6500 ft]. Wild red pepper grows as individuals or groups in gardens, fields, vacant lots, river flood plains, abandoned fields, roadsides, and early secondary forest." (Francis 2004: 147)

General uses: Red chili peppers are used in many parts of the world, mostly as a spice added to fresh, dried, or refined food, and as the "principal or incidental ingredient in sauces" (Francis 2004: 148). The chief chemical source of the common "biting sensation" of these peppers are the capsaicinoids (alkaloid compounds), mainly capsaicin. Serving as a condiment, chili peppers add flavor and diversity to everyday dishes, and *Capsicum* fruits are a first-rate source of vitamins A and C (e.g., see Bosland and Votava 2000; Lim 2012). Furthermore, chili peppers can add variety to contemporary diets in a number of geographical regions, including island areas within Micronesia where daily food consumption is currently based on only a few staple crops (e.g., see Valenzuela 2011; Yamamoto 2014). Chili peppers have also had numerous medicinal uses, especially as a "salve to relieve muscle, joint, and toothache pain, to treat cough, asthma, and sore throat, as a stimulant, and to treat stomach ache, seasickness, and flatulence" (Francis 2004: 148). Besides being responsible for the spiciness in chili peppers, capsaicin is also largely responsible for its effectiveness in relieving pain from arthritis and migraines and treating coughs or stuffy noses, and it also has "anti-clotting properties" (Valenzuela 2011; also see Lim 2012 for a comprehensive review of the chemical contents and phytochemical effects of chili pepper fruits).

Uses in Pohnpei: Yamamoto (2011) recorded the local uses of *Capsicum* plants on Pohnpei and the atolls of Mwoakilloa and Pingelap in order to identify the relationships between people and chili pepper plants on these islands. Yamamoto noted that the local names appear to be related to "chile" in Spanish, suggesting that some *Capsicum* peppers were introduced into the region from the tropical New World by the Spanish during the Manila galleon trade, perhaps when the Spanish colonized Pohnpei. However, Ragone et al. (2001) assumed that *C. frutescens* was introduced to Pohnpei during the early period of European or Western contact (1830–1886), and was also brought into this island during the American Trust Territory period (1945–1986); on the other hand, Yamamoto (2011: 92) suggested that additional "research on *C. frutescens* specimens collected early in the 20th century in Oceania will reveal more about its dispersal."

During his recent research on Pohnpei, Yamamoto (2011: 90) identified two species of *Capsicum* being used in Pohnpei:

> " *C. annuum* (two types: a pungent long fruit and a nonpungent bell pepper or sweet pepper) and *C. frutescens* (several pungent types exhibiting green or greenish-yellow immature fruit color and a variety of fruit sizes). People who cultivated the long fruit cultivar of *C. annuum* on Pohnpei said that it had been introduced there within the past five years. Cultivation of sweet peppers seemed unknown in three regions; although the peppers were sold in markets, some were confirmed to be imported from the United States of America. Unlike *C. annuum*, *C. frutescens* is often used for seasoning foods in the daily diet and has become an indispensable condiment on Pohnpei and [Mwoakilloa] and Pingelap atolls."

Balick et al. (2009: 501) referred to some of the uses of *C. frutescens* on Pohnpei, such as the use of chili pepper fruit as a flavoring to "spice fish," to treat skin rashes or wounds to prevent infection and stop bleeding, and as a medicine for chickens with an "unspecified illness." Students in ethnobotany classes at the College of Micronesia on Pohnpei have reported that peppers have been used to flavor food, as well as to treat "worms" and "viral colds."

Yamamoto (2011: 87) also reported on the varied use of *C. frutescens* today in Pohnpei:

> " ... as a condiment (on fresh or salted fruit and fruit soaked in unboiled water or the water of mature coconuts), as a vegetable (the leaves), and as a medicine (the fruit as an anthelmintic drug [used to eliminate parasitic worms], the leaves for curing boils and wounds, and the flowers for promoting childbirth or blood expulsion in pregnant women). With modernization, however, residents of Pohnpei have been eating fewer *C. frutescens* leaves in recent years. Rediscovering and maintaining knowledge of plants already naturalized on each island are important to food security."

Englberger et al. (2009: 123, 129) listed "chilli pepper leaves (*se sele*)" as among the highest-rated vegetables in their documentation of the traditional Pohnpeian food system. Their survey of the Mand community on Pohnpei indicated that these leaves are generally easy to cultivate and desirable for "traditional" consumption, especially as a source of vitamin A; Englberger and her colleagues also listed chili pepper leaves as a high priority for promotion as a vegetable crop on Pohnpei.

"Even though *Capsicum* did not originate in FSM, it has been cultivated and consumed there for an estimated 300–400 years and thus it is reasonable to consider it a traditional local food. The importance of this plant as a nutritious and local food in this region has been largely overlooked despite the fact that the fruits and leaves are rich in carotenoids and vitamins… In addition, information on *Capsicum* usage itself is very limited." (Yamamoto 2011: 88)

- **Memiap** or **Moamiap** (P), **Kaniahpo** (Mw), **Kaeiniap** (Pi), **Mami** (Nu), **Memeabu** (Ka) *Carica papaya*

Description: This common tropical fruit crop, *Carica papaya*, is a soft-stemmed large herb (non-woody plant) or small tree reaching heights of 2–5 m (5–25 ft), with a gradually tapering trunk and light-colored bark which contains a caustic milky juice. It produces broad, smooth leaves up to 50 cm (2 ft) across with deeply incised lobes that are clustered at the top of the plant. The flowers of this cultivated species are quite variable; in fact, *C. papaya* is dioecious, i.e., individual plants produce either all female (fruit producing) flowers or all male (pollen-bearing) flowers. However, some male plants change to fruit-bearing female plants after being cut off at the top, and some plant types produce male and female sex organs on the same flower. In addition, a fourth variety produces all three types of flowers (i.e., those that have only female sex organs, others that have only male sex organs, and still others that have both male and female sex organs in the same flower). The female flowers of *C. papaya* are five-petaled, creamy white, and fragrant, and the edible fruits are large, grooved, and yellow or yellow and green.

Distribution: Known in English as "paw paw" in some areas, but mostly as "papaya," which is also part of its scientific name, *C. papaya* is native to tropical America. It probably first arrived in Micronesia in the 19th century or earlier, after European and Asian contact in the region. It may have come from the Philippines, where it was introduced by the Spanish in the 16th century. It grows quickly from seed spread by humans or animals. It is cultivated extensively throughout the tropics and also grows as a weedy plant that does well in sunny places with relatively nutrient-rich soil.

General uses: Papaya is one of the more commonly eaten tropical fruits in the world today. Papaya fruit are best picked when they are half-ripe, still firm, and have a yellow-green color. The fruit is ripe when it is soft and easily bruised, and some varieties of ripe papaya will have a more pinkish or reddish than yellowish flesh. The fruit pulp is sweet, tasty, and nutritious since it contains calcium, natural sugars, and vitamins A and C. It can be eaten raw, baked, stewed, cooked with other fruits in jam, or made into candy, and the juice can be extracted for a refreshing drink or sweet soup. In various islands of Micronesia, this relatively recently introduced plant is widely cultivated and it grows wild for use as food, medicine, and adornment.

Uses in Pohnpei: According to Christian (1899: 339), *C. papaya* was brought into Pohnpei "about 1840 by an old French settler." Approximately a half century ago, *C. papaya*, known on Pohnpei as **Memiap** or **Moamiap**, which appears to be derived from an English name "mommy apple," was said to be "never cultivated but grows wild everywhere" but is "well liked because it is plentiful and requires no work and its fruit is flavorful" and "the fruit is sometimes used as chicken feed and the leaves are eaten by pigs and cows" (Lessa 1977: 167). There are also a number of medicinal uses for this plant on Pohnpei:

"The leaf petioles are used by children as toys to make small peashooters… To treat bloody diarrhea… To treat stomach ulcers, the ripe fruits are eaten… To treat stomach worms… Alternatively, to treat worms and parasites, a ripe fruit is cut in half and eaten, along with all the seeds inside, once daily for four days… To treat headaches, a handful of flowers are pounded together with a rock, put in a cloth, and squeezed on the wound… To treat a fungus infection… To treat skin rashes… To treat back pain, a root after being cleaned, is boiled in as much water as needed, then consumed daily…

Mature and young papaya plants along with other vegetation in the village on Mwoakilloa Atoll.

With a papaya and a knife, it's easy to enjoy a quick and healthy snack.

Rows of papayas arranged for sale at a local market in Kolonia.

Sioahk, a Pohnpei starling, feeding on a ripe papaya.

To treat an asthma-type cough... To treat *kilinkutuwa* (a fungal infection of the skin), the latex from the skin of the green fruit is placed on the affected area)." (Balick et al. 2009)

A common local remedy in Pohnpei for a jellyfish sting involves rubbing unripe papaya on the affected area. Unripe papaya has qualities that inactivate the jellyfish nematocysts and sooth pain. According to Lee et al. (2010: 9), sap of the unripe papaya fruit is collected on Pohnpei and rubbed over the wounds of stonefish and scorpionfish stings to remove poison substances from the skin and to relieve distress and accelerate the healing process:

> " This is prepared by slicing the top of the [papaya] fruit and draining the sap onto a brown leaf of [banana] that can either be placed directly on the wound as a poultice or the sap can be rubbed directly on the affected area."

Papaya is also reportedly used to bring relief from a burning rash caused by contact with "fire coral" (**Rarkapwudong**); in this case, ripe fruits are cut into halves and applied to the rash to treat the pain and induce healing (Lee et al. 2010: 9).

Uses in Mwoakilloa and Pingelap: Papaya is known as **Kaniahpo** on Mwoakilloa and **Kaeiniap** on Pingelap. On Mwoakilloa, it has been reported as "the leading fruit tree," but "not a very important source of food" because it does not thrive in normal atoll soils (Murphy 1950: 73; also see Lessa 1977: 167). In contrast, papaya appears to thrive "especially well on Pingelap and the fruit often reach the length of 18 in or more," with fruits reportedly, more or less, "saved for presentation at feasts" (Damas 1994: 56). According to Levin et al. (2018), her interviews with 14 adult members of the Pingelapese community—11 men and 3 women between the ages of 29 and 80—indicated that the great majority of plants grown on Pingelap by people interviewed "focused on Western Pacific cultigens" (such as breadfruit, bananas, coconut, swamp taro, and pandanus), with papaya being the only widely grown cultigen from outside this region.

▸ **Likotokotasahu** (P), **Kosoakos** (Mw), **Kosokos** (Pi), **Hiisoa** (Nu), **Uihoo** (Ka)
Cassytha filiformis

Description: *Cassytha filiformis* is an indigenous, thin-stemmed, tropical coastal twinning vine. It is a parasitic, herbaceous, and leafless species. The string-like, pale green to yellow or orange stems of *C. filiformis* attach themselves to other plants by suckers (known as haustoria). These plant organs take energy and nutrients from other plants. The many stems form dense, tangled mats that can climb over and almost completely cover a host plant. *C. filiformis* produces tiny greenish white flowers clustered on short spikes, and its small white fruits are round and fleshy.

Distribution: There are about 20 species of *Cassytha*, of which "at least three species are locally important parasites of woody vegetation in some locations worldwide: *Cassytha ciliolata* Nees; *C. filiformis* L.; and *C. melantha* R. Br." (Nelson 2008: 6). One of these, the tropical coastal vine *C. filiformis*, is widespread, affecting a broad variety of coastal or littoral plants in Micronesia as well as many other tropical shoreline areas of the world. *C. filiformis* is regarded as a native species in Micronesia. It spreads out over the ground and low-lying plants in the strand vegetation, and it also sprawls over taller plants near the shore and somewhat inland; occasionally, it is found in the upland vegetation. It may cover shrubs and even some small trees with its dense masses of orange stems.

General uses: Nelson (2008: 1) referred to *C. filiformis* as being either a "blessing or a bane, depending on one's perspective." In some areas of the tropical Pacific, *C. filiformis* is traditionally well regarded, with "diverse healing applications and with promise for modern medicine." It also provides some food resources, as well as vine stems used in thatched roofing. It has also served as a "potential biological control agent for invasive plants." However, this relatively common, parasitic vine can be regarded as a "pestiferous and pathogenic weed that is a threat to agriculture and endangered plants and is capable of transmitting other pathogens between plants." Thaman (1994: 168) referred to *C. filiformis* as one of those tropical coastal plant species found on many Pacific islands that is "associated with revelry or ceremonial dress."

Uses in Pohnpei: *C. filiformis* is known as **Likotokotasahu** in Pohnpeian. This native, thin-stemmed, twinning vine has been used in making of head garlands (*mwaramwar*) "to be worn as a badge of shame after not having caught any fish during a fishing contest" held twice a year on Pohnpei. One of these fishing contests is religious, and the other is "held by the municipal government in combination with the breadfruit festival" (Balick et al. 2009). *C. filiformis* is also used medicinally to treat high fevers, pains in body, and to grow long hair. In addition, parts of **Likotokotasahu** are reportedly pounded and mixed in with water on Pohnpei to use as a shower, which is intended to restore memory or cure other mental problems.

Uses in Mwoakilloa and Pingelap: Glassman (1953: 293, 300) provided the first botanical record for *C. filiformis* on Mwoakilloa and Pingelap, indicating that this climbing parasitic plant was commonly found in the coastal strand vegetation; Glassman (1953: figure 2) included a photograph of *C. filiformis* growing on *Guettarda speciosa*. No uses for *C. filiformis* were provided by Glassman.

▸ **Koatun** (P)
Ceiba pentandra

Description: *Ceiba pentandra* are large to immense trees that may grow to heights of about 60 m (~200 ft), adding up to 4 m (13 ft) a year, and can live for as long as two centuries. These trees develop wide buttresses at their base and produce large, more or less level crowns on top. Their trunks have gray bark and are usually spiny, cylindrical, and smoother far up. Their leaves are large and palmately compound with 5–8 leaflets radiating from a common point on each leaf. *C. pentandra* trees are deciduous, dropping all their leaves each year. Their large showy flowers are bell-shaped with five white to pinkish petals. These inflorescences produce a pungent odor that attracts pollinating bats. The thick, woody fruit pods contain about 200 small brown seeds. With a multitude of flowers, a single tree may produce up to about 800,000 seeds each year. The seed pods open while hanging on the tree and release their seeds along with many soft, silky, long, cotton-like fibers which are dispersed by the wind.

Distribution: *C. pentandra* is reportedly native throughout a large area of the American tropics, from Mexico through Central America and south to Peru, Bolivia, and Brazil, and possibly in West Africa. This large tree has also been introduced to other regions of the tropics by people and is now found in various areas of Micronesia, including Palau, Yap, the Mariana Islands, the FSM, and the Marshall Islands. It occurs in a range of habitats from savannah to forest in a variety of soils from rich to poor, but it requires a dry season and sheds its leaves during periods of drought.

General uses: *C. pentandra* has been referred to as a "super-tree" because of its ability to reach immense size and its multiple uses. It is a valuable tree in many tropical countries, providing wood, oil, and food, and the long seed hairs are used for insulation and stuffing. In Micronesia, *C. pentandra* has been planted as an ornamental tree, along roadsides and as a boundary marker; in addition, its lightweight wood can be used in light construction, its fibers are used to stuff pillows and mattresses, its young seed pods can be eaten, and it has medicinal uses.

Uses in Pohnpei: *C. pentandra*, known in Pohnpeian as **Koatun** (i.e., "cotton"), was introduced to Pohnpei "during the German and Japanese periods," apparently "for its potential value as a fiber crop." However, this tree has been cultivated primarily as an ornamental on Pohnpei, although its trunks have been used to make canoe hulls (see Brosi et al. 2007: 878) and its young leaf shoots are reportedly pounded and used to treat unopened boils, *umwpwos sawawi*:

> "The shoots from the young leaves are pounded together and placed on the unopened boil like a poultice. This treatment is repeated each time the leaves fall off or until the boil opens and drains." (Lee et al. 2010: 4; also see Balick et al. 2009: 426; Glassman 1952 and Ragone et al. 2001)

▸ **Liwadawad marer** (P), **Maler** (Mw)
Centella asiatica

Description: *Centella asiatica*, the Asian pennywort, is a small, low-lying herbaceous plant with creeping stems and many small, kidney-shaped, serrated leaves 2–5 cm (0.75–2 in) wide. New individuals of this plant are produced both by seeds and by runners (clones) at the nodes. The inconspicuous flowers are small, white, and clustered in groups of 3–4.

Distribution: This sometimes weedy plant is widely distributed in many tropical areas of the world, including many islands in the Pacific region. It is common in shady, often swampy areas near the coast, and to a certain extent in inland locations up to about 300 m (~1,000 ft) above sea level. It sometimes occurs in open places such as roadsides, cultivated land, and pastures. Fosberg et al. (1979) and others have listed this creeping plant as native in Micronesia.

General uses: *C. asiatica* has been used in folk medicine in some places for hundreds, perhaps thousands, of years and has become relatively important in scientifically oriented medicine during more recent times. Parts of this herbaceous plant are said to lower the blood pressure and are reputedly rejuvenating. Eaten raw or finely cut, the leaves of *C. asiatica* can be roasted with scraped coconut, and some believe these prepared leaves can purify the blood and relieve indigestion, nervousness, and dysentery. Indeed, *C. asiatica*, with its long history of widespread therapeutic use in Asia, is now becoming popular in other areas of the world for a variety of medicinal purposes:

> "Triterpenoid, saponins, the primary constituents of *Centella asiatica* are manly believed to be responsible for its wide therapeutic actions. Apart from wound healing, the herb is recommended for the treatment of various skin conditions such as leprosy, lupus, varicose ulcers, eczema, psoriasis, diarrhea, fever, amenorrhea, diseases of the female genitourinary tract and also for relieving anxiety and improving cognition." (Gohil et al. 2010: 546)

C. asiatica has long been commonly used traditionally by Pacific islanders in Polynesia (e.g., Tonga, see Yuncker 1959: 208) as well as in Micronesia.

Uses in Pohnpei: The prepared leaves of *C. asiatica* (**Liwadawad marer** in Pohnpeian) are reportedly used as food in soups and as medicine for babies and others for a variety of ailments, including diarrhea, dysentery, "stomachache when feeling nauseous," as well as for burns and a type of fungal skin infection (see Balick et al. 2009: 281, 328–329; for a remedy to treat diarrhea in infants, see Lee et al. 2010: 30). In addition, placing drops of juice extracted from pounded leaves of *C. asiatica* is reportedly used to make infants "less afraid of hiccups," because, in both Pohnpei and Kosrae Islands, hiccups are a sign of a healthy baby (e.g., see www.comfsm.fm/~dleeling/ethnobotany/centella_asiatica.html).

▸ **Madeu** (P), **Madu** (Mw)
Cinnamomum carolinense

Description: *Cinnamomum carolinense* is a large tree that can grow to heights of about 60 m (~180 ft). Its woody, aromatic bark is about 2.5–4 cm (1–1.5 in) thick, and its alternately arranged leaves are about 8 cm (~3 in) wide and 15 cm (6 in) long. During its seasonal blooming, the tree produces small, white flowers, and its round fruit become a deep red color as they ripen. A single fruit of this tree can contain three or four brown seeds.

Distribution: The large endemic *C. carolinense* tree is only known from Pohnpei, Kosrae, and Palau, where it occurs in volcanic soils in the wet montane forests, centered at about 250 m (~800 ft) in elevation. Exploitation of this species for a medicinal tea on Pohnpei (discussed below) has raised concern about its possible toxicity, and about its reportedly shrinking population in the upland wet forests of that island (Balick 2014: 141). Costion and Lorence (2012: 62) indicated that "it is rare" in Palau (e.g., Kitalong et al. 2014 recorded the "limited presence" of this tree in his survey research on Palau; also see Kitalong et al. 2011: 69).

Uses in Pohnpei: *C. carolinense* is a native cinnamon tree species on Pohnpei, where it is known as **Madeu** in Pohnpeian. Kanehira (1931: 280) reported that **Madeu** was found in "primary forests at low altitudes" on Pohnpei, where its bark "boiled in water is used for drinking." In the middle of the 20th century, Glassman (1952: 15) reported that local people consume a drink "prepared from the bark of native species of *Cinnamomum*." More recently, research reported by Reynertson et al. (2005) noted that numerous species of *Cinnamomum* produce safrole, a known hepatocarcinogen (causing cancer of the liver), which has raised concern about the common consumption of this beverage by people on Pohnpei. Scientific laboratory analysis confirmed the presence of the carcinogen safrole in the bark shavings of the endemic *C. carolinense* cinnamon tree, but not in the prepared tea. Thus, it appears that the "traditional preparation method, which boils the bark shavings, degrades the safrole," making the tea safe to consume. Traditionally, the hot *Cinnamomum* beverage was used to treat backaches, excessive menstruation, and a culturally specific disease called *soumwahu en naniak*, which is associated with the mangrove forests. This cinnamon drink also has a tradition of being used for the following ailments: "to treat syphilis; as a tonic to boost the immune system; to enlarge a urine stream that has contracted perhaps as the result of a prostate problem; and to treat something locally known as 'men's sickness' that involves swelling and pain in the groin area that is said to come from overwork" (Lee and Balick 2005). It has also been reported that bark scraped off *C. carolinense* is combined with cleaned roots of **Aiau** (*Ficus prolixa*) "to treat colds and flu." The combination of *C. carolinense* bark and *F. prolixa* roots is prepared and consumed as follows, with a strong cautionary statement:

> "Together these are boiled in a kettle of water. This tea is consumed once per day until symptoms improve. The same tea is brought to a boil each time it is consumed. **Caution**: Due to the potential toxicity of this plant, consult with a local expert and medical practitioner before using this remedy." (Lee et al. 2010: 81)

The bark of this tree is aromatic and thus sometimes used as seasoning. *C. carolinense* is harvested by stripping the bark. This girdling of a section of the tree can cut off part of its energy and nutrient transport system and may ultimately result in the tree's death. Due to this harvesting method, the cinnamon trees have not been traditionally managed adequately, at least in modern times, and so this species is no longer common in Pohnpei today. However, ways to harvest the trees sustainably have been developed for scraping "the stem in a way that allows the bark to regrow within 6 months" (see Lee et al. 2010: 81, fig. 79); in addition, with the establishment of protected areas where seedlings of this native cinnamon tree can be planted as a way for local people to profit from *C. carolinense*, it is hoped that this natural resource and cultural knowledge will persist into the foreseeable future (see Lee and Balick 2005; also see Balick et al. 2009; Lee et al. 2010; Balick 2014).

▸ CITRUS SPECIES

There are many species in the genus *Citrus*, several of which are important and much-desired fruit crops cultivated in various tropical or subtropical regions of the world. From the scientific point of view, *Citrus* taxonomy is complex and unsettled. Although modern genetic science is attempting to sort out the origins of the genus *Citrus* and the "true" species, many citrus fruit are from hybrids of two or more *Citrus* species. It is generally assumed that there were three original *Citrus* species, and all others arose from hybridization events (horticultural crossings between different *Citrus* species). Which of the hybrids is considered to be true species varies widely according to the classification scheme used. Therefore, the *Citrus* species described and discussed here must be considered as taxonomically tentative. In the Pohnpeian language, the general name for all *Citrus* is **Karer**, a word which, as an adjective, means "sour," but as a noun can stand for lime (*C. aurantifolia*), lemon (*C. limon*), sour or wild orange (*C. aurantium*), tangerine (*C. reticulata*), and sweet orange (*C. sinensis*). We will look at these five species in detail. All are introduced species, brought to Pohnpei after Western contact, but some probably quite early on. References to *Citrus* species can be found in the writings of Luelen Bernart (1886–1946). Bernart was a member of a well-known Pohnpeian family and became widely recognized for the depth of his traditional Pohnpeian knowledge. His highly regarded work, "The Book of Luelen," is commonly considered the most complete account of Pohnpeian history ever written by a native of that island (e.g., see Fischer et al. 1977).

▸ **Laman** or **Laim** (P), **Lɛmɛn** or **Laim** (Mw), **Karaer** (Pi), **Laim** (Nu), **Beleni mmala** (Ka) *Citrus aurantifolia*

Description: Lime (*Citrus aurantifolia*) is a bushy, vigorous shrub or small tree, about 4–8 m (12–24 ft) in height. This woody cultivated species produces sharp spines, variable but alternately arranged leaves, and fragrant, pink to white-petaled flowers. The smooth, globose, green to light yellow-skinned fruit are 3.5–6 cm (~1.5–2 in) wide and contain aromatic, generally acidic pulp and juice which can be sour or sweet depending on the variety.

Distribution: The lime tree is probably indigenous to southern Asia or Indonesia and appears to have been introduced into the Pacific Islands relatively soon after European contact (Thaman and Whistler 1996). Today, lime trees are commonly cultivated in home gardens on some Pacific islands; they are also sometimes planted in monospecific plantations, e.g., in Fiji, Niue, and the Cook Islands. *C. aurantifolia* occasionally becomes naturalized in Micronesia and is the only *Citrus* species well established in Kiribati and some other atolls in the region.

Uses in Pohnpei: Englberger et al. (2009: 119) listed the fruit of *C. aurantifolia* as a "traditional" crop on Pohnpei. It is known in Pohnpeian as **Laman** or **Laim**, as well as by the general term **Karer** noted above. In general, *C. aurantifolia* is cultivated for its juice, which is "used to marinade raw fish, make drinks, and for flavoring food and desserts" (Clarke and Thaman 1993). In Pohnpei, lime juice is used to enhance the taste of food, especially of fish sashimi, which gives it another local name, ***Karer en sasimi***. Yet another name, ***Karer en lopwolopw***, is reserved for lime used to whiten the fibers when making grass skirts. There is also a wide variety of medicinal uses for different parts of the lime tree. Some are referred to below:

> "The young leaves are used for coughing. You have to boil the leaves and drink the boiling water. **Karer** can also be used for headache. You can take the young leaves, pound them and put them in oil. When you have a headache, apply [it] to [the] skin. The green leaves are used for *umwulap* (steam bath) when you have a cold or flu. You can also mix the leaves with other leaves and boil the mixture, drink to help the blood vessels, the muscles, and other bodily organs. The juice of the fruit, drunk undiluted, is [also] good for coughing. Any cut not caused by a knife can be treated with undiluted juice. This will hurt, but the wound will heal more rapidly. The skin of the tree (*kelupwur*) can be cut into

A close-up of prized limes from Mwoakilloa, renowned for their exceptional quality.

Using a fruit-picking pole to gently shake limes loose from the tree.

Children eagerly gathering limes to present as a gift for a visiting guest from off-island.

Limes from Mwoakilloa on the coral-strewn ground where they grew.

Trays of limes for sale in Kolonia, Pohnpei.

pieces, put into a cloth, placed in boiling water, squeezed into a cup, and [drunk]. This will cure *soumwahu en lusulus*, [which is] characterized by high fever and convulsions. Locals tend to refer to this as "jumping" sickness" (www.comfsm.fm/~dleeling/botany/1999/vhp/citrus_aurantifolia.html).

Balick et al. (2009: 490–491), who referred to the consumption of lime fruits as food or condiment, also provided more information regarding the medicinal uses of **Laman** in Pohnpei:

> "The plant is used to make a local tea... To treat shortness of breath... To treat a wound such as stepping on a rusty nail or other sharp rusty objects, the fruit is cut in half, placed over the fire for two to three minutes, and placed on the wound to extract the rust... To treat fever... To treat *lusulus* (seizures)... To calm the *soumwahu en rerrer* [epilepsy]... To treat influenza..."

Uses in Mwoakilloa: In 2007, a student from the College of Micronesia reported that the bark is used "in water to cool a very high fever accompanied by violent chills," locally known as *lujluj* (www.comfsm.fm/~dleeling/ethnobotany/healing_plants_presentations.html).

▸ **Orens** (P), **Karɛr** (Mw), **Karaer** (Pi), **Aalonsese** (Nu), **Beleni maangala** (Ka)
Citrus aurantium

Description: The sour orange (*C. aurantium*) is an erect tree that ranges in height from less than about 3 m (10 ft) to about 9 m (30 ft) and has a more compact crown than the sweet orange (*C. sinensis*). *C. aurantium* produces smooth, brown bark and green twigs that are angular when young, and flexible, with thorns about 2.5–8 cm (~1–3 in) long but not very sharp. The fragrant evergreen, compound leaves with minutely toothed edges are about 6–14 cm (~2.5–5.5 in) long and about 4–10 cm (1.5–4 in) wide; they are usually ovate in shape, with a short point at the apex, and alternately arranged on broad-winged petioles that are considerably longer than those of the sweet orange. In addition, the leaves of *C. aurantium* are dark green above, pale beneath, and dotted with tiny oil glands. The bisexual flowers are highly fragrant and are borne at the axils. They are solitary or clustered and about 3.75 cm (1.5 in) wide with five white, slender, separated petals surrounding a mass of up to more than 20 yellow stamens. The useful fruit is round, oblate, or oblong-oval and about 5–10 cm (~2.5–5 in) wide. *C. aurantium* has a rough surface and a relatively thick, aromatic but bitter peel with minute, sunken oil glands. The peel becomes bright reddish-orange when mature, and the 10–12 segments of the acidic, pulpy fruit contain a few to numerous seeds. Because the center of the fruit becomes hollow when mature, the sour orange fruit is "able to float" (Stone 1970).

Distribution: *C. aurantium* is probably indigenous to southeastern Asia and is generally assumed to have been a relatively early post European-contact introduction into the Pacific Islands, although natives of some tropical South Pacific islands, especially Fiji, Samoa, and Guam, believe the tree was brought to their homelands in before European contact. On the other hand, because the mature fruit becomes hollow and so can float, this species of *Citrus* might have dispersed to some islands without human help.

General uses: A species of multiple uses, the sour orange, *C. aurantium*, is mostly used to make beverages and to flavor food. It is fruit sold as a minor cash crop in some countries, and parts of the tree are widely for used medicine (e.g., see Clarke and Thaman 1993). Where it is cultivated on Pacific Islands, including some islands in Micronesia, the crushed fruit and macerated leaves of sour orange trees are used to make a lather in water which is used as soap for washing clothes and shampooing the hair.

Uses in Pohnpei: *C. aurantium* is known in Pohnpeian as **Orens** or, like all other citruses, **Karer**. According to Balick et al. (2009: 491), **Orens** is used medicinally in Pohnpei to treat high fever (*soumwahu en rerrer*) "when the **Oahs** palm leaf (*Metroxylon amicarum*) is not available, the skin [of *C. aurantium*] is peeled off, squeezed on a cloth, and put into the mouth three times daily."

▸ **Laman** (P)
Citrus limon

Description: The "true" lemon (*Citrus limon*) is a relatively small woody species reaching heights of about 3–6 m (10–20 ft), and it normally has sharp thorns on the twigs. The leaves are arranged alternately, and are oblong, elliptic, or long-ovate in shape, about 6.25–11.25 cm (2.5–4.5 in) in length with fine-toothed margins and slender wings on the petioles. Although the leaves are reddish when immature, as they age, they turn dark green on the upper side and light green below. The flowers are fragrant and may be solitary or in clusters of two or more flowers in the leaf axils. *C. limon* flowers have four or five petals, each about 2 cm (0.75 in) long, that are white on the upper surface (inside) and purplish beneath (outside), with 20–40 stamens with yellow anthers that are more or less united. The true lemon fruit is about 7–12 cm (2.75–4.75 in) long and more or less oval in shape with a nipple-like bulge at its apex; the peel or skin is usually light yellow with aromatic oil glands, and the juicy, acid fruit pulp is pale yellow with 8–10 segments. Some fruits are seedless, but most have a few elliptic or ovate, pointed, smooth, seeds (e.g., see Morton 1987). Lemon fruits usually have either a smooth or rough green to yellowish skin; fruits with the smoother skin type are generally juicier, while the rough or bush type typically has more flavor (Matanane et al. 2014: 114).

Distribution: Although the native geographical range of the *C. limon* is uncertain, the true lemon "may have come from southeastern Asia" (Smith 1985). It is now grown commercially worldwide in tropical, semi-tropical, and warm temperate countries, including the Mediterranean region. Its range of cultivation in Micronesia and the greater tropical Pacific region is relatively limited; however, "the lemon tree has the reputation of tolerating very infertile, very poor soil" (Morton 1987: 160–168).

General uses: Roots, bark, fruits, and seeds of lemon trees have been used traditionally for a variety of reasons, especially for medicinal purposes and refreshing beverages, over an extensive global range, including within some islands in Micronesia. Safford (1905: 229) provided a summary of the uses of this common lemon fruit:

> "This fruit is valuable for its acid juice and for the oil obtained from its rind, known as the "essence of lemon." The latter may be obtained by scraping and pressing or by distillation. The former, together with lime juice, is the source of citric acid. Lemon oil is of a pale yellow color, fragrant, and aromatic. It is used for flavoring and in the manufacture of perfumery, especially of eau de Cologne."

Uses in Pohnpei: *C. limon* is known in Pohnpeian as **Laman**. According to Sproat (1968: 75), the only *C. limon* variety of consequence found in Micronesia a half century ago was the "Meyer's variety," which, like other citruses, was introduced to Pohnpei. Meyer lemons are named for Frank Meyer, the agricultural explorer who found these "special lemons" in their reported country of origin, China, where they were being used as decorative houseplants. Meyer identified this plant as a variety of the lemon tree and brought it back to the U.S. in the early part of the 20th century. The Meyer lemon variety is said to grow well and produce "large crops of medium to large juicy fruits." This lemon variety is reportedly a hybrid of the lemon, orange, and mandarin trees, and, as Meyer suspected, was apparently developed originally in China. The fragrant fruit peel or skin of a Meyer lemon has a deep yellow color, and the fruit itself is rounder than a common lemon. Meyer lemon fruits are said to have a sweeter, less acidic flavor than fruits of other well-known varieties such as the 'Lisbon' or 'Eureka' varieties, which are typically sold in supermarkets in Pohnpei and around the world (e.g., see Christman 2018). The pulp of the Meyer lemon is dark yellow and contains up to ten seeds per fruit. In Pohnpei, the first juices extracted from lemon fruits have been used "as a meat tenderizer or marinade" as well as a taste enhancer on "raw fish, sashimi, or other meat," and parts of lemons are used in the treatment of "unspecified children's illness" as well as to "treat colds and fevers..." (Balick et al. 2009: 491; also see Herrera et al. 2010; Lessa 1977: 160). In 2005, a college student reported that the leaves of "Pohnpei **Laman**" are used for coughing and the boiled bark tea of the lemon tree "is used to treat *soumwahu en lusulus* (high fever with chills)" (www.comfsm.fm/~dleeling/ethnobotany/healing_plants_presentations.html).

▸ **Manterihn** (P), **Mansorihn** (Mw)
Citrus reticulata

Description: The mandarin orange (*Citrus reticulata*) is a small, sometimes spiny tree that grows up to 9 m (~30 ft) high with leaves that have scalloped edges and slightly winged petioles. It produces flowers with white petals, while its fruits are somewhat flattened and roundish, and the orange pulp is sweet to somewhat sour. *C. reticulata* fruits are generally smaller in size than those of the true orange, *C. sinensis*. The skins of this *Citrus* tree, known commonly as the mandarin orange or tangerine, may turn lighter green or yellow to orange when ripe, and these skins are loose and peel easily.

Distribution: *C. reticulata* is indigenous to East Asia and Southeast Asia. It is a post-European-contact introduction to the Pacific Islands, but now it is rather commonly found in garden areas, fallow forests, and tree groves surrounding villages on some Micronesian islands, but not usually on atolls.

Uses in Pohnpei: This introduced species is known in Pohnpeian as **Manterihn**, and its fruit is eaten as food after removal of the skin. Its leaves are prepared and used as a body bath to "treat a headache" (Balick et al. 2009: 493).

▸ **Orens** (P)
Citrus sinensis

Description: The sweet orange (*Citrus sinensis*) is a small to medium-sized tree which can grow up to 12 m (~35–40 ft) in height. Normally, it develops slender, flexible, axillary spines. Its ovate to elliptic leaves are arranged alternately and are acute at the tip with winged petioles. The white flowers of *C. sinensis* are 5-parted, fragrant, and subglobose in shape. Its smooth fruit has yellow, or sometimes red, pulp that is sweet to somewhat sour. According to Matanane et al. (2014: 114), the two most common cultivars of *C. sinensis* in the Pacific region are the "navel orange" which is seedless, and the "Valencia orange," which has a dark orange rind and juicy fruit. Under tropical climatic conditions, such as in Micronesia, the skin color of *C. sinensis* remains green or can change to yellow, but not orange.

Distribution: Native to South China or Southeast Asia, *C. sinensis* has been introduced to a variety of Pacific islands after European contact (Clarke and Thaman 1993; Smith 1985). Today, in the tropical Pacific islands, including Micronesia, it is cultivated in subsistence garden areas, fallow forests, village tree groves, and home gardens. *C. sinensis* occasionally becomes naturalized in disturbed areas and is normally protected when clearing new food gardens. In addition to its juicy fruit pulp, this plant has other uses in the Pacific Islands where it occurs; its timber can be used in light construction, and it is considered to be a good firewood. Parts of *C. sinensis*, including the fruit, have been used medicinally is some of these islands as well as for a snack food.

Uses in Pohnpei: *C. sinensis* is cultivated on Pohnpei for its fruit and "was introduced [into Pohnpei] from the Philippines, Borneo, and Hong Kong during the German and Japanese periods" (Balick et al. 2009: 493; also see Ragone et al. 2001).

▸ **Koahi** (P)
Claoxylon carolinianum

Description: Endemic to Pohnpei, *Claoxylon carolinianum* is a relatively small understory tree up to about 10 m (~30 ft) tall, known in Pohnpeian as **Koahi**. *C. carolinianum* generally develops a roundish crown. Its mostly whitish to greenish flowers are small and rather inconspicuous; they are borne on spikes emerging from the axils of the leaves. The oblong-elliptical leaves are about 8–16 cm (~3–6 in) long with slightly serrated margins.

Distribution: The genus *Claoxylon* was first described in 1824 and is now known to be found in paleotropical areas from Madagascar to South Asia, Southeast Asia, Melanesia, and other parts of the tropical Pacific. There are reportedly somewhere between 80 and 100 species in the genus *Claoxylon*, which range from dioecious subshrubs to small trees. *C. carolinianum* is a tree species known only from the lower regions of the upland native forest on Pohnpei and is reportedly one of three endemic species in Micronesia; the other two Micronesian *Claoxylon* species, according to Costion and Lorence (2012: 77), are *C. longiracemosum*, endemic to Palau, and *C. marianum*, endemic to the southern Mariana Islands.

General uses: *C. carolinianum* is a species in Euphorbiaceae, the spurge family, which is characterized by producing milky latex that often contains a large variety of phytotoxins (toxic substances produced by plants). These potentially harmful substances include diterpene esters, alkaloids, glycosides, and ricin-type toxins. However, several Euphorbiaceae species, including *C. carolinianum*, also have medicinal associations.

Uses in Pohnpei: Many decades ago, Kanehira (1931: 289; 1935: 348) identified the Pohnpei endemic *Claoxylon* species as *C. longifolium* (Bl.) Miq., and reported that it was a "medium-sized tree," with poisonous latex. Kanehira indicated that it was common in the Pohnpei "forests from low to medium altitudes" and listed its native Pohnpeian name as '*Koy*' or '*Kôee*'. This unique Pohnpeian woody species, **Koahi**, is now recognized as *C. carolinianum*, an understory tree endemic to Pohnpei which is believed to have special attributes and supernatural properties on this island. For example, it has reportedly been used magically to strengthen machetes and knives when these cutting tools are put into cut stems of *C. carolinianum* trees. Parts of this small tree (cut pieces of the branches and stems) are also applied to fishing nets to make them more efficient. Balick et al. (2009: 372) referred to a number of therapeutic uses of *C. carolinianum* (**Koahi**) on Pohnpei:

> "To treat *katieni* (a form of ghost sickness)... To treat *soumwahu en naniak* (mangrove sickness)... To use as a *wini en kamwakel* (a plant to cleanse the body)—for example after drinking too much *sakau* or for a bad stomach... To treat headache... To treat a skin virus (possibly caused by the *Molluscum contagiosum* virus)... [and] To reduce high fever..."

▸ **Ilau** (P) (Mw) (Pi), **Manu mmala** (Nu), **Hia** (Ka)
Clerodendrum inerme

Description: *Clerodendrum inerme* (accepted as *Volkameria inerme* by some botanists) is a sprawling or climbing shrub that grows to heights of 2–8 m (6–28 ft), with woody, smooth stems up to about 3cm (>1 in) wide. The simple leaves are oppositely arranged and elliptical in shape, with an entire margin and an acute to acuminate tip; they are about 5–10 cm (2–4 in) long, more or less glabrous, with lateral, pinnate venation. *C. inerme* produces attractive, whitish flowers 3.5–5.5 cm (1.5–2 in) in length, each with four long, reddish purple stamens extending beyond the petals of the flowers; they are usually comprised of three flowers joined at a common base point. Its fruits are oval-shaped, about 1–1.5 cm (0.5 in) long, and split into four single-seed nutlets when mature.

Distribution: Seaside clerodendrum, as one of its common English names suggests, grows well along tropical beaches where it tolerates the salt spray of the ocean and the harsh rays of the sun. Indeed, this widespread woody plant is an indigenous species found in the coastal vegetation from tropical Asia to western Polynesia, and in all the major island groups of Micronesia, although it may have been introduced by people into some of these Pacific islands areas. It grows at the landward edge of the mangroves, in some swampy areas, along lowland streams, and sometimes in secondary vegetation, as well as in rocky and sandy areas near the shore. *C. inerme* is used in many places in landscaping, as a ground cover or a hedge plant, especially near the sea. However, it should be noted that *C. inerme* may develop "an aggressive growth characteristic, and is likely to become invasive in areas where it does not naturally occur, but where it has been planted," such as in parts of the Hawaiian Islands; in other words, "it has the potential to form dense brambles smothering other plants, and would probably be difficult to remove" (Chethana et al. 2013: 39).

General uses: As a cultivated hedge plant, *C. inerme* has been used traditionally for ornamental purposes in home gardens, largely because of its profuse white flowers that form in clusters and are accented by delicate purplish protruding stamens. *C. inerme* is also used for a variety of medicinal purposes in various areas of tropical Asia; for example, it is used widely in India in Ayurveda and Sidha healing traditions, and where it has an incredible diversity of local names (e.g., see Quattrocchi 2012: 1013). It is also used medicinally and planted as an attractive ornamental shrub in several tropical Pacific islands, including many in Micronesia. Some of its traditional uses, such as in wound treatments, are to some degree supported by experimentally observed effects of scientific extracts; however, there are "very few biological studies available on bioactive components of this plant" (e.g., see Chethana et al. 2013: 39). Several medicinal applications in Micronesia are associated with use of *C. inerme*; e.g., www.comfsm.fm/~dleeling/ethnobotany/text/ethnobotany.html.

Uses in Pohnpei: *C. inerme* is known in Pohnpeian as **Ilau**. Its attractive flowers are used in body decoration, and parts of the plant are used in many traditional medicinal preparations on Pohnpei. For example, a Pohnpeian student in an ethnobotany class at the community college on Pohnpei reported in 2004 that **Ilau** "is used in medicine, decoration, and the black arts." More specifically, this student reported that flowers and/or leaves of *C. inerme* have been used "to treat fever, skin rash, flu, headache, infected umbilical cord, eye infections, evil spirit prevention." As an ornamental which can grow from cuttings (cloning), *C. inerme* is used for decorative purposes because it is "very tolerant of shaping and trimming to a particular shape" (www.comfsm.fm/~dleeling/botany/1998/vhp/geenilau.html). Furthermore, Balick et al. (2009: 405) provided the following discussion of the uses of *C. inerme* (**Ilau**):

> "The leaves and flowers are used in *mwaramwar* (head garland), put on a women's comb to be worn in the hair, or worn directly in the hair... To use as a charm to keep a husband and wife together forever... to treat *soumwahu en eni* (ghost sickness)... To treat uterine infections... Alternatively to reduce high fever... To treat headache... To treat headache and weakness... To treat a newborn baby's belly button... to take away the evil spirit that holds the baby back during labor... To treat coughing or a sore throat... to treat earache... To treat *riphwin karakar* (characterized by rash that children get between the legs, under the arms, and/or on the buttocks, with a fever)... To cool and treat *inenkil* (cellulites)..."

More recently, another student from Pohnpei shared an additional Pohnpeian use of **Ilau**: "to treat dark spots on the skin, **Ilau** is mixed with copra scrapings and rubbed on the skin daily." Furthermore, yet another student from Pohnpei "noted the use of a mixture of crushed **Ilau** and coconut milk to produce a hair conditioner that yields long, silky black hair" (see danaleeling.blogspot.com/2016/09/plants-that-heal.html).

Uses in Pingelap: *C. inerme* is also known in both Pingelap and Mwoakilloa atolls as **Ilau**, as it is in Pohnpeian. A Pingelapese student in an ethnobotany class at the community college on Pohnpei reported in 2004 that Pingelapese people use *C. inerme* "for fever and pink eye," and more specifically that the leaves of this plants are used for treating "fever and cough." This student also indicated that the leaves of *C. inerme* "are also used in conjunction with other plants leaves for ghost diseases" (www.comfsm.fm/~dleeling/botany/1998/vhp/geenilau.html).

▸ **Kotop** or **Koatop** (P), **Koasop** (Mw)
Clinostigma ponapense

Description: Despite its small size, Pohnpei has a diverse native palm flora. It is home to five native palm species, two of which are endemic (found nowhere else but on Pohnpei); these unique Pohnpeian species are *Ponapea hosinoi*, known in Pohnpeian as **Kedei** (discussed later in this chapter), and *Clinostigma ponapense*, known in Pohnpeian as **Kotop** or **Koatop**, which is described here. *C. ponapense* is a rapidly growing, tall, attractive tree with leaves that look somewhat like those of the introduced betel nut palm (*Areca catechu*, **Pwuh**). The native *C. ponapense* palm tree produces a solitary whitish-colored trunk as an emergent woody species in the upland vegetation and can grow to heights of about 20 m (65 ft) or more. It produces a light green crown shaft, and its fronds have pinnately compound, fine, drooping leaflets. When mature, *C. ponapense* develops a mass of stilt roots surrounding the trunk, as well as distinctive flowering stalks (panicles) that grow below the crown of fronds. The flowering stalks of *C. ponapense* are yellowish and branched, with the youngest flowers towards the tip. The black fruits are oblong and strongly compressed. *C. ponapense* palm trees are generally larger than the two native species of *Ponapea* palms (the endemic *P. hosinoi* and the indigenous *P. ledermanniana*). The black fruit and tapering leaflets of the **Kotop** (*C. ponapense*) palm trees differentiate them from the native **Kedei** (*Ponapea* spp.) palms.

Distribution: *Clinostigma* is a genus of about 11 species endemic to the tropical Pacific islands (Costion and Lorence 2012: 64). *C. ponapense* is only found on Pohnpei so is classified as an endemic species on this island. This relatively common palm is often the dominant plant in some areas of the humid Pohnpei mountain forests, especially in wetter, foggy areas. *C. ponapense*'s biogeographical range in Pohnpei is between elevations of about 450 m and 600 m above sea level in the wet, upland "palm forests" (i.e., uncultivated areas in primary forest, the *nanwel*).

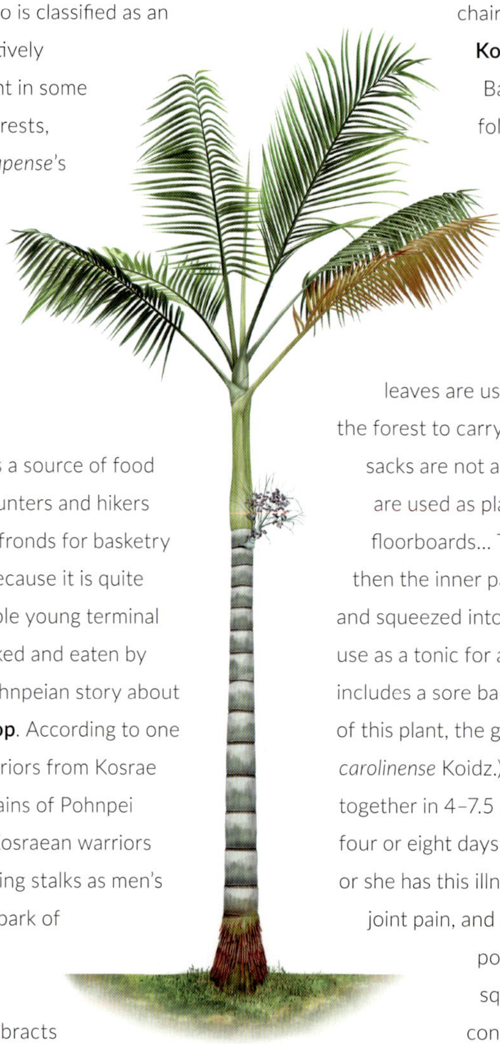

Uses in Pohnpei: The **Kotop** palm tree is a source of food and medicine. Today, it is also used by hunters and hikers in the upland forest as a source of palm fronds for basketry and for roofing on temporary shelters because it is quite plentiful. The heart of palm (i.e., the edible young terminal bud or "cabbage") of **Kotop** is often cooked and eaten by forest visitors. There is a well-known Pohnpeian story about the light-colored flowering parts of **Kotop**. According to one interpretation of this story, invading warriors from Kosrae saw many palms flowering in the mountains of Pohnpei as they approached the island; but the Kosraean warriors mistook these many light colored flowering stalks as men's skirts made from the fibers of the inner bark of **Keleu** (*Hibiscus tiliaceus*); consequently, the Kosraeans retreated. Traditionally, Pohnpeians have used the large, fibrous bracts that hold the infloresences of *C. ponapense* for carrying things, e.g., wild yam. Children use the crown shaft as a toy; for example, one child will get onto the large bract and the other kids will pull it, or if it is on a hill, they will just sit in it and use it as "skate" to slide downhill. Pohnpeians make a shelter of the **Kotop** fronds by simply gathering the large leaves of this palm and placing them over a line of cordage or over a long piece of wood placed about 1–2 m (4–5 ft) above them and horizontal to the ground. In the past when people went into the upland areas of Pohnpei, and even today when some people go into the forest to work, they would not bring any food or drinks with them because they knew that their needs would be met by the natural resources of this rich environment. The **Kotop** palm is still one of those plentiful plants that people in the Pohnpeian forest depend on for constructing shelter from rain, providing food, and for drink. It is also a food source for some of the animals of the upland forests, such as introduced pigs and deer. In addition, Pohnpeians use the wooden inner part of the **Kotop** trunk for flooring in their local huts, and this same trunk wood has been used to construct walls, shelves, chairs, tables, and many other things. Finally, **Kotop** reportedly has some medicinal uses. Balick et al. (2009: 261) provided the following description of many uses and cultural associations of this important endemic palm on Pohnpei:

> "The palm heart (the apical growing point of the stem) is cooked and eaten as a food… The leaves are used to weave baskets… that are made in the forest to carry products being gathered when other sacks are not available… The leaves of juvenile plants are used as plates… The stems are split and used as floorboards… To guard a child's soul, four roots are split, then the inner part is ground, put into the hand scrubbed and squeezed into a cup, and consumed by the child… To use as a tonic for an unspecified man's sickness, which includes a sore back, sore joints, and no energy, four roots of this plant, the ground back of **Madeu** (*Cinnamomum carolinense* Koidz.), and other unspecified plants are boiled together in 4–7.5 L (1–2 gal) of water and consumed for four or eight days (do this yearly) whether a person feels he or she has this illness or not… To treat back pain, headache, joint pain, and muscle pain, four or eight roots are pounded, wrapped in a piece of cloth, squeezed in 1 L (0.25 gal) of water, and consumed (1 L daily) for four of eight days…"

▸ COCONUT PALM

The following section discusses the coconut palm, undoubtedly one of the most important plants for people living in Micronesia. This large species also has a great array of uses in the rest of the Pacific and the world's tropical regions in general. We provide significantly more information about the botanical, geographical, and ethnobotanical characteristics of the coconut palm tree because its importance in many islands is so huge that it is often referred to in English as the "tree of life."

▸ **Nih** (P), **Ni** (Mw) (Pi), **Nui** (Nu), **Niu** (Ka)
Cocos nucifera

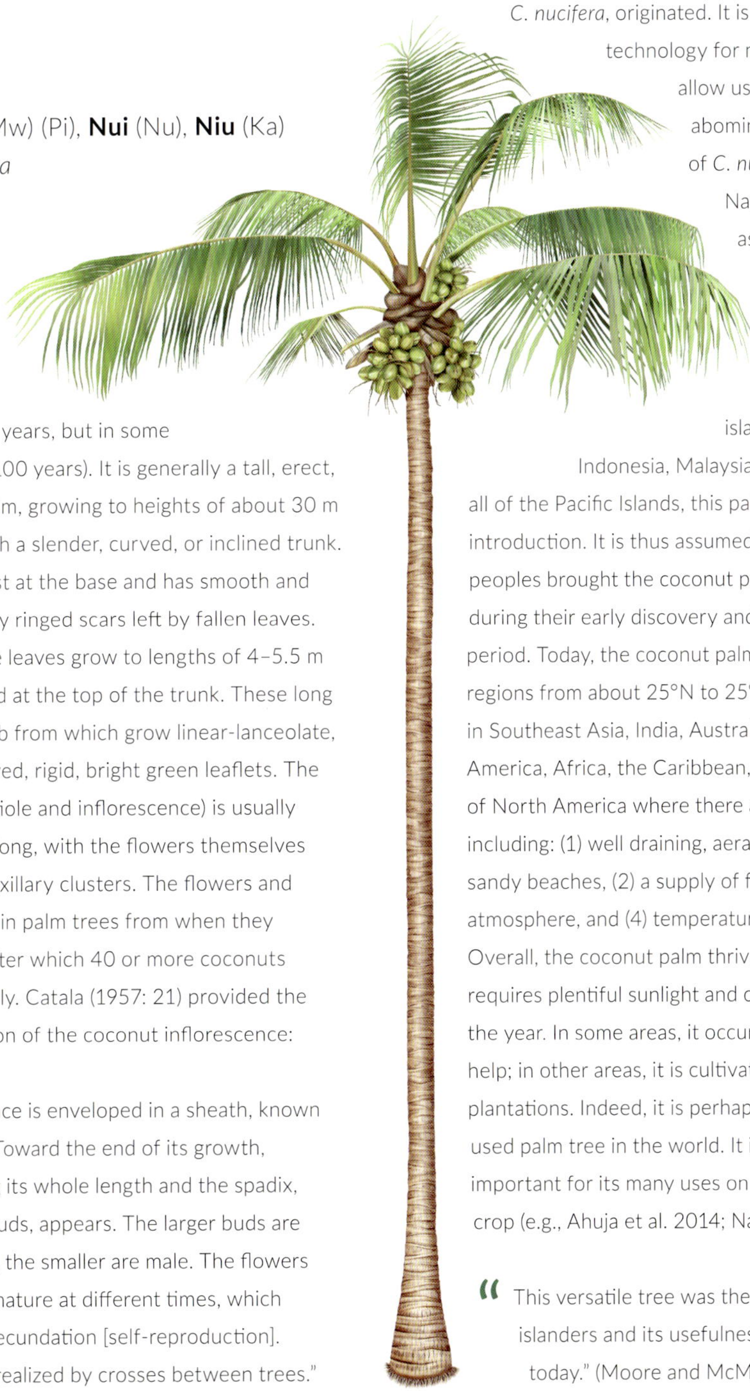

Description: The coconut palm (*Cocos nucifera*) is a long-living plant (normally about 70 years, but in some cases up to about 100 years). It is generally a tall, erect, single-stemmed palm, growing to heights of about 30 m (90–100 ft) tall, with a slender, curved, or inclined trunk. The trunk is thickest at the base and has smooth and gray bark marked by ringed scars left by fallen leaves. The palm frond-like leaves grow to lengths of 4–5.5 m (12–15 ft), clustered at the top of the trunk. These long leaves have a midrib from which grow linear-lanceolate, more or less recurved, rigid, bright green leaflets. The flowering stalk (petiole and inflorescence) is usually about 2 m (6–7 ft) long, with the flowers themselves produced in large axillary clusters. The flowers and fruits may develop in palm trees from when they are six years old, after which 40 or more coconuts are developed yearly. Catala (1957: 21) provided the following description of the coconut inflorescence:

> "The inflorescence is enveloped in a sheath, known as the spathe. Toward the end of its growth, this splits along its whole length and the spadix, covered with buds, appears. The larger buds are female flowers, the smaller are male. The flowers of both sexes mature at different times, which prevents autofecundation [self-reproduction]. Fertilization is realized by crosses between trees."

Coconuts generally ripen in 9–10 months. A large, roundish coconut fruit (technically a drupe) can be as big as a man's head and 1–2 kg (2–5 lb) in weight, with a thick, fibrous husk surrounding a hard endocarp (nut). Many cultivated varieties of this important crop plant have been recorded from Micronesia; some of those in Pohnpei and the outer islands are identified and briefly discussed below.

Distribution: For many decades, scientists and scholars have attempted to determine where in the world the coconut palm, *C. nucifera*, originated. It is possible that "ever-advancing technology for molecular genetic analysis may allow us to once and for all solve the abominable mystery" of the origins of *C. nucifera* in this century" (see Nayar 2017). The coconut palm is assumed by many, but certainly not all, to be a native of the biogeographical region known as Malesia, which encompasses most of the islands of Southeast Asia, specifically Indonesia, Malaysia, and the Philippines. In most or all of the Pacific Islands, this palm was probably a pre-European introduction. It is thus assumed that early Pacific island voyaging peoples brought the coconut palm to the Micronesian islands during their early discovery and colonization in the prehistoric period. Today, the coconut palm is found in many tropical regions from about 25°N to 25°S. The palm can be found in Southeast Asia, India, Australia, the Pacific Islands, South America, Africa, the Caribbean, and the southern extremes of North America where there are ideal growing conditions, including: (1) well draining, aerated soils normally found on sandy beaches, (2) a supply of fresh groundwater, (3) humid atmosphere, and (4) temperatures between 27°C and 30°C. Overall, the coconut palm thrives on sandy, saline soils and requires plentiful sunlight and consistent rainfall throughout the year. In some areas, it occurs spontaneously without human help; in other areas, it is cultivated in gardens or large scale plantations. Indeed, it is perhaps the widest grown and most used palm tree in the world. It is very common and especially important for its many uses on atolls, often as the only cash crop (e.g., Ahuja et al. 2014; Nayar 2017; Small 2017).

> "This versatile tree was the stuff of life for most Pacific islanders and its usefulness is still greatly appreciated today." (Moore and McMakin 1979: 119)

General uses: *C. nucifera* is one of the most useful plants in the tropical Pacific. It is featured in traditional mythology, legends, songs, proverbs, and riddles throughout the region and has great ceremonial and high-rank importance in Polynesia and Micronesia. The famous 19th-century American author Herman Melville referred in awe to the "Cocoa-Palm" in his second book, Omoo, published in 1847. According to Melville, this palm "is the most important tree by far" to the Polynesians, claiming that it is "emphatically the Tree of Life: even the bread-fruit in the multifarious uses to which it is applied" is less significant to Pacific islanders than the coconut palm. Melville (1847: 271) provided an articulate description of this palm's imposing growth and innumerable uses based on his experiences in the Polynesian islands of the Society and Marquesas groups:

> "Its very aspect is imposing. Asserting its supremacy by an erect and lofty bearing, it may be said to compare with other trees as man with inferior creatures. The blessings it confers are incalculable. Year after year, the islander reposes beneath its shade, both eating and drinking of its fruit; he thatches his hut with its boughs, and weaves them into baskets to carry his food; he cools himself with a fan platted from the young leaflets, and shields his head from the sun by a bonnet of the leaves; sometimes he clothes himself with the cloth-like substance which wraps round the base of the stalks, whose elastic rods, strung with filberts, are used as a taper; the larger nuts, thinned and polished, furnish him with a beautiful goblet: the smaller ones, with bowls for his pipes; the dry husks kindle his fires; their [fibers] are twisted into fishing-lines and cords for his canoes; he heals his wounds with a balsam compounded from the juice of the nut; and with the oil extracted from its meat embalms the bodies of the dead. The noble trunk itself is far from being valueless. Sawn into posts, it upholds the islander's dwelling; converted into charcoal, it cooks his food; and supported on blocks of stone, rails in his lands. He impels his canoe through the water with a paddle of the wood and goes to battle with clubs and spears of the same hard material."

The coconut palm, with more than 300 uses including the provision of food for millions of people, has been referred to by many as the "world's most useful plant" (e.g., Chan and Elevitch 2006: 277; also see Ahuja et al. 2014, Nayar 2017, Small 2017). We may add here to what the 19th-century author Melville described so well about the usefulness of this palm: for example, the large roundish coconut fruit (drupe) has a thick, fibrous husk surrounding a hard endocarp (nut) which contains the firm edible pulp (endosperm, "coconut meat") that is white and oily. When the fruit is young, the nut contains sweet water, which can sometimes be life-saving when long droughts occur. In addition, the coconut husk is a dense mass of fibers called "coir," which can be woven into strong twine or rope and has been used for many things such as padding mattresses, upholstery, and life preservers. The fibers are resistant to seawater decay and therefore can be used in cables and rigging on ships, as well as for production of mats, rugs, bags, brooms, and brushes. Historically, coconut palms have been one of the main sources of income for producing countries, with many different products made out of fruits that are utilized and appreciated among various cultures around the world. In addition, its leaves have long been used make baskets, roofing thatch, etc. and the apical buds of the adult plants are a first-rate palm-cabbage; furthermore, an alcoholic beverage known as "toddy" or palm wine is extracted from its sugar sap, which is tapped from its inflorescences:

> "In Micronesia the sweet sap, or toddy, which runs from the young, broken inflorescence of the [coconut] palm is an important item of diet. The inflorescence, while still in the spathe, is shaped into a long cylinder by biding of plaited coconut fibers. The end is cut off daily to allow the sap [*sikaliwi* in Pohnpeian] to drip from the cut surface into a coconut shell or bottle suspended from the extremity. The yield is approximated [to] two pints in 24 hours. The fresh toddy, which contains 15 to 16 percent sugar, is diluted with water for a beverage. If allowed to stand, [the] toddy ferments rapidly, and within about 15 hours contains 6 percent alcohol... When concentrated over a low flame, [the] toddy becomes a type of molasses, known as [*malahsis* in Pohnpeian]. This is used in a number of dishes."
> (Barrau 1961: 30–39; also see Clarke and Thaman 1993)

In the following discussion about the multipurpose coconut palm (*C. nucifera*) in Pohnpei and the outer islands, we first focus on the traditional uses of this remarkably useful tree. Then we present a short history of the origins, development, and decline of commercialized coconut plantations in Micronesia, and how this history relates to Pohnpei. Finally, we briefly describe recent conservation and revitalization efforts in this region that are focused on this keystone biocultural species.

Uses in Pohnpei: The coconut palm is known as **Nih** in the Pohnpeian language. Its leaves are used to make baskets and other woven implements for everyday use and handicrafts. For instance, these leaves are used in the making of a local purse called a *kopwou*. The body of the purse consists of woven leaves of **Nih**, the wood sticks at the top of the purse are made from the wood of **Ketieu** (*Ixora* spp.), and the braided outer bark of **Keleu** (*Hibiscus tiliaceus*) is used to make the strap for the bag. Coconut palm leaves are also used for house thatch on Pohnpei, even though the leaves of **Parem** (nipa palm) and **Oahs** (ivory nut palm) provide more durable leaves for this purpose. In addition, coconut palm leaves are also used to cover the stone oven (*uhmw*), as well as for shading young plants, and as a torch at night (e.g., see Balick et al. 2009).

Coconuts, the fruits of the palm, provide copra (dried coconut "meat"), a major commercial product of Pohnpeian farmers. In the past, the main cash crop in most of Micronesia was copra, which was shipped out of the islands to be processed into extracted coconut oil. During the latter part of the 20th century, low prices worldwide led to a huge reduction of copra production in the islands of Micronesia, and it has remained low even with government subsidies. This recent history of massive declines in copra production has been especially distressing for local farmers on the outer islands for whom coconuts were an important source of income. Even today on Pohnpei Island, because of the relatively low price of copra, the nuts are more commonly used to feed pigs (additional relevant copra history is presented on pp. 195–198). On the other hand, grated coconut meat is pressed for cream that is still an important cooking ingredient on Pohnpei, especially with fish, breadfruit, and various root crops:

> "Coconut cream is prepared by scraping the meat of a mature coconut out of the shell, then squeezing it through a cloth or plant fibers. Both *Hibiscus* bark and coconut husk can be used as the bast fibers through which coconut milk is squeezed from the grated coconut." (See Balick et al. 2009: 78).

Interestingly, coconut "oil" was reported to be a special, traditional ingredient in the basic marriage ceremony in Pohnpei:

> "The formality of marriage between young people is singular. The girl is brought into the house, and sits down whilst here future mother-in-law rubs coconut-oil vigorously into her back and shoulders. This is called *keiiehdi* or anointing. A garland of flowers is placed on the head, and the ceremony is concluded by a feast." (Christian 1899: 113; note: *marekeiso* = to anoint specifically with coconut oil; and *keii* = to have coconut oil applied on skin as a medicine)

Other examples of the significant use of the coconut palm on Pohnpei highlight the diversity of this species' resources, which include the following: the *pohndahl* (shell or endocarp of the coconut) is used as a source of charcoal for cooking, and it is also utilized to make the *ngarangar* (cup) for drinking the traditional beverage of *sakau*; in addition, the key cordage, i.e., the traditional, multipurpose coconut sennit rope or twine, is made from fibers of mature coconut husks. Among the various other uses for coconut palms in Pohnpei are its diverse medicinal applications. Some practical and therapeutic uses of the coconut palm on Pohnpei have been provided by Balick et al. (2009: 263–265):

> "The liquid inside the fruit is used as a beverage... Toddy, a fermented beverage, is also collected from the young inflorescences of this palm by cutting them before the bracts open and collecting the liquid, then allowing it to ferment. The endosperm of the fruit is harvested and dried in the sun to make copra, to extract oil. The fiber from the leaf base is used as a cloth in which to put medicinal herbs (*inipal*), and squeezed or rubbed on the skin. As a hex to help remove the teeth of a dog that bites a person, the flowers are pounded and rubbed on the wound where the dog has bitten the person... To treat dysentery, four flowers are pounded and wrapped in a piece of cloth, squeezed in the liquid of the fruit, and consumed. This treatment is repeated daily until better... To treat 'uterine cancer', the younger, very small fruits are chewed and inserted into the uterus to help fight the 'cancer'... To use as an eyewash to relieve a painful eye caused by sunlight or some dirt in the eye, the inner (beige) bark of the leaf stalk (the ones that just grow to about 1 m tall) is scraped into a piece of cloth and squeezed into the eye... To treat *lusulus* (seizures), the roots are pounded and squeezed in 1 lit (0.25 gal) of water and consumed..."

Coconut beverage uses on Pohnpei have been significant traditionally. Coconut "water" seems to have been the first "food" that newborn Pohnpeian babies received from their mothers after birth. According to Englberger et al. (2003c: 8):

> "... in the past, mothers did not give their breast (and colostrum [i.e., the first secretion from the mammary glands after giving birth, rich in antibodies]) immediately to the newborn, thinking that it was inadequate. Young coconut water was given until the milk came so that coconut oil lubricant was often given as the first feeding." (Also see Miller 1953: 19–23).

Young coconuts (*uhpw*) are gathered for popular coconut juice drinks, and even sold in some places in recent times. In the middle of the 20th century, Glassman (1952: 15) referred to the coconut

A child cuts green coconuts from the top of a palm at Pohnrakied village, Pohnpei.

Grinding coconuts on Mwoakilloa Atoll. This activity is essential to traditional cuisine, subsistence practices, and cultural activities across Micronesia.

Helping a younger sibling stay hydrated and enjoy fresh coconut water on Pingelap Atoll.

Locally produced coconut oil sold in Kolonia, for cooking as well as cosmetic use, including some infused with flower scents.

palm as a source of alcoholic beverages on Pohnpei, with notable commentary regarding the previous prohibition of distilled brews on this island during the mid-20th century: "Several kinds of beverages are drunk... Fermented drinks are used in only a few localities; and distillation of beverages has been declared unlawful since the American occupation." Interestingly, he wrongly assumed that drinks made from *Rahmedel* and *Rahmwanger* (main varieties of **Sakau**) are "fermented drinks prepared from the sap of the coconut inflorescence..."

Cultivars in Pohnpei: The generic name **Nih** refers to the many cultivated varieties developed locally over many centuries. Additionally, it also includes varieties introduced to Pohnpei in more modern times. For example, at least one cultivar was reportedly brought to Pohnpei from Sulawesi (Indonesia) by the Japanese in the early part of the 20th century for food and beverage (Ragone et al. 2001). Please see p. 426 for our list of coconut palm varieties grouped in several cultural categories.

Uses in Mwoakilloa: The coconut palm is known as **Ni** in Mwoakilloa. Murphy (1950: 61–62), in his "economic geography" written in the middle of the 20th century, noted that coconut palms were "by far the most common tree" on the islets of this atoll, and where "coconut land" made up "by far the greater proportion" of the two major agricultural land classes. Although, during much of the 19th and 20th centuries, coconut palms dominated most land areas on Mwoakilloa because of their commercial value for copra production, Murphy (1950: 65–70) argued that these palms were a crucial plant resource well before the economic shift on this atoll from a strictly subsistence economy to one involving considerable reliance on commerce. Essentially, trade focused mainly on copra production and money exchange to secure a variety of foreign products. Murphy described in some detail how and why coconut was still "the most important tree" on the atoll when he visited, despite the changing economy and cultivated landscape after foreign contact. He included notes on the various phases of palm and fruit production along with processing of the coconuts for trade with visiting commercial freighters. Murphy (1950: 76) also referred to declining but still existing traditional uses of the coconut palm on Mwoakilloa; for example, he indicated that parts of native sailing canoes, as well as the best local-made hats, fans, and belts, were still being made on Mwoakilloa using parts of this multipurpose tree.

Glassman (1953: 298), in addition to declaring that the coconut palm was by far the most common and useful tree on Mwoakilloa Atoll with "all parts of the plant being used," indicated that most of the coconut palm tree varieties "were planted by the early inhabitants." Glassman (citing Christian 1899) noted that consumption of coconut "toddy" (an alcoholic beverage created from the sap of the large coconut inflorescence) on Mwoakilloa had been "prohibited by the American missionaries." More recent commentary regarding the importance of coconut liquid in the lives of people living on Mwoakilloa Atoll is provided by Taboroši and Sánchez Collazo (2011: 95–96):

> "Water content of fruits used in the local diet are an additional source of hydration. Juicy fruits, such as papayas and citrus, provide water and nutrients when regularly consumed. Coconuts, however, are by far the most important. They grow year-round, bear abundant fruit, and [each] contain around 300 ml (10 fl. oz.) of liquid that is an excellent source of water and electrolytes. Ample supply of coconuts can sustain proper hydration even if no other sources are available. Coconuts are plentiful on Mwoakilloa and are regularly but not intensively used for drinking. There are different cultivars (varieties). Some are sweeter than others, and some grow on shorter trees. Residents report consuming less than one coconut per individual on an average day and perhaps 4–5 per week. The actual number depends on personal preference and also ability to climb trees. In the words of one of our informants, "There is no limited amount of coconuts, it's just that it's hard to get them." Coconuts are, therefore, not conserved and palms are said to produce just as many nuts as in the past. Children may consume more coconuts than adults."

Taboroši and Sánchez Collazo (2011: 95–96) also referred to the use of coconuts on Mwoakilloa in many "celebrations and meetings" and noted that they are "readily brought to visitors." However, these authors also noted how droughts and storms affect the availability of coconut liquid on this atoll: "... although coconuts can be an important component of usual hydration and even cooking needs of islanders, it cannot be considered an alternative to safe drinking water and definitely not a reliable emergency source." Furthermore, Taboroši and Sánchez Collazo (2011: 96) described the periodic problems associated with scarcity of rainfall, gale winds, and overall the reliability of coconuts as a source of drinking water:

> "Prolonged droughts are known to gradually reduce quantity or quality of nuts. Major storms can virtually eliminate standing crop of coconuts in an instant. Residents report that major storms damage almost every coconut palm and drinking nuts cannot be found for several months afterward."

Recent documentation at the College of Micronesia, by a student from Mwoakilloa, refers to the continued use of the *kiam*, a basket made from coconut leaves and utilized for a wide variety of purposes (see www.comfsm.fm/~dleeling/ethnobotany/63material_culture.html). Another student from Mwoakilloa reported the following medicinal usage on that atoll:

> "The brown spots on the petiole [of the coconut palm leaf] can be used to stop bleeding of small cuts... scrape these off of the petiole. Another treatment for cuts can be obtained from *koalo*, [which are] aerial roots that descend from the external base of a coconut palm. The tip end of this root has a white spot. Clean it, remove it, smash it, apply to the cut. This works on larger cuts than those treated by the brown spots on the leaf stem base." (www.comfsm.fm/~dleeling/ethnobotany/healing_plants_presentations.html)

Cultivars in Mwoakilloa: Murphy (1950) and Glassman (1953) recorded the existence of a number of coconut varieties in Mwoakilloa. We have checked them against information from the Mokilese-English Dictionary (Harrison and Albert 1977) to create the comprehensive list on p. 426.

Uses in Pingelap: The coconut palm is known as **Ni** in Pingelapese. In the middle of the 20th century, according to St. John (1948), coconut palms were abundant on Pingelap Atoll, where he observed them being cultivated and growing spontaneously (i.e., self-sown). Damas made research visits to Pingelap Atoll from 1975–1983, and based on these, he noted (1994: 54–55) that coconut palms could "be found almost anywhere" on that atoll, occurring spontaneously as well as cultivated. Damas also reported that the coconut palm was "the third important crop for food" on Pingelap (after giant swamp taro, *Cyrtosperma merkusii*, and breadfruit trees, *Artocarpus* spp.). In addition, Damas indicated that the "primary importance" of coconut in the Pingelap community was consumption of its liquid as "a natural beverage" and copra serving as "the only cash crop." Damas described standard planting practices for coconut plantations on Pingelap during the time of his research:

> "With regard to planting techniques, I had sometimes noticed fallen brown coconuts lying in neat rows of eight or ten, in groups of 70 or 80, with roots anchoring them to the ground, and with shoots beginning to sprout from the nuts [and after the coconuts had germinated] with the new plants feeding on the embryo within the mature nuts, and with roots then beginning to provide nourishment from the ground, the nuts are replanted at various places on the islands [islets of the atoll]."

Damas reported that coconut trees on Pingelap Atoll normally bear "copra nuts" within about five years and continue to produce effectively until they reach about "50 or 60 years old." Although coconuts were an important item in the Pingelapese diet in the past, by the time Damas made his visits to Pingelap, "store-bought foods and beverages had to a large extent replaced them," but coconuts did remain a source of "cream sweetened with sugar, which is applied to a number of recipes involving taro, breadfruit, or bananas." However, Damas also pointed out that even though introduced beverages such as coffee, tea, canned or powdered milk, and soft drinks had more or less replaced the use of coconuts for liquid refreshment, "the green [coco]nut is still important as a beverage... especially when people are working at some distance from the village, at which times the drinking nut supplies the chief source of fluid and the jelly of the green nut provides a daytime food ration." According to Damas, these "drinking nuts" are also used in "food-distribution ceremonies" on Pingelap. Numerous and diverse uses can be gleaned from the Pingelapese-English dictionary (Hattori 2012) and the many unique terms such as *dal* (coconut shell, container, plate or dish), *dahloang* (canoe roller, usually the primary rib of coconut leaf), *seinahk* (coconut leaf mat), *pwas* and *poa* (different basket of coconut leaves), *pwelekaelaekael* (coconut debris piles used as boundary markers), *penikae* (vinegar made from coconut, lime, and salt), *do* (coconut sennit soaked in the ocean to make rope), etc.

Cultivars in Pingelap: There are at least five coconut palm cultivars in Pingelap (see p. 427), including the exceptionally short ***Nihkusa*** and the unusual ***Adoahl***, a cultivar with a sweet, edible husk. In the words of Hattori (2012: 253):

> "People peel the husk with their teeth (*esies*) and chew [a] piece of the husk. They bite the husk by sticking

their lower jaws into the husk [and] peel the husk upward. Juice from ***Adoahl*** stains your clothes."

The Pingelapese language also has a rich vocabulary to distinguish coconut fruits by their maturity. St. John (1948) recorded three stages, while Damas (1994: 232–233) received from his Pingelapese informant, Albert, the names of nine stages. We used the Pingelapese–English dictionary (Hattori 2012) to expand the list and use up-to-date spellings, and then turned to other available dictionaries and unpublished research conducted by iREi with local informants and other sources (see p. 13) to find equivalent terminology from other Pohnpei State atolls, all presented in table 3.2.

Uses in Kapingamarangi: Hiroa (1950) documented a wide variety of traditional uses of the coconut palm for food, drink, and other aspects of the material culture related to this very important tree among the people inhabiting Kapingamarangi. Based on ethnobotanical information from a series of sources (Wiens 1956, 1962; Niering 1956, 1963; Emory 1965), Bayliss-Smith (2012: 110), listed the following crops (staple foods) cultivated as of 1953 to supply the basic energy needs (caloric value) for the atoll population "in rank order" on Kapingamarangi during the mid-20th century: (1) coconut, (2) breadfruit, (3) giant swamp taro, (4) taro, and Bayliss-Smith also provided the "probable pre-contact staple foods, ranked according to calories provided" as: (1) coconut, (2) breadfruit, (3) taro, (4) fish. The Kapingamarangi Lexicon (Lieber and Dikepa 1974) provides wonderful insights into the importance and local uses of coconuts through the rich vocabulary it contains. It includes the names of distinct maturity stages of coconut fruits (see table 3.2); names of cultivars (see p. 427); names of coconut parts: *agaanga / duuaganga* (coconut husk), *bulu* (part of the husk with soft fiber), *loi niu* (coconut leaf), *duaaniu* (leaf spine), *hahainiu* (midrib of leaf frond), *dila* (coconut leaflet), *ibu* (coconut shell), *golooloo* (coconut stalk), *bagaliha* (area of the coconut where stem is attached), *mugamuga* (soft part near the stem of a young coconut), *dagaluu* (sap from inflorescence), *doume* (flower sheath), *moigaga* (coconut husk fiber), and *gaga* (cloth-like bark); and many other terms including various objects made from coconut materials: *dabagau* (coconut leaf floor mat), *hakai niu* (coconut leaf twine), *ipmatu* (coconut cup), *lol maduu* (coconut oil), *na* (coconut leaves used for wrapping), *tanga imu* (coconut leaf mat used to cover the oven), etc.

Cultivars in Kapingamarangi and Nukuoro: Names for coconut varieties identified for these two atolls are based on the dictionaries (Carroll and Soulik 1973; Lieber and Dikepa 1974) and are listed on p. 427, but also see Anderson et al. 2013, Drummond et al. 2016, and Stone 1966a.

Historical review of the copra trade in Pohnpei and the outer islands: Over the past two centuries, major changes have occurred in the relationship between many Pacific islanders and their coconut crops, particularly on atoll islands but also on several high islands:

> "In the industrial developed countries of Europe and North America, the coconut had a golden phase for over 100 years from the mid-19th century. The end of the Second World War and its aftermath had signaled the fall of this crop plant. The coconut palm has continued as an orphan crop ever since." (Nayar 2017:xv)

The following historical overview of the rise and fall of the copra industry in Micronesia is based to a large extent on extensive research by F. X. Hezel:

> "The final and most enduring step in pre-colonial trade was the establishment of the copra industry throughout [Micronesia]. Shortly before 1850 coconut oil began to be widely used for the manufacture of soap and candles and its commercial value soared practically overnight. During the early 1850s whaling captains began stopping at islands to collect whatever coconut oil they could during their whaling voyages... By 1870 traders were beginning to collect copra rather than coconut oil, thus eliminating the tedious and wasteful process of pressing the oil before shipment. This had the added advantage of furnishing companies with the copra cake that could be used in Europe as animal feed or fertilizer. The market price of copra at this time was so high that the large European firms that were just beginning to expand into the Pacific largely abandoned the China trade to concentrate on copra... [Eventually, from about 1870 on] large German companies... came to dominate trade in eastern Micronesia. [They] recruited resident traders to supervise their trading stations and purchase copra from the islanders. These traders, generally Europeans or Americans, were provided with a simple home, a warehouse, and a supply of trade goods—at first largely ironware, cloth goods, tobacco—and were instructed to obtain as much copra as they could with their trade stock. The agents were paid a monthly salary or given a stipulated percentage of the profits, depending on the nature of their contract. A company ship would visit them periodically to pick up

the copra and deposit a new stock of trade goods, but irregularity of shipping was a problem then as now... As the market price of copra rose and competition among trading firms grew, price wars between traders often ensued [and then when] the world market price dropped suddenly the following year, the firms found themselves operating at a loss on islands such as Ebon [an atoll in the Marshall Islands]. The trading rivalry also had other marked results. As the usual trade goods grew more abundant and commonplace on the islands, agents were under increased pressure to sell goods that were formerly considered taboo: firearms and liquor especially... As an incentive to produce increasing amounts of copra, firms had to offer an ever more dazzling array of consumer goods. In the Marshalls, at least, simple iron goods and calico dresses were no longer sufficient inducement to make copra. Hence, the stores in Jaluit, reputed to be among the best-stocked in the Pacific, displayed canned foods, ship biscuit, sewing machines, and gold rings on their shelves." (Hezel 1984; also see Hezel 2001, 2013)

Thus, over time during the 19th century, the availability of "Western-made goods," such as canned foods, ship biscuits, modern equipment, personal accessories, and other dry goods encouraged the accumulation and utilization of imported merchandise desired by many Micronesians, including those in Pohnpei and the outer islands. Cash-driven consumerism and basic economic changes had a profound impact on local food consumption and normal everyday life in various areas of Micronesia. These significant lifestyle changes in the late 19th century occurred as societies in many tropical Pacific areas came under the governmental control of German administrations installed as European colonial rulers. Coconut plantation agriculture and its associated commercial copra trade became the economic "backbone." This shifted societies from their traditional agricultural subsistence to a new commercial consumerism economy in many of the tropical Pacific islands, including many in Micronesia. The Westernized copra trade and associated cash system would continue through the second half of the 19th century and most of the 20th century.

Following Spanish control on Pohnpei during the late 19th century, "severe laws were enforced by the German government, making it imperative, under threat of punishment, for each adult male to plant ten coconuts a month" (Migvar 1965: 4). Then, in the early 20th century, after the First World War, the German government surrendered its colonial control, and subsequently, in much of Pacific, the Japanese government usurped power. Under the Japanese occupation of many islands, including Pohnpei and the outer islands, large copra plantations were established in some eastern areas of Micronesia. For example, the Japanese South Seas Government leased land to the Japanese-owned Nan'yō Bōeki Kaisha (South Seas Trading Co.) for the cultivation of coconuts. This company began its operation in Micronesia in 1906, and in 1914 it inherited the extensive interests of the Jaluit Company, which continued to be the major trading company during the Japanese period. Around 1930, this key commercial company leased 2,500 acres of land in Madolenihmw, Pohnpei, and the entire atoll of Pakin and became directly engaged in copra production. Approximately 125,000 coconut palms were planted at Madolenihmw.

Then, when World War II began, the Japanese seized full control of these plantations and changed their economic focus. For example, the Etscheit coconut plantation of about 500 acres near Kolonia on Pohnpei (much if not all purchased from the estate of the Polish immigrant Jan Kubary) "was all cut down during the war and made into sweet potato gardens" (Migvar 1965: 4).

After WWII and the removal of the Japanese administration in Micronesia, copra production continued to be a key commercial activity in terms of tonnage and monetary value from the late 1940s until the early 1970s. According to Migvar (1965: 4–6), copra production in 1938 as recorded by Japanese administration officials for "Ponape District" amounted to 3,578 tons out of a total of 14,620 tons for all Micronesian areas under Japanese control, while after the war, under the U.S. Trust Territorial Administration, tonnage of copra production in 1963 amounted to 3,295 for "Ponape District," about 26% of the total 12,817 tons for all the U.S. Trust Territory of the Pacific.

Table 3.2. Local names of coconut fruit development stages in the Outer Islands of Pohnpei.

Pingelap	Mwoakilloa	Nukuoro	Kapingamarangi
kiripw	*kiripw* (smallest nut stage)	*doloinui* (youngest stage)	
uhpw	*upw* (unripe, all soft shell edible)	*mugamuga* (first stage of a green coconut)	*maadali* (immature coconut before it has any meat)
a (bottom of shell still soft and edible, top hard)	*pɛn pwulopwul* (young)	*daahaa mada balabala* (intermediate, with soft mushy meat)	*balabala* (green coconut with soft and thin meat)
pen (drinking nut)	*pɛn* (green, drinking)	*daahaa* (second stage of a green coconut) *daahaa mada makaga* (intermediate, before firm meat) *daahaa mada giigii* (next intermediate stage)	*luumada* (drinking coconut) *luumada halihali* (drinking coconut with little meat) *luumada hai ganniu* (meaty drinking coconut) *luumada gu mada* (green coconut reaching ripe stage)
kalaepaen	*pɛn mwangaj* (intermediate)	*daahaa modomodo* (third stage of a green coconut)	*mee ka* *mee modomodo* (nearly mature coconut)
mwaengas (nut is now brown, still on trees, can be drunk still, milk is effervescent)	*mwangaj* (almost brown)	*modomodo mada gii* (intermediate stage)	*maduu haaliu* (nearly mature)
aering (fallen brown coconut used in copra making, mature)	*oaring* (brown, mature)	*gaadinga gaha dea* (nearly mature) *gaadinga* (brown, mature)	*maduu* (fully mature coconut) *maduu dogi* (coconut selected for seedling)
woasaenporok (coconut plant that has started growing but no leaf yet)	*poad* (sprouting)	*somo* (sprouting coconut)	*tubu* (coconut sprout)
par (nut has grown embryo and germinated)	*sil* (young shooting) *par* (shooting)	*mee osi* (coconut seedling)	*homo* (coconut seedling) *homo dogi*
waeiahk (sprout now more than three feet tall, but still using nutrition from shell)	*waiahk*	*dilobuna* (sapling) *hai ganabu* (coconut tree in which the base of the trunk begins to develop)	

The copra production peaked by the mid-1960s, and Migvar (1965: 1) described the "great traditional significance of the coconut palm" to people living in regions where this tree was found:

> "The coconut palm is really the most universal plant that nature has ever provided. There is hardly any other plant in the tropics or the temperate areas that has so many uses, and of which every part of the plant, from the roots to the very tip of the leaves, can be used to make life more pleasant. For centuries the coconut palm has given life to thousands of people on atolls in the Pacific and Indian Oceans. These people have utilized the palm and its by products to the fullest extent that they know how."

Based on the "positive" economic growth in copra production in Micronesia during the post-world war period, Migvar (1965: 1) also expressed a common mercantile point of view about the changing lifestyle associated with the key economic crop of the time:

> "The coconut palm is the most important tree crop for the people of Micronesia and provides the major cash export—copra. It not only provides the money to buy sugar, salt, and clothes but also provides daily subsistence. If the palm were not present, most of the atolls would today be uninhabited. The coconut palm is life and life is the coconut palm."

Due to the rising tonnage and dollar value of the copra trade in the U.S. Trust Territory of Micronesia (for example, 5,150 tons produced in 1900 and 13,828 in 1964), Migvar (1965: 7) welcomed what he and others at the time saw as the bright future role of the coconut palm in the region:

> "The coconut will be the life line of Micronesian economy for the foreseeable future. The atolls are well suited to coconut cultivation, and thus far no economic plant has been discovered or developed which can readily take its place. On the high islands cacao, black pepper, ramie, pineapple, sugar cane, and vegetables will be strong competition for the coconut. It is good that many Micronesians will not, in the future, have to depend on the coconut alone as a source of cash income. However, the atoll inhabitants are destined to depend on coconuts for their livelihood." (Underlined for emphasis).

However, in the 1970s, the copra industry began to falter, and subsequently, coconut plantations in the tropical Pacific region were progressively neglected, leaving many islands, especially atolls with large numbers of coconut trees, without company ships arriving regularly to collect the copra harvests. The FSM Agriculture Policy report of 2012–2016 summed up the relatively short-term economic gains with long-term consequences for the islands:

> "Copra was a former ubiquitous cash crop throughout the FSM, but production has now ceased almost completely due to inefficiency, volatile commodity prices, and poor transportation linkages coupled with increasing senility [old age] of the coconut palms."

More recently, in some areas of the FSM, attempts to initiate a revival of the coconut industry have been undertaken. For example, on Pohnpei in 2018, the Micronesia Challenge and the Terrestrial Monitoring Program agreed to assist FSM Vital (formerly FSM Petrocorp) on the Coconut for Life Project (C4L). The C4L Project is an initiative to revive the coconut industry in the FSM. This industry had been stalling since the 1970s. Whether or not this activity will allow a revitalization of coconut palm farming for commercial products, people living on the outer islands will most likely continue to rely on the coconut palm for its remarkable set of traditional resources from its roots to its canopy of large and useful fronds.

Of course, although the coconut palm no longer retains its dominant role in the commercial agricultural economy of Pohnpei and the outer islands, it remains a vital, multipurpose resource, and an integral part of culture and island way of life.

This local setup on Pingelap Atoll uses fire to quickly dry copra for sale.

A shed on Pingelap Atoll where discarded coconut husks are stored for latter use as burning fuel.

A closeup view of dried chunks of copra, one of the Micronesian atolls' few cash crops.

Neatly arranged half-shell coconuts, drying in the sun on Nukuoro Atoll.

▸ **Sawa** (P) (Pi), **Jawa** (Mw), **Dalo** (Nu), **Dala** (Ka)
Colocasia esculenta

Description: *Colocasia esculenta* is one of the most important plants for many people living in the tropical Pacific islands. This very widespread cultivated aroid herb is known in English as taro, sweet taro, true taro, soft taro, or dryland taro. It is a perennial herb that can grow to a height of about 2 m (6 ft) or more in some tropical regions where it is cultivated; but in Pohnpei gardens, this taro species is usually much smaller in height, normally only up to about 60 cm (~24 in) (see Balick et al. 2009: 138; also see Sproat 1968: 16). *C. esculenta* produces a cluster of large leaves up to 30–40 cm (12–16 in) in length and normally pointing downward or horizontally. These leaves are smooth, heart-shaped, and peltate (i.e., the stalk or leaf petiole is attached at a point on the underside within the edges of the leaf or its margin). The true roots of *C. esculenta* develop from the large swollen underground part of the stem (known botanically as a corm). The flowers, which are rarely produced by the cultivated varieties of *C. esculenta*, are borne in a multi-flowered spadix surrounded by a spathe. The corm can weigh up to about 2 kg (~4.5 lb), and after sufficient cooking, it becomes edible, with its young leaves also becoming edible if well cooked.

Distribution: *C. esculenta* is a widely important aroid root crop that is believed to be native to the Old World tropics of South or Southeast Asia, where it is has been cultivated since ancient times; others have suggested that it may also be native to areas of northern Australia and Near Oceania (e.g., New Guinea). This taro is planted in both wetland areas (usually around the edges of natural or manmade swampy environments) or under "dryland" conditions where there is sufficient moisture available from rainfall (normally about 24 cm (60 in) per year). *C. esculenta* is grown from cuttings taken from the top of the corm, with all leaf stems except the center leaf petiole cut away at their bases. These sections from the mother plant are cultivated in either dry or wetland conditions. Propagation from cormlets (small side growths from the corm) are sometimes used, but commonly in most places:

> " In preparing land for planting, selection of site is the first step. Dryland varieties will grow in low spots or open fields or on sloping hillsides with fairly deep sticky soils which are intermittently irrigated. Level, partially swampy areas are also satisfactory for dryland variety propagation. Wetland varieties need to be planted in paddies or meadows with access to [a] steady flow of cool water during the 8 to 12 months required for development from planting to maturity. Regardless of type, soils should be at least 8–12 in deep, fertile and with a good percentage of humatic material [a highly purified organic fertilizer made of natural ingredients]." (Sproat 1968: 17–18)

It should be noted that in modern times, many varieties of *C. esculenta* taro have been heavily attacked by pest species. "In brief, the majority of Polynesian and Micronesian cultivars of *Colocasia* taro that were derived from a narrow genetic base are very susceptible to today's pests and pathogens" (Manner and Taylor 2011).

General uses: Manner (1993a: 88) commented on the availability of published knowledge about of *C. esculenta* in the atolls and low islands of Micronesia and noted that its use and status need more study and the "productivity of *Colocasia* taro in the atolls is [also] not well documented." From a broader regional perspective that included the high and low islands of Micronesia, Manner (1993b: 34) described the cultivation of the two most important taro species (sweet and swamp taros) in terms of the habitats, methods of preparation, planting, and mulching:

> " In Micronesia, the bulk of taro production of *Colocasia esculenta* and *Cyrtosperma chamissonis* [= *C. merkusii*] takes place in permanent to semi-permanent lowland patches. On the high islands, the favored areas for the wetland cultivation of taro are the freshwater swamps and marshes located

inland of the mangroves, and the alluvial bottomlands. Areas selected for planting are cleared of vegetation and drained. The soil is then dug up, and various leaves, twigs, and seagrasses are added as a mulch to increase soil fertility."

Manner continued:

> "In the atolls, by contrast, both *Colocasia* and *Cyrtosperma* taros are planted in pits located near the centers of the larger islets where the hydrostatic freshwater lens is the thickest, the water is low in salinity, and the possibility of wind-driven salt spray and water contamination from storm waves is low."

The cultivation of these two main taro species (*C. esculenta* and *C. merkusii*) in wetlands is one of the most distinctive agricultural systems devised by traditional Pacific islanders; indeed, these taro farming systems "can be found in a wide range of habitats, including freshwater swamps on both high volcanic islands and atolls, mangrove swamps, ponded lowlands, and irrigated slopes and terraces," with a great variety of methods devised and implemented "to transport water to the irrigated pond, including bamboo pipes, dams, and other stream diversions" (Dixon et al. 2012).

> "Taro is one of the few crops in the world (others being rice and lotus) that [can be] grown in flooded conditions. Large air spaces in the petiole permit the submerged parts to maintain gaseous exchange with the atmosphere. However, surrounding water must have enough dissolved oxygen. If the water is too stagnant and warm, the high temperature and resultant low oxygen content can cause the basal rotting and wilting of the taro." (Taboroši and Sánchez Collazo 2011: 103)

The primary food products from *C. esculenta* taro throughout much of the Pacific Islands today for both subsistence and commercial purposes include the swollen, mostly underground corm, the young leaves, and the petiole that connects the corm to the leaves. These parts of true taro must be cooked (a necessity for all parts of taro plants) to make them edible. The main food source of the *C. esculenta* taro plant is the corm, which is variously "boiled in water, baked, fried or steamed in underground earth ovens" (e.g., see Manner and Taylor 2011). If the corm and young leaves of *C. esculenta* are well cooked, they are both edible and nutritious, providing an excellent source of calories, and if large amounts are eaten, they also provide nourishing amounts of protein, fat, calcium, phosphorous, iron, and thiamine. However, the edible parts of true taro provide very little or no carotene and only small amounts of ascorbic acid and riboflavin, so these important substances must be consumed from other food sources (e.g., see Englberger et al 2009a; Murai et al. 1958).

In addition to its most significant uses as a food source, *C. esculenta* has been used widely for a series of traditional medicinal purposes in tropical Pacific islands. For example:

> "In New Guinea, the leaves are heated over a fire and are applied to boils. A decoction of the leaves is drunk to promote menstruation. A decoction of the leaves, together with some parts of other plants, is taken to relieve stomach problems and to treat cysts. The sap of the leaf stalk is used in treating conjunctivitis. The scraped stem, together with some parts of other plants, is used to create an appetite. The plant is also used to treat wounds." (World Health Organization 1998: 55)

Uses in Pohnpei: *C. esculenta* is called **Sawa** in Pohnpei. Augustine Primo (1992: 6), a former employee of the Pohnpei Division of Agriculture, and then an Extension Agent I at COM-Pohnpei Campus, referred to the following stories relating to its legendary origins:

> "According to Pohnpei legends, the first taro on Pohnpei was not introduced, rather it was found long, long ago in the Municipality of Nett located in the north of Pohnpei. The first people that found this taro named it **Eniseupadok**, meaning planted by ghosts or spirits. Anything that happens in Pohnpei without the presence or consent of humans is always referred to as a spiritual creation. This original taro variety was not edible. According to the local anthropologist Ioakim David (whose traditional title is *Kehlak Takaieu*), the ancient Pohnpeians used this plant only for medicinal purposes."

Primo further described how, after the original taro was discovered, another type was found in the northwestern part of Pohnpei at Palikir in Sokehs Municipality:

> "This variety was named **Sarawihpe**. Since this variety was edible, they gave praise to God before they ate it. In Pohnpeian, the name *sarawihpe* is a word that means the act of praise or worship of someone higher than you. Since this time, taro has been an important food on Pohnpei. Unfortunately, neither of these two original varieties exists today, so we do not know what they looked like. All the sweet taro varieties planted today are introductions during the post-European contact and colonial periods."

Although we do not know when *C. esculenta* first occurred on Pohnpei, it almost certainly arrived via humans as a "canoe plant" of agricultural importance, perhaps as early as 2,000 years ago (see below). Primo (1992: 6–8) described why he regarded sweet taro as a very significant plant in Micronesia as well as on Pohnpei:

> Sweet taro (*Colocasia*) is cultivated almost everywhere in the Federated States of Micronesia and in the other islands of the Pacific and it has sown important values where it is grown. Sweet taro is considered one of the major food crops on Pohnpei Island by local people. They prepare taro in many different ways, especially during special occasions in the community in which all are asked to prepare food."

According to Primo, one reason why sweet taro is a significant plant on Pohnpei is the length of time this crop remains productive during the year, especially for people who are economically dependent on farming:

> In Pohnpei, the main breadfruit season begins in May and slows down by late July, and the yam season only begins in earnest in December and usually ends in March or April. A good Pohnpeian farmer schedules his planting of taro so that it can be harvested during the gap between the breadfruit and yam seasons. This is why sweet taro is an essential component of the traditional Pohnpei agroforestry system."

Primo also referred to the importance of adaptation of sweet taro to several soil types on Pohnpei without commercial fertilizer, only needing locally available traditional fertilizer at planting time, along with some "weeding, desuckering, hilling-up and mulching." Utilizing these traditional farming practices provides the Pohnpeian farmer with a large enough taro harvest to satisfy family consumption and use for special events.

> Sweet taro is used a lot at parties and other special events because it can be prepared in eight to ten different ways. It can also be stored for a longer time than all other root crops, without much change in quality and taste. The market demand for sweet taro, especially the major varieties, is good, and sweet taro planted in large quantities can be fed to livestock if the market demand is very low at the time of harvest."

Primo stressed the importance of sweet taro as a food source, referring to it as a "nutritional treasure" for the people of Pohnpei and the outer islands.

> The taro corm and heart-shaped green leaves are packed full of the vitamins and minerals the body needs to stay healthy. The taro corm is an excellent source of energy, which the body needs to stay active. The taro corm also provides fiber which is needed in order for the intestines and other parts of the digestive system to work properly. When eaten in large quantities, taro corms are also a good source of calcium and iron. Calcium helps to make strong bones and teeth, and iron keeps the blood healthy. Women especially need a lot of iron in their daily diet. Taro leaves contain Vitamin A which is needed for proper growth, healthy eyes, and protection from disease. Taro leaves are a good source of Vitamin C which keeps the body tissue strong, helps the body use iron, and aids chemical reactions in the body. Taro leaves also contain Vitamin B1 (thiamin) which helps the body use energy foods and Vitamin B2 (riboflavin) which is needed for normal growth and healthy eyes." (Also see Murai et al. 1958; Englberger et al. 2009, 2011, 2013).

Finally, Primo commented on the "Future Research Needs" related to the "healthy" cultivation of sweet taro in the FSM and cited "Twelve steps to growing healthy sweet taro" (see Shresthra 1990). This government pamphlet, published by the Pohnpeian Agricultural Division about 30 years ago, highlighted the major research need of the time: "the control of *Phytophthora* leaf blight, which is the most serious problem to the island's farmers." More recently, Singh et al. (2012: 182) summed up the danger of this taro leaf blight disease and emphasized the need for modern breeding for disease resistance:

> Taro leaf blight (caused by the oomycete *Phytophthora colocasiae*) is a disease of major importance in many regions of the world where taro is grown. Serious outbreaks of taro leaf blight in Samoa in 1993 and in the last few years in Cameroon, Ghana and Nigeria continue to demonstrate the devastating impact of this disease on the livelihoods and food security of small farmers and rural communities dependent on the crop. The spread of the disease to new geographical areas also poses a major threat to neighboring countries and taro growing regions still free from the disease. Past research, particularly in the Pacific, has demonstrated that management measures such as chemical and cultural control are largely ineffective and that breeding for disease resistance is the most sustainable approach to manage the disease."

Verma (2014b: 13) repeated some of the same warnings and suggestions:

> Among the diseases that affect soft taro, leaf blight caused by *Phytophthora colocasiae* is the most prevalent. Its incidence is influenced greatly by the climatic conditions and is most serious during wet seasons. Its presence usually diminishes during the dry months of the year. Leaf blight can be recognized by the formation of purplish to brownish circular water-soaked spots on the surfaces of the leaves. A clear yellow liquid is exuded from the spot. This disease has devastated taro wherever it has occurred, the last outbreaks were in American Samoa and Samoa in 1993. To control leaf blight in taro, at present there is no fungicide cleared for commercial use. The only solution to control leaf blight is to cultivate leaf blight resistant taro varieties. Other diseases of soft taro are dry rot caused by *Sclerotium rolfsii* and leaf spot caused by *Phyllosticta colocasiophyla*. These can be serious in soft taro but seldom occur in well-managed and well-drained fields."

In addition to its use as food, *C. esculenta* has medicinal uses on Pohnpei. Its leaves "are used to make a cup for drinking medicines, such as an unspecified medicine for depression that is squeezed directly into the cup and consumed" (see Balick et al. 2009: 256). Primo (1992) indicated that the consumption of water collected in **Sawa** leaves by young boys and girls has been associated with good mental and physical development on Pohnpei:

> During the child's early growth, the parents will often instruct their child to go to a taro patch early in the morning and look in the upright leaves to see if some water remains in them. The children are then asked to drink this water every morning. If a child does [this] every morning, they predict that the child will be '*lolokong*' [*loalekeng*] (intelligent) and '*marara*' [*marahra*] (a fast runner)."

Lee et al. (2010: 12) reported the use of **Sawa** to treat stings inflicted by **Likendindep** (stingray); accordingly, a 30 cm (12 in) section of unbroken taro "root" is carefully wrapped in cloth, soaked in water, and applied to the area on the skin where the sting occurred in order "to relieve pain and draw the poison from the wound."

Historical perspectives of taro in Pohnpei: Just how important taro may have been in Pohnpei's history was revealed by a recent remarkable use of aerial remote sensing (known as LiDAR). This method for measuring distances (i.e., "ranging") uses laser light to illuminate the target and measure the reflection with a sensor; then the differences in laser return times and wavelengths can be used to make digital 3-D representations of the target. The recent airborne LiDAR survey of the UNESCO World Heritage Site of Nan Madol and Temwen Island uncovered an ancient irrigated cultivation landscape, which appears to have been large series of wet fields suitable for growing taro, located adjacent to the core area of Nan Madol on Pohnpei and described as a "vast archaeological landscape consisting of a network of cultivation features connected by a complex water management system" (Comer et al. 2019).

> Fieldwork verified the presence, across Temwen, of low, wet, cultivable areas, many of which are connected by water channels or separated by earthen berms. The berms themselves may also have been cultivated. In complexity, labor investment, and organization, the system is comparable to Nan Madol itself, the largest archaeological site in Micronesia, with structures on about 100 artificial islets built of stone and coral on a reef flat. Constructed over a millennium, Nan Madol was the seat of the Saudeleur Dynasty, which persisted from about 1200 to 1600 CE. The cultivation system appears to have been able to provide ample food for consumption, feasting, and redistribution or trade. If the landscape alteration described here proves to date to the time of the Saudeleur Dynasty, it will offer many avenues of research into the economic basis of Nan Madol's regional dominance."

This area has been used in living memory "as a field for growing swamp taro," though it is "now almost completely abandoned," with current residents in the area "say[ing] that many members of the family that once grew swamp taro there have left Pohnpei" (Comer et al. 2019). What species and cultivars of taro were grown here in the ancient times? When exactly did this farming take place and what techniques were being used? That can be answered by additional studies in the area, analysis of soils, pollen, phytoliths, and plant macroremains. If such research confirms the ancient taro farming system, it would suggest that taro production "would have resulted in an enormous and relatively secure store of food... [and that] supply of taro could have been tapped not only for redistribution or trade, but also for feasting that enhanced the prestige of the rulers" (Comer et al 2019; also see Haun 1984). Therefore, "the discovery of this landscape of cultivation features suggests ways in which Nan Madol's economic strength may have contributed to a position of great influence on Pohnpei and in Micronesia," perhaps encouraging peoples on geographically accessible atolls "to maintain regular economic, social, and cultural links with Nan Madol via sophisticated sailing technologies and navigational techniques" (Comer et al. 2019).

Cultivars in Pohnpei: The use of **Sawa**, compared with other taro species in Pohnpei, has varied over the last decades. Lessa (1977) indicated that the corms of **Sawa** were being "baked or boiled and eaten as a substitute for breadfruit and yams [and that] the leaves are sometimes cooked." However, Lessa also pointed out that the cooked stem portions of the giant "dryland" taro (*Alocasia macrorrhizos*) were still being consumed more commonly than **Sawa**, but not as much as the cooked **Mwahng** (the huge swamp taro, *Cyrtosperma merkusii*). Today, *A. macrorrhizos* is considered a famine food and rarely eaten, while **Mwahng** and **Sawa** are still cultivated and consumed today on Pohnpei, but less so than on the outlying atolls. Lessa observed that **Sawa** is "far less important [on Pohnpei] than on atolls, ranking sixth in order of preference and use among the starchy foods." Lessa (1977: 142) also noted that "at least 17 varieties" had been recorded in Pohnpei and indicated that these varieties are "distinguished by whether they grow in wet or dry places, and whether they are cultivated or wild or both, as well as by morphological characteristics." The Pohnpeian–English Dictionary (Rehg and Sohl 1979) provides a list of 15 names of Sawa cultivars, which probably includes most varieties that Lessa had found and also the legendary and no longer extant taro that Primo (1992: 6) wrote about. We provide the full list on p. 420.

Uses in Mwoakilloa: Bentzen (1949: 32) described "taro" history and contemporary cultivation on Mwoakilloa during the first half of the 20th century. He referred to the significance of cultivation of both *C. esculenta*, known on Mwoakilloa as **Jawa**, and *C. merkusii*, known on Mwoakilloa as **Mwɛhng**. Lessa (1977: 142–143) referred to **Jawa** as "the second crop in importance in the wet gardens" on Mwoakilloa Atoll, behind **Mwɛhng**, the most important crop on this atoll. Even though **Jawa** was "considered tastier" than **Mwɛhng**, Lessa indicated that it was "a luxury crop and not a main source of food" on Mwoakilloa. Today, **Jawa** is still cultivated on Mwoakilloa, but **Mwɛhng** remains more significant, largely because it is better adapted to salty soils and the harsh, dry atoll climate, including strong winds.

Cultivars in Mwoakilloa: Taro cultivars on Mwoakilloa were recorded by St. John (1948), Bentzen (1949), Glassman (1953), and Damas (1994). See p. 421 for our summary cross-checked with the Mokilese dictionary.

Uses in Pingelap: Sweet or true taro (*C. esculenta*) is called **Sawa** in Pingelapese. During his visit to Pingelap in the middle of the last century, St. John (1948) observed it cultivated in small numbers and referred to it as a Pingelap crop "of very minor importance." More recently, Damas (1994) agreed with that observation and indicated that it is being grown around the edges of the main patch of giant swamp taro, *C. merkusii* (**Mwaeiang** in Pingelapese), or planted "in minor rows in the otherwise empty spaces between the major divisions of the [giant swamp taro] patch." According to Damas, **Sawa** is "a very minor source of food in view of the small numbers observed" and noted that it "is fertilized by coconut husks, which are seen lying on the surface at the base of [sweet taro] plants, a fertilizing medium not considered suitable for most wet taros, where the mulching is done with various grasses and leaves." In addition, Damas (1994) pointed out that the majority of **Sawa** varieties "require special preparation before they can be eaten." This involves removal of the corm's "skin," and then slicing and drying these starchy corms before cooking them in earth ovens. After cooking them, the Pingelap people grind up the corms "into a coarse flour which is then formed into a cone mixed with sugar or into cakes flavored at times with bananas."

Cultivars in Pingelap: Hattori (2012) lists eight varieties of "soft taro" in Pingelap, though it is presently not known which of them are *Colocasia* and which *Xanthosoma*, and one appears to be *Alocasia*. Additional cultivars were noted by Bentzen (1949) and Glassman (1952); see p. 422.

Uses in Kapingamarangi: *C. esculenta* is known on Kapingamarangi Atoll as **Dala**. According to Hiroa (1950), this species of taro *C. esculenta*) was introduced to this atoll "in ancient times and is esteemed as the best of the local foods." Hiroa described the excavated, swampy pits in which this taro species was being grown on this atoll:

> "The taro is planted in cultivations which are excavated down to subsoil water. The sand and coral gravel is loosened with a point[ed] stick, scooped up with half coconut shells into coconut-leaf baskets, and deposited around the borders of the excavations. The soil thus forms ridges in the vicinity of the gardens. These cultivations are fairly extensive on some of the larger islands [of the atoll]."

According to Hiroa (1950: 43–44), these pits support the cultivation of the giant swamp taro (*C. merkusii*), known in Kapingamarangi as **Bulaga**, but before *C. merkusii* was brought to the atoll for the first time, reportedly from Nukuoro Atoll in the late 1880s, all the sunken gardens of Kapingamarangi were actually filled with sweet taro, **Dala** (*C. esculenta*). However, the introduction and subsequent cultivation of **Bulaga** to a large extent replaced **Dala** because the former grows more vigorously and has greater resistance to arid atoll conditions.

> "On Kapingamarangi Atoll, the taro pits are found on islets greater than 3.8 ha in size (Wiens 1962), and are absent on the smaller islets as the freshwater lens is poorly developed or absent. On these islets, coconuts and breadfruit are the principal tree food crops." (Manner 1993b: 34).

Referring to mid-20th century reports, Bayliss-Smith (2012: 113) noted that *C. esculenta* was the "dominant staple" in earlier times in Kapingamarangi but that it was only cultivated on one islet of this atoll in 1953. Hiroa (1950: 17–19, 43–48) described four main types of food preparations made with taro: *'hakaroi'* [*hagalloi*], *'hakate'* [*hagadee*], *'putara'* [*budala*], and *'hokooto'* [*hagoodo*], which involve various mixtures of cooked sweet taro, swamp taro, and breadfruit, the major food energy sources for the residents of Kapingamarangi.

Cultivars in Kapingamarangi: Although many varieties of **Dala** were formerly recognized on Kapingamarangi, not many are still known (see our list on p. 423). It should be noted again that **Dala** (*C. esculenta*) was the only taro cultivated on this atoll until **Bulaga** (the giant swamp taro, *C. merkusii*) was introduced to Kapingamarangi by a Nukuoro woman, wife of the resident trader, in the late 19th century; **Bulaga** subsequently replaced **Dala** as the primary staple on Kapingamarangi (Hiroa 1950: 44; Emory 1965: 7).

▸ **Kahil** (P)
Commersonia bartramia

Description: *Commersonia bartramia* is a shrub or small tree. It can reach heights of about 20 m (65 ft), but often is smaller, with white to gray mottled bark and conspicuous lenticels on the bark that function as part of the plant's gas exchange with the atmosphere. The leaves are ovate, sometimes lobed, about 6–15 cm (2–6 in) long and 4–10 cm (1.5–4 in) wide with serrated margins. The underside of the leaf is densely covered with yellow to gray hairs. The white flowers are produced in small clusters and are radially symmetrical, as is typical for the mallow or hibiscus family (Malvaceae). The fruit is a dry, bristly capsule that grows up to 2 cm (1 in) in diameter that splits into five segments, each of which contain 2–6 seeds.

Distribution: *C. bartramia* is reportedly native to areas of Southeast Asia, Australia, New Guinea, the Bismarck Archipelago, New Caledonia, Fiji, the Society and Marquesas Islands, and the Caroline Islands of Micronesia. It is a common woody species occurring in secondary and dry forests as well as in patches of forest or grassland thickets; it is also found occasionally in garden and fallow areas, where it is an important pioneer species.

> "This is a one of the fast-growing set of secondary "bandage trees" that grow up in disturbed areas, covering soil and reducing erosion. It is adapted to harsh conditions and frequent fires in savannas, resprouting from the base when burned and often flowering and fruiting while shrub-sized." (Falanruw 2015: 218)

C. bartramia is common in disturbed places, including for example the savanna areas of Palau and Yap. Savanna grasslands with associated trees occur on volcanic soil substrates where the primary forest has been removed; most, but not all, of these open grassland areas are human-induced, largely due to repeated man-made fires. There are some endemic savanna species on islands such as those in Palau and Yap, which suggests there may have been some isolated pockets of native or natural savanna. The loss of native forest, much of it due to human impact, has resulted in soil erosion and degradation; this has reduced the probability of forest regeneration, at least over the short term. As previously suggested, these savanna grassland systems persist because of repeated burning caused by humans (see Hunter-Anderson 2009 for a counter-argument). Savanna is found throughout the island of Babeldaob, in areas of Koror on Palau, and over extensive inland areas of Yap. The vegetation consists of grass, sedges, ferns, shrubs, and a few scattered tree species, including some endemic species. Trees and shrub species, including *C. bartramia*, that grow in the harsh savanna ecosystem are potential reforestation species.

General uses: Throughout much of the western tropical Pacific, *C. bartramia* has a wide variety of traditional uses:

> "Wood used in light construction, for fishing floats, and as a fast-burning firewood; light wood used to start fire by friction (e.g., in the Solomon Islands); small poles and sticks used to stake yams; fibrous bark used as cordage for fishing lines, nets, baskets, belts, girdles, headbands (and bark cloth in Melanesia); strips of bark used as a crude rope to carry produce and firewood, and for lashing in construction; roots used medicinally in Fiji." (Clarke and Thaman 1993)

Uses in Pohnpei: *C. bartramia*, known as **Kahil** in Pohnpeian, is indigenous to the island. It produces wood that is sometimes used on Pohnpei as construction material, and the stem has been used to make trellises for supporting plants, such as yam vines, as well as for firewood and lumber. In addition, the fruit is bound together and placed where the rats often eat to poison those pests (see Balick et al. 2009; also see Raynor 1989).

▸ **Ikoik** (P) (Pi), **Kanau** (Mw), **Ganava** (Nu), **Laagau mmee** (Ka)
Cordia subcordata

Description: *Cordia subcordata* is an attractive, medium-sized tree with a dense crown that grows to heights of 10 m (35 ft) or taller with pale-grayish, slightly fissured bark. *C. subcordata* produces broad, light green, ovate, pointed leaves up to about 30 cm (12 in) in length on long petioles; the attractive flowers are orange and trumpet-shaped. The clustered fruits are broadly egg-shaped, 2–4 cm (1 in) long, turning pale-grayish brown and woody when mature, and containing 2–3 corky seeds.

Distribution: *C. subcordata* produces floatable fruits and durable seeds, so has apparently experienced long-distance dispersal throughout much of the Tropical Pacific; indeed, it is considered to be indigenous from the Indian Ocean and tropical Asia in all the islands of Micronesia and Polynesia. However, it may have also been introduced to some islands by people because of its varied cultural utility (e.g., see Mueller-Dombois and Fosberg 1998; Thaman et al. 1994: 25; Burney 2011). This tree thrives in coastal forests and thickets, often near the shoreline, but can grow at elevations up to about 150 m (~500ft). It is not common on some of the larger Micronesian islands (e.g., on Yap proper or the main islands of Palau), but is frequently found on outer islands, particularly on atolls. It is planted through both seeds and cuttings. It was formerly common in coastal strand forests, especially in disturbed sites, but is now "increasingly scarce due to overexploitation of wood for carving" and can sometimes be found "in some home gardens on atolls or occasionally as a roadside tree" (Clarke and Thaman 1993; also see Friday and Okano 2006b).

General uses: *C. subcordata* has many uses by the Pacific Islanders, including the Micronesians. Apart from its use as a shade tree, windbreak and soil stabilizer in coastal areas, its wood is highly esteemed for making handicrafts, and house posts made from the timber of this tree reportedly can last up to 100 years. The wood will not split when a nail is driven through it. Commonly, the soft but very sturdy, chocolate brown and blond-colored wood of *C. subcordata* is:

> "Used in general construction and is among the most favored carving woods; also used for making canoe hulls, thwarts, rudders, weather platforms, outrigger booms and paddles, furniture, headrests, bowls, trays, plates, combs, food stirrers, food containers or boxes, airtight reef boxes for fishing equipment, fishnet floats, fishing rods, tools, coconut-climbing sticks, toys, drums and slit-gongs, tobacco pipes, images of gods (*tiki*), and other carved objects for sale to tourists; young saplings used for fishing rods and flutes." (Clarke and Thaman 1993; also see Friday and Okano 2006b)

Uses in Pohnpei: Known as **Ikoik** in Pohnpeian, this tree provides wood for small boat construction and the making of paddles on Pohnpei. Although many assume this species is native to Pohnpei and the nearby atolls, it may also have been introduced, at least in some areas or islands, by the early Micronesian settlers. Balick and his colleagues (2009: 342) provide information about the medicinal and some additional uses of *C. subcordata* on Pohnpei:

> "The ripe fruits are eaten as food... The wood is used as firewood... The flowers are used in *mwaramwar* (head garlands)... To add fragrance to body oil, as many flowers as needed are placed in coconut oil and boiled or heated in a pan... To stimulate a child's appetite... To close an open cut, the pounded bark is put on the cut... To protect a child from any type of sickness... To treat a baby's mouth rash... To stop the bleeding from a cut and to relieve the pain, the bark scrapings are pounded, wrapped in a piece of cloth, and squeezed onto the cut..."

Uses in Mwoakilloa and Pingelap: *C. subcordata* is known as **Kanau** in Mwoakilloa and **Ikoik** in Pingelap. According to Glassman (1953: 296), its "trunk is used in making canoes" on Mwoakilloa Atoll. For Pingelap, he indicated that this coastal strand tree could reach heights of about 50 ft (~15 m), occurring frequently near the shores of this atoll.

It should be noted here that *C. subcordata* is a multipurpose tree species that has a special ecological role as it is one of several woody species that are especially adapted to help protect the people and local agriculture located near the ocean environment, and this applies especially to atolls. Manner (1993b: 33) summarized this special ecological sheltering function of *C. subcordata* and other typical coastal shrubs and trees in the tropical Pacific, including Micronesia:

> "On atolls, the pattern of agroforestry is arranged in zones and reflects the distribution of natural vegetation and the severity of environmental constraints (wind-generated salt spray, wave damage, saline ground water, and drought). The shores and beaches contain a sparse, salt tolerant herbaceous cover, backed by a fringing vegetation of shrubs and low trees which serve as a windbreak and buffer against hurricane-generated waves and salt spray. Species commonly found in this zone are *Scaevola taccada, Cordia subcordata, Heliotropium arboreum, Pandanus tectorius, Soulamea amara,* and *Guettarda speciosa.*"

▸ **Dihng** (P), **Manu lauligi** (Nu)
Cordyline fruticosa

Description: *Cordyline fruticosa* was formerly classified in Liliaceae and more recently in the families of Agavaceae and Laxmanniaceae which are now both subfamilies of the Asparagaceae in the Angiosperm Phylogeny Group III system of flowering plant classification (the third version of a modern, mostly molecular based, system of plant taxonomy). The plant is an attractive, erect "palm-like" shrub that grows from thickened, tuberous roots to heights of 1–3 m (3–10 ft) or more. Its stems are simple, or somewhat branched and marked with leaf scars. Generally, the large and useful dark green to rust or red leaves are 30–50 cm (12–20 in) long, smooth, lanceolate to oblanceolate, arranged spirally, and located mostly near the apex of the stem. The clusters of small, pinkish, six-lobed tubular flowers are borne in terminal panicles. The fruit, when produced, is fleshy, globose, and yellow or reddish. In English, this shrub is known as the "ti plant" after its eastern Polynesian name. Today, a large number of multicolored variations of this plant are grown in tropical areas, where this species has become an important landscape plant in urban and rural areas.

Distribution: Probably indigenous to Near Oceania or possibly somewhere in insular Southeast Asia, the ti plant is another species that was probably brought to Micronesia by early Pacific island voyagers. In the tropical Pacific region, this shrub is a widespread, important plant cultivated from stem cuttings. "Common in garden areas, in home gardens, in ceremonial grounds, and as border plantings throughout the non-atoll Pacific... commonly naturalized in fallow vegetation; rare to occasional on atolls" (Clarke and Thaman 1993). On some islands, it can be found in the forests as well as in food gardens and near homes. It appears to be a relatively recent introduction to some of the smaller islands, particularly atolls, of Polynesia and Micronesia, and is occasionally naturalized in poorly drained areas, along streams, and in other disturbed sites on some of the higher islands.

General uses: *C. fruticosa* is a very widespread, important multipurpose plant cultivated in the ground or in pots from cuttings. It traditionally has medicinal, ceremonial, and aesthetic associations with Pacific Islanders throughout much of the tropical Pacific region and may have been cultivated first as a root crop by pre-Austronesian people. Indeed, it may have been one of the initial crops to be managed and eventually cultivated in Near Oceania (e.g., in New Guinea; see Merlin 1989; Ehrlich 1989: 53–54). Some cultivars of this species produce a large, edible, tuberous root which traditionally has served as a supplementary food plant and famine food, and on several tropical Pacific islands, the sugar-rich roots were baked for a long period of time (up to four days) in an earth oven and then "consumed as food, sweets, refreshment, or confectionery, or, in Hawai'i, made into an alcoholic beverage known as *okolehao*" (Clarke and Thaman 1993). The leaves of *C. fruticosa* are used as a food wrapper and also for storing and cooking foods, as well as for dancing skirts and body ornamentation. In some Pacific islands, it has long served as a boundary marker or as an attractive ornamental species. *C. fruticosa* also has strong medicinal and spiritually protective associations with many Pacific Islanders. It is often planted in ritualistic and burial grounds, and on many tropical Pacific islands it is or was a "very important ceremonial and magico-religious plant, commonly planted as a garden marker or to ward off evil spirits or to discourage black magic or sorcery in garden areas" (Clarke and Thaman 1993). Brower (1974) referred to Micronesian magicians chanting to ti plants (*C. fruticosa*), "naming various causes of death and expecting the plant to tremble in response to the right cause" (see Ehrlich 1989: 52; also see Ehrlich 1999, 2000a, 2000b).

Uses and cultivars in Pohnpei: *C. fruticosa* is known as **Dihng** in Pohnpeian (a local name also applied with modifiers to the introduced *Costus speciosus*); in addition, there are two names, ***Dingmwahn*** and ***Dingpein***, for different cultivars of *C. fruticosa* on Pohnpei according to the Pohnpeian–English Dictionary (Rehg and Sohl 1979). *C. fruticosa* has several kinds of uses in Pohnpeian culture, including those that are for medicinal or ritualistic use, minor product development, and ornamentation (Herrera et al. 2010: 53; also see Glassman 1952: 107). Among the traditional medicinal uses of this plant on Pohnpei, Balick and his colleagues (2009: 283–284) reported that **Dihng** has been:

> "Used to decorate a *nahs* (feast house)... To protect a new building from sorcerers trying to destroy the building, a handful of leaves is pounded and places in the four corners of the new building... To treat anemia in a woman after she has given birth... To treat upset uterus... To treat severe chest pain or stomachache... To treat back pain, muscle pain, headache, and joint pain... To treat *soumwahu en ohl* (men's sickness)... To treat stingray stings, apical meristems from two to four plants are mashed on a rock, wrapped in a cloth, and the juice is squeezed out of the wound immediately and as long as needed to remove the poison and to help in the healing of the wound [see more on this therapeutic use below]... To treat depression... To treat *soumwahu en naniak* (mangrove sickness)."

On Pohnpei, **Dihng** has also reportedly been used to treat venereal disease and abscesses, and medical practitioners may chant with leaves of this plant held over a sick person for healing purposes. Lee et al. (2010: 6) also described the treatment of a stingray sting using the "young central leaves [apical meristems] from three or four different [*C. fruticosa*] plants" which "are pounded with a rock and wrapped in a piece of cloth." From this mass of pounded leaves bound in cloth, the "pulp is squeezed and the juice is applied to the wound." If this medicine is appropriately "applied immediately upon injury... it is said that the treatment removes the venom, relieves pain and aids in wound healing." In addition, Lee et al. (2010: 130–131) described the use of pounded leaves of *C. fruticosa* "to treat abscesses and deep boils" as well as "to stop heavy menstrual bleeding (menorrhagia)."

Uses in Nukuoro: *C. fruticosa* is used as decorative plant "to build boundary walls" (Drummond et al. 2016).

▸ **Kiepw** (P) (Mw) (Pi), **Giebu** (Nu), **Daladala** (Ka) *Crinum asiaticum*

Description: *Crinum asiaticum* is known in English as the "grand crinum" or "spider lily" (because of the shape its flowers). It is a large, bulbous herb that actually belongs in the Amaryllis family, not the Lily family. *C. asiaticum* produces a rosette of 6–30 long, narrow or oblong, acute, stemless, evergreen leaves, up to about 2 m (~6 ft) long, and clasping at the base. A thick, more or less flattened flower stalk grows out of the center of the rosette of leaves to heights of about 60 cm (~2 ft). The flowers consist of a narrow, straight or curved, white to reddish tube, from which six long petals, up to about 10 cm (4 in) long, spread widely or in a funnel shape. The yellow-green fruits are fleshy, irregularly rounded, 2.5–6.5 cm wide, and have 1–3 seeds.

Distribution: *C. asiaticum* is native to southern Asia, where it is normally found along seashores and other sandy places near riverbanks; it is used traditionally as a source of medicine in many regions where it grows. It is now cultivated in numerous tropical areas, mainly as an ornamental. Early voyagers probably brought this plant to Micronesia; indeed, this plant, and its introduction on numerous Micronesian islands, is referred to in a well-known traditional navigator story from Yap State. Today, it is sometimes found as an ornamental plant growing near houses, as well as on beaches, along village paths, and in coastal areas. Another related species, also known in English as "spider lily" or "beach spider lily," *Hymenocallis littoralis* (see p. 264), is native to tropical America and was probably introduced to Micronesia more recently. It is sometimes confused with *C. asiaticum*.

Uses in Pohnpei: *C. asiaticum* is known in Pohnpeian as **Kiepw** and is sometimes integrated into house gardens as an ornamental plant; its flowers are sometimes included in *mwaramwar* (head garland). Balick and his colleagues (2009: 253) offered the following concerning the medicinal uses of **Kiepw** on Pohnpei:

> "To protect wounds from flies, stem pieces are places on wounds as a cover... To treat the *kilin kidi* (impetigo), one leaf is heated over a fire, mixed with coconut meat, and rubbed into the skin... To treat sore nerves, the longest root of one plant is added to scraped coconut meat, then placed in a piece of cloth with a hot stone, and rubbed into the sore area... To treat unopened boils, eight crushed fruits are placed in a piece of cloth and gently rubbed into the boil until the boil is gone... To treat arthritis, fruits are pounded into a piece of cloth, squeezed into a cup of water and consumed (one cup daily) for four or eight days... To treat centipede bites, an unspecified amount of one clean tuber is scraped off and placed onto the..."

Uses in Pingelap: St. John (1948) reported that *C. asiaticum* was seen on Pingelap only as a cultivated ornamental plant. Damas (1994: 234) said *C. asiaticum* and a related decorative species, *Zephyranthes rosea* (the rain lily) from the same family, Amaryllidaceae, are both known in Pingelapese as **Kiepw**. Damas indicated that the two ornamental species are "used only in decoration" and are cultivated near houses on Pingelap.

▸ **Auleng** (P)
Curcuma australasica

Description: *Curcuma australasica* is a rhizomatous herbaceous perennial plant in the ginger family, Zingiberaceae. It can be found growing wild and is also cultivated for ornamental purposes due to the attractive flowers. Its long leaves are about 1 m or more (~4 ft) in length, developing from its underground rhizomes. Its eye-catching inflorescence comprises colorful pink to purplish, slightly hairy and sterile bracts, along with its smaller bright yellowish flowers within its lower, greenish bracts. Its rhizomes do not elongate and are yellow to orange within. The fruit capsule develops within a persistent inflorescence, with its top third pubescent or hairy. *C. australasica* is a close relative of *C. longa*, the cultivated turmeric plant.

Distribution: *C. australasica* is probably a native plant of tropical northern Australia, where today it is listed as a threatened species; beyond its geographical range in Australia, it is also reported from areas in Indonesia, New Guinea, the Bismarck Archipelago, the North Solomons, and other Pacific islands including some in Palau, the FSM, the Mariana Islands, and French Polynesia). *C. australasica* grows wild in some forested areas of the high islands of this region such as Chuuk, Pohnpei, and Kosrae with an altitudinal range from mainly near sea level up to about 500 m (1,500 ft), or higher. It is also cultivated for ornamental purposes near sea level on several Micronesian islands, including some of the low-lying atolls. *C. australasica* is dormant during the dry season, and when cultivated, it requires plenty of water during the growing season. Its habitat adaptation is normally in shady rainforest margins, often on rocky terrain, but it can also be found in disturbed areas or grasslands by the edge of forests, as well as alongside forest streams and secondary forest regenerating after fires.

General uses: In addition to its ornamental use on several Micronesian islands, *C. australasica* is becoming popular with plant collectors in various parts of the world as another attractive member of the ginger family. Some local people in Micronesia reportedly use *C. australasica* for therapeutic and other purposes, such for "grass skirts," *mwaramwar* (garlands), a source of yellow-colored body paint, and even magical uses (e.g., see Manner and Mallon 1989: 42).

Uses in Pohnpei: *C. australasica* is known as **Auleng** in Pohnpeian and is an introduced species, not a "native turmeric" plant, but it does have a long history in Micronesia and has several uses on Pohnpei, including for food, food preservation, skin care, and as a medicinal source. For example, its leaves, along with banana or breadfruit leaves, are used "to line pits" to preserve pieces of breadfruit (**Mahi**) via fermentation: "... preserved breadfruit, or *mahr*, lasts for a year or longer, with fresh batches made each breadfruit season" (Ragone and Raynor 2009: 80–81). According to Balick et al. (2009: 313), the leaves of *C. australasica* are also reportedly used for wrapping up fermented food such as *mahr*. In addition to smoothing a person's skin, the rhizome of *C. australasica* is reportedly scraped and mixed with grated coconut and squeezed onto the body before taking a shower. In addition, Balick et al. (2009: 312) referred to a Pohnpeian remedy for an infection that involves pounding and then squeezing "a clean tuber" of *C. australasica* onto the infected area.

▸ **Oahng** (P), **Oang** (Mw) (Pi), **Gaalau ango** (Nu), **Dala kala** (Ka) *Curcuma longa*

Description: Turmeric, *Curcuma longa*, is a member of the ginger family, Zingiberaceae. It is a smooth, erect herb up to about 1 m (3 ft) in height, and develops from fleshy, yellow to orange-red, aromatic rhizomes. These finger-like rhizomes usually grow horizontally, and propagate vegetatively by branching or breaking. Its leaves are formed upright and are mostly lanceolate to elliptic in shape with fine, parallel veins. The leaves can grow to about 40 cm (16 in) long and about 7.5 cm (3 in) wide on a stem of similar or shorter length. The flower stalk is cylindrical, growing within numerous, overlapping, broadly lanceolate, whitish cream to greenish yellow bracts (modified leaves).

Distribution: The origin of the turmeric plant (*C. longa*) cannot be accurately reconstructed because of the longevity of its trade and use over a large geographical range in Eurasia. However, this multipurpose ginger plant is probably native to somewhere in South Asia or possibly Southeast Asia, more specifically in areas with a monsoonal climate, i.e., where there is a regular dry period followed by a moderate to very wet (monsoon) season. The fact that very few if any seeds are produced by *C. longa* indicates that this predominantly sterile plant is an ancient cultigen (a plant that has been intensively selected artificially by people over a long period of time); indeed, the turmeric plant is primarily reproduced via clones, i.e., by planting sections of its valuable rhizomes. Humans have apparently planted rhizome cuttings for so long that the plant has lost its ability to produce seeds; thus, it is fully domesticated and therefore completely dependent on human help for its reproduction. In fact, an old, outdated scientific name for the turmeric plant, "*Curcuma domestica*," is a scientific binomial synonym that is sometimes still used for this species. The various varieties of the turmeric plant (*C. longa*) are widespread from tropical West Africa to eastern Polynesia, and some of varieties are traditionally very important in Micronesia. Although it is not known to occur in a truly wild state, we can assume, as noted above, that *C. longa* originated in cultivation somewhere in South or Southeast Asia, and from there it was eventually transported to many tropical Pacific islands by early voyagers. It is found mostly in cultivated places and areas of tropical secondary forest, from the coast to upland regions.

General uses: Turmeric is a useful and significant plant among people in many cultures from the heights of the Himalayas to Easter Island in the farthest reaches of the Pacific. It has had a very lengthy ethnobotanical relationship with many different cultures as a condiment, cosmetic, and cure (e.g., see Sopher 1964). Its roots have been used since antiquity for food flavoring (the grated root is an essential ingredient in Asian curry), medicine, and dye. Religious and ceremonial functions of *C. longa* have been very important traditionally in many Pacific cultures, including on Pohnpei and its associated outer islands. Islanders utilize turmeric now to color and perfume coconut oil, which they wear to church, for traditional dance, and at other occasions. It may be used to dye cloth, to paint the body, or to act as a protective charm of a magical nature. Indeed, *C. longa* is also an important source of medicine in many Pacific cultures. It is quite significant that many Pacific peoples have traditionally anointed babies and deceased persons with coconut oil scented and colored with turmeric. Thus, turmeric has traditionally been among the first and last plant substances a human encounters in life.

General information regarding flavoring and medicine: Turmeric powder prepared from the rhizomes of *C. longa* has been used for centuries for food flavoring, providing a mustard-like, earthy aroma and pungent, slightly bitter flavor to foods. Turmeric has long been used in Asian traditional medicine as a stomach tonic and blood purifier, and for the treatment of liver ailments, skin diseases, and wounds. It has also been used externally to heal sores and as a cosmetic. The yellow-orange color of turmeric comes from the yellow pigment called curcumin found in the rhizomes. Today, turmeric is widely believed to be potentially beneficial in treating or reducing symptoms associated with a wide range of health conditions due to its antioxidant, antitumor, anti-inflammatory, and antibacterial effects. However, although turmeric has long been used in Ayurvedic (South Asia) and other traditional systems of medicine, "no high-quality clinical evidence" has been documented for use of turmeric or its main constituent, curcumin, as a therapy:

> With respect to curcumin/curcuminoids [main active ingredients in *C. longa* rhizomes] in vivo studies and clinical trials, we believe there is rather "much ado about nothing." Certainly, the low systemic exposure levels reported in clinical trials do not support its further investigation as a therapeutic." (Nelson et al. 2017: 1631)

Nevertheless, curcumin supplements are standardized and are now found widely in the marketplace. Indeed, a caveat in Nelson et al (2017: 1631) is worth noting: "Of course, we do not rule out the possibility that an extract of crude turmeric [raw rhizome material] might have beneficial effects on human health." In any case, turmeric has a very ancient history in India, where it is still a very important spice; indeed, India produces the great majority of the global crop and uses most of this domestically. Turmeric also has a quite lengthy history of importance in China as well as mainland and insular Southeast Asia. For example, in Thailand, turmeric is one of the most important folk remedies; its rhizomes have been used to treat dizziness, gonorrhea, and peptic ulcers and as an appetite stimulant, carminative, astringent, and antidiarrheal, and externally the rhizome is used to treat insect bites, ringworm, wounds, bleeding, and the teeth and gums. Among the herbs often classified as spices, turmeric is one of the best researched for pharmacological effects. Turmeric also adds flavor, color, and aroma to foods, such as in soups with taro, breadfruit, yams, and bananas. On some islands, the leaves of the plant are also used in garlands and "grass" skirts.

Uses in Pohnpei: More than a century ago, Christian (1899: 113) described what he observed as the desired apparel and appearance of the Pohnpeian man: "The men's dress consists of a long thick native kilt or girdle of split coconut filaments ('*kol*' [*koahl*]), [with] the upper part of the body generally bare." Furthermore, Christian noted that women produced the preferred clothing and beautifying cosmetics for a Pohnpeian man: "They make the leaf-girdles of the men, and compound the coconut-oil and fish-oil, and mix the cosmetic turmeric, without which no Ponapean dandy's toilet is complete." (Underlined for emphasis).

Turmeric (*C. longa*) is called **Oahng** in Pohnpeian. Balick et al. (2009: 313) reported its uses on Pohnpei:

> The leaves are used as food and put into soaps. On Pohnpei, there is a legend about a woman who fell from heaven and was trapped between two rocks. She could not get out. One person used a turmeric root potion to free her, allowing her to slip out from between the rocks, and then he married her. The two rocks still remain [in] place and can be seen today... The leaves are used to make *mwaramwar* (head garland) [and preparations of the grated yellowish rhizome are used] as a food additive... to treat a headache, back pain, muscle pain,

and joint pain... to clean out the contents of babies' stomachs after births, the rhizome is ground and mixed with coconut oil, squeezed through coconut fiber from the base of the leaves so it is well mixed, cooked until it is hot, then fed to the baby (mostly easily using a sponge), waiting for four days after the mother begins breastfeeding... To treat acne and pimples, a rhizome is scraped, chopped into a cloth, and rubbed on the face... To treat the redness of skin around a babies' skin folds... To treat heat rash, the root is mashed into coconut oil, and used to rub on skin... [and it is used as a] skin lotion..."

Lee et al. (2010: 56) further described the treatment of skin rashes using turmeric (*C. longa*) in Pohnpei:

> "Four tubers are collected from the soil. The tubers must be cleaned first and then pounded together with a rock. The pulp is wrapped into a piece of cloth and the juice is squeezed into the hand. The juice is rubbed on the rash twice daily until it has healed."

Lee et al. (2010: 8) also reported on the use of turmeric in Pohnpei as a remedy for the sting effects of a sea urchin:

> "The tuber is scraped and wrapped in a piece of cloth. The juice is squeezed on top of the wound to relieve the pain and is said to draw the poison from the wound. This medicine is applied immediately upon injury. It is believed the wound will heal once the poison has been removed."

In addition, Lee et al. (2010: 123) noted that turmeric is combined with other plant species "to treat *soumwahu en naniak* or mangrove sickness" and recorded that turmeric material is "added to oil, used on skin burns, used in oil for skin, powdered applied to skin for dancing (Yap), spice added to curry [and young] leaves cooked with meat for flavoring." Also see Dana Lee Ling (www.comfsm.fm/~dleeling/ethnobotany/text/flora.html), who indicated that the leaves of *C. longa*, as well as leaves from other ginger plants, "are popular in cooking with locally raised chicken" on Pohnpei (www.comfsm.fm/~dleeling/angio/curcuma_longa.html).

Cultivars in Pohnpei: Lee et al. (2010: 8) reported that on Pohnpei, turmeric (*C. longa*) "is often referred to as ***Oangitik*** to describe the variety with the bright yellow tuber, while ***Oangalap*** refers to the variety that is paler in color." These two varieties have distinct uses: ***Oangalap*** is a large-leafed cultivar, the roots of which are used for medicine, making yellow dye, and flavoring soup, whereas ***Oangitik*** (also known as ***Kisinioang***) is a small-leafed cultivar whose roots are used for a red dye in tattooing and for medicine and as an ingredient in Japanese curry. According to Lessa (1977: 148), on Pohnpei, the former is used in "flavoring soups" and "for dyes," and the latter for food purposes and in curing gonorrhea. He also mentions varieties he spelled as '*Luh-aroo*' and '*Ongkol*', which we could not identify. The Pohnpeian–English dictionary also lists cultivars called ***Oangen pahsed*** and ***Oangen palau***. In contrast, **Oangen pele** (also known as **Oangen pahnle**), with a leaf of intermediate length and a slightly less yellow root, is actually a different species, "wild ginger" (*Zingiber zerumbet*) (see p. 376).

Uses in Nukuoro and Kapingamarangi: Turmeric (*C. longa*) rhizomes are reportedly used to treat bruises in Nukuoro (Drummond et al. 2016). In the early 20th century, the plant was introduced from Nukuoro to Kapingamarangi (Emory 1965), where it was given the name **Dala kala**, which means "fragrant taro" (Bayliss-Smith 2012: 114).

▸ **Kemeu** or **Koamoau** (P)
Cynometra ramiflora

Description: The genus *Cynometra* includes perhaps more than 100 species which belong in the large botanical family Fabaceae (known in English as the legume, pea, or bean family). *Cynometra ramiflora* is an indigenous legume tree in Pohnpei that is small to medium-sized, branched, and evergreen. It produces a rounded crown, grows to heights of 8–15 m (up to ~50 ft), and has a well-developed tap root system. *C. ramifora* has no aerial roots but sometimes develops buttresses. Its bark is grayish and its leaves are compound, with leaflets 2.5–4 cm (1–1.5 in) long and 1–1.5 cm (~0.5 in) wide; new foliage is more or less pinkish to beige in color. *C. ramiflora* produces small, white flowers in relatively inconspicuous clusters, and its brown, indehiscent dry fruits are scaly, roundish, usually wrinkled, and are 1.2–5 cm (0.5–2 in) long by 1.3–4 cm (0.5–0.8 in) wide.

Distribution: *C. ramiflora* is native from Malaysia to the Western Pacific, including Pohnpei. It is found along streams and in swampy areas and at the landward edge of the mangrove, often near *Heritiera littoralis* and *Barringtonia racemosa*. It is also occasionally found inland in lower undisturbed forests up to 400 m (~1200 ft) above sea level. In Micronesia, *C. ramiflora* occurs in the high islands of Palau, Yap, the Mariana Islands, Pohnpei, and Kosrae.

General uses: *C. ramiflora* has a number of traditional uses where it occurs naturally. The dark brown, heavy, hard and close-grained wood of *C. ramiflora* is used to make door-posts or for other construction purposes, and although it is "not very durable," it is reportedly useful for decorative veneers, inlaid work, as a source of firewood, and to make charcoal. The leaves and oil extracted from the seeds are reportedly used in some tropical areas to make lotions for treating skin diseases, and the tree is occasionally cultivated along roadsides and in gardens as an ornamental plant due to its attractive foliage, flowers, and shade value (e.g., see Tan et al. 2010).

Uses in Pohnpei:
C. ramiflora is called **Kemeu** or **Koamoau** in Pohnpei, where it is found along streams and in swampy areas (*nanpwoaki*); it is also found commonly at the landward edge of the mangrove on this island, often near other trees adapted to this wetland environment. *C. ramiflora* is also occasionally found inland in undisturbed forests, e.g., "from 1000–2000 feet in elevation" in Pohnpei (Haws 2006: 98). In Pohnpei, *C. ramiflora* is reportedly a source of medicine, construction, fuel, and oil (Herrera et al. 2010: 116).

▸ **Sapasap** or **Use** (P), **Moahdong** (Mw), **Saepasap** (Pi), **Helii baalanga** (Nu), **Dimuudi** (Ka)
Cyperus javanicus

Description: In this text, we only refer to one of the many sedge species in the genus *Cyperus*, *Cyperus javanicus*, which is relatively common on Pohnpei. *C. javanicus* is a perennial, coarse-tufted sedge that is short to medium-sized, up to 40–100 cm (~2–4 ft) tall with slightly triangular culms (stems). It is a short-lived perennial species, living less than five years. *C. javanicus* produces 12–24 spikelets with pale yellow to brownish inconspicuous flowers in the form of compound umbrella-like clusters up to about 15 cm (6 in) across; these small inflorescences have branched rays up to 10 cm (4 in) long. The small fruits of this sedge are 3-angled, ellipsoid, dark brown or black achenes.

Distribution: There are more than 4,000 species of sedges (family Cyperaceae), with about 70 genera worldwide; *Cyperus* is one of these genera. Sedges are most common in the tropics and are mostly associated with wetlands or areas near streams with relatively poor soils: "They are well represented in Micronesia where there is no shortage of poor soils and wet areas" (Costion and Lorence 2012). The name of *C. javanicus* indicates that this sedge is native to Java in Indonesia. Although it may sometimes be regarded as a weedy plant, it is also assumed to be native (indigenous) to some Micronesian islands and many other tropical areas of the world. In its natural, widespread habitat, *C. javanicus* (commonly known as the "marsh sedge") is found frequently in sandy coastal areas, mud flats, coastal wetlands exposed to brackish or salty water, at the landward edge of the mangroves, and sometimes in lowland taro plantations (e.g., see Stemmermann 1981).

General uses: *C. javanicus* has been used among Pacific Islanders for a variety of traditional and more modern purposes. For example, it was cultivated and extensively utilized by native Hawaiians for cordage, garlands (*lei*), and medicine (in Hawaiian this sedge is known as *'ahua'wa*).

C. *javanicus* can also be used in environmental conservation and restoration projects; e.g., it is tolerant of flooding and can be planted near stream banks to control erosion:

> "It is an excellent natural feature for restoration projects and used as food, nesting material and shelter by native waterfowl. This sedge tolerates salty water and soils. The plants will naturally reseed themselves for regeneration without becoming invasive." (www.nativeplants.hawaii.edu/plant/view/Cyperus_javanicus)

Uses in Pohnpei, Mwoakilloa, and Pingelap: In Pohnpei, *C. javanicus* is known as **Sapasap** or **Use**. Glassman (1953: 305) collected *C. javanicus* on Ahnd Atoll near (eight miles from Pohnpei) where this sedge was reportedly known as 'Kusakus' and commonly found along beaches and in coconut groves. Glassman (1953: 298) also indicated that this sedge is known on Mwoakilloa Atoll as 'Mordong' (= **Moahdong**) and could be found occurring commonly in the coastal strand and in coconut groves. Damas (1994: 232) lists the native name for *C. javanicus* as 'Sapasak' (= **Saepasap**) on Pingelap Atoll, and refers to St. John (1948) who described *C. javanicus* on Pingalap as "growing at edge of fresh swamp and being used to remove the [odor] from the body [with inflorescences that] were used for perfume."

▸ **Mwahng** (P), **Mwɛhng** (Mw), **Mwaeiang** (Pi), **Daogoli** (Nu), **Bulaga** (Ka)
Cyrtosperma merkusii

Description: *Cyrtosperma merkusii* is a very large, cultivated "swamp taro" that in some cases can reach heights up to 4–6 m (15–20 ft). In fact, *C. merkusii* is the largest species in the aroid or taro family (Araceae), and it lives for a relatively long time, with some varieties surviving for as much as 15 years. *C. merkusii* produces 6–8 very large sagittate leaves carried on long, thick petioles (stalks) up to 4 m (12 ft) long in a spiral arrangement with a relatively short leaf sheath. The leaves can grow up to 2 m (6 ft) long, but are usually shorter. The tips of the large leaves point upward. There is no above-ground stem, but the underground section develops into an enormous edible swollen corm which varies in weight according to cultivar type and age. Commonly, the corm of weighs about 15–25 kg (33–55 lb) when harvested, and rarely it may weigh 90 kg (200 lb) or more after about 10 years' growth. The corm shape varies among the varieties of *C. merkusii* from almost spherical to cylindrical or conical: "The giant swamp taro is believed to be the largest plant in the world which produces edible corms," and small young corms, known as cormels, are produced by a fully developed corm as side shoots after about three years. The cormlets are the vegetative reproductive units. They develop from "eyes at the base of the stem of older plants" and will eventually send up leaves and develop into "suckers" (NRI 1987: 308; Sproat 1968: 24). *C. merkusii* is generally propagated from these cormlets (or suckers), which are separated from the side of the mother plant and planted as clones to form new swamp taro plants. The tiny flowers of *C. merkusii* are produced on a multi-flowered spadix surrounded by a yellowish spathe. Numerous cultivars have been listed, with as many as 50 identified in the FSM (Rao et al. 2014). *C. merkusii* was formerly, and in some cases still is, a very important starchy, staple food source on many islands in Micronesia.

Distribution: Throughout the tropical Pacific region, edible aroid species, including *Alocasia macrorrhizos*, *Colocasia esculenta*, *Cyrtosperma merkusii*, and more recently *Xanthosoma sagittifolium*, are or have been important subsistence food crops. However, there are significant differences in the distributions and uses of these large herbaceous plants in this region's three major cultural regions. The huge swamp taro (*C. merkusii*) is generally the major aroid in Micronesia and Melanesia, while sweet taro (*C. esculenta*) is commonly the major aroid in Polynesia. Currently, giant taro (*A. macrorrhizos*) is only important as a staple crop on a few islands in western Polynesia and Micronesia. Although the distributions of swamp taro (*C. merkusii*) and sweet taro (*C. esculenta*) match up with "cultural regions," their adaptation to ecological factors rather than the cultural preferences of the people cultivating the plants has determined their traditional geographical patterns of use. For example, the traditional importance of *C. merkusii* "on the low islands and atolls of Micronesia is due to its ability to withstand brackish water and outcompete *C. esculenta* under such conditions" (Thompson 1982: 185). In fact, giant swamp taro is one of the very the few subsistence food plants which will grow on atolls.

Giant swamp taro is commonly grown in freshwater swamps or marshy areas on both high and low islands in Micronesia; however, it has long been assumed that this taro originated in Southeast Asia and/or possibly Near Oceania (e.g., New Guinea). During prehistoric times, prior to contact by Europeans, a number of useful plants were transported by people on sailing canoes throughout a vast area of the tropical Pacific islands. Up until recently, most scholars and scientists believed this giant swamp taro was one of those useful species introduced by humans to various islands in Remote Oceania. These isolated Pacific islands include most of those in Micronesia, and many in Polynesia, which have long histories of traditional cultivation of giant swamp taro.

A half century ago, the French ethnobotanist Barrau (1961: 42) referred to the similarity of local names for giant swamp taro over a wide geographical range: "From the Palau Islands to Samoa and the Cook Islands, via Yap, Ulithi, Truk [Chuuk], and Kapingamarangi, the vernacular names of this taro are very similar and can be compared with some native names in the Philippines." Although the origins of the domesticated *C. merkusii* and its relationships with other species in the genus *Cyrtosperma* are still not clear, we now know that swamp taro, *C. merkusii*, has long been an important starchy, staple food source in Micronesia as well as elsewhere in tropical Southeast Asia and the Pacific. For example, Weisler (1999a/b) documented the ancient existence and cultivation of giant swamp taro in the Marshall Islands, beginning about 2,000 years ago, and Hather and Weisler (2000) documented its early occurrence and possible cultivation on the very remote Henderson Island, near Pitcairn Island in eastern Polynesia, as far back as about 500 years ago (also see Horrocks and Weisler 2006).

More recently, especially remarkable archaeological and paleobotanical research has uncovered ancient evidence that swamp taro appears to have reached Pohnpei on its own, and this appears to have occurred thousands of years before the first people arrived on these islands! Athens and Stevenson (2012: 35) summarized the assumptions about human transported "canoe plants" into the islands of Remote Oceania, including Micronesia:

> "The legacy of this earlier research [led] most paleoenvironmental investigators to believe, first, that the presence of major cultigens [e.g., swamp taro] in their coring [archaeological excavation] records constituted undisputed evidence for human presence in the landscape, and second that the timing of initial human settlement could be determined by the presence of introduced plants in the pollen coring records, be they cultigens or other economically useful plants. While these premises are generally true, we have learned that there can be important exceptions." (Underlined for emphasis).

An exception that Athens and Stevenson (2012) discovered was ancient pollen grains of swamp taro recovered from two mangrove swamp cores from opposite sides of Pohnpei (Nanitipw-2 and Lewetik-4). The oldest prehistoric pollen from the excavated cores on Pohnpei dates back at least 7,000 years ago, which is thousands of years before humans arrived on this island!

> "Pollen data from the Nanitipw and Lewetik cores are unambiguous in their documentation of the presence of *C. chamissonis* [= *C. merkusii*, giant swamp taro] on Pohnpei since the early Holocene. The early part of this time frame places its presence on Pohnpei over 5000 years prior to the likely advent of humans on Pohnpei ca. 2000 years ago as documented by archaeology. This leads to the obvious conclusion that the natural distribution of [*C. merkusii*] was far greater than previously realized, extending north and eastward from Melanesia to at least the Eastern Caroline Island of Pohnpei."

The long-distance natural dispersal of *C. merkusii* may have occurred via birds that could have eaten some of the small reddish or orange berries that truly wild swamp taro naturally produces for dispersal and reproduction. These berries probably contained fertile seeds and were carried to distant islands, such as Pohnpei, in the birds' guts (e.g., see Hay 1988: 429). Another possible way the salt-tolerant swamp taro may have be able to reach Pohnpei on its own would be by drifting as part of a raft of "vegetative parts within clumps of soil protected within the root mass of drifting uprooted trees" (Athens and Stevenson 2012: 45).

It is interesting to note that Sproat (1968: 23) referred to giant swamp taro being "found growing wild in natural swampland" on Pohnpei and in some places in Chuuk and Palau. Thompson (1982: 189–190, citing other authors) commented on the "ecology and distribution" of *C. merkusii* taro in relationship to the other major aroids in the tropical Pacific region, with reference to how and when swamp taro became important to humans on Pohnpei:

> " The herbaceous habit of aroids and the fact that they are usually grown in more protected interior swamps provides greater resistance to wind damage by tropical storms. Thus, when high winds destroy other crops, aroids typically remain to provide sustenance. For example, in 1957 and 1958 when typhoons destroyed most of the staple foods on [Pohnpei], Cyrtosperma became an important food (Mahony and Bikajle 1960). The aroids also contrast with pandanus and breadfruit in their non-seasonality. Aroid tubers [such as those of swamp taro] can be harvested year-round while breadfruit and pandanus are chiefly available only in summer and winter." (Murai 1954)

Thompson (1982: 191) continued, pointing out how atolls such as Mwoakilloa (in Pohnpei State), Ulithi (in Yap State), and Arno (in Marshall Islands) sometimes suffer from long dry periods when food can still be harvested from swamp taro, while "the typically larger and/or higher islands of Melanesia and Polynesia provide environments more favorable to the growth of other aroids, particularly sweet taro..." Furthermore, weeds in the giant swamp taro patches are usually shaded out by the very large leaves of this plant, thus reducing work efforts in cultivation: "The *Cyrtosperma* are among the rare subsistence plants which will thrive on atolls" (Massal and Barrau 1956: 8).

General uses: Although use of *C. merkusii* has declined in several areas of the Pacific due to the introduction of new crops and modernization (see below), it is still cultivated in many high islands as well as in the low islands of Micronesia for its swollen, starch-rich, underground parts. It has long been grown on atolls in human-constructed, muck-filled, fresh-water swamps (pits dug to just above the fresh water lens) in the interior areas of the larger islets, and this farming practice appears to have been an ancient activity in these islands (e.g., see Weisler 1999a/b; Horrocks and Weisler 2006). Barrau (1960: 3) provided a perspective of the importance and adaptation of the giant swamp taro for human cultural development on tropical Pacific atolls:

> " *Cyrtosperma* [*merkusii*], an aroid with [an] edible tuber, to give only one example, was, and still is, grown on the Micronesian atolls [where] the Micronesians developed specialized techniques. Pits are dug in the central part of the islets until the level of the relatively fresh ground-water is reached. On the mud, the *Cyrtosperma* grows in baskets filled with compost and thus finds an artificial environment rather similar to its natural one."

Cyrtosperma swamp taro is commonly interplanted with *Colocasia* sweet taro in or near wetland areas, with *C. merkusii* commonly cultivated "as a fringe species between an open field or patch of *Colocasia* taro and the forest as giant swamp taro tolerates shade quite well," and on atolls, "the smaller taro depressions are heavily shaded by adjacent trees" (Manner 2011b). For more information regarding taro production for the two main aroid species, *C. esculenta* and *C. merkusii*, cultivated traditionally in Micronesia, see Manner 1993a/b, 2008, 2011a/c; Clarke and Thaman 1993; and Thompson 1982).

In many areas of the Pacific, cultivation of traditional taro species, i.e., both *C. merkusii* and *Colocasia esculenta*, has declined, with two major factors influencing this decline:

1. Shifts from subsistence to cash economies, e.g., the rise and then fall of the coconut copra industry;

2. The introduction of new, "easier to grow" crops. For example, the New World "taro," *Xanthosoma sagittifolium*, has replaced the native taros because of its ability to grow well on "dry land" (i.e., unirrigated or non-swampy areas).

3. In a recent review of giant swamp taro (*C. merkusii*) and its many cultivars, Rao et al. (2014: 96) described how contemporary economic pressures and desires have affected the Pacific islanders' choices of food according to a new set of influences from beyond their home islands:

> " Modernization has played a pivotal role in shaping the present trends in lifestyle preferences with nutritional Pacific staple crops being replaced by nutritionally poor imported foods. Giant swamp taro is no exception and has also 'fallen' a victim to modernization. It was once a major food source in the

Indo-Pacific region, but it is now being replaced by western food at an accelerating pace. Foods such as instant noodles, rice, flour, biscuits are not only replacing giant swamp taro but also other local food crops such as pandanus, taro, yams, breadfruit and many others. Moreover, not all the cultivars are cultivated as preference varies according to taste, use, ease of cultivation and so on (Iese 2005). So apart from the effects of climate change and modernization, current preferences may also contribute to the extinction of some of these cultivars."

However, Rao et al. (2014: 96) also focused on the worldwide effects and future consequences of global climate change, especially for the low-lying islands, including those in the FSM:

> "The production levels of this giant aroid in the atolls are now under threat from the effects of climate change and related sea level rise. The increased frequency and intensity of king tides and sea swells bring huge waves inundating the low lying atoll islands... This inundated sea water percolates into the groundwater lens increasing salinity in the fresh groundwater lens... Along with this, the disturbance caused by rise in sea level to the sensitive balance of the groundwater lens results in seawater intrusion which adds to the increasing salinity levels..."

In summary, *C. merkusii* is presently a minor crop from a worldwide perspective, but it has played an important role in human adaptations to relatively inhospitable Pacific island environments. Where giant swamp taro is specifically adapted, it continues to be an important component of the subsistence food-producing system of many remote island societies. Even with the decline of active cultivation of productive cultivars, wild, and semi-wild populations of *C. merkusii* may continue to be important as emergency and famine foods throughout the Pacific.

Uses in Pohnpei: The Pohnpeian name of giant swamp taro is **Mwahng**. As noted above, we now know that this taro species has been cultivated on Pohnpei and other islands in the region for a very long time. However, Sproat (1968: 25) indicated that more than 50 years ago on Pohnpei, giant swamp taro had already become known "generally as a reserve food and not cultivated extensively," while others indicated that **Mwahng** has remained one of the most important famine food staples on Pohnpei which "can be used when other foods are scarce, thus contributing to food-security" (Rehg and Sohl 1979). Indeed, giant swamp taro is still, more or less, an important staple food crop of Pohnpei, along with breadfruit, bananas, coconuts, and yams, while it continues to serve the people of Pohnpei as a traditional "famine food" when needed. Giant swamp taro is slow growing and may take several years to mature, but Sproat (1968: 24–25) referred to "some early maturing" Pohnpeian varieties, "notably ***Sounpwong weneu***" (see cultivar section below), which can produce a "good-sized corm in a year's time under good growing conditions." Generally, giant swamp taro is planted in Pohnpei and the outer islands "so that there are mature plants available at all times of the year."

Sproat (1968: 25–26) provided a summary of preparation and cooking of **Mwahng** for human consumption:

> "*Cyrtosperma* [**Mwahng**] is usually prepared for eating by peeling, cutting into sections and boiling or grating fresh and mixing with banana or coconut milk and then it is baked in an underground oven. The grated and mixed material is made up into one or two pound cakes, wrapped with leaves then placed in the oven. *Cyrtosperma* tubers are sometimes peeled, scalded, then chopped into pieces and dried in the sun. After proper drying the pieces may be stored in cans for a month or two. Slow, oven-baked pieces of *Cyrtosperma* tubers can be kept and eaten for about a week or more."

Lessa (1977) also commented that the large corms of *C. merkusii* are normally "baked among hot stones or boiled," and that the petioles (leaf stalks) were said to be occasionally cooked and consumed in Pohnpei: "the plant is used primarily during the fall and spring lean periods in place of yams and breadfruit, but it has no season and is often allowed to grow for several years before being harvested." The cooked corm of the giant swamp taro is a nutritious food with some yellow-colored corm cultivars containing high levels of vitamin A, calcium, potassium, and zinc (see Englberger et al. 2003b, 2008, 2009a). A number of documented food preparations involving cooked giant swamp taro corms, alone or mixed with other plant foods, are noted by Rehg and Sohl (1979): *mwahng pwoil* (boiled), *mwahng uhmw* (baked), *mwahng idihd* (baked with coconut cream), *pihlohlo en mwahng* (grated with mashed bananas and coconut cream), and *rotama* (grated and mixed with bananas and sweetened).

Cultivars in Pohnpei: There are many recognized cultivars of **Mwahng** in Pohnpei. Many were documented in the last century (e.g., see Bascom 1965). Lessa (1977) referred to 29 recorded cultivars of giant swamp taro being planted in Pohnpei. Most are among the 26 varieties identified in the Pohnpeian–English dictionary (Rehg and Sohl 1979). We list them all on p. 420. More recently, agricultural scientists (Rao et al. 2014) provided an extensive discussion and description of many cultivars identified in the FSM, including Pohnpei and the outer islands. These scientists used an analyses of variable plant characteristics for giant swamp taro, which helped them identify "36 distinctive cultivars" that they "characterized" out the "50 cultivars present in the FSM Pilot farm."

Uses on atolls in Pohnpei State: Global climate change should be kept in mind as we focus on the use and importance of agriculture crops such as those grown on low-lying atolls, including the outer islands of Pohnpei State; this is especially significant for those key species such the giant swamp taro, *C. merkusii*, which have been cultivated over many generations and about which the local atoll-dwellers have a remarkable breadth of traditional knowledge.

> "Sea level rise and shifting weather patterns are affecting food and water security. Freshwater lenses on outer islands are increasingly vulnerable to salt water intrusion from storms and sea level rise. The effects of king tides are particularly evident in the country's outer islands leading to salt-water inundation in taro patches." (Fletcher and Richmond 2010)

Giant swamp taro is grown intensively in the high rainfall atolls of Pohnpei and Chuuk States (FSM) where it has long been "prized as a staple food" (Sproat 1968: 25). More specifically, on the atolls of Pohnpei State, giant swamp taro has, at least historically, been considered the most significant local calorie-rich food and otherwise very important culturally: "*Cyrtosperma* withstands harsh, dry atoll climates, sandy saline soils, and strong winds and hurricanes, and thus it is the most important food crop on atolls" (Englberger et al. 2009: 132–134). In the middle of the 20th century, Mahony (1960: 112) underscored the significance of the durability of giant swamp taro on these atolls:

> "On Ngtik [Sapwuahfik], Mokil [Mwoakilloa], and Pingelap, *Cyrtosperma* is the stuff of life. It is by far the most important crop and people pride themselves on their knowledge of its cultivation. And they should, for they spend a goodly portion of their lives in cultivating [and] caring for *Cyrtosperma*. It is the mainstay of the diet, their life's blood." (Also see Bascom 1965; Bentzen 1949; Bradbury and Holloway 1988; Damas 1994; Emory 1965; Manner 2011c; Murphy 1949; Murai et al. 1958; Pollock 2006).

Extensive and centuries-old taro patches surrounded by coconut palms on Pingelap Atoll.

Research carried out more than 30 years ago by Bradbury and Holloway (1988) focused on the mineral content of the giant swamp taro; their research identified significant cultivar variation due to differing soil mineralogy among those plant varieties derived from the atolls compared with those developed on the volcanic island of Pohnpei:

> "They showed that the iron and zinc levels of cultivars grown on the volcanic island of Pohnpei were higher than those same cultivars grown on the Ngatik Atoll (Sapwuahfik). By contrast, cultivars grown on the atoll had a higher calcium content." (Manner 2011c: 12)

A half century ago, Sproat (1968: 25) recognized the importance of cultivating swamp taro to atoll-dwellers on islands with high rainfall. He also warned that "its culture is beginning to disappear; the younger generation prefers making copra and using cash to purchase imported rice and flour." That was true, but it did not take into account that declining interest in swamp taro cultivation on atolls such as Mwoakilloa may have been a short-sighted mistake, given the collapse of the copra economy in the 1970s (the history of the rise and fall of copra industry in Micronesia, including in Pohnpei and the outer islands, is discussed in the section of this book dealing with the coconut palm, *Cocos nucifera*).

Uses in Mwoakilloa: Recent archaeological research on Mwoakilloa has revealed an ancient excavated swampland associated with the cultivation of *Cyrtosperma* that dates back many centuries. The evidence for ancient taro cultivation on Mwoakilloa includes a prehistoric volcanic stone pounder that must have been introduced from beyond this atoll, probably from the high island of Pohnpei; in addition, land snail subfossil analysis and identification of pit cultivation have also been established:

> "Recent excavation on the island of Mwoakilloa has provided a date of 1750 BP for the initial settlement of the smallest atoll in the East Caroline archipelago lying between Pohnpei and Kosrae. XRF analysis of stone found at these levels suggests that the early settlement came from Pohnpei and was characterized by a marine foraging culture but was later influenced by cultures from Kosrae. Land snail analysis suggests influence from Kosrae and the Marshall Islands introduced

Swamp taro thriving in centuries-old pits on Mwoakilloa Atoll, with a community groundwater well adjacent.

pit cultivation practices for the growing of swamp taro (*Cyrtosperma*)." (Thompson 2011; also see Poteate et al. 2016)

Cultivation of the giant swamp taro and its use in Mwoakilloa were documented in the years following WWII, in the middle of the 20th century, as part of the region-wide "Coordinated Investigation of Micronesian Anthropology":

> "Taro [*C. merkusii*] far outranks the other foods in importance as it is the one staple that can be drawn upon in predictable quantities all through the year. More time is spent on the cultivation of taro than on all the other foods combined. Such cultivation is laborious work involving the digging of pits deep enough to reach below the water level five to twelve feet deep, and the building up of artificial humus which take generations of endless toil. The importance of taro in the economy of [Mwoakilloa] has been increasing as the population increased. In the pre-typhoon days, taro was raised only in limited quantities. There were three natural swamps on the atoll suitable for taro cultivation, two on Urak [islet] and one on Manton [islet]. At that time taro probably had little other than a ceremonial value... Without taro cultivation it is doubtful whether the people would be able to support themselves on this small infertile atoll. For this is the only type of intensive cultivation that is carried on in [Mwoakilloa]." (Bentzen 1949: 30)

Bentzen (1949: 31) provides a fairly broad discussion of taro history and contemporary cultivation on Mwoakilloa during the first half of the 20th century, noting the significance of cultivation of both sweet taro, *C. esculenta*, and especially giant swamp taro, *C. merkusii*, on Mwoakilloa:

> "Digging taro pits is recognized as the most difficult work. But the security that possession of plenty of taro provides as a constant food supply and the prestige that has been attached to its cultivation is sufficient incentive for every man to expand his taro land holdings. Not only can wet land taro be planted and harvested at any time of the years, but some varieties will mature sufficiently in six months to be eaten while other varieties [*C. merkusii*] can be left in the ground for at least fifteen years. It is said that there are [giant swamp taro plants on Mwoakilloa] which are as much as twenty years old."

In reference to the cultivation and importance of giant swamp taro on Mwoakilloa, Murphy (1950) emphasized the impact of the highly porous soils on this atoll:

> "While low rainfall produces the dry conditions that have led to the importance of [*C. merkusii*] and the extreme cultural elaboration surrounding its growth on the Gilbert Islands [Kiribati], drought of a different sort is the apparent reason for its importance on [Mwoakilloa] Atoll... Water might seem an unlikely problem on [Mwoakilloa] where rainfall averages over 250 cm per year. Murphy (1950: 61) notes, however, that "the extreme porosity of soils makes the problem of water supply serious." This extreme porosity leads to a faster rate of loss of fresh water from the fresh-water lens, resulting in more brackish lens water and therefore more brackish water in taro pits. Reflecting the brackishness of taro pits is the fact that [*C. merkusii*] comprises more than 85 percent of the taro grown on the atoll [Mwoakilloa]..." (Thompson 1982: 190–191)

According to Lessa (1977: 143), about a half century ago, the most important crop in wet gardens of Mwoakilloa Atoll was the giant swamp taro, while Colocasia esculenta taro was the second most important crop in the wet gardens, and although it is considered tastier than *Cyrtosperma* [*merkusii*], it is a "luxury crop" and thus was "not a main source of food."

On Mwoakilloa Atoll today, the giant swamp taro is grown for its "barrel-shaped" root stock, which may weigh as much as 22 kg (~50 lb); it is prepared for eating by being boiled or baked, or ground with pandanus, arrowroot, coconut meat, or molasses before cooking; it is by far the most important crop of the wet gardens, and traditionally appears in some form in most meals (e.g., see Murphy 1950; Lessa 1977; Taboroši and Sánchez Collazo 2011).

It should also be noted that *C. merkusii* on Mwoakilloa Atoll remains highly useful to the local people, but it is not as important as in the past due to the importation of food stuffs from outside sources:

> "The inner part of Kahlap [main islet] contains a large taro pit divided into many smaller patches maintained by individual families. These water-filled depressions are excavated and maintained by people in order to cultivate wetland taro, which requires ample water supply. In the past, smaller pits have been dug in the interior of the other two islets. They do not appear to be regularly maintained nowadays but residents indicate that some people still use them." (Taboroši and Sánchez Collazo 2011: 104)

Taboroši and Sánchez Collazo (2011: 104) also commented on agricultural problems on Mwoakilloa Atoll, related to the wetland taro cultivation in modern times:

> "The leaves of some large taro plants turn yellow and brown. At the same time, the root is observed to become rotten, described as *'perrer'* [*pɛrɛr*] in [the] Mokilese language. This is happening more in the part of the taro patch that faces the ocean and the residents feel that it is caused by intrusion of salty water into the patch." (Note: Underlining is added here to draw attention to the severe impact that global warming and resulting rising sea levels are having now, and most likely in the future, on atoll environments and the human populations that live on them).

Cultivars in Mwoakilloa: The generic name for giant swamp taro (*C. merkusii*) on Mwoakilloa is **Mwɛhng**. According to Bentzen (1949: 32), **Mwɛhng** existed in 15 varieties on Mwoakilloa in 1947, and he classified four of these cultivars as indigenous to Mwoakilloa and 11 as introduced from Pingelap, Ngatik, and Pohnpei in the first half of 20th century. We were able to match most from Bentzen's list to the up-do-date spellings provided in the Mokilese–English Dictionary (Harrison and Albert 1977), which does not clearly distinguish between different taro species but lists 18 cultivars that are likely *Cyrtosperma* (see p. 421).

Uses in Pingelap: St. John (1948: 108) observed the cultivation of giant swamp taro on Pingelap. He indicated that it was the main food plant on Pingelap Atoll (his spellings *'lepuel'* and 'Muiăng' correspond to Pingelapese *lɛhpwɛl* [taro patch] and **Mwaeiang** [*C. merkusii*]):

> "Near the center of the island was a large swamp that possibly had been enlarged by digging. It seemed to be 300 feet wide and more than 600 feet long. Within this low, wet garden, or *'lepuel'*, each family controlled a plot and each year fertilized it with leaves and trash. The *'lepuel'* was thickly planted to 'Muiăng'. The natives reported that the crop matured in 3 years, but could be harvested any time after 1 year. If allowed to grow to maturity, the corm attained large size, as much as 2 m [6.5 ft] in length and 60 cm [~25 in] in diameter and a weight so heavy that two men were needed to carry it."

Lessa (1977: 143), a half century ago on Pingelap, also named giant swamp taro as the most important food crop. Near the end of the last century, Damas (1994: 233) reported that this taro was still the "major food crop" and certainly the main cultivated aroid species on Pingelap Atoll. According to him, giant swamp taro corms are harvested at variable intervals depending on the variety:

> "Many corms are left in the ground for up to 10 years, while five years is perhaps a general average for harvesting when all species are considered. The fast growing plants like the ***Sounpwong waenau*** [*C. merkusii* variety], and [*Colocasia esculenta*] are most important in the periods of recovery after greater or lesser flooding of the main patch. My informants also indicate that as plants mature the greater part of the leaves, branches, and stems actually shrink in size. Therefore, one expects larger root growth after this process is observed. Fertilizer is applied when rain water is high in the patch so that the grass and leaves strewn on the surface will deteriorate in the water and then sink to the bottom where the resultant product will then work effectively."

Damas (1994: 54) also provided a valuable description of how the Pingelapese farmers responded to the major flooding associated with the major typhoon storm of 1957, as well as other periods of less dangerous conditions, but still troublesome minor flooding involving saline water. The indigenous Pingelapese methods of draining the salty floodwaters out of the main agricultural area is ingenious and instructive for present and future farmers on atolls, and it is relevant for revitalizing taro cultivation following natural disasters.

> "Atoll islands are the diversity hot spots for *C. merkusii* (giant swamp taro) and its traditional knowledge, especially in the Federated States of Micronesia (FSM). In these atolls and islands the giant aroid is given high importance in the food, cultural and traditional systems. Despite this, a significant lack of knowledge and the threat of salt water intrusion in the face of climate change exists for the aroid." (Rao et al. 2014: 95)

Recently, Yamamoto (2018: 6) carried out a "detailed study of household food consumption for every meal over a period 31 months" for one family on Pingelap, which revealed (at least for that household) that *C. merkusii* remains, along with breadfruit, one of the two major staple crops on this atoll. See the section on breadfruit earlier in this chapter for more of Damas' discussion of Pingelapese cultivation and use of this important agroforesty tree.

Cultivars in Pingelap: St. John (1948: 108) recorded five varieties of giant swamp taro (*C. merkusii*) cultivated on Pingelap. Damas (1994) listed the same varieties (with somewhat changed spellings) and added seven others. Interestingly, he noted that his informant Albert referred to three *C. merkusii* varieties, specifically named after their discoverers, Aikem, Seria, and Silington—grandfather, son, and grandson—who discovered one each of these varieties, which are all planted from the flower rather than from the stalk and cap [i.e., asexual vegetative reproduction, using a plant cutting that includes an upper section of the corm and part of the main stem]. Moreover, Damas noted that these three varieties are said to attain:

> "... the largest sizes of any of the aroids with a record corm of the '*Silington*' [***Naein silingidaen***] variety being marked on the wall of a municipal building said to be 10 feet above ground. The largest roots of these types extend well above the level of the mud and coconut husks and other fertilizer is piled upon them."

Damas also commented on the great size of the ***Pisep*** variety of giant swamp taro, noting that it is used in annual competitions and does not need fertilizer. Most of these varieties of *C. merkusii* on Pingelap are listed with up-to-date spellings in modern Pingelapese orthography by Hattori (2012). We discuss and consolidate this information on p. 422.

Uses in Kapingamarangi: The giant swamp taro, locally known as **Bulaga**, is apparently a major staple food on Kapingamarangi today, though it was historically not present there until the latter part of the 19th century, when it was introduced from Nukuoro. Eventually, this newcomer species displaced the previous keystone aroid, *Colocasia esculenta*, locally known as **Dala**, on Kapingamarangi in the 20th century. **Bulaga** is now grown in pits (*wai*) located on 11 islets of the atoll (e.g., see Bayliss-Smith 2012: 113). Most of about 500 Kapingamarangi islanders reside on Werua and Touhou islets, "but they utilize the other islands for cultivation (mainly *Cyrtosperma* taro) and for harvesting other plant materials for food, fuel, handicrafts, and housing construction" (Buden 1998: 1; also see Niering

Harvesting swamp taro on Sapwuahfik Atoll.

1956, 1963; Wiens 1956). Manner (1993b: 34, citing Wiens 1962) noted that "the taro pits [where giant swamp taro is cultivated] on Kapingamarangi Atoll are located on islets greater than 3.8 ha in size" because the freshwater lenses on smaller islets are insufficiently developed or completely nonexistent.

Thompson (1982: 191) described how environmental differences between geographical regions in the Pacific, and not cultural differences, explain why giant swamp taro is the main aroid crop in Micronesia, while sweet taro dominates in Polynesia and Melanesia:

> "This is illustrated by the situations on Kapingamarangi Atoll (Eastern [Caroline] Islands) and Makatea (Tuamotu Archipelago). Although located in Micronesia, Kapingamarangi is inhabited by Polynesians (Moore 1950). As Polynesians, it might be expected that their major aroid would be [sweet taro] and both Niering (1956) and Moore (1950) report that it was the principal aroid on Kapingamarangi in the past. Both state, however, that [giant swamp taro] is clearly dominant today and is the major food plant in the diet. Rainfall within the Eastern Carolines [FSM] is highly erratic from year to year as well as during the course of a year and extended droughts occur (Wiens 1956). Fischer and Fischer (1957: 6) note that within the Eastern Carolines "Kapingamarangi is the island most disturbed by occasional prolonged drought." Particularly during these droughts, [giant swamp taro] outcompetes [sweet taro] (Miller 1953; Niering 1956). Environment, rather than cultural expectations, determines the importance of [giant swamp taro] on Kapingamarangi."

Cultivars in Kapingamarangi: Unlike **Dala**, which was known to occur in at least 5–6 cultivars on Kapingamarangi, **Bulaga** on the atoll appear to be all the same, except for one named cultivar recognized as distinct (see p. 423).

Women cutting taro roots with knives and preparing them for sale in an outdoor kitchen on Pohnpei.

▸ **Ulung en kieil** (P), **Manu mangamanga** (Nu)
Davallia solida

Description: *Davallia solida* is a fern species found growing on the ground or on other plants (i.e., it is both epiphytic, growing on other plants, and epilithic, growing on bare rock). *D. solida* is a highly variable species, which, according to some botanists, is subdivided into three varieties: var. *solida*, var. *pyxidata*, and var. *fejeensis*. Generally, this fern species produces scaly, prostrate, creeping rhizomes (specialized roots) about 9 mm (~0.2 in) thick. Its rhizomes are covered by fine, rusty-brown hairs. The feathery-like fronds are variable in length and often pinnate; they are produced on erect stipes (stalks) that are tough but flexible, triangular, pinnately divided 2–4 times, and up to 50 cm (20 in) long. The specialized foliage dieback of species in the genus *Davallia*, in particular *D. solida* and another related fern, are described below:

> "The leaves of *Davallia* are articulated to the rhizome and are deciduous when old. They do not usually fall off at a definite season, but in *D. denticulata* and *D. solida* all leaves on a plant drop together, and this is followed rapidly by the growth of new leaves. In strongly seasonal climates the leaves fall as soon as the dry season is well established, and the plants rest in a leafless condition until the next rainy season. In non-seasonal climates plants are never bare of leaves for long. The deciduous habit is of great value for adaptation to seasonal climatic conditions. Environmental conditions such as relative humidity, light intensity, soil moisture content, soil nutrients affect the population density of the fern and the size of leaves. In the rainy season the number of leaves increases. Ferns growing under shade produce bigger leaves but fewer in number compared with those exposed to sunlight. Fertile leaves may occur the whole year round." (www.uses.plantnet-project.org/en/*Davallia*_(PROSEA)).

Distribution: The fern *D. solida* is reportedly native, or at least naturalized, from Malaysia to parts of eastern Polynesia. It is found growing on trees and on the ground, sometimes in relatively dense stands, but especially as an epiphytic climber up on woody plants. *D. solida* commonly grows on plants from mangrove and other coastal forests to upland areas, especially in dry forests near the shore, or on low scrubby vegetation on mountain ridges, often on lava flows and in association with other epiphytic ferns. *D. solida* has also been widespread as an ornamental species and even as a houseplant in places where it does not do well outdoors throughout the year.

General uses: *D. solida* is an eye-catching fern species that has been broadly cultivated as a decorative hanging plant or air plant:

> "They are often attractive indoor ornamental plants, grown in pots or in hanging baskets. The leaves are used fresh or dried in floral arrangements. The term "ball ferns" originates from Japan, where the flexible rhizomes of *D. trichomanoides* are tied into balls and other figures; when these figures are hung and watered, new leaves will subsequently appear on the surfaces, and they become self-contained hanging baskets." (Baltrushes 2006)

The fronds and rhizomes of *D. solida* are also reportedly used widely as ingredients in the medicinal mixtures of local healers in parts of Polynesia (e.g., see Baltrushes 2006 for use of *D. solida* on a number of tropical Pacific islands, e.g., in cleansing baths for newborns and for inducing menstruation on Mo'orea in the Society Islands). More generally, the rhizomes of *D. solida* have been used in its native or naturalized range as an herbal tonic to treat osteoporosis, arthralgia, and arthritis, and relatively recent research has identified antioxidants in the rhizomes (Chen et al. 2008).

Uses in Pohnpei: The Pohnpeian name for *D. solida* may be **Ulung en kieil**, which means "pillow of the **Kieil** [a species of lizard]." This fern is reportedly used in Pohnpei for treating cuts, open wounds, and bruises, as well as for "mangrove sickness" (Balick et al. 2009). Lee et al. (2010: 4, 73, 129) describe uses of *D. solida* and refer to mixtures involving other plant substances collected in Pohnpei for topical medicine and treatment of "culture-bound" syndromes of illness; they also provide cautionary comments about hygiene and the potentially toxic dangers of caffeic acid, a compound found in *Davallia* species.

▸ **Kehmelimel** or **Likaritikitik** (P)
Dendrolobium umbellatum

Description: *Dendrolobium umbellatum* is a common shrub of the beach forest and is often found on rocky areas near the sea, growing horizontally out towards the sea. The leaves are compound with three leaflets (i.e., trifoliate), and it produces flowers and fruits all year. The flowers of *D. umbellatum* are creamy white, and the fruits are quite unique, resembling a "beaded chain." This woody species is sometimes planted to improve soil conditions because, as a member of the bean family, Fabaceae, it is a "nitrogen fixer," and it is noted for its ability to help limit soil erosion (e.g., see Selvam 2007).

Distribution: *D. umbellatum* is found in various areas of tropical Africa, East Asia, the Indian subcontinent, Southeast Asia, Australia, and on some Pacific islands. *D. umbellatum* is sometimes known as "the horse bush" in English. "The seeds stick to clothing, helping the plant to be spread" (Falanruw 2015: 104).

Uses in Pohnpei: *D. umbellatum* is known in Pohnpeian as **Kehmelimel** or **Likaritikitik**. It has been reportedly used traditionally on Pohnpei for culturally bound uses as well as for some therapeutic purposes:

> "For use in sorcery (magic) to make a storm in the ocean, the stem is cut, a local prayer is said to the plant, and then it is thrown into the ocean... To treat arthritis, eight leaves are pounded, placed in a piece of cloth, squeezed into a cup of water, and consumed once daily for eight days..." (Balick et al. 2009: 384)

▸ **Peinuhp** or **Uhp** (P)
Derris elliptica

Description: The genus *Derris* belongs in Fabaceae, the bean or legume family, and includes many species of woody vines or lianas and trees native to Old World (e.g., see Bailey and Bailey 1976; Sirichamorn et al. 2012). *Derris elliptica* (recently revised by botanists to *Paraderris elliptica*, e.g., see Lichvar et al. 2016) is assumed to be an alien species in Micronesia. It is a close botanical relative of *Derris trifoliata*, which is native in parts of Micronesia, and of *Derris malaccensis*, which originates in New Guinea but is naturalized on many other islands in the Pacific (see PIER 2002). *D. elliptica* is a large, sprawling woody vine that can reach over 10 m (30 ft) in length. Its branches are densely covered with brown hairs, and its compound leaves are about 15 cm (6 in) long with 4–6 pairs of leaflets that are obovate-oblong (i.e., they have an elongated form but are teardrop shaped, with the stem attached to the tapering point). The undersides of the leaves are brownish and silky. The flowers are borne in stalked clusters of three racemes up to about 30 cm (~1 ft) long, with pinkish to bright red petals (racemes are flower clusters with separate flowers attached by short equal stalks at equal distances along a central stem). The fruits (pods) of *D. elliptica* are oblong or oblong-elliptical and up to about 7.5 cm (3 in) long with narrow "wings." Each pod contains 1–3 olive, brown, or black seeds that are flat and reniform (kidney-shaped; or roughly circular in shape with a notch). An aggressive species, *D. elliptica* reproduces both vegetatively (clones) and by seeds. Its roots contain the pesticide known as rotenone, which is used widely as a fish poison and is cultivated in some areas for the horticulture industry.

Distribution: Apparently from early times, the root of *D. elliptica* has been used as fish poison in much of mainland and insular Southeast Asia. *D. elliptica* is believed to be native to some areas in the eastern part of South Asia and much of mainland and insular Southeast Asia. For example, Merrill (1923: 299) reported it growing abundantly in the Philippines as a rambling climber amongst "thickets along streams, secondary forests, etc., at low and medium altitudes" (also see Sirichamorn et al. 2012; Bailey and Bailey 1976). Throughout Micronesia, *D. elliptica* is likely to be an alien or introduced sprawling, woody plant species. As in many other tropical areas, it was introduced to the region by humans, especially during the period of Japanese government control. Notice the close similarity of the Micronesian names for this species: **Duub** in Palau; **Yuub** in Yap; **Bagin** or **Tuba** in the Mariana Islands; **Uup**, **Wúúp**, or **Wúún** in Chuuk; **Uup** on Namoluk Atoll; **Up** on Namonuito Atoll; **Uhp** or **Peinuhp** on Pohnpei; **Op**, **Oop**, or **Up** on Kosrae; and **Te obu** on Banaba. *D. elliptica* is found growing wild near wetlands, as well as in forested areas inland, and can be an aggressive vine that smothers other plants and other objects as it spreads. For example, as an invasive species it has been reportedly observed climbing over small trees and shrubs on Rota and it has been recorded as a problematic weedy species on many other islands in the Marianas and elsewhere in Micronesia, especially on the high islands (e.g., see www.hear.org/pier/species/derris_elliptica.htm).

General uses: This *Derris* species is cultivated or collected as a wild-growing plant in some tropical regions for its "fish poison." Some fishermen have practiced a form of fishing in which they crush the roots of *D. elliptica* and throw them into the water. Fish that are stunned or dead float to the surface, where they can be easily reached and harvested. The toxic pesticide rotenone is found in the roots of the *D. elliptica* vine, with its concentration varying from 1.6–10.0% (e.g., Thothathri 1982: 17). These roots are commonly known in English as "derris or tuba root of commerce" (e.g., see Sirichamorn et al. 2012: 432). Rotenone is a respiratory inhibitor which kills fish by depriving them of oxygen exchange (e.g., Eldredge 1987). Although rotenone affects most species of fin fish, many aquatic invertebrates (e.g., shell fish) appear to be unaffected by it (e.g., Des Rochers 1992: 11). *Derris* roots have been used as a fish poison traditionally and in modern situations over a wide geographical range; the white milky sap exuded from the pounded roots is the raw material for stunning and catching fish, but today it is banned in many areas because all, or at least most, of the fish in areas where it is applied are disabled and harvested in bulk. Thus it has often been exploited widely for commercial fishing purposes. On the other hand, the native people in parts of Southeast Asia have also traditionally used *D. elliptica*'s exuded sap as an arrow-poison for hunting terrestrial animals, as well as for fish poison (see Iwasa 1975, Etoh 2001; also see www.stuartxchange.com/Tubli.html).

Generally in Micronesia, the use of *D. elliptica* root extracts has been strongly condemned due to their impact on viable fishing; for example, "Destructive and unsustainable fishing methods—e.g., dynamite, chlorine, fish poisoning plant (*Derris elliptica*), and small mesh gillnets" are listed as a key threat in the "Over Exploitation and Unsustainable Harvesting Methods and Practices" for governmentally protected natural areas of the FSM (see the FSM National Action Plan for PoWPA, the Programme of Work on Protected Areas; www.cbd.int/protected/implementation/actionplans).

Uses in Pohnpei: Ragone et al. (2001: see Appendix 1) listed *D. elliptica* as one of those introduced plants that was brought to Pohnpei before the Japanese administrative takeover in 1914, but the exact date of introduction is not known. These authors also listed the use of this plant as an "ornamental." Today, there are a number of reported local names for *D. elliptica* in Pohnpei (see above). As in several other Micronesian island areas, the roots of this vine have been used as a source of effective, but potentially destructive, fish poison.

▸ **Kehoaroahr** (P), **Kainipil** (Pi)
Derris trifoliata

Description: *Derris trifoliata* is an indigenous species in Micronesia and a close botanical relative of *D. elliptica* (notorious fish poison plant that is now an alien invasive species on a number of Micronesian islands); the botanical genus *Derris* is classified in Fabaceae, the bean or legume family. *D. trifoliata* is usually a slender, woody climber or liana that can grow to be as much as 15 m (45 ft) or longer, and under certain conditions, it can blanket the ground and surrounding vegetation (Stemmerman 1981: 87). *D. trifoliata* also sometimes develops into an erect shrub, producing smooth or glossy dark red branches from which compound or pinnate leaves with 3, 5 or 7 leaflets are borne (*trifoliata*, the specific or epithet name for this species, refers to its leaves normally having three leaflets). The inflorescence of this scrambling woody species is a slender raceme (i.e., a flower cluster with separate flowers attached by short equal stalks at equal distances along a central stem). The individual flowers are small, whitish to light pink or rose colored, and pea-like in form. The slightly wrinkled fruits each have a disk-like orbicular to oblong shape that is about 3–4 cm (1–1.5 in) long and contain 2–3 seeds.

Distribution: As an indigenous climbing woody vine or liana, *D. trifoliata* has a wide native range in the tropics from parts of Africa, including some islands in the Indian Ocean and regions of South Asia, Southeast Asia, and the Western Pacific. *D. trifoliata* can creep up trees in the strand vegetation that occurs near the coast in swampy scrub vegetation, as well as in forests behind beaches, on muddy foreshores, in or near estuaries, and along the edges of mangroves, sometimes in pure stands.

General uses: *D. trifoliata* has been reportedly used medicinally as a stimulant, antispasmodic, and counter-irritant, as well as for rheumatism relief, indigestion, and painful menstruation that typically involves abdominal cramps. Somewhat similar to its close relative *D. elliptica*, *D. trifoliata*'s bark and roots are sometimes used as a "fish poison" (e.g., Yusuf et al. 2009) and occasionally for cordage. Because of its attractive pea-like flowers and coin-like fruits, the plant has been touted for its horticultural potential, and the leaves are reportedly fed occasionally to cattle and other livestock.

Uses in Pohnpei: Because of its well-developed ability to hold the shore or soil together, the ritual use of *D. trifoliata*, known in Pohnpeian as **Kehoaroahr**, is believed to improve the efficiency of fishing gear to catch and hold fish. According to Pohnpeian tradition, *D. trifoliata* holds the shore, and therefore the land, of Pohnpei together, as well as having several other uses:

> "To use as a fish bait, leaves are pounded and placed with a fishnet in the water to bring more fish to your net... The [long stem] of this plant was used to make cordage to tie things. It is reported that it is not used for this purpose anymore, except as a yam trellis... To use as a charm to protect one's working area (farm and building) from sorcerers or from a sorcerer trying to make a person's crops grow poorly, the vine is pounded and buried in the ground around the working area... To use as a charm for buildings or houses that are incomplete in construction (caused by sorcerers)... To use as a charm to help catch more fish, the bark is placed in the canoe before fishing... To use as a charm to dispel other sorcery on your canoe or net... To use as charm to harm a person who steals (often applies to crops such as stolen *sakau*...)" (see Balick et al. 2009: 384–385)

Uses in Pingelap: *D. trifoliata* is known as **Kainipil** in Pingelapese. St. John (1948: 111) described it as a "vine, climbing on trees near beach," but did not record any use for it. Damas (1994: 237) added that it is a "vine which is pounded and [the exuded sap is] used on fresh wounds."

▸ **Dipwinmal** or **Mwedilinmal** (P)
Dicranopteris linearis

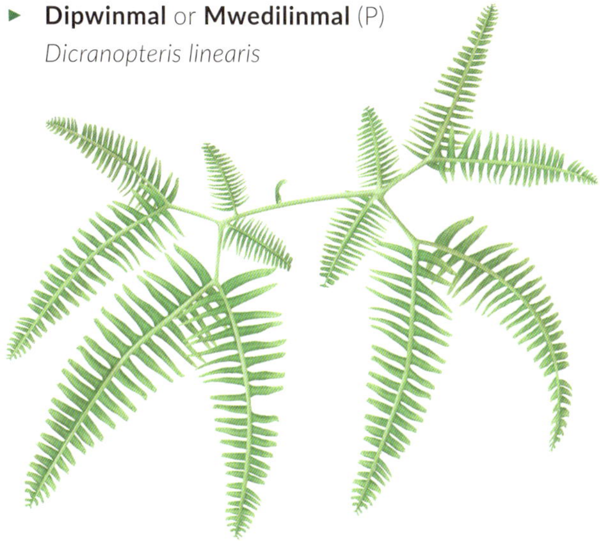

Description: *Dicranopteris linearis* is a vigorously creeping fern that spreads laterally by extensive rhizomes to form large clumps. The rhizomes have tips covered with dense scales and develop at or just beneath the soil surface, where they intermittently send up fronds that are large, stiff, and dichotomously branched (i.e., dividing into two parts or sections). *D. linearis* sometimes grows as a large, thick mat of strong stems and wiry fronds, forming a tough barrier. In fact, *D. linearis* can quickly develop into an almost impenetrable thicket up to 2 m (6 ft) or more in height, shading out all other plants. It can grow also up onto trees, reaching heights of 7 m (22 ft) or more, with its slim fronds meshing together and making it virtually impossible to penetrate. These fern clumps burn relatively easily due to an accumulation of dead stems and leaves; this creates a potential fire danger.

Distribution: *D. linearis* is an indigenous native fern common in many disturbed areas or secondary forests in tropical or subtropical areas ranging from Africa to the Indian subcontinent, as well as in tropical China, Southeast Asia, and parts of Australia, New Zealand, and the volcanic Pacific islands. It is generally an indicator of very acidic soils. It is adapted to colonize forest margins or open ground, especially in poor soils where heavy erosion or landslides have occurred, and almost always in full sun. This aggressive fern facilitates ecological succession because it rapidly covers areas cleared of the vegetation by natural disturbances or human activities, such as burning vegetation. It can spread by spores, or by its rhizomes which clone, scramble along the ground, and climb on other vegetation. "The stem grows from the rhizome, branches at a 45° angle, and forms fronds that continue to bud and branch... [and in this manner the fern's] growth can continue for a long distance as the plant forms a mat, grows over itself in layers, and spreads" (Romanchak et al. 2005). From a positive point of view, this vine-like fern plays a significant role in stabilizing exposed soils, especially in areas of steep slopes and high rainfall. In fact, *D. linearis* forms a distinct indigenous vegetation type, designated as *Dicranopteris* fernlands (e.g., see Mueller-Dombois and Fosberg 1998; Merlin 1985; Maxwell 1982), and it plays an important role in successional structuring of the plant community in many humid tropical ecosystems. This includes the high islands of Micronesia, where it may develop into "very dense thickets" in areas that are consistently kept more or less open by people setting fires; this "in turn prevents forest regeneration but also increases plant diversity in the savannas as it is hard for other plants to compete with *Dicranopteris*" (Costion and Lorence 2012: 58):

> This fern species is... able to colonize and maintain dominance due to its shallow rhizomes, highly effective leaf area, and phosphorus use efficiency. [*Dicranopteris linearis*] fern is also known to contribute to soil organic matter in the form of leaf litter as soil nitrogen and phosphorus [are] returned to the poor soils common in tropical ecosystems." (Romanchak et al. 2005: 517)

As noted above, the rapid, multilayered, clonal fern growth and specialized nutrient requirement of *D. linearis* also appear to help dampen the impact of alien invasive species by crowding them out, and thereby it can play a significant conservation role in preserving native species while enhancing ecosystem restoration over time:

> Growth form and nutrient use in *Dicranopteris* [*linearis*] render this species unrivalled in terms of its ability to vegetate harsh landscapes and maintain dominance by blanketing a site with leaves. Carbon gain in *Dicranopteris* [*linearis*] under nutrient-poor conditions can be substantial, resulting in important impacts on regeneration, ecosystem-level productivity, nutrient uptake, soil genesis and, consequently, ecosystem development." (Russell et al. 1998: 778)

Slocum (2000) suggested that fern patches could create a suitable condition and pathways for secondary species to colonize and help successional processes proceed, thus leading to the restoration of heavily disturbed or destroyed forest areas in the tropics. For example, Cohen et al. (1995) suggested "that disturbing (clean weed, root removal and till) *D. linearis* fernlands soil could accelerate the germination of herb, shrubs and trees seeds," however the seedlings recruited from these *Dicranopteris* "disturbed fernlands were dominated by herb, sedge, and grass species," not forest species (see Florentine and Westbrooks 2004). More research is recommended, focusing on the role and value of *D. linearis* fernlands in independent or related soil stabilization and renewal, as well as its putative associations with tropical forest restoration.

General human uses: Because *D. linearis* can spread rapidly in open disturbed areas, its growth is sometimes culturally encouraged or integrated into soil stabilization schemes, especially in steep landscapes. In addition to its ability to slow down rates of erosion, there are also a number of ethnobotanical relationships associated with the *D. linearis*. For example, in some tropical areas of its natural, widespread distribution, the pliant fibrous leaf stems (petiole and leaf axis) of *D. linearis* are traditionally woven together to make matting, fish traps, chair seats, thatching for walls and roofs, pouches, caps, and ropes. Indeed, the lignified peripheral fibers of the stem are used widely in Asia for plaiting. Traditional uses have also been noted in Remote Oceania. According to Pukui (1983), a kind of Hawaiian tea brewed from the fronds of the *uluhe* fern (*D. linearis*) was imbibed traditionally by Hawaiians as a laxative; and "Shore-dwellers of Ka'u on the island of Hawai'i applied the derisive term *pala uluhe* (ripened in *uluhe* leaves) to poor upland farmers, who ripened bananas in pits lined and covered with *uluhe* leaves, rather than allowing them to ripen in the field" (also see Cambie and Ash 1994 for traditional medicinal use references for *D. linearis* in Fiji, as well as a listing of the chemical compounds identified from samples of this fern; in addition, see Baltrushes 2006 for a reference to traditional medicinal use on Moorea Island).

Uses in Pohnpei: *D. linearis* is known in Pohnpeian as either **Dipwinmal** or **Mwedilinmal**. It has cultural significance on Pohnpei. For example, perhaps because of its natural toughness, it is used as a source of spiritual protection for a house when it is being built. It is also used "at local dances in *mwaramwar* (head garland), a stem (with leaves) the circumference of the person's head is wrapped in a circle and tied around the head" (see Balick et al. 2009). According to the online Pohnpeian–English dictionary, it is "traditionally placed in the home and inserted in thatch to prevent roach infestation.

▸ YAMS

Yam species, particularly *Dioscorea alata*, are among the most important plants for people living in the high islands in the Pacific, and in Micronesia nowhere more so than Pohnpei. Because of this, we first present the botanical, geographical, and ethnobotanical characteristics of the yams in general found in Pohnpei, then discuss individual species.

> "The genus [*Dioscorea*] comprises 603 species, distributed throughout tropical regions and independently domesticated on three continents. Over 50 species are edible, and without doubt these were important food for the first humans." (Walter and Lebot 2007)

Indeed, yams have been an important food source for humans for millennia and may well have been actively sought by our ancestors as they moved out of Africa and spread about the Earth as hunters and gatherers. In parts of West Africa, **Nyame** (from which the name "yam" is derived) is a sky god, which may be a reflection of this very ancient source of wild food for people. Many yam species have spread far and wide from their places of origin and have become basic food crops and important cultural species for millions of people in tropical and subtropical regions all over the world, especially in three principal areas: West Africa, Central and South America, and Southeast Asia. It is from Southeast Asia that some of these cultivated yam species were brought to and utilized in some of the Micronesian islands.

Yams are among the major root or tuber crops that have sustained the lives of many Micronesian islanders for centuries or even millennia. Certain yams not only provide essential food but are also important components of a prestige economy (for example, as offerings during communal feasts associated with funerals, weddings etc.). Thus, on a number of high volcanic Micronesian islands, the conservation of the diversity of yam species and varieties or cultivars within species is important economically, ecologically, socially, and culturally (e.g., see Krishnapillai 2005a/b).

Species in Pohnpei: All members of the yam family, Dioscoreaceae, in Remote Oceania are assumed to be human introductions (e.g., Whistler 1991, 2009a). Five main yam species were first introduced long ago and are now found in the high islands of Micronesia. These include three cultivated species: *Dioscorea alata*, *Dioscorea esculenta*, and *Dioscorea nummularia*, while two relatively minor "wild" species are used primarily as famine food sources: *Dioscorea pentaphylla* and *Dioscorea bulbifera*.

Multiple yam varieties for sale at a local market in Pohnpei.

All of these yams apparently originated in Southeast Asia or Near Oceania (especially New Guinea). From there, yams were purposely transported by humans in prehistoric times to various Pacific islands, over many centuries during the prehistoric period, or perhaps in some cases after European contact. Pohnpei is the center of yam diversity in Micronesia, and all five and more species have been recorded, but only some of these five yam species are found and used on other islands where yams grow in Micronesia. Four were brought in by the early voyagers: *D. alata* (the most important traditional food and prestige yam), *D. esculenta* and *D. nummularia* (two other more or less widely cultivated species), and *D. bulbifera* (a wild or semi-wild yam used as a source of traditional medicine or, if cooked for a long period of time, a source of famine food).

Additional edible yam species have been introduced to Pohnpei since European contact began in 1826. Among them are *Dioscorea pentaphylla*, which has compound leaves with five leaflets and may no longer be present on the island, and *Dioscorea hispida*, which has compound leaves with three leaflets and is said to have been introduced from Hawai'i in the 1950s (Balick et al. 2009: 43).

Warning: Like many wild yams, *D. pentaphylla* and especially *D. hispida* contain toxic compounds and require special processing before consumption. In 2015, PestNet.org reported that several families on Pohnpei fell seriously ill with diarrhea and vomiting after eating the tuber of what was likely *D. hispida*.

Cultivars in Pohnpei: Based on a research project started in Pohnpei in the late 1980s, Raynor and his colleagues (1992: 25) investigated the production and significance of cultivated yams (*Dioscorea* spp.) on the island. They described yams as Pohnpei's most significant series of crops and acknowledged their prominent role in the traditional prestige system:

> “One hundred seventy-seven locally-recognized cultivars of yam representing five species were reported by informants. Yam production is seasonal, complementing the breadfruit season. The yam season is delineated by a series of feasts and ceremonies. In order to insure a steady supply of yams during this time, farmers plant several cultivars, use different trellising methods, and employ other techniques. Cultivation of yams is done mainly under live breadfruit trees, although several other species of trees are used, and are often ring-barked to improve yam growth. Taboos and magic are still employed by many farmers. Although production per unit land area is low under the traditional system, yams of large size are produced which are highly suitable for prestige purposes." (Also see Raynor et al. 2009).

It is important to understand that Pohnpeian classifications of yams and their cultivars do not match the genetic classification and scientific nomenclature. In our following detailed discussion, we recognize there is some confusion concerning the overlap of local and scientific names for the cultivated *Dioscorea* species of yams on Pohnpei, which greatly complicates the identification of the botanical species to which a particular cultivar belongs. Local names are not a reliable indicator because **Kehp**, the name of the most commonly cultivated species, *Dioscorea alata*, is also the generic name for all yam species. Two other widely cultivated species are *D. esculenta* (cultivars known as **Kehpineir** and others) and *D. nummularia* (cultivars known as **Kehmmwas** and others). However, these names sometimes overlap, being used for certain cultivars that may belong in any one of these three yam species. In contrast, the wild-growing **Kehmpalahi** (*D. bulbifera*) is more distinct and less likely to be confused with the rest. The relatively recent introductions, *D. pentaphylla* and *D. hispida*, are uncommon, with the former having no local name and the latter known as **Kehp en awai**. Their occurrence and use on Pohnpei remain inadequately documented. Therefore, all or the vast majority of yam cultivars mentioned in this book are those selected by farmers over time from the main cultivated yam species, *D. alata*, and to some extent *D. esculenta* and *D. nummularia*.

As noted above, a study of crop cultivars on Pohnpei carried out over 30 years ago (Raynor and Fownes 1991: 151) indicated that there were more than 170 different cultivar names for yams in Pohnpei, which is the most named varieties of any crop on this island. Despite difficulties in establishing the species of different yam cultivars in the field, it is known that the major cultivars "were of *D. alata*, one of which, [***Kehp en dol***], made up 18% of all yams recorded." Raynor and Fownes also pointed out that many of the "commonly-occurring yam cultivars" have been introduced since European contact (also see Bascom 1965). Unfortunately, the prime ***Kehp en dol*** cultivar "is extremely susceptible to yam dieback or anthracnose, and since 2001 its cultivation has declined to the point where it is now only rarely encountered on farms and in the local market" (Raynor et al. 2009: 41). In any case, today on Pohnpei there are assumed to be almost 200 locally named cultivars reported by various informants. The full list we provide on pp. 427–429 includes 194 cultivar names we gathered from the most authoritative sources, with duplicate names and multiple spelling variants resolved as much as possible.

Yams and people in Pohnpei—a warning for the future: In summary, Pohnpeians should pay special attention to the future status and traditional use of all yams species and also to the survival of the great many yam cultivars developed in Pohnpei, including the most important traditional food and prestige varieties. Why? Because Pohnpeians, as well as other Micronesians who cultivate and depend upon yams for sustenance and cultural traditions, must be aware of the progress and impact of global warming and related climate change.

Yam species, like most other plants, are under threat due to global warming and how this climate change affects the production of these important crop plants. The 5th National Report by the Pohnpei Government to the Convention on Biological Diversity (2014: 15) referred to evidence showing a steady increase in annual mean air temperatures in Pohnpei and Yap between 1950 and 2010, and how the climate change is affecting the health of **Kehp** and **Sakau** (FSM 2012: 23).

> "In Pohnpei stakeholders reported that changing weather patterns are affecting the soil and reducing crop yields. Changing soil conditions may also be behind recent outbreaks of fungus affecting more than 100 varieties of yams in Pohnpei and **Sakau** plants in Pohnpei and Kosrae. In each State, stakeholders reported that changing weather patterns have already resulted in different harvesting patterns than previously known. For example species of breadfruit are ripening at different times during the year, which is impacting traditional and cultural practices that revolve around harvest seasons. These observations are in line with scientific findings about changes in soil and weather patterns."

▸ **Kehp** (P), **Koahp** (Mw), **Kaehp** (Pi), **Iaama** (Nu), **Eeme** (Ka)
Dioscorea alata

Description: *Dioscorea alata* is a commonly cultivated yam species in various tropical areas on the Indo-Pacific. It is known in Pohnpeian as **Kehp** and in English by various names such as the greater yam, winged yam, or water yam. *D. alata* is a-high-climbing, herbaceous or somewhat woody, tuberous vine. Its storage organs or tubers are club-shaped or globose, often lobed with blackish or brownish skin covering its white or purplish flesh. The lengthy stems of this vine are 4-angled and winged, about 5 mm (~0.25 in) thick and often produce small aerial tubers (bulbils) at the leaf axils. The long modified stems, up to about 9 m (30 ft) in length, have no thorns and differ in color among the various cultivated varieties of this species. The stems grow upward, twisting counterclockwise as they climb up on to woody plants, sticks, or available props placed in the ground. The leaves are about 15 cm (6 in) long or more, are smooth with 7–9 distinct veins, and are arranged in opposing pairs. They have an elongated heart shape and an extended, pointed tip. According to Bradbury and Holloway (1988: 27), cultivated yams are vegetatively propagated from the tubers (about 20% of the previous crop is stored for planting), and perhaps as a result, many cultivars now only flower rarely. The small flowers, although forming infrequently, are borne in leaf axils and in clusters at ends of stems. *D. alata* is a dioecious species, i.e., the male and female flowers occur on separate plants, with male flowers forming as panicles up to about 30 cm (1 ft) long and the female flowers in smaller spikes. The underground tubers of *D. alata* can become enormous, especially under attentive cultivation; some tubers of this species can attain weights of more than 50 kg (100 lb)!

Distribution: *D. alata* is one of the foremost cultivated yam species and is now "the most widely distributed species globally" (Muimba-Kankolongo 2018). In spite of "its economic and cultural importance, very little is known about its origin, diversity and genetics" (Arnau et al. 2017). However, we do know that *D. alata* was brought from its original home in Southeast Asia or Near Oceania to various tropical Pacific islands for its starchy tubers, probably starting more than 2,000 years ago (e.g., Coursey and Martin 1970). In lowland New Guinea, yams, especially *D. alata*, are a staple of most traditional peoples. Yams are still cultivated intermittently throughout insular Southeast Asia, and there is increasing evidence that they were cultivated well before the expansion of Austronesian-speaking peoples into Remote Oceania including Micronesia. For example, Bourke (2013) refers to Ille Cave on the island of Palawan in the Philippines, which would have been connected to Borneo by land prior to the rise in sea level in the Holocene, and reported that there was "crucial evidence" for cultivation of yams more than 8,000 years ago. Bourke quotes Barker et al. (2011: 6):

> "The presence of parenchyma [cellular tissue] identical to those of the modern domesticated yam [*D. alata*] at Ille Cave in secure stratigraphic contexts provides extremely strong evidence that it was being intensively exploited by people in Palawan thousands of years before the supposed Austronesian expansion."

D. alata yams are still an important staple food "in the cuisine of some islands of the Philippines, [where] a popular dish, *ube halaya*," is prepared from grated tubers of *D. alata* (Bourke 2013). What characteristics have contributed to the widespread growth and consumption of this yam?

> "Traits of *D. alata* make it particularly valuable for commercial cultivation. These include high yield potential, ease of propagation, early growth vigor for weed suppression, and long storability of tubers. Tubers possess a high nutritional content with an average crude protein content of 7.4%, starch content of 75–84%, and vitamin C content ranging from 13.0 to 24.7 mg/100g." (Arnau et al. 2017)

General uses: Some have argued that *D. alata* has its ancient domestication origins in Melanesia:

> "It is certainly one of the most ancient of cultivated plants. Domesticated in New Guinea, which is also its center of diversity, and then distributed throughout the tropical world, it is the best represented of the species of yams. It reached Vanuatu at a time that is still uncertain, probably with the people who made the Lapita pottery 2,500 years ago, but it may even be older than that. The winged yam [*D. alata*], a seasonal plant, is used everywhere in customary exchanges, and is the staple food of coastal populations. The vegetative cycle of this yam sets the rhythm for the village life, for certain rituals and for the cultural calendar." (Walter and Lebot 2007)

In Melanesia, as in some areas of Micronesia (such as Pohnpei), *D. alata* assumes unique "extra-culinary status" as a component of the celebrated "big man" complex, which among other characteristics involves ceremonial efforts of conspicuous consumption or display by farmers to obtain prestige:

> "In part this behavior can be seen as a product of the growing requirements and storage capabilities of the tuber itself. The plant requires rich well-drained soil which is present in abundance on the high islands of Melanesia and absent in atoll environments: it crops prolifically and tubers can grow to enormous size; and finally, unlike taro, it keeps well in storage. Because of these characteristics it can be grown in excess of food requirements and, like money, in our own economy becomes available in effect as a collectable product." (McLean 2014: 121–122, also see Ross et al. 2008: 258)

A recent guide to traditional foods of Pacific islands describes the quality and preparation of *D. alata*. Their ground tubers can become large and of excellent food quality; under some conditions, the tubers may be stored for 12 months or longer. Some of the more importance aspects of preparation include the following: it should be washed well before use; it can be baked, roasted, or steamed whole with skins; it can also be boiled (after peeling), stewed, or made into soup; possible poisons should be avoided by washing them in cold water, cooking them repeatedly, and adding lemon juice to cooking water; cooked yams should be peeled and cut to dry in the sun or in a drier (e.g., see Matanane et al. 2014).

Uses in Pohnpei: *D. alata* has long been an important prestige food in Papua New Guinea, Fiji, Tonga, Vanuatu, Samoa, and in some islands of the FSM, including Pohnpei.

> "Of all the subsistence root crops on Pohnpei, the yam is by far the most important, although its importance to the prestige economy as a major component of feast and competitions far outweighs its value as a food crop." (Raynor et al. 2009: 40)

D. alata yams are definitely among Pohnpei's most important traditional cultivated food plants. Their cultivation can bring much pride to a Pohnpeian farmer, and traditional relationships between yams and people on the island have ancient roots in the Pohnpeian culture. Yams are normally planted near tall trees, usually breadfruit, for support and to encourage growth, and sometimes poles extending to the higher branches of the trees are put in place for the vines of the yam to climb up on. The cultivation of very large yams is a way in which Pohnpeian men can gain prestige and manifest their farming expertise, so many of their cultivation techniques are often not revealed openly.

In the middle of the 20th century, the French ethnobotanist Jacques Barrau (1961: 45) commented on the difficulties he encountered in Pohnpei when he inquired about yam cultivation because of the enigmatic nature by which men produced the yam crop on this island: "For them, yams possess a certain ritual value; and they are reluctant to take an outsider into their yam gardens, which are usually hidden in the forest, or to supply any information on cultivation methods or the species and varieties grown." Bascom (1970: 88–89) accounted for this mysterious behavior as part of the Pohnpeian "prestige economy." Bascom described how a Pohnpeian man can rise in local status by successfully participating in prestige competitions in which he contributes large yams and quantities of pit breadfruit to feasts given by others to honor chiefs. Meanwhile, individual regional chiefs carefully observe the men to see which of them consistently deliver the biggest yams, and then accordingly they choose which of these men should fill open titles, and if they already possess a title, the chiefs may promote them to higher ranks. During this prestige competition, it is important for the yam farmer to follow a protocol of behavior:

> "... a man is not allowed to show open pleasure in success but must display appropriate humility. He must not act proudly or boast openly about his achievement. When others discuss the merits of his yam, he pretends not to listen. When they come up to tell him that his yam is the largest, he protests that it really isn't. Finally, as Barrau found out, it pays a man to keep quiet about yams, kava plants, or pit breadfruit he has in his possession or is keeping in reserve for future occasions. Information of this kind is concealed in the hopes that the element of surprise may enable the owner to surpass his neighbors at future feasts." (McLean 2014, citing Bascom 1970: 88–89; also see Barrau 1961)

More recently, Raynor and Fownes (1991: 151) reported:

> "More than 15% of yam plants recorded in the field were unidentified as to variety, due to the secrecy regarding yam cultivation and the reluctance of farmer to discuss them with us" (see Raynor et al. 2009: 40–63 for a detailed discussion of yams and their traditional cultivation in Pohnpei; especially note the massive prestige yams displayed in Fig. 2, p. 42, with their top decorations including leaves of **Tehnlik**, the bird's nest fern, *Asplenium nidus*)."

Standal (1979: 4) described the prestige value of giant yam production by local farmers in Pohnpei and the dimensions of these very large *D. alata* "prestige" yams:

> "Some varieties of perennial yam grow to gigantic sizes in [Pohnpei], and several people are required to carry a single piece of tuber. In [Pohnpei], the size of yam is known as a 2-man, 4-man, or 5-man yam, which describes the number of men needed to carry the yam pieces. A single 6-man yam tuber can weigh up to [270 kg (600 lb)] and a long variety of yam tuber grown on hillsides may [grow] to [1.6 m (6 ft)] long and about [30 cm (1 ft)] in diameter [for comparative purposes]. Usually the average annual variety of yam weighs [6.8–9 kg (15–20 lb)] and is the most common size of yam in the islands of the Mariana and Yap."

While yams are a fundamental prestige component of most feasting events, for which they are grown to sizes up to 100 kg (220 lb), edible cooked yams also play a role as a staple food in the Pohnpeian diet (Raynor et al. 2009). Yam production is seasonal, complementing the breadfruit season. The yam season is defined by a series of feasts and ceremonies. In order to ensure a steady supply of yams during this time, farmers plant several cultivars, use different trellising methods, and employ other techniques. Cultivation of yams is done mainly under live breadfruit trees, although several other woody species are used, and they are often ring-barked to improve yam growth. Taboos and magic are still employed by many farmers. Although production per unit of land area is low under the traditional system, yams of large size are produced which are highly suitable for prestige purposes.

Uses in Pingelap: Damas (1994: 234) referred to a yam (*D. alata* or *D. esculenta*?) known in Pingelapese as **Kaehp**, which was either being cultivated or growing wild in some areas of this atoll. Damas compared his observations with those recorded by St. John (1948):

> "Edible yams said to be grown and cultivated in village... but I think at the times of my visits [yams were] chiefly growing wild as vines in the forest. Albert [his informant] says its tubers can grow 'as long as man's arm'."

▸ **Kehmpalahi** (P)
Dioscorea bulbifera

Description: *Dioscorea bulbifera* is a wild or semi-wild herbaceous vine that is high-climbing and tuberous, producing smooth, vigorously twining stems that twine to the left (clockwise). Its large leaves are distinctly veined, heart-shaped, and grow up to 20 cm (8 in) or more in length. The long-stemmed herbaceous vine may arise from an underground tuber, although often tubers are inconspicuous or absent. The stems have no spines or wings. The flowers are small and abundant in long, slender, spike-like inflorescences, and the seed capsules have wings about 20 mm (0.8 in) long and 9 mm (0.4 in) wide. The conspicuous aerial tubers (bulbils) are roundish, green, or reddish and up to about 13 cm (5 in) wide; they are borne in leaf axils (the angle between a leaf and stem), and *D. bulbifera* is sometimes called the "air potato" because of these aerial bulbils. When full-sized, the bulbils are called "offsets" and fall off (or are removed and planted to produce new plants).

Distribution: *D. bulbifera* is found from East Africa to Polynesia (Stone 1970) and is thus said to be "the most widespread of all yams" (Burkill 1951). It was probably an early and intentional introduction to some Pacific islands, then purposely transported by humans in prehistoric times to other islands over many centuries during the prehistoric period, or in some cases after European contact. Despite this association with people, *D. bulbifera* largely remains a "wild yam" (i.e., a self sown, naturalized yam species). It is a common, sometimes locally abundant adventive species in cultivated areas and along roadsides, as well as being naturalized in secondary and even sometimes in primary forests. It may become common in thickets and open areas. It is an aggressive alien weed in some tropical or subtropical areas; for example, it is listed as a "Category I" invasive alien plant in Florida and is currently altering native plant communities and ecological functions there by smothering native species: "Asexual propagation by means of bulbils that drop off parent vines to the ground in abundance ensures that eradication of the plant is exceedingly difficult once it has invaded an area" (Masterson 2007; also see Langeland and Burks 1998).

General uses: The tubers of *D. bulbifera* are globose or pear shaped and sometimes lobed, and some are edible (in cultivated races); however, others (in wild races) can cause nausea or even be fatal when consumed (Stone 1970: 127–128). Nevertheless, in a number of areas of the world, the tubers and bulbils of *D. bulbifera* are used as food, and the plant is cultivated as an agricultural crop:

> " In the Pacific [*D. bulbifera*] is found mostly as a wild or weedy plant with aerial tubers which are poisonous unless specially prepared. However, there are a few cultivars which produce edible (not poisonous) aerial tubers, and even some cultivars where both the aerial tuber and the underground tuber can be eaten." (Wilson and Hamilton 1988)

In addition to its occasional use as a food source, this yam species "has also long been used as a folk medicine, including use as an analgesic, aphrodisiac, diuretic, and a rejuvenative tonic." Some of the natural compounds produced by *D. bulbifera* and identified by modern chemists include steroidal diosgenin, an active component of birth control pills, and dihydrorodioscorine, which has anti-fungal qualities (Masterson 2007). *D. bulbifera* is the most common and widely distributed wild species of yams in Micronesia, but it is seldom planted. It is sometimes found growing wild in the Marianas but is very common on most islands in the FSM (Fosberg and Sachet 1987b: 134).

Uses in Pohnpei: This yam is generally known in Pohnpeian as **Kehmpalahi** or **Palahi**; however, "when in the ground" (i.e., tubers in the soil rather than hanging on the vine), this yam is referred to in Pohnpeian to as **Uhn palahi**. Another older name for this wild-growing yam is **Ohdek**. Lessa (1977) noted that the "roots" (tubers) of this yam were extensively processed and cooked twice and used "principally as a diet (not a medicine) for invalids, although a few individuals like to eat them occasionally before and after meals." Lessa also indicated that "wild pigs eat both the roots and the vine seeds but domesticated pigs do not like either." Victor (2000), a student at the College of Micronesia, provided some locally observed notes on the invasiveness of this introduced "canoe plant" in Pohnpei, along with references to traditional uses for *D. bulbifera* on Pohnpei:

> "This certain herbaceous vine likes to climb on everything that is in its way. It has several medicinal uses locally, like [to] help [with] easing the pain on sprained ankles, and certain other uses that is in combination with other plants. History or stories of the past depicts that the plant was used as yam, way before the [other] yams found [their] way to the island of Pohnpei. It was one of the major food productions before that is why one the island of Parem there is so many... First, in healing the sprained [ankle], the fruit [bulbil] of the vine, which is brownish in color is cut in half and the insides are scraped out and put into a cloth or something that will easily let the fluid out of it. [Then] we massage a sprained ankle with it. Always massage down toward the ground and outwardly of the foot. It is also used for some medicinal [purposes], but the identities of the other ingredients are [reserved among the] family... In talking about the usage of this plant as food, in the ground after a certain time, lies something just like the yams today. It can be put in an uhm [earth oven], boiled with coconut oil, and it can also be used to feed the pigs and they certainly like it."

Balick et al. (2009: 276–278) provided additional insight into the variety of the traditional edible and medicinal uses of *D. bulbifera* in Pohnpei:

> "The tuber is ground and boiled until soft and fed to ill people as a comfort food... To use as a food for very old people, a tuber is cleaned and scraped, cooked, and given to the person to eat... To open and treat boils, the tuber is cut in half and the inside scraped out onto the boil... To treat *nankaokon medek* (illness characterized by headache and joint pain), four or eight leaves are boiled in a pot of water, then the person, sitting over the pot and covered with a sheet, breathes in the steam from the boiled leaves... To treat abscess and boils, especially those under the arm and for hard boils that are growing, the fruit is cut, chopped, placed in a piece of cloth, and then juice exuding from the cloth is rubbed around the abscess or boil four times daily for four days..."

Unsurprisingly for a wild yam, *D. bulbifera* does not seem to have many named varieties on Pohnpei, though some do exist. Lessa (1977: 145) referred to "two varieties" of wild-growing *D. bulbifera* on Pohnpei. The Pohnpeian–English Dictionary refers to ***Palahirot*** as a variety of wild yam and to ***Weisoahn*** as an inedible variety of **Palahi** which has been "used an indicator of when to plant yams; when ***Weisoahn*** begin to sprout, it is time to plant."

▸ **Kehpineir** (P) and other names (which may overlap with *D. nummularia* cultivars) *Dioscorea esculenta*

Description: *Dioscorea esculenta* has many scientific binomial names (synonyms), several common English names, and is one of the more important cultivated yams in the world. This yam is a strong, high-climbing, tuberous, annual vine, which can reach up to about 2.4 m (8 ft) high and spread up to about 1.8 m (6 ft) across. It often has prickly-armed stems that are somewhat woody and hairy, but these are usually found on "wild" plants and normally not on cultivated ones. *D. esculenta* is the only commonly cultivated species in the Pacific Islands which twines to the left and has spines on the stem (Wilson and Hamilton 1988). Leaves are oval to heart-shaped (or nearly round to kidney shaped), pointed at the tip, up to 15 cm (6 in) long, and arranged alternately on the stems. Very young leaves are covered with so many hairs at the tips of the vines that they have a velvet-like appearance. The flowers are unisexual, and the fruit is a very rare reflexed capsule about 27 mm (1 in) long; the seeds are winged all around. Several, sometimes many, underground tubers develop downward from a corm lying close to the soil surface; the tubers have relatively thin, brown or gray-brown skins with a bitter layer just under its skin, and the edible flesh is white. Some have described the flavor of these yams as being sweet and agreeable and similar to that of sweet potato or chestnuts.

Distribution: The *D. esculenta* yam was probably transported by humans in prehistoric times to various Pacific islands over many centuries during the prehistoric period, or in some cases after European contact (e.g., see Horrocks and Nunn 2007). *D. esculenta* is adapted to humid and sub-humid tropical habitats, growing best where the rainfall is fairly well-distributed throughout the year with a total range of about 875–1750 mm/year (~35–70 in/year); it probably evolved in regions with a dry monsoon season climate, somewhere in tropical Asia. Indeed, some believe *D. esculenta* to be native to Southeast Asia. It also occurs as a "wild" plant beyond that region, such as in northern India and New Guinea, but it not known whether these are plants that have escaped from ancient cultivation or "real relicts of its natural distribution area." From its putative Southeast Asian homeland, the major direction of prehistoric human dispersal has been out of the continent of Asia via the Philippines to the Pacific Islands, and after about 1500 CE, throughout the tropics. Today, cultivation of *D. esculenta* is the most significant in parts of Nigeria, India, Southeast Asia, China, New Guinea, Oceania, and the Caribbean. Of all edible yams, *D. esculenta* ranks third in production and utilization, with only *D. rotundata* (West African yam) and *D. alata* more widely used (Martin 1974: 1; also Coursey 1967, 1969, 1972).

General uses and uses in Pohnpei: Yams are among the most important plants in Pohnpeian culture (e.g., see Bascom 1948, 1965; Petersen 1977; Hunter-Anderson 1991; Balick et al. 2009). Yams form a central component of most feasting events, for which they are grown to sizes up to 100 kg (Raynor et al. 2009) and also play a role as a staple food in the Pohnpeian diet.

> "In many traditional yam-growing communities [such as Pohnpei and other islands, especially in Melanesia], the cultivation of yam is highly ritualized, attempting to optimize these agronomic influences and to appease the supernatural forces held responsible for further unexplained variability." (Melteras et al. 2008)

Presently, we have only small traces of archaeological evidence for ancient "thorny" yams in Micronesia. For example, Moore (2005: 101) reported yam thorns, probably *D. esculenta*, excavated from a site on Guam that have been radiocarbon dated roughly to about 1,000 years ago (calibrated range of A.D. 986–1210) and "a single yam pollen grain, associated with a calibrated radiocarbon date of A.D. 670–880," identified in a soil sample from an archaeological site on Tinian (see Hunter-Anderson 2005). In addition, there is ancient evidence from Fiji for both introduced sweet taro (*Colocasia esculenta*) and lesser yam (*Dioscorea esculenta*), dated to about 2,500 to 3,000 years ago (Horrocks and Nunn 2007).

D. esculenta is a species of cultivated yam that produces small egg-shaped tubers which lie shallowly in the soil. The tubers are quite sweet and whitish in color, and the skin is very thin. This species is suitable for traditional garden planting and is one of the tastier and more nutritious of the cultivated yam species, with a long history of cultivation and use in some islands of Micronesia. The starch of *D. esculenta* is more easily digested than that of other yams, and thus this species is used in special diets for persons with digestive problems. The raw tubers are used as a poultice on swellings, as well for ailments such as piles and syphilis. Burkill (1966, original edition 1935) referred to *D. esculenta*'s medicinal use for swellings, including those associated with tumors. According to Olayemi and Ajaiyeoba (2007), the tuber of *D. esculenta* has been used to cure ulcers, boils, and abscesses:

> "It contains allantoin, a cell-proliferant that speeds up the healing processes. It has been used traditionally as a contraceptive, in the treatment of menopausal symptoms and various disorders of the genital organs. It has been suggested for ethnomedicinal uses as anti fatigue, anti-inflammatory, anti-stress, anti-spasmodic and immune deficiency remedies in various ethno-medicines. The peel has been reported to possess anti-cancer and anti-fungal properties.

The main constituent of *Dioscorea* species is the well-known saponin dioscin. The aglycone, diosgenin, is the major starting material used in the industrial production of steroidal hormones. Diosgenin has also been utilized for hundreds of years to treat rheumatism and arthritis-like ailments."

▸ **Kehmmwas** (P) and other names (which may overlap with *D. esculenta* cultivars) *Dioscorea nummularia*

Description: *Dioscorea nummularia* is a robust, high-climbing, perennial spiny vine that has relatively thick stems up to 1 cm (0.4 in) wide or more at the base, usually with many spines; under low vegetation, the stems of this vine scramble over the ground, or twine into the surrounding plants. The leaves, which are distally alternate to proximally opposite, are oval to triangular in shape, narrowing at the tip. One to four male flowers are bunched together on downward leafless branches up to 4 cm (1.5 in) long; one or two female flowers together are up to 15 cm (6 in) long. Tubers are usually long and cylindrical with white flesh; however, the tubers of some cultivars are difficult to dig out because they are relatively large, very long, branched and irregular in shape. "But other cultivars produce clusters of approximately 5–15 small tubers which are easy to dig" (Wilson and Hamilton 1988; also see Stone 1970: 128)

Distribution: *D. nummularia* is typically a Melanesian cultivated yam species. It is found in East Malaysia, the Philippines, and Indonesia to Polynesia, including areas in Micronesia; *D. nummularia* has also been introduced to several Caribbean Islands and India. According to Fosberg and Sachet (1987b: 136), *D. nummularia* is widely cultivated for food in Micronesia, occurring "on all inhabited high islands," and is undoubtedly "an aboriginal [prehistoric] introduction."

General uses: *D. nummularia* is cultivated in New Guinea and other islands of South Pacific as a food and medicinal plant, and the tough stems are used in some areas for making cordage. Although the tubers are never poisonous, they are not in high demand due to their saponin (amphipathic glycosides) content. Recent microfossil archaeological evidence (starch grains) indicate, if valid, that *D. nummularia* was an ancient cultivated plant about 2800–3100 years ago on Vao Island, "one of a string of small, low-lying islands (each <2 km^2) located off the north-east coast of the much larger [Melanesian] Malekula Island in northern Vanuatu" (Horrocks and Bedford 2010). Indeed, in Vanuatu, *D. nummularia*, sometimes referred to as the "wild yam," is very popular; baked or boiled, it is an important traditional food in these Melanesian Islands "during the transition between the two major yam harvests" (Walter and Lebot 2007: 84).

Uses in Pohnpei: *D. nummularia* is known in Pohnpeian as **Kehmmwas** and by other names depending on the cultivar. Many of these names overlap with *D. esculenta* cultivars. According to Balick et al. (2009: 278) *D. nummularia* is used medicinally on Pohnpei:

> "To reduce fever, the leaves of [*D. nummularia*] are pounded (young leaves are used in the treatment for a child, older leaves for adults), placed in *inipal* (coconut fiber from the base of the leaves, *Cocos nucifera*), squeezed into a cup of water, and consumed... To treat uterus pain, four leaves are pounded, squeezed into 1 lit (0.25 gal) of water and consumed (1 lit daily) for four or eight days..."

Cultivating yams with *paledou* method, where several plants are grown under a trellis structure for the vines to climb.

Men carrying a large yam as an offering at a community ceremony. Traditionally, a set of contributions to a *kamadipw* (traditional feast) includes a yam or a bundle of breadfruit, *sakau*, and a pig).

Carrying an offering of *kehi en kehp*, a large yam tied to a pole.

Using *ahk tuhke*, a pointed wooden stick to dig up yams.

Yams being baked in an *uhmw*, a traditional oven of heated stones placed on the ground with the food.

▸ **Sadak** (P), **Jadak** (Mw)
Elaeocarpus carolinensis

Description: *Elaeocarpus carolinensis* is considered to be an endemic species to Pohnpei and Kosrae (Costion and Lorence 2012: 75–76). This species was first described botanically by the Japanese botanist Gen'ichi Koidzumi (1883–1953). It is a large, native tree on Pohnpei that may grow to heights of 35–40 m (100–130 ft) under prime conditions. It produces pointed, evergreen, simple leaves which are alternately arranged on the stem and are elliptic in shape, reaching 10–15 cm (4–6 in) long and 3–5 cm (1–2 in) wide. The leaves have margins that are either entire or serrated with a swollen petiole, and they almost always turn yellow to red before falling to the ground. Some of its distinctive cream colored, cup-shaped flowers turn red with age, and its fruit (drupes) are bluish when ripe. All Micronesian *Elaeocarpus* species, listed below from Costion and Lorence (2012: 76), are considered endemic to the areas indicated in parentheses: *E. carolinensis* Koidz. (Eastern Carolines), *E. joga* Merr. (Micronesia), *E. kerstingianus* Schltr. (Pohnpei), *E. kusaiensis* Kaneh. (Eastern Carolines), *E. kusanoi* Kaneh. (Eastern Carolines), and *E. rubidus* Kaneh. (Palau).

Overall, the genus *Elaeocarpus* contains between 150 and more than 500 species, depending partly upon whether it is classified as belonging to the family Elaeocarpaceae or Tiliaceae. Obviously, this genus needs much more study to determine the accurate amount of species variation in *Elaeocarpus*.

Distribution: The many species in the genus *Elaeocarpus* are widely distributed, found variously in Asia, Australia, Madagascar, Mauritius, New Zealand, Papua New Guinea, and the Pacific. *E. carolinensis*, and a few other species of *Elaeocarpus*, occur in some of the high islands of the FSM and Palau; e.g., *E. carolinensis* occurs on the islands of Pohnpei, Kosrae, and reportedly on Palau in the rainforests of these islands, from sea level to about 400 m, where it is a canopy tree considered critical for rainforest survival on some islands. These *Elaeocarpus* trees can be found in both the lowland and upland forests, including tropical montane cloud forests on Pohnpei and Kosrae. As noted above, *E. carolinensis* can develop into a large tree in some places, especially those that are protected from strong winds and where there is deep fertile soil. Unfortunately, due to upland deforestation for clandestine commercial cultivation of **Sakau** (*Piper methysticum*), or subsistence farming on some islands, *E. carolinensis* is less common than in the past in its montane Eastern Caroline, Micronesian range.

Uses in Pohnpei and Kosrae: *E. carolinensis* is known as **Sadak** on Pohnpei, and as **Nahnek** on Kosrae; on both of these high volcanic Micronesian islands, these native trees are important components of the rainforests and are used to make canoe hulls (much like the trunks of the breadfruit tree, *Artocarpus altilis*). The wood of *E. carolinensis* is lighter than *Campnosperma brevipetiolata*, which is known as **Dohng** on Pohnpei and as **Elahk** on Kosrae. *C. brevipetiolata* is also a common native rainforest tree on both Pohnpei and Kosrae; however, even though **Sadak** has lighter wood, it has been less desirable than **Dohng** for making canoe hulls because its wood is not as strong. On the other hand, in more recent times, both of these native tree species have been considered as being a potential source of commercial timber because they are relatively common species in the upper forests of Kosrae (e.g., see Laird 1983: 16).

▸ **Rehtakai** (P), **Pwuroar in doakoa** (Mw), **Roah saekai** (Pi), **Geinga dolo** (Ka)
Eleusine indica

Description: *Eleusine indica* is an annual or short-lived perennial grass that forms clumps of culms (the above-ground or aerial stems of grasses and sedges); these culms branch out from the base, erect or prostrate. *E. indica* is distinguished by its flattened culms. It can grow up to 1 m (3 ft) or more with leaf blades that are flat or sometimes folded and 15–30 cm (6–12 in) long. Flowering spikes, 2–6 (usually 5), grow from the apex of the culm, with one slightly below the apex. These spikes of flowers are up to 10 cm (4 in) long. Individual spikelets have a few tiny flowers without awns (a hair- or bristle-like appendage) and are dark green. The tiny seeds of *E. indica* are ridged and striated (for more detail, see Stone 1970: 191–192). A single *E. indica* plant is said to be capable of producing up to 150,000 small seeds!

Distribution: Reportedly native to India or Africa, this weedy plant now has an immense geographical range, from the tropics to temperate environments. It is quite common on roadsides and on wasteland (disturbed or open areas near settlements or buildings), especially in sandy soil. *E. indica* is considered to be a major weed in all relatively open tropical areas where it has become established, including cultivated cropland, gardens, and roadsides up to elevations of about 900 m (3,000 ft). It also occurs in plantations and nurseries, growing best in wet, fertile, cultivated soil in full sunlight, and once it becomes established, it is hard to eliminate (Swarbrick 1997a: 16; also see Swarbrick 1997b). Although this species is native to somewhere in the Old World tropics, it is now naturalized in many tropical and temperate areas, including many, if not all, islands of Micronesia. Although some have considered *E. indica* to be a native grass on some islands in Micronesia, others believe it is an alien plant introduced by humans to these high volcanic islands (e.g., Kosrae) and low coralline inlands (e.g., the Marshalls). Some have suggested that *E. indica* was an ancient introduction to the Pacific Islands as far east as Polynesia, but was a European introduction to Hawai'i, where it was first recorded in the middle of the 19th century. In any case, this weedy grass is now common in various areas of Micronesia, including Pohnpei.

General uses: *E. indica* is a notorious weedy grass species in many tropical areas of the world today, apparently being poisonous to some mammals, and its seed is reportedly a potential contaminant of cultivated grains. Nevertheless, this plant is also used in some areas as a source of medicine, food, and materials for weaving, making paper, and other products. In addition, all parts of *E. indica*, especially the root, have been reportedly used in traditional medicine in various places within its vast present-day range of human-mediated distribution (e.g., see Lim 2012: 230–234). Reputed uses of *E. indica*, such as decoctions of the boiled plant or otherwise, include as a diuretic, to expel parasitic worms and other internal parasites from the body, and to induce sweating and reduce fever, as well as for treating coughs and other ailments (e.g., see Kulip 1997); furthermore, the seeds of *E. indica* are sometimes said to be consumed "as famine food and also used in the treatment of liver complaints" (see Lim 2012: 230; Iqbal and Gnanaraj 2012).

Uses in Pohnpei: *E. indica* is known as **Rehtakai** in Pohnpei (also the name for another invasive grass, *Chrysopogon aciculatus*). *E. indica* grass is very hard to pull out of the soil because its roots develop a strong grip in the ground. In the Pohnpeian language, *takai* means "stone," and the meaning of **Rehtakai** in the Pohnpeian language indicates that this alien grass species is hard to pull out of the ground. *E. indica*'s vernacular name on Namoluk Atoll in Micronesia is **Fätilimwech**, which also refers to the difficulty of pulling this grass out of the soil. In Pohnpei, *E. indica* has been used for a variety of purposes; for example, Balick et al. (2009: 298) referred to a variety of these applications:

> "To treat unspecified baby's sickness... To use as a charm to protect a person... For use as a stimulant or perhaps as a charm to never become tired when running during a track and field competition, four of the soft parts of the leaves are chewed and swallowed... To use as a stamina boost, the stems are chewed before races... To treat continuous menstrual bleeding... To maintain healthy hair... Alternatively, as a hair conditioner, one plant is collected, cleaned, mixed with scraped coconut, put into a cloth and squeezed on the hair..."

▸ **Oaloahdilap** (P), **Woaloahd** (Mw), **Waelahd** (Pi), **Helii lausedi** (Nu), **Miiamia** (Ka)
Enhalus acoroides

Description: *Enhalus acoroides* is a native seagrass or saltwater herb commonly found in the marine wetlands of Pohnpei. It is a relatively large "seagrass" and the only species in the genus *Enhalus*. It is also a dioecious species, having individuals that produce only male flowers and others that produce only female flowers. The leaves can be remarkably long, and the female plants of this seagrass species have longer leaves than the male plants. The average leaf blade is about 1–2 cm (~0.5–1 in) wide and may be up to about 1.5 m (~5 ft) long, with in-rolled leaf margins that emerge from a solid creeping rhizome (root-like plant stem) which has tough black bristles. In some tropical

saltwater regions, *E. acoroides* seagrass meadows may cover wide marine areas. Male flowers are tiny (1 cm or 0.4 in) long, and when these float on the water surface, they resemble small pieces of Styrofoam. The male flowers are produced from a cup-shaped inflorescence formed at the base of the plant. The female flowers are relatively large and are attached on a long stalk; they have three large-ribbed white petals about 2–3 cm (1 in) long which usually drop off relatively soon after blooming. The fruit of *E. acoroides* is round to teardrop shaped, and about 4–6 cm (~2 in) in diameter with an outer surface or skin that is dark, ribbed, and hairy. As the fruit ripens, it opens and releases 6–7 white seeds. These seeds float for approximately five hours and then they begin sinking, which ensures that they are not dispersed far from the mother plants. Roots of the newly germinated seeds develop quickly. However, similar to other seagrass species, *E. acoroides* spreads mostly by vegetative reproduction. *E. acoroides* is the only seagrass species that releases pollen to the surface of the water when reproducing sexually. Flowers are water pollinated (hydrophilous). *Enhalus* seagrass is pollinated by male flowers that detach from the plant and then float on the surface until they reach a female flower, where pollination can occur. This feature restricts its distribution to intertidal and shallow sub-tidal areas. *E. acoroides* is a slow-growing, persistent species with a poor resistance to disturbance, suggesting that areas where *E. acoroides* is found are quite stable over time. The bristles at the base of the plant are the fibrous remains of old leaves. Indeed, a distinctive characteristic of *Enhalus* seagrass is "the tough vascular bundles which stiffen leaf margins and remain after leaves are broken or gone" (Falanruw 1992).

Distribution: Seagrasses are generally important primary producers (the base of a food web) in tropical Indo-Pacific shallow saltwater communities. The number of seagrass species decreases (attenuation) across Melanesia from west to east (Tsuda et al. 1977; Kock and Tsuda 1978); for example, Palau has ten described species, which is the greatest number of seagrass species in Micronesia, and going eastward, Yap has seven, Chuuk has five, Pohnpei apparently only has three, and the Marshall Islands has only one. Among the few seagrass species to reach Pohnpei without human help is *E. acoroides*, which is very widespread in the tropical Indo-Pacific, where "it is found from southern Japan to Taiwan to the Philippines, Indonesia, and Malaysia extending throughout the Gulf of Thailand and the coast of Vietnam to Hainan, China [and] occurs across insular Southeast Asia to New Caledonia, northern Australia and across Micronesia to the Northern Mariana Islands" (Short and Waycott 2010). Dispersal studies indicate that the fruits of *E. acoroides* can float in seawater for as much 10.2 days, and within that amount of time they could spread out as much as 63.5 km (about 40 miles; see Lacap et al. 2002); however, *E. acoroides* might occasionally be able to disperse much longer distances (>1,000 km) when influenced by strong currents (see Nakajima et al. 2014). Within the many areas of its native range, *E. acoroides* is common and extensive, especially in embayments. It grows submerged in shallow coastal bays, with its roots embedded in sand, mud, or in pools of tidal flats. However, its population trend is declining in a number of locations within its range due to local threats and is most likely declining overall in its areal extent. Threats for *E. acoroides* include coastal development, dredging and marine construction, minor damage from boating and shipping activities, and coastal runoff (for example, impacts in Pohnpei and Kosrae). *E. acoroides* is especially threatened by the creation of fish pens for farming. The species is also susceptible to shallow-net trawling and eutrophication. As this species is primarily found in estuaries, shallow lagoons, muddy bays, and inside marine basins, it is often disturbed by vessels and anchorage. Furthermore, *E. acoroides* is found in a narrow depth range. Therefore, this species is especially sensitive to a variety of disturbances, including the impacts of global climate change, such as sea-level rise.

> "In the Western Pacific islands, it is recognized in the need for sanctuaries and protected areas and the concept of traditional or community management of these areas. There are NGOs focused on its conservation and environmental protection integrated with traditional leadership and government agencies, suggesting that conservation measures and the acceptance of enforcement will continue to improve." (Short and Waycott 2010; also see Green and Short 2003)

General uses: The literature documenting the traditional knowledge of seagrasses is relatively limited, but some important sources of information describe the unique value and varied uses of seagrasses, including *E. acoroides*, within coastal indigenous societies. In fact, there are several traditional uses of the *E. acoroides* seagrass among tropical cultures, including some of those in Micronesia, where this species has been a resource for a number of purposes, including as an extractable source of a durable fiber.

Uses in Pohnpei: *E. acoroides* is known in Pohnpeian as **Oaloahdilap**. In addition to the traditional use of the large edible seeds of this seagrass as a customary delicacy, its inner stem fibers have been used to make Pohnpeian fishnets, and its leaves and lower parts are used for medicine. The leaves of **Oaloahdilap** are produced from a solid, creeping rhizome (root-like plant stem) which has tough black bristles. These bristles at the base of the plant are the fibrous remains of old leaves. According to Balick et al. (2009: 282; also see Herrera et al. 2010: 53), **Oaloahdilap** has had a number of traditional uses for food, fiber, medicinal, ritualistic, and miscellaneous other purposes on Pohnpei:

> " To use as food like a green onion, the base of seagrass is chopped together with sea cucumber… To use for weaving grass skirts and fishnets, the outer stem is scraped off and the inner plastic like fiber is then processed into grass skirts and fishnets… [and medicinally parts of this seagrass are used] to treat *soumwahu en eni* (ghost sickness) [and to] dispel *enihn nansed* (an evil ocean spirit) that has possessed a woman…"

The strong and durable **Oaloahdilap** fibers have long been an important resource in Pohnpeian culture. However, today on Pohnpei as well as other Micronesian islands, modern monofilament nets are used instead of nets made from traditional materials. In the past, fibers extracted from **Oaloahdilap** seagrass were traditionally used to make nets to catch reef fish on Pohnpei; **Oaloahdilap** is said to produce the strongest natural fibers in Pohnpei. Formerly, Pohnpeian women made their own nets, and fishing was a community activity. Now, people fish more individually. Previously, community fish herding was also a common activity practiced by men, women, and children. Monofilament nets are now used by men and women to surround a school of fish (Lambeth 2001: 18). It is interesting to note there is an aquatic herb found in the fresh waters of Pohnpei known as **Oaloahdenpil** (= *Blyxa aubertii* Rich.) which is used for medicinal purposes (e.g., see Herrera et al. 2010: 53).

▸ **Pahr** (P) (Mw) (Pi), **Ulu davage** (Nu), **Laagau hai wanii** (Ka)
Erythrina variegata

Description: The genus *Erythrina* in Fabaceae, the bean or pea family, comprises about 110 species of trees and shrubs. The name "coral tree" is used as a collective term for all these *Erythrina* plants. *Erythrina variegata* is a fast-growing, spreading tree with a wide distribution, including many tropical Pacific islands. Note: Falanruw 2015: 106 and others have identified the coral tree found in Yap and elsewhere in Micronesia as *Erythrina variegata* var. *orientalis* (L.) Merr., while Herrera et al. (2010: 120) list two distinct varieties of *E. variegata* (i.e., *E. variegata* var. *variegata* and *E. variegata* var. *orientalis*) as occurring on Pohnpei. If the environmental conditions are favorable, *E. variegata* trees may grow at "rates greater than 1.5 m (5 ft) per year" (Whistler and Elevitch 2006b). As a full-grown tree, *E. variegata* can reach up to 20 m (65 ft) in height and has a thorny trunk and branches. The trifoliate leaves are arranged alternately and are 6–25 cm (2–10 in) in length on long petioles. Clusters of dark to scarlet red flowers with very unequal petals are produced in long, dense stalks (racemes) at the leaf axils. The fruits are curved, linear to oblong pods 12–22 cm (5–11 in) long and contain 3–10 hard, kidney-shaped seeds about 1.5 cm (0.5 in) long. Another much less common *Erythrina* species found in some areas of the tropical Pacific is *Erythrina fusca* Lour., which can be differentiated from *E. variegata* by the rounded tips of its leaflets and the darker, nearly purple flowers; *E. fusca* is reportedly found in swampy areas in some parts of Micronesia on the islands of Palau, Yap, Kapingamarangi, Pohnpei, and Kosrae, as well as in North Malaya (Falanruw 2015: 106). The endemic species *E. sandwicensis* is restricted to the Hawaiian Islands.

Distribution: The coral tree, *E. variegata*, is found from Zanzibar, off East Africa, to eastern Polynesia. It occurs on most of the high islands of Polynesia and Micronesia, especially on atolls, although it may have been introduced on some islands, e.g., Pohnpei, where there is a variety sometimes referred to as the "wild type" (e.g., see Whistler and Elevitch 2006b), with leaves that are mottled with a yellow color (*E. variegata* var. *orientalis* (L.) Merr.). The coral tree, *E. variegata*, is usually found in coastal forests, more commonly on the outer atolls, but also on the rocky shores of high islands, and occasionally inland in lowland and ridge forests; it normally grows in sandy or limestone soils up to 250 m (800 ft) in elevation.

General uses: Cultivated widely in tropical regions of the world, *E. variegata* is a valuable, multipurpose tree, often grown as an ornamental tree and for soil improvement as a nitrogen-fixing species (typical of species in the bean family, Fabaceae). As a fast-growing species, *E. variegata* is also used to shade understory tree crops such as coffee and cacao. In addition, it is used commonly as a boundary marker, living fence post, trellis support, windbreak, animal fodder, and outrigger float; the latter use is due to its light, buoyant wood. In addition, in some areas, parts of this tree are used for medicinal purposes (e.g., see Whistler and Elevitch 2006b). It is even reported that the "young, tender leaves and young sprouts [of *E. variegata*] are eaten as a vegetable" (e.g., see Fern 2014–2018). In the Pacific Islands, *E. variegata* is considered a valuable multipurpose tree used medicinally and planted to serve as living fences.

Uses in Pohnpei: Glassman (1952: 25) referred to "*Erythrina*" as one of the "woody strand plants most likely to be seen" in the vegetation near the shorelines of Pohnpei. There are two closely related species of *Erythrina* found on Pohnpei today: *E. fusca* and *E. variegata*. Both are locally known as **Pahr**, and are further distinguished from an indigenous perspective as being "male" and "female" versions: **Paramwahn** (*E. fusca*) is male (*mwahn*) and **Parepein** (*E. variegata*) is female (*pein*). **Paramwahn** and **Parepein** trees have interchangeable uses. Their leaves are reportedly pounded and used to make a drink to cure curses. The leaves, bark, and roots of these *Erythrina* trees are burned on red-hot stones and then the cursed person, with a cover over their head, breathes in the smoke from the smoldering parts of these trees. The leaves are also wrapped around rock and placed where a new house is to be built to make the work easier (an association with the light wood of these trees) and to protect a new house structure (see Merlin et al. 1992: 29–30). Also see Balick et al. (2009: 386–387) for additional therapeutic uses for *E. variegata* on Pohnpei; for example: "To make a heat treatment for nerve damage that may have been caused by sorcery (a treatment known as *umwulap en kau*)... To treat *rihpw* (itchy swelling) on arms..." Glassman (1952: 21) noted that wood from "*Erythrina* (**Pahr**), *Hernandia* (**Pingiping**), and *Thespesia populnea* (**Pone**)" was sometimes used in canoe construction Pohnpei, but these tree sources "are not so highly prized" as the most important ones, which he identified as "*Campnosperma brevipetiolata*, *Elaeocarpus carolinensis*, *Xylocarpus granatum*, *Parkia korom*, and breadfruit, in that order."

▸ **Seir** or **Pwuhr** (P), **Jeir** or **Pwur** (Mw)

Fagraea berteroana

Description: The genus *Fagraea* includes a diverse group of trees, shrubs, scrambling lianas, and epiphytic stranglers that are well known for having hard wood and fragrant, showy flowers. *Fagraea berteroana* is a shrub to large tree that grows to variable heights of 1.5–20 m (5–65 ft). It produces many branches and is commonly epiphytic (grows upon other woody plants or stone surfaces). It bears rather thick leaves, is quite fragrant and produces white tubular flowers that turn yellow as they age. The succulent, ellipsoid to globular fruit change from yellowish-orange to bright red and contain many very small black seeds immersed in pulp when ripe. Botanists have long been puzzled about whether the *F. berteroana* trees found in Micronesia belong to one or more species, and if it is only one species, whether they belong to more than one variety. For example, in his comprehensive flora of Guam,

Stone (1970: 42) referred to these taxonomic questions when describing the distribution of *Fagraea* in Guam:

> "... although the species found in Guam is here called *F. galilai*, there is considerable evidence to show that it has at least a close relationship with *F. berteriana* [which is a synonym for *F. berteroana*] Gray [ex. Benth.], a species originally described from Tahiti. What the real boundaries of the species are may well require an intensive study of the whole genus, a matter which is beyond the scope of a flora, but would form a monograph. But, until such a study is performed, we will have to remain in some doubt as to the correct name of the Guam plants of *Fagraea*."

A decade later, Fosberg et al. (1979) argued that *F. berteroana* trees vary enough from one island or island group to another (primarily on the basis of the length of the flower "throat") that these botanists divided the Micronesian populations of the species into a number of varieties; however, more recently, Whistler and Elevitch (2006c: 3) commented on these putative varieties of *F. berteroana*:

> "... var. *galilai* (Palau), var. *kusiana* (Kosrae), var. ladronica (Guam and the Northern Marianas), var. *pogas* (Chuuk), and var. *sair* (Pohnpei and Kosrae),... are of questionable value because this kind of separation, based mostly on minor and variable characteristics, has not been done elsewhere over the range of the species."

For now, here we recognize *F. berteroana*, and possibly *F. ksid*, as the only *Fagraea* species occurring in Micronesia. Some aspects of this taxonomic confusion are discussed in more detail below.

Distribution: The diverse assemblage of approximately 70–75 species in the genus *Fagraea* is distributed from Southeast Asia to the archipelagos of the Pacific Ocean, according to Costion and Lorence (2012: 78). The woody plants of *Fagraea* are found growing from sea level to altitudes of 3000 m (about 9500 ft), mostly in forests gaps, forest edges, along stream beds in wet tropical forests, and less commonly in mesic forests, savannas, and mangrove swamps. The center of *Fagraea* species diversity is "Malesia and Papuasia" (i.e., much of Southeast Asia, tropical Australia, New Guinea, and tropical islands in the Southwest Pacific). Members of this genus are also apparently indigenous to regions of Sri Lanka, India, southeastern Asia to southern China, Hainan, and Taiwan, northern Australia, New Caledonia, and throughout the Pacific Islands from the Marianas to the Marquesas. One of these species, *F. berteroana*, is found more or less commonly in lowland to montane forests, including those on some tropical Pacific islands, especially on lowland lava flows, as on Savaii in Samoa, and on narrow ridges, as on Rarotonga in the Cook Islands (Merlin 1985; Whister and Elevitch 2006c: 2; Costion and Lorence 2012). *F. berteroana* is now a popular ornamental in the tropical Pacific, particularly in Hawai'i, but is only native to parts of Australia, New Guinea, Micronesia, and east to the Marquesas. Birds and bats commonly eat the fruit of *F. berteroana* and therefore have probably played an important role in dispersing its relatively small seeds. Indeed, Motley (2004: 397) pointed out that the seeds of *F. berteroana* are spread by birds and bats "internally or externally attached to feathers and fur by a sticky latex present in the mesocarp," and the naturalist Guppy (1906) referred to the fact that *Fagraea* fruits can stay afloat for several days, which suggests that some *Fagraea* species growing near moving freshwater could be stream dispersed. Motley (2004: 407) described how the Pacific island biogeography of *Fagraea*, the distribution of native fruit doves and pigeons, and the voyages of Pacific Islanders might be connected:

> "The widespread island species *F. berteroana* is interesting with respect to island biogeography and Polynesian voyages. This species, which has similar uses throughout Polynesia and Melanesia, occurs naturally on most islands except the Hawaiian archipelago. It has been reported that the fruits of this species are a favorite food of [native] fruit-doves and pigeons (Brown 1935; Parham 1972), which are present on most of the islands in the Pacific with the exception of Hawai'i. This correlation in distributions may account for the absence of *Fagraea* in Hawai'i. However, this is probably only a partial explanation for the dispersal of *F. berteroana*. This species was important to early human voyagers who no doubt aided its dispersal [because of the "showy, fragrant flowers, hard wood, habit, and overall attractiveness of the tree species" (Guppy 1906)]. *F. berteroana* is an ideal ethnobotanically important plant for studying the genetic diversity of populations between and among islands. This type of study may help elucidate the voyaging patterns of Pacific Islanders. Since the plants were important to these cultures, but not [as] important as crop plants like taro (*Colocasia esculenta* (L.) Schott) or banana (*Musa* L. spp.), there may have been less reticulate [complex, network-like] movement of plants between islands, a phenomenon that confounds genetic investigations. In addition, *Fagraea* was not as important to the European settlers in the past or present, so there has been little or no immigration of

'foreign' genotypes into the island populations during modern times. This should make *F. berteroana* an ideal candidate for studying human/plant migration patterns across the Pacific."

General uses: "The nearly constant profusion of fragrant flowers is the hallmark of this tree" (Whistler and Elevitch 2006c: 8). *F. berteroana* is relatively easy to cultivate and can produce many flowers with its attractive appearance and scent, but flowering can sometimes be uneven throughout the year. In parts of the Pacific, the inner bark and leaves of related *Fagraea* species are used for medicinal purposes, and in some parts of Polynesia, such as the Cook Islands and Tahiti, *F. berteroana* has had sacred traditional associations; Motley (2004: 401) reviewed the literature focused on the past and present highly regarded Pacific island status of this species in Polynesia:

> " In Polynesia the flowers of *F. berteroana* were so highly regarded for use in leis (flower necklaces) and for scenting coconut oil to perfume the hair and body that a Tahitian legend states that the first pua tree was brought to earth from the tenth heaven by Tane (god of forests) [**Pua**, the Polynesian name for *F. berteroana*, means "flower," essentially referring to it being the archetypical Polynesian flower]. As a result of this legend any images (*tiki*) carved of Tane were made of *Fagraea* wood (Neal 1965). In the Cook Islands, Mangaian legend states that the *Fagraea* tree (**Pu pua** or **Puapua**) guards the entrance to the underworld (Neal 1965) and that the branches are regarded as the road by which the spirits of the dead descend to the land of spirits, called *po* (Whistler 1990). Brown (1935) considered *F. berteroana* to be so highly valued by the Marquesan people that he thought it should be their national flower. The yellow color of aging *Fagraea* flowers may also have contributed to their value in Polynesian culture since yellow and red were colors usually reserved for use by royalty. In modern times the floral fragrance of *F. berteroana* is considered useful for aroma therapy. The Hawaiian Paradise Trading Company (1996) reports that *Fagraea* essence reduces cravings for drugs (especially *Cannabis sativa* L.) as well as urges toward sexual promiscuity. The essence is also used to restore mental clarity and purposefulness."

Whister and Elevitch (2006c: 7) summarized the diversity of uses in Pacific for *F. berteroana*:

> " It is esteemed mostly for its fragrant, white flowers, as a garden ornamental, and for its timber. The long-tubed blossoms are widely worn, singly over the ear or strung into fragrant leis. They are also highly favored for making scented coconut oil. The durable, light brown, fine grained wood is of good quality and can be smoothed into a high polish. The wood was much valued for making tool handles, tools, house posts, furniture, drums, and in modern times, handicrafts. Its relatively small size does not give it much potential for timber. Today it is mainly cultivated as a garden ornamental and for its fragrant flowers used in personal adornment."

On the other hand, Motley (2004: 402) also commented on the potential toxicity of *Fagraea*:

> " No recorded poisonings have been attributed to the consumption of *Fagraea* fruits or decoctions made from the plant parts. However, this by no means verifies that there have not been unrecorded poisonings and deaths as the result of improper consumption. This point can be driven home by the report of Grant, Fosberg, and Smith (1974), which stated the Tahitians used the fruit to treat bronchitis, but that in large quantities the fruits are poisonous. The delicate balance of folk medicine is that it can be either a cure or a curse to human health. For this reason, casual experimentation or use of these plants internally is not recommended."

Uses in Pohnpei: Motley (2004: 404) tells us the local name **Seir** attributed to *F. berteroana* in Pohnpei, but the most common name used on this island in reference to *F. berteroana* derives from the word *pwur*, which literally means "to return." That is because the "flowers are commonly used in the necklaces given to visitors upon their departure from the island as a symbol of the host's wishes that they will return again." Balick et al. (2009: 400) recorded additional information regarding the uses of *F. berteroana* in Pohnpei:

> " To use as a flytrap, the peeled fruit is placed near the toilet or on a table to draw flies away from the food (the flies are attracted to the sticky flesh of the fruit but get stuck when they land on it)... The flowers are used for three types of *mwaramwar*: *elin sohpeidi* (or *elin soupeidi*) for *Nahnmwarki* and *Nahnken*; *elin soupwur*, to give a person who is not expected to come back, as if in war; and for the standard *mwaramwar*... To use as a charm for protection against sorcery... To treat *kilsarawi* (shingles, *Herpes zoster*)... To treat *kanahria* (gonorrhea, causing painful urination in men)... To treat *soumwahu en mwidihlik* (kidney stones, including pain in the back and numbness in the legs)... To treat an unspecified kidney ailment... To treat swollen testicles associated with *soumwahu en ohl* (men's sickness)... To treat boils..."

▸ **Tuhke karisihn** (P), **Koarsihn** (Mw), **Karesihn** (Pi)
Falcataria moluccana

Description: The large tree species of *Falcataria moluccana* has undergone considerable taxonomic change in its nomenclature over the years; it has been previously placed in other genera, such as *Adenanthera*, *Albizia* and *Paraserianthes*. *F. moluccana*, as it is correctly referred to today as a scientific species, has commonly been known in English as "albizia." *F. moluccana* is a big, introduced, fast-growing tree, up to about 40 m (125 ft) tall and up to as much as 100 cm (40 in) or more in diameter, with a spreading, flat crown, along with layered subcrowns. Its leaves are bipinnately compound, about 23–30 cm (8–12 in) long and alternately arranged, and its bark is white, gray, or greenish, mostly smooth or slightly warty, with its young parts covered densely with rusty, pressed hairs. Its compound leaves have a multitude of small leaflets. On the leaves, a large nectary develops below the lowermost pair of pinnae (a leaflet or primary division of a pinnately compound leaf), and a smaller nectary develops between or below most pairs of pinnae. (Some alien insects, such as "the invasive and biting fire ants" may feed on these nectaries and increase the problematic naturalization and spreading distribution of these fast-growing, but relatively short-lived and shallow-rooted, large *Falcataria* trees.) Flowers develop into panicles about 20 cm (8 in) wide, often with two serial branches from one bract scar. The relatively small seeds are ellipsoid, laterally flattened, and are transversely arranged in straight seed pods about 10–13 cm (4–5 in) long (see Wagner et al. 1999: 690; Falanruw 2015: 108).

Distribution: *F. moluccana* is sometimes referred to as an elegant, beneficial, large, deciduous tree because of its aesthetic appearance, including attractive gray bark and feather-like flowers (Wagner et al. 1999; Starr et al. 2003). Indeed, because of its fast growth with wide-spreading branches, *F. moluccana* has been widely used as a shade plant for coffee and other cash crops in many parts of the world; it has also been planted due to its adaptation to nutrient-poor soils from sea level to about 1,500 m (~4,500 ft) in the tropics. In modern times, *F. moluccana* has been introduced throughout much of the tropics as a rapid-growing plantation tree, or in small numbers for shade and as an ornamental capable of increasing in height by about 7 m (~20 ft) in its first year of growth. Its native range is probably in parts of Indonesia and Melanesia, since this is where the greatest diversity within the species is found. Ecologically, *F. moluccana* is a pioneer species occurring sometimes in primary, but more typically in secondary, lowland rainforest as well as in light montane forest, grassy plains and along roadsides near the ocean. It is adapted to fairly wet and monsoonal climates with a dry season of up to 2 (4 maximum) months. It is important to recognize that this large tree species is sensitive to fire and easily damaged by strong winds, which has consequences for human settlements, road networks, and plantation agriculture. Still widely cultivated and exploited, it has escaped and become invasive, especially in natural lowland humid forests on Pacific and Indian Ocean islands, where it alters ecosystem functioning through nitrogen fixation and eliminates native species. This tendency could also occur in other countries where it is present but not yet widespread, such as tropical America and Africa, and the dangers of possible invasion by this huge tree should be highlighted prior to making any further introductions! In Micronesia, *F. moluccana* has been cultivated on some islands, including Palau, Yap, Guam, Pohnpei, and Kosrae. Some naturalization has been reported from some of these islands. Although some have suggested that this species does not spread rapidly, it appears able to extend its invasive range gradually, if not faster, through native forest ecosystems. Alien populations of *F. moluccana* in the Hawaiian Islands, where it has been listed on the State of Hawai'i Noxious Weed List (Ostertag et al. 2009), and elsewhere should be monitored carefully (e.g., see Hughes and Denslow 2005; Allison et al. 2006; Hughes et al. 2011; Tummons 2013). In his ecological study of the floral impact of *F. moluccana* tree populations on Mo'orea in French Polynesia, B.J. Wong (no date) summed up the problematic future for this large alien woody species in the tropical Pacific and elsewhere, even given its multipurpose usage history, including use as a biofuel:

> " *F. moluccana* is an enormous presence on the island of Mo'orea, and, due to its value as a timber species and ability to spread and persist, it is not likely to disappear. Future

monitoring will be critical to ensure that *F. moluccana* does not begin to affect the ecosystem on a larger scale. Temporal studies showing effects of *F. moluccana* over time, as well as more intensive sampling, would both serve to better inform us of the possible threat posed by this invasive."

General uses: *F. moluccana* (albizia) has been used for reforestation and as an ornamental due to its fast growth and nitrogen-fixation capacity (e.g., see Hughes and Denslow 2005), and reportedly its wood has been "used for a variety of purposes including canoe building and furniture making" (Starr et al. 2003; Little and Skolmen 1989). Other reported uses of the wood include lightweight construction, packing materials, paneling, cabinets, toys, wooden shoes, musical instruments, lightweight veneer and plyboard, low-density particleboard, pulp, paper, and matchsticks. As noted above, although *F. moluccana* has been described as an "elegant, deciduous tree with wide-spreading branches" and was introduced into Micronesia and Hawai'i as a timber and reforestation species, it has also been widely recognized as a potentially dangerous invasive species. Indeed, it is gradually—and in some areas rapidly—spreading throughout parts of the tropical Pacific region, including areas on some Micronesian high islands (e.g., see Smith 1985; Meyer 2000). It is used as a shade plant for coffee and other cash-crop plantations in many tropical parts of the world, and it is one of the faster growing trees in the world, with some importance as a pulp species in Asia; furthermore, it fixes nitrogen and can improve the soil fertility, especially in disturbed areas, so may not necessarily be considered a "bad tree," but it can become a very aggressive invasive woody species as it invades and overtakes forests, shading out other trees. Albizia trees (*F. moluccana*) grow very rapidly, even on nutrient-poor soils, and are adapted to areas from sea level to about 1,500 m (~4,500 ft) elevation in some tropical regions, but they are most common in moderately wet, lowland areas. These trees have been often used in ornamental and forestry plantings, especially for reforestation projects and more recently as a source of biofuel (e.g., see Chimera et al. 2010); however, it must be remembered that *F. moluccana* trees also have brittle branches that break easily in storms, especially due to hazardous winds.

Uses in Pohnpei: *F. moluccana* was introduced to Pohnpei in recent times for timber and reforestation purposes. As a large, fast-growing tree which does well in full sunlight, it has already spread rapidly in some lowland areas of Pohnpei, especially in open areas such as grasslands. The very light, soft wood of *F. moluccana* does not offer high-quality lumber. However, it is used as a source of firewood, thus its Pohnpeian name, **Tuhke karisihn**, which means "kerosene tree" or "kerosene plant." Seeds of this tree have sometimes been broadcast in open, degraded grasslands and fernlands. As a legume plant (in the bean family), **Tuhke karisihn** can improve the soil (bacteria on its roots fix nitrogen in the soil). As it grows tall, to heights of 25–30 m (80–100 ft), it eventually shades out the grasses and helps prevent repeated burning of the open areas. After this tree has been growing for some years in open areas, and has helped improve soil conditions, farmers can plant crops such as **Mahi** (breadfruit) and **Sakau** (kava) in the understory of **Tuhke karisihn**. It should be hoped that eventually farmers will remove the majority of these large and helpful but potentially dangerous "nurse" trees by girdling them. Then the breadfruit and other agroforest species in the recovered agricultural areas can grow to their full potential. However, it must be stressed again that *F. moluccana*'s tendency to take over disturbed areas relatively quickly with its heavy limbs and trunk should be considered when planting this tree near human settlements and transportation networks. East of Pohnpei on the high island of Kosrae, by the end of last century, some *F. moluccana* had been planted in Kosrae, and warnings about its problematic environmental impact as a possible invasive, large tree species were published:

> "This species has the potential, over time, of spreading widely through the forest (as can be seen on Pohnpei). Further planting should be discouraged. If this is not a desired species, emphasis could be given to cutting the existing trees."
> (Space et al. 2000; also see Space and Falanruw 1999)

▸ **Aiau** or **Oaioa** (P), **Au** (Mw), **Aiau** (Pi), **Aoa** (Nu), **Iaiau** (Ka)
Ficus prolixa

Description: There are reportedly more than 750 species of fig plants (genus *Ficus*) in the world. *Ficus* species are known from many tropical and subtropical countries, and all of them are woody. Some are trees; others are shrubs or woody climbers. Many fig trees begin their lives epiphytically on other trees and then send down aerial roots to reach the ground; others climb up trees by means of clasping roots; and still other have extensive root systems, either above or in the ground, or both. *Ficus prolixa* is a large, sometimes huge, member of the fig family that is an indigenous native species in Pohnpei. This spreading "banyan" or "strangler fig" tree can attain heights of 30–40 m (100–125 ft) or more. It produces a milky sap and bears mostly elliptical leaves, which are 6–15 cm (2–6 in) long, and have small hard spots (i.e., cystoliths) only on their lower surface. On the other hand, a related fig tree species, *Ficus microcarpa*, which is also found on Pohnpei, has cystolithic cells in both the upper and lower surfaces of the leaf (Stone 1970: 250; also see Falanruw 2015: 156). The flowers of *F. prolixa* are enclosed by a spherical receptacle, and the fruits are globular berry-like and hollow (technically a synconium or a cluster of many ingrowing flowers), i.e., according to Stone 1970: 247, the "Fruit [is] compound, the receptacle invaginated so that the flowers and matured fruits are entirely enclosed [and its] external appearance [is] fruitlike even when in flower." The compound fruits become reddish when ripe and contain many small seeds. This tree frequently starts life as an epiphyte growing on the branch or in a notch of another tree. As it grows older, it sends down aerial roots. If these reach the ground, they quickly form roots in the soil and become much thicker and more vigorous. The aerial roots supply nutrients to the epiphytic fig tree, allowing it to grow faster than the original host tree upon which the strangler fig germinated as an "air plant." The aerial roots of the fig gradually surround the host tree, preventing its main trunk from expanding, while at the same time its foliage smothers that of the host. Eventually, the host tree dies, allowing the *F. prolixa* tree to continue growing without competition. The long aerial roots of *F. prolixa* that grow down from branches sometimes create an almost impenetrable barrier for humans and other large animals. On some Micronesian islands, these tangled masses of branches, prop roots, and trunks create dark shadows and recesses in which spirits are thought to make their homes.

Distribution: The Pacific banyan, *F. prolixa*, is found on many high, tropical volcanic islands in the tropical Pacific region, including those in New Caledonia, Fiji, Cook Islands, Tonga, Samoa, French Polynesia, and Micronesia; it also occurs in some limestone areas of these islands, such as Guam in the Mariana Islands. Although this tree can germinate on rocky ground or outcrops on mountain ridges, it often starts life as an epiphyte growing on another plant, sometimes enlarging to point that it strangles the host plant.

General uses: Thaman (1992a: 51) summarizes the cultural significance and use of *F. prolixa* in Melanesia and Polynesia:

> "[*F. prolixa*] is regarded as having spiritual importance in many areas, and is the focus of creation mythology and cosmogony in Polynesia and Melanesia; [it is] important in rituals and ceremonies in Melanesia; is found in and serves as the focus for the ceremonial meeting place (*nakamal*) in Vanuatu; [its] timber and aerial roots used in construction and tool making in Melanesia; occasionally used for canoe hulls in Papua New Guinea;... [its bark bast fiber is] used to make bark cloth in Vanuatu, Fiji, Niue, and Tahiti, and for making very large seine nets in Tahiti; sap used as chewing gum and, in Vanuatu, for putty and caulking and to dye ceremonial sashes and belts; [its] sap is used for chewing gum; roots, aerial roots, bark, fruit latex, and leaves used medicinally; leaves used in garlands and in ceremonial dress; latex used in waterproofing in Tahiti; fruit important food of birds and fruit bats."

In Micronesia, there are also a number of recorded traditional uses of the bark, leaves, and other parts of the large native *F. prolixa* banyan tree. These uses include production of cordage, starting fires, practical and magical medicine, and as a hair conditioner. In regards to its use to condition hair, its leaves are reportedly "added to oil to promote thick, lush hair" (www.comfsm.fm/~dleeling/ethnobotany/text/flora.html).

Uses in Pohnpei: This large indigenous tree, whose Pohnpeian names are **Aiau** and **Oaioa**, is also known as **Likatengensapw** ("that which holds the land") and is rarely cut down. On Pohnpei, *F. prolixa* is believed to serve as the *mwoalen eni* ("home of spirits"). It is not cultivated, but remains as a wild component of the native Pohnpeian forest vegetation:

> "The **Aiau** (*F. prolixa*) also strengthened the land. This tree grows on the mountain ridges. It is said to be one of the first plants to appear after the origin of Pohnpei. Early Micronesian settlers observed **Aiau** growing in Kiti. The trunk grew taller and taller and the branches fell over and apart toward what is now Sokehs and Kiti, and that is why that area is now fruitful with food (fruits coming from branches). The roots of the **Aiau** spread toward Madolenihmw and sprouted. It thus held the land together from north to south." (See Merlin et al. 1992: 1).

Rehg and Sohl (1979), in their Pohnpeian dictionary listing of **Aiau** (*F. prolixa*), refer to the use of the leaves, fruits, bark, and roots of this large banyan fig tree in local medicines. Lessa (1977: 133) noted that a medicine made with the leaves of this banyan tree (*F. prolixa*) and parts of the bird's nest fern, *Asplenium nidus*, has been used on Pohnpei to treat *mai*, "a serious disease believed to be caused by a tree spirit." Lessa (1977: 152) also noted that in Pohnpei "the bark and branch roots [of *F. prolixa*] are utilized in the treatment of tetanus and as a hemostatic in menstruation." More recent ethnomedicinal research indicates that scraped bark of *F. prolixa*, in conjunction with parts of other plants including lime (*Citrus aurantifolia*), is used to treat the sickness characterized by high fever and convulsions ("jumping sickness" [tetanus?]) known locally as "*soumwahu en lusulus*" (see Balick et al. 2009: 445; also see Riesenberg 1948: 417).

▸ **Nihn** (P), **Kuwɛhn** (Mw), **Nin** (Pi)
Ficus tinctoria

Description: *Ficus tinctoria* is a fig species indigenous in Pohnpei and known from many tropical and subtropical Pacific islands. It is a very variable evergreen woody species, occasionally seen as a scandent shrub with long leaning limbs only a meter tall, while in other situations it can develop into a small, pendulous tree up to 18 m high. It may also grow to become a large tree with a spreading crown that can reach heights of up to 25 m (~80 ft); indeed, sometimes the trunk of a **Nihn** tree can be as much as 3 m (~10 ft) in diameter (e.g., see A.C. Smith 1985). However, it usually does not reach the size of *F. prolixa*, which is a similar but usually a much larger native tree. The leaves of *F. tinctoria* are alternately arranged, oval, and glossy dark green above, pale green below, with entire margins and up to 20 cm (8 in) long and 9 cm (3.5 in) wide. *F. tinctoria* bears red flowers and has thorns near the leaves; it also produces copious milky sap, and its fruits are yellow to bright orange or even reddish, about 17 mm (~0.5 in) in diameter. The fruits are arranged in the axils, usually produced in pairs, and commonly eaten by birds.

Distribution: *F. tinctoria* is considered by various biogeographers to be a native species in numerous tropical areas ranging from Southeast Asia to many tropical Pacific islands. In fact, both *F. prolixa* and *F. tinctoria* are supposedly indigenous fig tree species in Remote Oceania, widely found on high, tropical volcanic islands in the Pacific all the way to the Marquesas. Although *F. tinctoria* is typically found in beach thickets or in rocky coastal areas, it can germinate on outcrops on mountain ridges, and it often starts its life as an epiphyte growing on another plant. As this fig tree increases in size, it produces many slim aerial roots that grow downward from the branches. When these adventitious roots reach the ground, they root into the soil and further enlarge, providing extra nutrients to the fig plant, which subsequently develops much more robustly. Over time, the roots of the fig plant surround and squeeze the trunk of the host tree, and at the same time the emerging canopy growth minimizes the light that can reach the host tree, over time leading to the death of the host tree.

General uses: Although A.C. Smith (1985) referred to *F. prolixa* and *F. tinctoria* as indigenous in Fiji and Polynesia, he also suggested that these two fig tree species might not be native plants in Remote Oceania islands but rather were aboriginal introductions (i.e., "canoe plants" brought to remote islands by early Pacific voyagers) because their fruits and young leaves could have been an emergency food. Corner (1997) noted that *F. tinctoria* was the most widespread banyan fig species found in India, southern China, Indonesia, northern Australia, and several islands in Remote Oceania, including Micronesia. Thaman (1992a: 51–52) described some of the traditional applications of this useful fig tree in the tropical Pacific region, pointing out that *F. tinctoria* is widely cultivated and propagated vegetatively as a minor staple food plant in Micronesia and some nearby atolls. For example, green or ripe "fruits" are cooked or processed numerous ways and used to produce a minor staple food or made into puddings and sometimes dried to serve as a preserved food in some areas of Micronesia and Tuvalu. In addition, the wood of this fig tree is occasionally used in light construction for "digging sticks, canoe connectives, fishnet frames, fishtrap parts, earth sieves and for firewood… [and its fiber-rich] branches used to clean teeth" (also see Clarke and Thaman 1993).

Uses in Pohnpei: **Nihn**, as *F. tinctoria* is known in Pohnpeian, does indeed produce an edible fruit, but it is also used as a source of medicine on Pohnpei. In addition, during earlier times, its bark bast fibers were used for making cloth; for example, Christian (1899: 113) noted that Pohnpeian women "used to wear a sort of deep petticoat (*likau*), woven of the [fibers] of a tree called 'Nin' [**Nihn**]." Although **Nihn** likely refers to *F. tinctoria*, a rendering of a Pohnpeian chant about the divine origin of this island in a "creation myth" is said to refer to the Pohnpeian women of old "wrapping their hips in a *likau* made of cloth pounded from the supple inner bark of the breadfruit tree" (Flood et al. 2002: 144). In any case, on Pohnpei, the wood of the **Nihn** tree, along with woody parts of **Keleu** (*Hibiscus tiliaceus*) and **Topwuk** (*Premna obtusifolia*), was reportedly used to start fire by rubbing pieces of their woods together. Balick and his colleagues (2009: 446–448) provided more information about traditional uses for **Nihn** on Pohnpei, such as local hunters' knowledge that "black birds" (perhaps the native Micronesian starling known in Pohnpeian as **Sioahk**) and other forest birds consume its yellow to orange fruits, and that such knowledge motivates them to hunt in areas where **Nihn** trees are found. Balick et al. (2009) also noted that in the past, when food was scarce and people were hungry, they could eat the **Nihn** fruits. In addition, the fibrous bark of this type of fig tree has been used to make a *liep* cord which is "looped around ankles to climb a coconut tree," as well as to produce a strong rope that could be used in ways similar to that made from *H. tiliaceus* fibers; this included fabrication of **Nihn** cordage used to tie up firewood to be carried home. **Nihn** is also said to be used to make fish bait and a fishing lure used specifically to catch certain fish such as **Ahl** (skipjack tuna), and the stems of **Nihn** are used as construction material for small structures, such as cookhouses (*imwen kuk*). When dead, **Nihn** is said to be a good source of firewood. Furthermore, Balick et al. (2009) reported that this tree has been used in a variety of practical and magical remedies for diverse ailments on Pohnpei; for example, **Nihn** is said to be used "in sorcery to chase ghosts away from a sick person because it is a powerful plant; and to treat nervousness, four leaves are chewed and swallowed to calm a person in a big crowd of people and in playing games or sports." Moreover, **Nihn** is said to be used to treat a "possessed person" and those who become affected by *soumwahu en naniak* (mangrove sickness); additional therapeutic uses attributed to **Nihn** include the following:

> To treat joint pain in the arm... To make a baby strong... To use to treat sick dogs with a skin rash and falling hair... To treat cuts and dry wounds... To treat earache... To treat *saleng medek* (earache caused by insects crawling in the ear, water getting in, a blow to the head, or just proneness to earache)... To treat a skin rash or skin disease..." (see Balick et al. 2009: 446–448 and Lee et al. 2010: 92 for more detail regarding the use of **Nihn** in Pohnpei)

▸ **Idahnwel** (P)
Flagellaria indica

Description: *Flagellaria indica* is a member of the small family Flagellariaceae, which is comprised of woody vines or lianas. These vines, including *F. indica*, climb upward to reach tropical rainforest canopies, such as those in Pohnpei, using long tendril-tipped leaves. *F. indica* often grows up to be about 15 m (45 ft) tall. Although *F. indica* produces relatively slim stems, it is a tough but flexible woody vine or liana that ascends trees. The cane-like stems of *F. indica* are up to about 2 cm (0.8 in) wide, and its leaves are about 9–30 cm (4–12 in) long and 2–3 cm (1 in) wide with the tip originating from the midrib and coiled upwards to form a slender tendril. In addition, *F. indica* has very short or absent petioles which, when present, clasp the stems at the base of the leaf blades. Its tiny, creamy, white flowers are produced in large terminal panicles, and its small, pea-sized, greenish-red or light brown fruits are about 6–8 mm (0.2 in) in diameter, normally with only a single seed in each fruit.

Distribution: *F. indica* is one of four or five species in the genus *Flagellaria*, and is a climbing vine or liana native to tropical and subtropical regions including Africa, Indo-Malaysia, Australia, and a variety of islands in the tropical Indian and Pacific Oceans. This rattan-like plant is found occasionally in thickets and woods, more often in lower valley regions of some forests on the high islands of Micronesia.

Uses in Pohnpei: According to Balick and his colleagues (2009: 279–281), **Idahnwel**, as *F. indica* is known in Pohnpei, has many uses on this island. For example, the whole plant is sometimes used as thatch for roofing buildings and "as cordage to hold a thatched roof together," in which case "the leaves are taken off the stems and the stems are used to tie leaves of [the coconut palm or the nipa palm] onto a roof frame." Moreover, its stem is sometimes "used as a crossbar in the roof thatch," and the stem is traditionally "split and used to weave house walls, baskets, cabinets, and floors." **Idahnwel** is also used to support the vine growth of cultivated yams (mainly *Dioscorea alata*, **Kehp**); in this case, the yam plants are tied to the split stems of **Idahnwel**. In addition, **Idahnwel** is used in Pohnpei to make a special medicine (*winien koaloalamwahu*) to help treat sadness or depression (*koaloalamwahu*) in a variety of ways as a "calmative" and "charm." Often in combination with plant parts from other species, leaves of this climbing plant are used on Pohnpei to treat arthritis and to wean a baby from breastfeeding; for babies, parts of this vine are also used medicinally to treat infant shock, excessive crying, diarrhea, sleeplessness, loss of appetite, and feverishness (for more details about the traditional uses of **Idahnwel** in Pohnpei, see Balick et al. 2009: 279–281).

▸ Rahra (P)
Freycinetia ponapensis

Description: *Freycinetia ponapensis* is a woody vine (liana) native in Pohnpei. It is a member of the family Pandanaceae, which includes the very important *Pandanus tectorius* tree (**Kipar**). *F. ponapensis* is a coarse woody climber that can also form a thick low-lying vegetative cover when it spreads out over open ground. Its widely branching stems are about 2–3 cm (~1 in) wide and produce long, tough rootlets that become attached to shrubs and trees. These grasping parts of *F. ponapensis* lianas allow it to grow up into the forest canopy. The lanceolate-shaped leaves are sheathed at the base, and are long, linear, or elongated and have a pointed tip. The base and apex of the leaf margins are serrated with spine-like (toothed) structures. *Freycinetia* species, including *F. ponapensis* (like *Pandanus* species), are dioecious, which means that the male flowers, including the pollen-bearing anthers (male sex organs), and the female flowers, including the pistils (female sex organs), are located on separate plants. The fruits of *F. ponapensis* are integrated into a syncarp (compound fruit), with individual carpels developing into fleshy berry-like structures united in spiraling rows, and collectively, numerous small seeds are produced in this compound fruit. The column-shaped fruits develop at the terminal ends of branches. Balick and Cox (1996, and others) have pointed out that pollination of the flowers of *F. ponapensis*, like its close relative *F. reineckei* (which is native to the high Micronesian islands of Palau, the Marianas, and in Chuuk lagoon), relies on flying foxes, which are actually bats in the genus *Pteropus*. These bats are attracted to the large, fleshy bracts of the *Freycinetia* flowers (also see Cox 1982, 1990); bracts, such as those of *F. ponapensis*, are leaf-like structures that are often brightly colored in plant species lacking true petals and are sometimes mistaken for floral petals.

Distribution: *F. ponapensis* is a common, woody vine endemic to Pohnpei and Kosrae. *F. ponapensis* grows up around several types of trees in the rainforests of both these islands, in cultivated areas as well as in native forest areas.

Uses in Pohnpei: According to Raynor and Fownes (1993: 47), *F. ponapensis* is a relatively common, native liana in the Pohnpeian rainforests, where it is known locally as **Rahra** and has several cultural uses. Glassman (1952) referred to the use of an infusion of the leaves of *F. ponapensis* as a contraceptive on Pohnpei. More recently, Balick and his colleagues (2009: 290–291) provided the following information about the magical and medicinal use of this liana on Pohnpei:

> "To use as a charm to find out who stole one's **Sakau** plant, the shoot of young leaves [of **Rahra**] is pounded and placed where the plant was stolen. This will spiritually make that person sick, and that person will come and ask for forgiveness... To use as a charm to find a person who steals from you, eight young leaves are cut into pieces, pounded, placed in the location where the property was stolen; then the thief's hand will become swollen and he or she will

come back and apologize... To stimulate a child's appetite to help him or her grow up fast... To make *wini en sarwel* for helping to diagnose a baby's illness (used for babies who have been sick for a long time and are losing energy, which has a gradual onset and is sometimes referred to as shock when the child loses its soul due to a loud noise when the child is sleeping; the baby cannot blink, its color is similar to that of a child who is very cold, its temperature is increased, and diarrhea occurs, sometimes with blood in it)... [and **Rahra** is used to]... treat weakness and body aches..."

▸ **Kehnpwil**, **Koahnpwil**, or **Nikenpiri** (P)
Garcinia ponapensis

Description: *Garcinia ponapensis* is a native endemic woody species on Pohnpei that has opposing leaves and produces abundant latex or sap, which is typical of species of the genus *Garcinia* that are commonly referred to as sap trees, mangosteens, or garcinias. *Garcinia* is the mangosteen genus, named after the esteemed fruit from Southeast Asia, *G. mangostana*. Several other species are also cultivated for their fruits, and others have medicinal compounds that are utilized in traditional and modern medicine. *G. ponapensis* needs further taxonomic and other botanical study.

Distribution: *G. ponapensis* is an endemic tree species reportedly only found on Pohnpei, although a closely related species (or variety of *G. ponapensis*) is found only in Chuuk, and classified long ago as *Garcinia trukensis*. Unique trees in the genus *Garcinia* are found in the native mixed broadleaf forests of a number of the high islands in Micronesia, including Pohnpei. In any case, more research is needed to sort out the taxonomy of these native endemic Micronesian *Garcinia* trees.

Uses in Pohnpei: *G. ponapensis* is known as **Kehnpwil** or **Nikenpiri** in Pohnpeian, and has been important traditionally in Pohnpeian myth and medicine. For example, it is said to be used to stop menstrual bleeding, and an infusion of inner bark is used as a beverage for treatment of dysentery. Furthermore, this *Garcinia* species' sap makes it a good source of kindling wood to start fires in forest. Therefore, people often camp near these trees when staying in the *nanwel* (i.e., wild forest lands or lands outside of human authority often in the uplands). The trunk of **Kehnpwil** is also sometimes carved into house timber on Pohnpei:

> "The wood is used for fire kindling and is even good if the plant is still alive or if it is raining, as it is rich in flammable resin... [The latex sap is also] used as glue for diving goggles, to seal cracks in wood and canoes[. T]he bark is cut in many places and allowed to bleed for four or five hours; then the gum is collected and used directly as glue [and the resin is used medicinally to] treat menorrhagia..." (Balick et al. 2009: 352).

Note: *G. mangostana*, the common mangosteen fruit tree which is native to Southeast Asia and a close relative to the Pohnpeian endemic *G. ponapensis*, was introduced to Pohnpei during the time of the German influence and then Japanese administrative control. It is now cultivated in some areas of Pohnpei (e.g., see Ragone et al. 2001).

▸ **Luwekidinloal** (P)
Glochidion marianum

Description: *Glochidion marianum* is a woody plant, described as "a bushy shrub with shiny, light green, alternate leaves on short petioles... [and] male and female flowers occur separately in leaf axils... [with the fruit forming into] globose, flattened capsule containing several orange-brown seeds" (Moore and McMakin 1979: 60).

Distribution: According to Wagner and Lorence (2011), *G. marianum* should be reclassified as *Phyllanthus mariannensis*, which they described as: "Endemic to the Mariana Islands (Guam) where it occurs on limestone and basaltic soils in old fields and grasslands and disturbed native and secondary vegetation at ca. 60–150 m elevation." Moore and McMakin (1979: 60) also referred to this species as endemic "on Guam," but classified it as *G. marianum* (with local names of **Chosgu** or **Abas Duendes**); furthermore, Moore and McMakin described it as a shrub which "seems equally at home on both limestone and volcanic soils," and indicate that it is "common in the southern half of Guam."

Uses in Pohnpei: The Pohnpeian name for this native woody plant in upper forests is **Luwekidinloal**, which implies "return to normal." This refers to its use as medicine for curing "heartbroken" emotions after the break-up of a romantic relationship. Balick and his colleagues (2009: 462) discuss the use of this species to treat a broken heart, relieve depression, bring one's wife back, decrease anxiety, and to stimulate a baby's appetite (also see Balick et al. 2009: 462 for a description of the closely related *G. puberulum* Hosok., and its uses in Pohnpei). Interestingly, in Pohnpei, **Luwekidinloal** plants with red fruit [*G. ramiflorum*?] are considered to be female; those with whitish fruits are considered to be male [*G. marianum*?].

▸ **Mwehk** (P)
Glochidion ramiflorum

Description: *Glochidion ramiflorum* is a small to medium-sized, variable woody species up to 10 m (35 ft) or more in height. This shrub to small tree produces simple, alternate, elliptical to lanceolate-shaped leaves that are mostly 4–15 cm (1.5–6 in) long with entire margins, and the petiole connecting the leaf to the stem is often pinkish. Its flowers are tiny and yellowish. The small, greenish, spherical, and compressed or grooved fruit capsules split into separate units with reddish seeds when mature.

Note: Based on identified plant material, Herrera et al. 2010: 160, suggested that "*G. ramiflorum* likely represents one or more possibly undescribed species and needs further study," and according to Wagner and Lorence (2011), "Recent molecular phylogenetic studies [DNA research] and reevaluation of morphological characters have led to the inclusion of *Glochidion* within a broader delimitation of *Phyllanthus*." Therefore, Wagner and Lorence offered a new species name, *Phyllanthus cleistanthoides* (Fosberg) W. L. Wagner & Lorence, comb. nov., to replace *G. ramiflorum* J.R. Forst. & G. Forst and *G. puberulum*. However, these two *Glochidion* species' names have been used by Balick et al. (2009: 461–464) in their book dealing with ethnobotany in Pohnpei (also see Falanruw 2015: 86). Here, we retain the genus name *Glochidion* for a series of populations of shrubs or small trees in Micronesia that are discussed below, and we list *Phyllanthus cleistanthoides* as a synonym.

Distribution: *G. ramiflorum* is found from tropical Asia to the Marquesas Islands in Polynesia, including its occurrence in Vanuatu in Melanesia and the Mariana Islands and Caroline Islands in Micronesia. This relatively small tree is found in both limestone and basaltic soils in primary and secondary forest areas from near the coast to the upland areas of the higher islands; it is also found in grasslands, and fallow vegetation, occasionally in garden areas and sometimes in grazing areas, "often volunteering in fallow vegetation; commonly protected when clearing for new gardens, and is occasionally found in home gardens" (see Clarke and Thaman 1993). Birds are fond of the fruit and thus disperse the small seeds; for example, "The Micronesian Starling is known to feed on the fruits" of *G. ramiflorum* in Palau (Kitalong et al. 2008: 81).

General uses: The sturdy wood of different *Glochidion* species is "used in light construction, in tool making, and for firewood;

leaves and bark used medicinally throughout the Pacific," and a reddish dye is "made from the bark in New Guinea" (Clarke and Thaman 1993; also see www.comfsm.fm/~dleeling/ethnobotany/text/flora.html). Throughout Micronesia where this woody species is found, the terminal buds of *G. ramiflorum* are said to be commonly used medicinally to treat appetite, and according to Quattrocchi (2012: 1849–1850), parts of *G. ramiflorum* have been used to "prevent conception of a child" as well as "treating food poisoning." In addition, its wood is commonly used as a source of firewood throughout the region.

Uses in Pohnpei: *G. ramiflorum* is called **Mwehk** in Pohnpei. It can be found in places from near sea level to higher upland areas of the island. As on other Micronesian islands where this species occurs, on Pohnpei it serves as a source of firewood, and the terminal buds of the flowers are used to increase the appetite of young children. Other parts of the plant have been used as a remedy for urination ailments. Although **Mwehk** sprouts on its own in the *nansapw* areas, and is sometimes allowed to grow, yams (**Kehp**), when cultivated near **Mwehk**, are believed to do poorly, even on occasion becoming diseased. Balick and his colleagues (2009: 463–464) referred to several medicinal uses for **Mwehk** on Pohnpei, including the following:

> "To bathe and be "fresh," the leaf tips are squeezed into bath water before bathing the body with it… The plant can be used to make medicine taste good in an unspecified way… To use as a charm for women who are nervous about leaving the house… To stimulate a baby's appetite… To increase appetite… To treat ulcers… To treat *pwuhs en seri* (a newborn's belly button)… To treat a cut or infection, the terminal buds are pounded, placed in a cloth, and squeezed onto the cut or infection. This treatment serves as an antibiotic and a styptic. It will sting when applied… To treat depression… To treat depression or stress… To treat abdominal pain… To treat *mwopw* (asthma)… To treat conjunctivitis (pinkeye)… To treat *nan mas pwudong* (itchy eyes)… To treat an upset uterus… To cleanse the uterus…" (Also see Lee et al. 2010: 68, 101, 138, 139, 150).

▸ **Ihd** (P), **Id** (Mw), **Eles** (Pi), **Bua laamodu** (Nu), **Bua** (Ka)
Guettarda speciosa

Description: *Guettarda speciosa* is a shrub or small tree that may reach heights of about 15 m (50 ft) or more. *G. speciosa* is very bushy with short thick branches and multiple twigs about 1 cm (~0.3 in) across. It has large dark green, glossy, ovate leaves 12–30 cm (5–12 in) long, each with a prominent midrib and 7–10 pairs of lateral veins. The attractively scented, white, tubular flowers are up to about 5 cm (2 in) long and grow from the leaf axils. The roundish, faintly ribbed, stalkless fruits are green when young, but turn whitish with age; the leaves are also somewhat flattened and fibrous at maturity, by which time they are about 2–3 cm (~1 in) wide and usually contain 4–6 seeds, each in a separate cell or cavity. *G. speciosa* is pollinated at night by moths, and its fruits are eaten by hermit crabs. Because of their corky interiors, the fruits of *G. speciosa* can float and are durable, meaning they can disperse over long distances via ocean currents.

Distribution: *G. speciosa*, with its floating fruits, is a native indigenous coastal tree found throughout Micronesia. Like most of the plants occurring naturally in the strand vegetation or low coral island forest areas, it is also found on many other tropical Pacific islands beyond Micronesia. Besides growing naturally in tropical lowland coastal vegetation, *G. speciosa* can also be found "in places at considerable elevation." For example, it occurs infrequently in upland areas in the Mariana Islands (Fosberg et al. 1993: 59). The native geographical range of *G. speciosa* includes parts of tropical East Africa and throughout tropical Asia and the Pacific Islands region from the Marshall Islands to southeastern Polynesia (Elevitch 2006a/b).

General uses: The wood of *G. speciosa* is especially important on tropical low islands and atolls, where it has many uses, including "hard and durable wood used in light construction, for pilings, fish-trap stakes, stakes to hold garden mulch in place, coconut huskers, fishing poles, floats, spears, thatching needles, fishing rods, fishnet and bird-net handles, stilts, eel traps, fruit harvesting sticks, bowls, slit-gongs, for canoe hulls, supports, steering paddles, bailers, poles for poling canoes, and floats" (Clarke and Thaman 1993). *G. speciosa* is also known to be one of the most important tropical Pacific coastal plant species for use by islanders living on the low reef islands or atolls due to the very widespread medicinal use of its bark and leaves (e.g., see Thaman 1992a: 6). For example, *G. speciosa* is reportedly widely used to reduce fever and as a remedy for diarrhea, muscle aches, and wounds. It is probably most important in the Marshall Islands, where the blossom of **Utilomar** is unofficially regarded as the national flower. *G. speciosa* is also a very important plant in the coral island nation of the Republic of Kiribati, where it is a very common tree and is prominent in local legends and mythology (e.g., see Clarke and Thaman 1993; also see Catala 1957). In addition, *G. speciosa* is one of the most favored sources of nutritious leaves used widely as a vegetative mulch added to the soil to enhance soil moisture conservation, to improve the fertility and health of the soil, and to reduce the growth of weeds (e.g., on Kiribati, see Manner 1993b: 34).

Uses in Pohnpei: *G. speciosa* is known in Pohnpeian as **Ihd**. This useful tree has a number of uses, especially in the atolls of Pohnpei State, as well as atolls elsewhere in Micronesia. For example, its trunk has been used for house timber, especially on the low-lying islands and atolls where the number of different woody species is much more limited than on the high islands, such as Pohnpei. Lessa (1977: 180) referred to the use of parts of *G. speciosa* on Pohnpei to ease the pain of bruises, and more recently, Balick and his colleagues (2009: 481) provided a number of uses for this lowland coastal tree on Pohnpei:

> "The seed kernels are eaten fresh as food [but mainly by children because it is a lot of work] and adults do not have the time... To use as a fire starter... The wood makes excellent firewood... It is also used for lumber for house posts... and to make *ipew* (spear handles)... The flowers are picked and tied together for *mwaramwar* (head garland)... To make perfume coconut oil... To treat bedwetting in babies (male or females)... The leaves are used for problems with pregnancy in women in an unspecified way..."

Uses in Mwoakilloa and Pingelap: *G. speciosa* is known as **Id** in Mwoakilloa and **Eles** in Pingelap. Glassman (1953: 296) indicated that it was a common coastal strand tree that could reach up to about 30 ft (~10 m) tall. St. John (1948) noted that the logs of this tree are used on Pingelap to make canoe hulls, and the white, fragrant flowers are utilized as ornaments in the hair and to perfume coconut oil.

Uses in Kapingamarangi: On the Polynesian outlier atoll of Kapingamarangi in Micronesia, *G. speciosa* is known as **Bua** (*pua* or *bua* generally means "flower" in the Polynesian languages). The dried branches of *G. speciosa* reportedly provide the best fuelwood when they are not needed for other uses: "The [b]ranches are broken into short lengths of a couple of feet or so and tied in a bundle for transport to the home," and **Bua** wood is also used as the common source to make the many types of bowls that "are in everyday use in all families" (Te Rangi Hiroa 1950: 9, 14). The wood of *G. speciosa* is also reportedly used for construction and to make implements, while the root is used to make shark hooks, and the flowers are used in garlands on Kapingamarangi (Niering 1956).

▸ **Akenwel** (P)
Gynotroches axillaris

Description: *Gynotroches axillaris* is a special endemic woody species on Pohnpei where it is known as **Akenwel**. Botanically, *Gynotroches* is a monotypic genus of trees in the family Rhizophoraceae. In other words, the genus *Gynotroches* contains a single species, *G. axillaris*. *Gynotroches* means "woman wheel" in Greek and refers to the shape of its stigma, which is the pollen-receptive surface of a carpel (the basic female reproductive organ in flowering plants), while *axillaris*, the second name of the binomial, is derived from the Latin and refers to the position in the plant's axils. An axil is the upper angle location where a leaf is attached to a stem, and it is in the axils of *G. axillaris* that the flowers of this Pohnpeian species are produced. *G. axillaris* can grow up to about 35 m (~110 ft) tall, with a trunk diameter of up to 70 cm (28 in). Its smooth bark is gray to blackish. The flowers are greenish white. The roundish fruits are green, turning red to black, and measure up to 0.3 cm (0.1 in) in diameter.

Distribution: The natural range of *G. axillaris* is apparently restricted to some areas of Southeast Asia, Melanesia, and one island in Micronesia: Pohnpei. Its natural habitat throughout much of its range is near rivers and in sandy areas from sea level to 2,200 m (7,200 ft) altitude. *G. axillaris* and *Cyathea* tree ferns are the typically dominant plants in the tropical montane cloud forests of Pohnpei (e.g., see Merlin and Raynor 2005: 249). Its putative unique Pohnpei-only distribution in Micronesia needs further study.

Uses in Pohnpei: As a member of the family Rhizophoraceae, **Akenwel** is related to the **Akelel** and **Akapah** mangrove trees, but grows in the upland wet forest, as opposed to the lowland swamp habitats. It is very common in several areas above about 2,000 ft (~615 m) on Pohnpei, and reputedly has uses in traditional Pohnpeian medicine.

▸ **Sinter** (P), **Jinjir** (Mw), **Sintia** (Nu), **Tindia** (Ka)
Hedychium coronarium

Description: The ginger genus *Hedychium* includes about 40–50 herb species with robust rhizomes that are mostly native from Malaysia to India, and possibly Madagascar. In addition, numerous *Hedychium* species are cultivated by people because of their eye-catching, often attractively aromatic flowers, and in some cases for the rhizomes that are used to make perfume. Here, we focus on *Hedychium coronarium*, known in English as white ginger or butterfly ginger. *H. coronarium* is an erect, perennial ginger species that is unbranched, with its leafy shoots emerging from the rhizomes. The leafy shoots can grow to be about 1–2 m (3–6 ft) in height. These leaves are simple, alternate, and two-ranked, with elliptic to lanceolate blades 20–30 cm (8–12 in) long and 3–10 cm (1–4 in) wide; the lower surfaces of the leaves have hairs. *H. coronarium* produces many fragrant, white or yellowish, butterfly-shaped flowers throughout the year on Pohnpei. The fruits capsules are oblong and many-seeded, but form infrequently under cultivation.

Distribution: The white or butterfly ginger, *H. coronarium*, is probably native to the Himalayas and southwestern China. It is widely cultivated in the tropics and frequently escapes cultivation to become naturalized (growing wild) in various places, including in some Micronesian islands, where it was only introduced in recent times. It is well adapted to wet habitats and can be found in moist forests, roadsides, open areas, and along streams. Unfortunately, the creeping growth of this aromatic herb "overwhelms low-growing plants in pastures and forests" and therefore is capable of becoming invasive, forming dense, single-species stands because of vegetative spread by the expansion of its dense rhizomes (Motooka et al. 2002).

Uses in Pohnpei: *H. coronarium* was introduced into Pohnpei in relatively recent times and planted as an ornamental; however, this decorative white ginger plant has escaped cultivation and become an invasive species in some areas of this island (Meyer 2000: 90). Its Pohnpeian name is **Sinter**, and it has two local cultivars: ***Sinter pwetepwet***, typically with white flowers, and ***Sinter oangen waii***, typically with yellow flowers (see Rehg and Sohl 1979). Balick and his colleagues (2009: 314) referred to the use of the attractive budding flowers of **Sinter** in head garlands on Pohnpei; furthermore, they reported that these same flowers are used to condition human hair, making it sweet-smelling and gleaming: for example, unopened flower buds of this ginger plant on Pohnpei are "collected before the sunrise, pounded with grated coconut "meat" (copra from *Cocos nucifera*), and massaged into the hair in the morning and washed out in the evening." Balick and his colleagues (2009: 315) also described therapeutic use of the cleaned rhizome of this ginger, which is scraped into a piece of cloth, and drops are then squeezed into the eyes of people suffering from irritations such as "conjunctivitis (red eye or pinkeye)."

Note: Rehg and Sohl (1979) referred to *H. flavescens*, a synonym for *H. gardnerianum*, which is a close relative of *H. coronarium* that was also introduced to Pohnpei and is "used as an ornamental." However, *H. gardnerianum*, which is reportedly known on Pohnpei as **Sinter weitahta**," can be a very invasive and troublesome plant pest:

" [*Hedychium gardnerianum*] is a showy ornamental [ginger-type plant] which grows over a meter tall in wet climates and grows from sea level to an altitude of 1700 meters. It displaces native plants, forms vast, dense colonies and chokes the understory vegetation. It can also block stream edges, altering water flow. It is dispersed by birds over short distances and by man over long distances (in garden waste or via the horticultural industry). Even small root fragments will re-sprout, making it difficult to control." (www.iucngisd.org/gisd/species.php?sc=57)

▸ **Mworopwinsed** or **Merepwinsed** (P)
Heritiera littoralis

Description: *Heritiera littoralis* is an indigenous tree in Pohnpei and grows to heights of approximately 15 m (~50 ft). It has a slightly buttressed trunk and is often covered with lichens and mosses. The pointed, simple leaves of *H. littoralis* are ovate to elliptic with entire margins; they are arranged alternately and are 8–35 cm (~3–14 in) long and 3–12 cm (~1–5 in) wide. The flowers are inconspicuous, unisexual, and are produced from the leaf axils. *H. littoralis* produces broad, winged, and oval-shaped, green to brown woody nuts with a tall keel on one side. These nuts are about 8–9 cm (~3–4 in) in length and 3–5 cm (1–2 in) in diameter. In English, this tree is called the "looking-glass tree" because the undersurfaces of its leaves are grayish or silvery white.

Distribution: *H. littoralis* is found along river channels and at the landward edge in some areas of mangrove forests. "The lightweight seeds fall into the water and float along like little sailboats… to be distributed along coasts" (Falanruw 2015: 220). Its geographical range includes various mangrove areas in East Africa, South Asia, Southeast Asia, Northeast Australia, and a number of tropical Pacific islands, including Palau, the Mariana Islands, and the FSM. Although this mangrove-associated species is still widespread and common, "it is threatened by the loss of mangrove habitat throughout its range, primarily due to extraction [for timber use] and coastal development, and there has been an estimated 20% decline in mangrove area within this species range since 1980." This loss of mangrove area is due mainly to "coastal development and extraction at the extremes of their distribution," where the mangroves "are likely to be contracting… more than in other areas." Changes in climate due to global warming are also likely to further affect these areas in the native range of this species (IUCN 2010).

General uses: *H. littoralis* is a high-quality timber source and is also harvested for food and medicine in some parts of its wide natural range. The wood of this mangrove-associated species "is valued for its toughness, durability, and resistance to saltwater," and as a result, it is regularly used in coastal areas, such as those in the tropical Pacific, for "shipbuilding and in making pilings, bridges, and wharves" (e.g., see Thaman 1992a).

Uses in Pohnpei: Known as **Mworopwinsed** or **Merepwinsed** in Pohnpeian, *H. littoralis* is a source of timber on Pohnpei where it is used for general construction, as well as for making canoes (Glassman 1952); Christian (1899), who identified the local Pohnpeian name of this species as "Chaiping," referred to its use in boat building. *H. littoralis* is also a source of traditional magic and medicine on Pohnpei, as reported by Balick et al. (2009: 428): "To take revenge on another person for doing something bad to you, an image of that person's body is drawn or carved on the tree [*H. littoralis*] and the bark containing the image is cut off and placed in boiling water, accompanied with an unspecified spell…" and the bark of the tree is used "To treat *mehn war mwoui* (menorrhagia)…"

▸ **Pingiping** (P) (Pi), **Pingping** (Mw), **Bingibingi** (Nu) (Ka)
Hernandia nymphaeifolia

Description: *Hernandia nymphaeifolia* is large indigenous coastal tree. It can develop a robust trunk with gray to whitish bark and grows to a height of 20 m (65 ft) or more. The large, cordate (heart-shaped) leaves have 7–9 conspicuous veins radiating from the point of attachment to the petiole. *H. nymphaeifolia* produces small white male flowers with three petals, while the female

flowers have four petals. The hard-seeded fruits of *H. nymphaeifolia* are surrounded by a spherical, inflated yellowish (or reddish) vesicle, which has an opening at the top, and because of its conspicuously shaped fruit, this species is sometimes referred to in English as the "lantern tree." Its wood is relatively light and soft.

Distribution: *H. nymphaeifolia* is commonly found along tropical sandy or rocky shores in the coastal woodlands from East Africa and the Indian Ocean to Asia and the Pacific Islands. It is found widely in the tropical forested areas of Micronesia.

General uses: *H. nymphaeifolia* has a variety of uses for the tree itself and its wood. It also has traditional medicinal uses in various areas within its native range, and modern science has produced some positive results from testing its potential healing properties (e.g., see Quattrocchi 2012: 1962).

Uses in Pohnpei: **Pingiping**, as *H. nymphaeifolia* is known in Pohnpeian, is commonly found along sandy or rocky shores. Its name means "confusion" or "disorder," and it is said that some have used this tree species to bring confusion to their enemies. According to Balick et al. (2009: 402), the stems of **Pingiping** are used to make outriggers for local canoes, and the leaves are utilized in the fertilizing compost for **Mwahng**, the giant swamp taro (*Cyrtosperma merkusii*) on Pohnpei. Furthermore, Lessa (1977: 157) indicated that **Pingiping** "has medicinal uses" on this island.

Uses in Mwoakilloa and Pingelap: Glassman (1953: 300) reported that "*Hernandia sonora*" [= *H. nymphaeifolia*] is a common tree in the coastal strand of Mwoakilloa, growing to heights of about 5 m (15 ft) tall with numerous seedlings found beneath the parent plants, and wrote that the trunks of this *Hernandia* species were used by the natives to make canoes. Hattori (2012: 260) described it as "tall strong tree, with good amount of [pliability]," with seeds used "in craft work" and wood used to make canoe hulls or paddles on Pingelap Atoll.

▸ **Keleun wai** (P), **Kɛlɛu in wai** (Mw), **Kaelaeu in wai** (Pi)
Hibiscus rosa-sinensis

Description: *Hibiscus rosa-sinensis* is a very widespread ornamental plant. Most ornamental varieties of this species are hybrids; indeed, *H. rosa-sinensis* is a highly polymorphic group of plants that includes a host of complex hybrids and their derivatives (often involving crossing with various other *Hibiscus* species). This popular, commonly cultivated hibiscus is a woody shrub or small tree, up to 4 m (~12 ft) high, with glabrous, dark green to variegated, serrate leaves; its flowers are large, conspicuous 5-petaled, and red, pink, white, or yellow in color. As a cultivated plant, mostly with cuttings, *H. rosa-sinensis* rarely produces the fruit capsule which is "ovoid rounded, 5-valved, 15-seeded (or less)" (Stone 1970: 411).

Distribution: The origin of *H. rosa-sinensis* is uncertain, but it is probably indigenous to eastern Africa or tropical Asia. It is most likely an aboriginal or very early post-European-contact introduction to Melanesia and parts of Polynesia, and a pre-World War II introduction into most of the other islands of Polynesia and Micronesia.

General uses: The common hibiscus, *H. rosa-sinensis*, is generally a highly desirable, showy shrub that is planted widely in relatively warm environments. It requires very little gardening care, flowers prolifically, and overall it is a very good plant for landscaping. It can be found occasionally in rural areas and is common to abundant in urban areas where it is an attractive plant that often serves as a hedge or living fence. In the tropical Pacific, its flowers are frequently used in garlands and for decoration, and in some places the leaves are "fed to goats" (Clarke and Thaman 1993). *H. rosa-sinensis* has many common English names; one is "shoe black" or "shoe flower" because its flowers were once used to polish leather shoes. Another important use of *H. rosa-sinensis* has been the widespread therapeutic applications of leaves, barks, roots, and flowers in traditional medicine. Many research studies have documented the use of different parts of this species to treat various diseases because of their antioxidant, antimicrobial, antidiabetic, antiulcer, hepatoprotective, antifertility, antigenotoxic, and anti-inflammatory properties, which may aid in the management of numerous diseases (e.g., see Kristi and Patel 2016).

Uses in Pohnpei: *H. rosa-sinensis* is known as **Keleun wai** in Pohnpeian and is closely related botanically to **Keleu** (*H. tiliaceus*), a very useful and widespread plant in tropical Pacific islands, including Pohnpei. **Keleun wai** (*H. rosa-sinensis*), with its beautiful reddish, pink, yellow or white flowers, was introduced to Pohnpei in Spanish or German times and is now commonly planted near buildings. The flowers are commonly used in *mwaramwar* (garland worn on one's head) and leis. According to Lee et al. (2010: 39), on Pohnpei, a preparation of **Keleun wai** flowers is used "to speed the healing of boils."

▸ **Keleu** or **Kolou** (P), **Kɛlɛu** (Mw), **Kaelaeu** (Pi), **Goehau** (Nu), **Hau** (Ka)
Hibiscus tiliaceus

Description: Coastal hibiscus (*Hibiscus tiliaceus*) is an evergreen, much-branched, small tree that typically grows to 3–10 m (10–33 ft) in height with a sprawling form. In some places, this tree forms an erect, gnarled, crooked tree of low or medium height, while in other places it spreads horizontally over the ground in thickets, thus forming an apparently impenetrable network of trunks and branches. The round, variable-sized, heart-shaped leaves are more or less leathery, about 5–25 cm (~2–12 in) wide with entire or scalloped edges. The tops of the leaves are mostly smooth, while the lower sides are whitish with small matted hairs. *H. tiliaceus* produces many flowers, each about 5–8 cm (2–3 in) long, that are produced at or near the branch tips. When the flowers open, they form bright yellow cups, some with dark red centers; however, as the day progresses, the flowers change color to a rather dull orange, and then by nighttime are dullish red. The calyx cup is 5-toothed, downy persistent, and about 2.5 cm (1 in) long, while the 10–12 associated bracts are shorter. The fruiting capsule is downy and ovoid in shape, about 2.5 cm (1 in) long, and has 5 valves which contain three smooth seeds.

Distribution: *H. tiliaceus* has a wide distribution and complex natural and cultural history; it is unclear if it originated in tropical Asia or tropical America; "the seeds [of *H. tiliaceus*] can remain viable after floating in seawater for several months, which partially accounts for the tree's wide dispersal," and because of "its many traditional uses, coastal hibiscus has doubtless been spread by humans" (Elevitch and Thomson 2006: 394; also see Thaman and Whister 1996). Indeed, *H. tiliaceus* is widely believed to be indigenous to many parts of the tropics and has been introduced to new regions by people. It is possible that some varieties of *H. tiliaceus* were artificially selected by humans and then transported to islands or other environments where this species already existed as a native species, and it was also possibly brought to other areas first by people.

H. tiliaceus is relatively common to abundant on many if not most tropical Pacific islands and is currently widespread in the islands in Micronesia where it is probably native. However, because of its multitude of traditional uses, it is also broadly cultivated, and in some areas it has become naturalized (escaped from cultivation and growing spontaneously). *H. tiliaceus* plants are highly variable in form, leaf and flower color, etc., and several taxonomic varieties have been recognized. The fast-growing *H. tiliaceus* commonly grows in habitats along the seashore and in the back of mangroves, where its growth indicates the high water mark and the boundary between the end of saltwater penetration and the beginning of freshwater swamp. *H. tiliaceus* does well in a wide range of soils, including inhospitable brackish swamps, waterlogged soils, and limestone, but is most at home in coastal and near-coastal environments. However, it has been introduced into agricultural environments up to 800 m (~2600 ft) elevation, but remains mostly in the lowlands, and thus can be found growing in profusion along some streams. Once established, *H. tiliaceus* often persists and spreads, especially in moist gullies, streambeds, and other wet areas where it can become a dense mass of tangled limbs and leaves; therefore, it can sometimes be considered an invasive species, especially in disturbed or open forest and in degraded upland areas. It tolerates salty ocean air currents, but very strong winds can cause trees to fall over, which contributes to its tangled and intertwined branching form (e.g., Elevitch and Thomas 2006).

General uses: Throughout most of the tropical Pacific region, coastal hibiscus is a culturally significant plant species with a multitude of uses. As a densely branched woody species, it can provide a coastal windbreak and be used as hedge or boundary plant, and it often serves as a living trellis or fence posts. In addition, this multipurpose species is widely appreciated "by Pacific peoples for its contributions to soil fertility" (Falanruw 2015: 140), and its light but relatively sturdy wood is used for canoe parts, crafts, short-term utility, for fuel wood, and to start fires. The bark of *H. tiliaceus* contains strong fibers used for making rope, to caulk ships, and a variety of other purposes including making traditional clothing. To extract the bark from the plant, the bark is stripped lengthwise from the wood and soaked in water to separate the outer part from the smooth cream-colored inner part. These fibrous outer bark strips are dried and then woven into rope and used for a number of other purposes. The cordage made from this plant has the unusual property of being stronger when wet. Coastal hibiscus cordage is used to make a wide variety of items, including handicrafts, fishing nets, hammocks, mats, slings, bow strings, net bags, and string for sewing or making flower garlands. Clarke and Thaman (1993) provided an extensive description of the diverse uses of *H. tiliaceus* in the Pacific region, including those especially in the Hawaiian Islands and in Micronesia:

> "One of the most useful trees in the Pacific; commonly planted as living fencing and animal pens and in coastal areas, near houses, in gardens, and as an ornamental or shade tree; a creeping variety planted as wind-break in Hawai'i; its presence in forested areas considered a sign of former cultivation in Hawai'i; features in eastern Polynesian legends and Hawaiian fire-making legends; commoners not allowed to cut branches without permission of chiefs in Hawai'i; branches borne in battle by priests as a good omen and allowed to fall in retreat in Hawai'i; born by attendants at presentation of first fruits to kings on Easter Island; branches used as tapu markers to delimit restricted areas in Hawai'i; used to make spears used in typhoon magic in Ulithi; soft wood used in light construction and wood carving, for house rafters, pig-tethering posts, for canoe outriggers, spreaders, bailers, booms and occasionally hulls, fishing rods, hoists and floats, fishnet frames and handles, bows, fruit-picking rods, tools and tattooing comb handles, kite struts, jackstraw sticks, pestles, breadfruit splitters, coconut huskers, net floats, spears, shore-line posts to delineate fishing zones, fishing gear containers, noddy bird net handles and frigate bird nesting platforms (Nauru), and other purposes; a decent firewood, especially for slow smoking; used in making fire by friction; wood dried for six months used for fireworks in Hawai'i; bast [fiber] used as canoe caulking and to make cordage for clothing and dancing skirts and kilts, coconut-climbing bandages or foot harnesses, mats, sandals, sewing tape, bark cloth paint brushes, making fishnets, fishing line and lures, slings, kava strainers, sandals, tying corpses in tape, and cordage for tying, lashing and binding canoes, housing, and other things; bark used to strain kava in Pohnpei to give it its preferred slimy consistency; leaves, terminal buds, unopened flowers, and bark used medicinally, with leaves being used to reduce hemorrhaging and for treating neurological disorders; leaves used to parcel food, especially seafood, as plates, and to line and cover the earthen oven; leaves widely used as toilet paper; flowers used in garlands in Hawai'i; bark, shoots, and sapwood eaten in New Caledonia and other parts of Melanesia; leaves occasionally added to compost in Kiribati and Tuvalu; a number of distinct varieties or cultivars recognized in Melanesia and Polynesia."

Uses in Pohnpei: *H. tiliaceus* is known as **Keleu** or **Kolou** on Pohnpei, where it is a very widespread shrub or small tree, just as it is on many other tropical Pacific islands. On Pohnpei, it is one of the foremost multipurpose trees, with virtually all parts of the plant being used by people for a variety of purposes. It is used as a source of firewood because its wood parts are easy to burn. Its wood is also used in light construction, including making canoe parts; for example, a Pohnpeian student's presentation in the ethnobotany course at the local community college briefly referred to the major plant components of a traditional Pohnpeian canoe: "The hull is made of the trunk of **Sadak** (*Elaeocarpus carolinensis*), **Mahi** (*Artocarpus altilis*), or **Dohng** (*Campnosperma brevipetiolata*) [and the] upper part of the hull ([the] running board) is made of a lighter wood such as that extracted from **Keleu** (*H. tiliaceus*)." Poles or whole trees of *H. tiliaceus* are used for yam trellises on Pohnpei, and its leaves are composted as green manure in soil pits dug for yam cultivation. In addition, the useful bark of this abundant plant has been used on Pohnpei to make traditional clothing, and is processed into rope or cordage. Furthermore, bark fibers of **Keleu** are said to have replaced the **Nih** (coconut) fibers used to strain pounded **Sakau** rootstock; indeed, stripped bark of **Keleu** now provides the favored fibers for this common use on Pohnpei (Raynor and Fownes 1991: 140; also see Lebot et al. 1997). Furthermore, fibers of **Keleu** have also been used in the construction of fishing nets, and according to the Pohnpeian-English Dictionary, ***Pahtakai*** is the Pohnpeian name for a cultivar of *H. tiliaceus* "used for making ropes and paddles," and whose terminal buds are utilized as medicine, while it is also "prized for its dense wood." Lee Ling referred to the use of the inner bark of *H. tiliaceus* as a wrapping for the mixture of water and stone pounded **Sakau** rootstock:

> The squeezing (*wengweng*) is done using the inner bark of the hibiscus tree. This layer is the phloem and contains polysaccharides (long chain sugar molecules) that are good for the "gut" yet can also cause "gassiness" the next day. The viscous and slimy sap is referred to as *meteitei* (*moatoaitoai*) [and] *rahr* is another name for [the inner bark of] **Keleu** [when used as a bast in the preparation of kava]." (www.comfsm.fm/~dleeling/ethnobotany/text/#chapter10)

Two Pohnpeian proverbs (*lepin kahs*) referring to **Keleu** (*H. tiliaceus*) on Pohnpei are presented below:

1. *Tehn keleu marahra*—"(He is) a light hibiscus leaf."

> Leaves of the wild hibiscus tree are fairly large, light and flutter in the wind. This phrase is applied in derision to a man whose opinions are fickle or who cannot follow through on things he has started." (Reisenberg and Fischer 1955)

2. *Keleun ieng soupeidi*—"A chief is a hibiscus in the wind."

> The wild hibiscus tree sends up tall, flexible shoots which bend easily in any direction. According to two informants, this proverb sets a standard of fairness for the chiefs, who are supposed to respond to the needs and requests of any of their subjects, just as the hibiscus tree bends with a wind coming from any direction. According to a more caustic informant, the proverb refers to the readiness of the chiefs to favor anyone who will bring them plenty of food at feast, regardless of the other requirements of the situation." (Reisenberg and Fischer 1955; also see Hanlon 1988: 113)

There is a Pohnpeian story about a raft made of **Keleu** that was brought in during the time of the Saudeleurs (during the time of Isohkelekel, who invaded from the island of Kosrae). According to this story, the **Keleu** raft was dragged in by canoe from somewhere to the west of Pohnpei. It is said to have gone by Saipan and, according to this traditional story, when this raft arrived in Pohnpei, it brought in the **Keleu** plant (*H. tiliaceus*) for the first time.

A student in the ethnobotany course offered at the College of Micronesia on Pohnpei reported the following information about the usefulness of this important plant on that island:

> [**Keleu**] is an important plant to the people of Pohnpei ever since a long time ago. Pohnpeians use almost every part of the plant for certain cultural purposes. A long time ago, there were no clothes on Pohnpei, so the people had to dry the inner bark in order to make grass-skirts. Back then, grass-skirts were used as clothes, but today they are made specifically for cultural dances and displays. Aside from being made into grass-skirts [e.g., the traditional *koahl* worn by Pohnpeian men; see Anson and Raynor 1993: 144], the slimy, inner bark of the plant is used for the sacred ritual of preparing the local *sakau*. This inner bark is separated from the outermost bark, rubbed with water, and then it is used to squeeze the *sakau*. The inner bark of the hibiscus and the *sakau* work together to form a very significant part of our culture which is the preparation and drinking of *sakau*. The hibiscus tree is also necessary for healthy *sakau* production. These trees are cut down on growing *sakau* for the *sakau* to grow faster and bigger. Furthermore, the bark is important for yam production. Pieces of the bark are dried and tied together for farmers to use. Farmers help the vines of their yams climb up trees by tying one end of the tied bark to the base of the vine, and the other end in a tree. As the vine grows higher, it twirls around the tied barks all the way up into the tree. The leaves of the *Hibiscus tiliaceus* [are] also important for food preparation. Smashed or pounded local food are wrapped in the leaves before being cooked. Some people also cover their cooking

pots with hibiscus leaves before cooking their food. The leaf buds too, are very important because they are used as some form of medicine by the people of Pohnpei. When a person has a swollen skin bruise, the leaf buds are crushed, wrapped in a piece of cloth, and put in a cup half-filled with water. The person then gently massages the swollen bruise with the wrapped medicine. It is kept wet, and used over and over until the bruise is no longer swollen. The leaf buds are also used for some superstitious beliefs in Pohnpei. For instance, if a group of people are going to do a cultural display before a large crowd filled with strangers, they usually take a shower with water that contains crushed leaf buds before the sun rises on the day of the display. It is believed that this shower helps take away nervousness, and any other feeling that may interrupt one's concentration. The yellow flowers of the hibiscus plant [are] used mainly for head leis or *mwaramwar* and other forms of decoration. The stems, when alive, are used for carrying things like yams, breadfruits and other heavy farm products. A long pole used for breadfruit climbing is also made from the hibiscus tree. However, when the trees are dead, the stems can sometimes be used as firewood. Finally, the *Hibiscus tiliaceus* is very important for the people of Pohnpei because almost every part of serves some special purpose." (www.comfsm.fm/~dleeling/botany/1998/vhp/bruno02.html)

The therapeutic usefulness of *H. tiliaceus* is also referred to in some detail by Balick et al. (2009: 430–431):

> " To make *wini en loh* (a mechanism to treat choking on a fish bone)... Alternatively, to dislodge fish bones that are stuck in the throat (one of the most important medicines in Pohnpei, the medicine is very effective because the sap is very slimy and acts as a lubricant)... [some of the other therapeutic uses of parts of *H. tiliaceus* including leaf buds on Pohnpei are] To treat diarrhea, several remedies are reported... To treat gonorrhea... To treat baby or adult fevers... For children's fevers... To treat boils that are taking a long time to heal... To treat a bruise... To treat a swollen, bruised face such as after a fight... To treat anxiety... To treat early or imminent labor, the bark is scraped, wrapped in cloth, squeezed into a cup of water, and consumed daily when the mother is six months pregnant until the baby is born and period returns..."

Another student in a College of Micronesia class dealing with use of plants reported an "unusual" use of *H. tiliaceus* bark to treat a skin bruise (*mwou*):

> " The treatment was unusual in that six inches of bark are boiled in water until the water turns brown, then the water is drunk as a tea. This is unusual as the bruise is external and treatments for external skin issues are usually also applied externally. This is an internal treatment." (See danaleeling.blogspot.com/2014_09_01_archive.html).

Uses in Mwoakilloa and Pingelap: *H. tiliaceus* is known in Mwoakilloa and Pingelap by the same name, spelled **Kɛlɛu** and **Kaelaeu** in the respective languages. Glassman (1953: 300, coll. 2653) reported it growing in the coastal strand vegetation, "up to 50 feet high," and noted that its fibers extracted from the bark of this *Hibiscus* species "are used in making rope, fish line, hats, and baskets, and the leaves are often utilized in washing clothes" on both atolls. It has also been reported more recently that *H. tiliaceus* has a number of additional uses on Mwoakilloa, including some that are medicinal; for example, similar to its use elsewhere (e.g., on Pohnpei), *H. tiliaceus* is reportedly used on this atoll to reduce the swelling of a bruise by using unopened leaves of the plant which are chewed, mixed "with a small amount of oil" and then applied repeatedly to the bruised area of the body for sustained amounts of time (www.comfsm.fm/~dleeling/ethnobotany/healing_plants_presentations.html).

Uses in Kapingamarangi: According to Hiroa (1950: 130, 203), *H. tiliaceus* is "one of the most useful plants" on Kapingamarangi Atoll. For example, Hiroa reported that *H. tiliaceus*, locally known as **Hau**, supplied poles for the outrigger parts of canoes and the mast, made from a straight pole "3 to 5 in in diameter." Hiroa also featured the important traditional use of **Hau** bark, which has long been used "for cordage, tying material and warps and wefts for weaving," giving this description of how the white inner bark is detached from the stems:

> " To obtain the bast for cordage, the bark [*gili hau*] is removed [ihi] from young trees or saplings in strips. The subsequent treatment is summed up in the words '*ka hati, ka kumi, ka ihi*' [*ga hadi, ga gumi, ga ihi*] (bend, pinch, strip apart)."

Hiroa (1950: 149–158) further referred to the subsequent lengthy treatment of the fibers to make them ready for the loom weaving of customary **Hau** fiber garments worn on Kapingamarangi. Two- and three-ply **Hau** cordage was traditionally used for fishing lines and sinker cords, and wooden bows, sometimes made out of **Hau** wood, were used to hunt fish on this atoll (Hiroa 1950: 231, 243, 269).

▸ **Tekitek** (P)
Hoya schneei

Description: *Hoya schneei* is an endemic vine species only found on Pohnpei. The Pohnpeian name for *H. schneei* is **Tekitek**, which means "to look after it." This evergreen perennial vine has bluish-green foliage and produces relatively small, whitish flowers that have a pleasantly fragrant odor.

Distribution: *H. schneei* is one of two recognized endemic species of *Hoya* vines in Micronesia; the other is *H. trukensis* Hosok. *H. schneei* can be found in disturbed areas on Pohnpei, such as those used to grow crops. For example, Balick and others collected a specimen of this vine in a non-flowering state (sterile) in "Primary forest, cleared for **Sakau** (*Piper methysticum*) cultivation" on Pohnpei at 371 m (1217 ft) elevation in Madolenihmw Municipality, and another specimen of this vine was collected in 2009 near Kolonia, Pohnpei, along road "in agroforest and residential areas" within mostly alien vegetation (see New York Botanical Garden specimens 02136993 and 192255).

Uses in Pohnpei: This endemic vine is used in the making of *mwaramwar* (garlands) and in medicine on Pohnpei. Balick et al. (2009: 331) provided the following ornamental and medicinal uses of **Tekitek** on this island:

> "To treat back pain, the whole plant is pounded, put into a cloth, squeezed into a bottle, and the liquid is consumed... To clean a uterus, as many leaves as needed are pounded, wrapped in a piece of cloth, squeezed in 2 lit (0.5 gal) of water, and consumed (2 lit daily) for four or eight days... To use as a contraceptive for women, gather four or eight buds, pound, and put them into a cloth, squeeze them into a cup of water, and drink one cup each day to prevent a woman from getting pregnant..."

▸ **Kiepw** (P) (Mw) (Pi), **Giebu lauligi** (Nu), **Bukeulehu** (Ka)
Hymenocallis littoralis

Description: *Hymenocallis littoralis* is a perennial herb that develops from an underground bulb, growing to heights of 60–70 cm (24–28 in). The bulb of this species is 7–10 cm (3–4 in) in diameter, and with age, the bulb develops a "neck" that reaches approximately 4–5 cm (~2 in) in diameter. The attractive flowers are relatively large and white and produce a strong vanilla-like scent; the flowers are positioned sessile (i.e., attached directly by their base or stem without a stalk or peduncle). The tepals (flowering parts that cannot easily be classified as either sepals or petals) are attached to the staminal cup. The tube of each flower in this species is 14–17 cm (5–7 in) long or longer.

Distribution: *H. littoralis* is native to tropical America and was probably introduced to Micronesia more recently than *C. asiaticum*, a larger "lily plant" with which it is often confused. *Crinum asiaticum* is native to southern Asia, where it is normally found along seashores and other sandy places near riverbanks. Both species are known in Pohnpeian as **Kiepw** and the same pattern can be observed elsewhere in Micronesia. For example, in the languages of Chuuk State, the names **Siipw** and **Kiyopw** both refer to both *H. littoralis* and *C. asiaticum* (discussed in detail earlier in this chapter).

Uses in Pohnpei: *H. littoralis* (**Kiepw**) is used on Pohnpei for a number of purposes. According to Herrera et al. (2010: 33), it is used as a decorative and ornamental plant, as well as for some ritual purposes and as a source of miscellaneous products (see Balick et al. 2009: 253). For beautification purposes, the leaves and flowers of *H. littoralis* are used on Pohnpei in head garlands as well as for decoration of buildings. In addition, a number of medicinal uses have been reported. Lessa (1977: 145) referred to use of its leaves on this island in the "treatment of rheumatism and centipede bites," and also noted that its roots are used in "a mixture to treat a black magic disease called *kaw*." According to Lee et al. (2010: 14), the leaves and the bulb of *H. littoralis* are used separately to relieve the bite of a centipede: "The leaves are cut into pieces and placed over the bite with a heated stone over the leaves to steam the wound." (Steaming, *umwulap*, is reportedly a common application of traditional botanical medicines on Pohnpei.) Lee et al. (2010: 40) also indicated that the pounded fruits of the beach spider lily, *H. littoralis*, are used to treat boils, and for back aches and general body pain, a tuber of this plant is extracted "from the soil, cleaned, pounded on a rock, and wrapped in a piece of cloth," and then the "juice is squeezed into a cup of water and consumed once daily for four days immediately at the onset of symptoms" (Lee et al. 2010: 89).

▸ **Mworopw** or **Mwoaroapw** (P), **Mwoaroapw** (Mw) (Pi), **Ihi** (Nu) *Inocarpus fagifer*

Description: *Inocarpus fagifer* is a tree introduced to Pohnpei and typically found in wetland areas. It is the only edible member of its genus (*Inocarpus*), with a large seed that is a staple starch food source. This leguminous, medium-sized, evergreen tree produces a gnarled, buttressed, and fluted trunk with a broad crown reaching heights over 20 m (65 ft). The leaves are about 15–30 cm (6–12 in) long and 8–14 cm (~3–6 in) wide, with short petioles. The leaves are arranged alternately along the stems and have a papery or leathery texture. The clustered, small flowers are produced in axillary spikes and have white to cream or pale yellow petals. The large, more or less kidney-shaped, non-splitting fruits of *I. fagifer* vary among local varieties, but are generally about 7–8 cm (~3 in) long and 5 cm (2 in) wide. Each fruit encloses a single large "nut" (seed) with a chestnut-like flavor that is edible when cooked. Indeed, this tree is one of the most important "nut" species in the tropical Pacific.

Distribution: *I. fagifer*, the so-called Tahitian chestnut, is assumed by some botanists to have a broad native range stretching from parts of Indonesia in Southeast Asia through most of Micronesia to the Marquesas Islands of Polynesia, and including the Melanesian countries of the Solomon Islands, Vanuatu, Fiji, and Papua New Guinea. For example, Thaman et al. (2012: 220) refer to *I. fagifer* as an indigenous species from "Malesia into Micronesia as far as Kosrae and the Marshall Islands, and as far east as French Polynesia." However, in many areas of Polynesia (e.g., Samoa, Tonga, Cook Islands, and French Polynesia) and Micronesia (e.g., Pohnpei, Kosrae, the Marshall Islands, and Kiribati), the species is believed to have been introduced by Pacific Islanders during pre-European times (e.g., see Pauka 2006). *I. fagifer* occurs in secondary forests, as well as home gardens and coconut plantations, but its favored locations are near riverbanks, swamps, marshes, and along coastal shorelines. It may have been cultivated more intensively on some islands, and today the "Tahitian chestnut" is found mostly growing spontaneously. Although *I. fagifer* is shade tolerant, heavy shading may lessen its yield. It has a dense canopy, which makes it unsuitable for close planting between light-demanding agricultural crops typical in the tropical Pacific, such as sweet potato (*Ipomoea batatas*), taro (*Colocasia esculenta*) and sugarcane (*Saccharum officinarum*). On other hand, it can serve as a border tree that provides shade and shelter for more shade-tolerant crops and is apparently non invasive. It is thus compatible with other commonly cultivated trees in Micronesia, such as the coconut palm (*Cocos nucifera*), breadfruit (*Artocarpus altilis*), and the betel nut palm (*Areca catechu*), often in areas of local subsistence agroforestry.

General uses: In addition to its widespread use in parts of Micronesia, this tree species has been important throughout much of Polynesia and Melanesia, where the "mature seeds of the Tahitian chestnut (*I. fagifer*) are a seasonal staple or snack food" (Thaman 1994: 172; also see Whistler 2009: 134–137). In addition, the boiled or roasted "chestnut-like seeds are an important seasonal minor staple and cash crop sold at local markets and a source of plant protein, comparable in nutritional value to sweet potato" (Thaman et al. 2012: 220). According to Barrau (1961: 55), "In Polynesia and some Micronesian islands, surplus pods [of *Inocarpus*] have been customarily buried in the ground for future use." Several additional uses and information for the Tahitian chestnut have been provided by Pauku (2006):

> "Almost every part of the plant has been used traditionally. Leaves and bark are mainly used for medicinal purposes, while fallen branches are used for firewood. Even green wood is burned to dry copra. The wood is also used for crafts, tool handles, canoes, and light construction... The kernel must be cooked to make it edible. The nutritious kernels have protein and carbohydrate contents of about 5% and 22% respectively. It is prepared in many different ways, including roasting, grilling, boiling, baking... Fruits are harvested either directly from the tree at maturity or from the ground after ripening. The kernels have been sold mainly in domestic markets."

Pauku (2006) also provided some interesting folklore concerning the Tahitian chestnut in the Solomon Islands:

> "A myth in Choiseul (Varisi), Solomon Islands [Melanesia], relates a Tahitian chestnut to the death of a man named Porana, who betrayed the Chief of a particular tribe. The people ate large quantities of the cooked kernels and the ensuing flatulence [passing gas] suffocated Porana in a packed meeting room from which he could not escape."

Uses in Pohnpei: *I. fagifer* is called **Mworopw** or **Mwoaroapw** in Pohnpeian. It grows in coastal locations and in the uplands to an elevation of about 200 m (~650 ft); however, it is most common along riverbanks and in giant taro patches (e.g., see Pauka 2006). People eat the fruits after they have been boiled for 1–2 hours, and pigs eat the fallen nuts. Sproat (1968: 80) indicated that **Mworopw** "is eaten in appreciable quantities" on Pohnpei; however "it is not carefully cultivated and regarded as a reserve food source." In addition to its use as a food source, the bark and leaves of **Mworopw** are used in various applications as traditional medicine for treating "*soumwahu en naniak* (mangrove sickness)" and "*soumwahu en lih* (characterized by stomachache and back and joint pain caused by hard work)," as well as for amoebic dysentery, upset uterus, muscle pain, joint pain, and headache (Balick et al. 2009: 388–389).

Uses in Pingelap: Glassman (1953: 300) recorded that *I. fagifer* is known in Pingelapese as 'Marrup' [**Mworopw**]. Glassman also reported that *I. fagifer* had been introduced into Pingelap from Pohnpei and was cultivated on that atoll where it could grow to be about "15 feet high."

▸ **Soiu** (P), **Kebuk** (Mw)
Intsia bijuga

Description: *Intsia bijuga* is a medium to large tree, in some cases growing to about 35 m (~100 ft) tall with small, strongly fluted buttresses at its base. The bark of this valuable tree is light pinkish to reddish brown, weathering over time to a light gray. The bark is also smooth overall, slightly spotted, and peeling in different-sized scales which exposes renewed coloring below. *I. bijuga* produces compound leaves with 1–3 pairs of broadly ovate but asymmetric leaflets. The flower clusters, produced in panicles, are pure white to pink with a red claw and red to purple stamens. The seed pods are thick, leathery, and oblong and contain orbicular black seeds.

Distribution: This tree is reportedly indigenous in eastern Africa, Madagascar, South Asia, and from Southeast Asia to many high and low islands in the tropical Pacific; however, in some of the more eastern Pacific islands where it is found today, it could be an early anthropogenic introduction. Whistler (2009: 137) offered a scenario for its spread independently or with human help to remote tropical Pacific islands: "Although it has seawater-dispersed seeds that allow it to spread to and survive on atolls, it is such a useful tree that it would have been introduced to Polynesia [as well as Micronesia] if it weren't there originally." *I. bijuga* tolerates a wide variety of environmental conditions, from rough limestone to sandy and muddy coastal areas, or on the inner margins of mangrove swamps. In addition, it survives in quite dry to seasonally wet windward situations, and it is found occasionally in coastal or lowland forests and thickets as well as inland. It is also protected where it grows in garden areas. In some countries, it is occasionally planted in villages.

General uses: *I. bijuga* is found widely throughout Southeast Asia and the tropical Pacific. "It is considered to be one of the premier timber species of this region" (Marler and Lorence 1994).Thaman et al. (2006) refer to multiple uses of the highly respected and cherished, but decreasingly available, *I. bijuga* tree in many tropical Pacific islands:

> "It is undoubtedly one of the most highly valued trees in the Pacific Islands, both in terms of its traditional cultural importance and its value for commercial timber. Its durable, easy-to-work, attractive dark red-brown wood is especially favored for use in house building, furniture, and woodcarving. It is in very high demand and decreasing in abundance in most locations due to overexploitation for house posts, canoe making, and for woodcarving of valuable cultural artifacts such as kava bowls and weapons, and due to indiscriminate modern commercial logging. In addition to the wide cultural use of its timber throughout its range, it also has potential as a boundary marker or living fence post, for the production of fuelwood (from its branches when harvested for timber or woodcarving), as well as for its considerable medicinal value... It is a tree that should be considered under threat. Human population pressure, extensive traditional use, and increasing logging and commercial production of kava bowls and other artifacts for sale to the expanding tourist market, coupled with the lack of a tradition of replanting, have rendered exploitation of current populations unsustainable. There is a critical need for systematic propagation and replanting programs to address the situation." (www.agroforestry.net/tti/Intsia-vesi.pdf)

Uses in Pohnpei: This useful tree is reportedly known on Pohnpei as **Soiu**. Balick et al. (2009) indicated that the trunk of this species "is cut and used as housing material."

▸ Kangkong (P) (Mw) (Pi)

Ipomoea aquatica

Description: *Ipomoea aquatica* is known by several names in English, such as swamp morning glory or water spinach. It is a creeping or floating, normally subaquatic, herbaceous vine only found in areas of open fresh water or swampy environments. It can be both a useful crop and an invasive species. Its stems grow up to about 3 m (9 ft) in length, and its leaves are alternately arranged, more or less arrowhead-shaped but commonly oblong-lanceolate. Normally, the leaves are about 5–15 cm (2–6 in) long and 2–10 cm (1–4 in) wide with petioles 3–10 cm (1–4 in) long. Its flowers are produced in several cymes with typical morning glory shape, and the fused corolla is usually 4–5 cm (~2 in) long and pinkish-violet in color but darker in the throat (and, rarely, whitish). *I. aquatica* produces spherical or ovoid shaped capsules, each containing four or fewer small seeds about 1 cm (~0.4 in) long and normally, but not always, with fine grayish hairs on the surface.

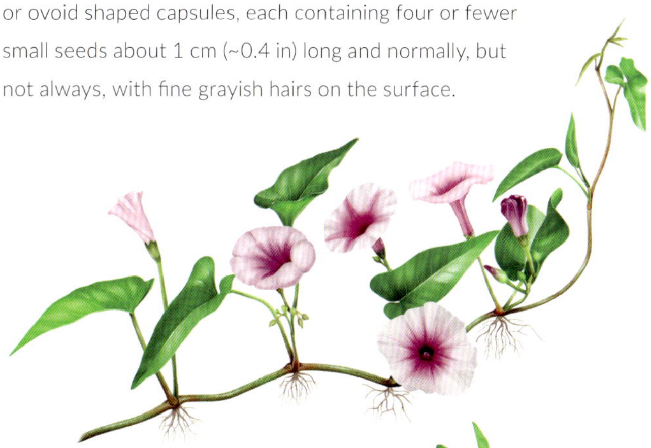

Distribution: Adapted to freshwater aquatic habitats, *I. aquatica* invades canals and ditches, but also becomes naturalized in lakes and marshes; in addition, it grows profusely in moist soil as well as in still or moving waters. Due to its abundant growth, this species can invade damp cultivated areas such as rice fields, sugarcane plantations, and other places with variable water levels. *I. aquatica* originated in tropical Asia (possibly India) and today can be found in South and Southeast Asia, tropical Africa, South and Central America, and Oceania; however, it is only in South and Southeast Asia that it is used as a significant leafy vegetable. It is intensively grown and frequently eaten throughout Southeast Asia, Hong Kong, Taiwan, and southern China. "Above all else, it is a... creeper on the banks of muddy streams and ponds, awaiting the opportunity to reach out over the water surfaces to form mats with other vegetation" (Holm et al. 1997: 412, 414). Since the name used for this plant in Pohnpei and throughout Micronesia is **Kangkong**, which is the same as in Tagalog, it appears that *I. aquatica* was introduced directly or indirectly from the Philippines (see www.comfsm.fm/~dleeling/ethnobotany/text/flora.html).

General uses: *I. aquatica* is considered to be a worthwhile vegetable crop in tropical regions (such as areas in China) where, in some cases, it has been cultivated for many centuries. Nevertheless, where it is an introduced species, it has the potential to damage natural wetland ecosystems, as well as harming subsistence or commercial rice and sugarcane production, critical irrigation systems, and even human navigation and recreation areas. These negative aspects associated with the naturalization of *I. aquatica* have stimulated its listing as a Federal Noxious Weed in the United States (1974). In other words, although *I. aquatica* might produce monetary or social benefits as a vegetable crop, these potentially positive aspects are offset by the likely damage it can produce and the cost of controlling its unwanted spread. On the other hand, where it has been under control, its food value is much appreciated; for example, its young stem and leaves are cooked or lightly fried in oil and can be eaten in a wide variety of dishes. The vines are also used as fodder for livestock in some places, and in parts of Malaysia it is grown extensively in fish ponds and fed to domesticated pigs. The very nutritious stems and leaves, which are high in protein, vitamin A, iron, calcium, and phosphorus, can be eaten raw, boiled, stir fried, steamed, or pickled as they are throughout much of tropical Asia. *I. aquatica* has also been used extensively as a medicinal plant, for example as a laxative and in treatment of ringworm. Clearly, this swamp morning glory species has both positive and negative aspects in terms of its use for food and medicine, compared with its potentially serious environmental impact.

Uses in Pohnpei: In Micronesia, according to Fosberg and Sachet (1977: 8), *Ipomoea aquatica* is "planted or spontaneous in marshy places or in open water," and has been found on Pagan, Saipan, Tinian, Rota, and Guam in the Mariana Islands, as well as on various other islands or island groups, including Palau, Yap, Ulithi, Chuuk, Satawan, Pohnpei, and Kiribati. In Pohnpei, this easily cultivated semi-aquatic leafy vegetable is called **Kangong**. Besides its edible leaves, the tuber can be "cooked in soup" and, as elsewhere in Micronesia, the whole plant is fed to pigs (Balick et al. 2009: 359).

▸ **Pedehde** (P), **Pidɛhdɛ** (Mw), **Pidaehdae** (Pi), **Budeedee** (Nu), **Bedeedee** (Ka)
Ipomoea batatas

Description: The "sweet potato" plant (*Ipomoea batatas*) is a member of the morning glory family, Convolvulaceae, and therefore it is not closely related to the "white" potato plant, which is a member of the deadly nightshade family, Solanaceae (also known as the tomato or potato family). Unlike the white potato crop, which is a thickened, underground stem (a tuber), the sweet potato crop is a swollen root, modified to store food reserves. Similarly, the sweet potato crop can look like some yams, but they are not closely related botanically. The sweet potato plant is a smooth, creeping, herbaceous vine with purple or green stems and milky sap. Its alternate, palmately lobed leaves have two prominent basal lobes and a pointed tip. The flowers are usually pale rose-violet in color, or, rarely, albino, 3–5 cm (1–2 in) long, and darker (purplish) in the throat, and its swollen roots are edible after cooking. The rarely forming fruit is an ovoid capsule with 1–4 seeds. The rarity of seed formation is a strong indication of the degree of artificial selection for vegetative reproduction of this domesticated species.

Distribution: Sweet potato plants have been introduced widely among the Pacific Islands, as well as many other regions of the world. It is a hardy crop and can be grown in poor soils with little fertilizer. The sweet potato plant is probably native to the lowlands in the northwestern region of South America, most likely in the Andes Mountains, where there are hundreds of varieties of this edible species. In Micronesia, it is cultivated on many islands, but it sometimes spreads as a semi-wild plant along roadsides and streams. The sweet potato was (or is) a major crop in the native cultivation systems of some areas in Polynesia and Melanesia, especially Hawai'i, New Zealand, and Easter Island [Rapa Nui], as well as in the New Guinea highlands. From its original homeland in South America,

sweet potatoes spread westward in the prehistoric period (pre-European contact) or historic period (post-European contact), probably in three major waves of movement (e.g., see Whistler 2009: 139–142 and Yen 1974). Two of these waves occurred during the early historic period some hundreds of years ago and involved Europeans, such as the Portuguese and Spanish, who brought "Camotes" or "Batatas" (*I. batatas*) to various areas of Southeast Asia and the Pacific, including parts of Micronesia. The local names for the sweet potato throughout Micronesia appear to be derived from the Mexican-Spanish name **Batatas** for the sweet potato: "In Micronesia, the sweet potato was probably introduced in the sixteenth or seventeenth century by the Spanish, via Mexico, the Manila galleons, and Guam" (Barrau 1961: 54).

General uses: Sweet potato is cultivated on the high islands of Micronesia and less successfully on many of the low coral islands and atolls. On the islands with soils derived from volcanic material, sweet potatoes are relatively easy to grow and are a good source of vitamins and fiber; indeed, the orange-fleshed tubers are rich in provitamin A carotenoids (see Englberger et al. 2003a). From a global perspective, the sweet potato is a very prominent food crop, and in some situations it is even a hardy life-saving resource:

> "Sweet potato has a long history as a life saver. The Japanese used it when typhoons demolished their rice fields. It kept millions from starvation in famine-plagued China in the early 1960's and came to the rescue in Uganda in the 1990's, when a virus ravaged cassava crops." (International Potato Institute CIP 2018)

In modern times, the sweet potato in all its varieties is the seventh most important food crop in the world after rice, wheat, white potatoes, maize, barley, and cassava, and in "developing countries" it ranks as the "fifth most important food crop," with well over 100 million metric tons being produced worldwide each year. Furthermore, the great majority of this sweet potato production takes place in the developing countries, with Pacific Islanders being "among the largest per capita consumers of sweet potato in the world" (e.g., www.cipotato.org). Verma (2014a: 5) provided information regarding the cultivation of sweet potato in the tropical Pacific region and emphasized its potential present and future importance to Micronesian farmers, whose traditional cultivation areas are under a variety of natural and man-made threats:

> "With the effects of global climate change and the increasing occurrence of natural disasters, farmers must be aware of these disasters and adopt preventative measures to minimize their impact on agricultural operations. Of all the cultivated root and tuber crops, sweet potato is probably the best at adapting quickly to new conditions. But still, its growth and development can be affected by adverse environmental conditions." (Verma 2014a)

Uses in Pohnpei: The sweet potato (**Pedehde** in Pohnpeian) is now considered to be among the "traditional" foods on Pohnpei, with six "cultivars" currently used as food. In addition to cooked tubers, the edible leaves are also marketed and are referred to as *se pedehde* in Pohnpeian (Englberger et al. 2009b: 18–19; also see Balick et al. 2009: 359).

▸ **Iol wah poh mehi** (P)
Ipomoea indica

Description: *Ipomoea indica* is a long-lived, widespread, native, sprawling morning glory vine with blue flowers. As a perennial twining climber, it can grow up to 15 m (50 ft) high upon supporting plants, but sometimes it scrambles over low vegetation or creeps along the ground. Its stems can form dense cover and occasionally exude a white milky sap when broken. The alternately arranged, heart-shaped leaves have three lobes and pointed tips and are hairy on both top and bottom. The leaves are borne on petioles 2–18 cm (~1–8 in) long. The large funnel-shaped or tubular flowers are 5–10 cm (2–4 in) long and 7–10 cm (3–4 in) wide; they are bright blue or bluish-purple with a paler pink or whitish-pink central tube. Flowering can occur throughout the year, and when this happens, the flowers are produced in clusters of 2–12 in the leaf forks or axils. The fruits are small, globe-shaped, papery capsules containing 4–6 dark brown or black seeds.

Distribution: The origin of *I. indica* is unclear as it appears to be pantropical in its present distribution. If so, it should be classified as indigenous in Pohnpei. In Micronesia, *I. indica* is known from several of the Mariana Islands as well as in Palau, Satawal, Chuuk, Pohnpei, and Kosrae. It is commonly found at low elevations at the back of the beach, especially in open and disturbed places; typically, it climbs on bushes and in plant thickets, forming dense mats on bare limestone soil even at elevations somewhat above sea level.

Uses in Pohnpei: According to Balick et al. (2009: 360–362), on Pohnpei, pigs feed on the leaves of *I. indica*. The leaves also have a number of medicinal uses, some involving the use of vine shoots, and some involving combination with parts of other plant species. These therapeutic uses of *I. indica* include the following: treatments for boils, women's sickness (known in Pohnpeian as *soumwhau en lih*), painful menstruation (known in Pohnpeian as *pahsoan mwekidida* or *kikond*), cleaning the uterus, curing a baby's sickness involving fever and loss of appetite (known in Pohnpeian as *soumwahu en seri*), as well as treating ulcers and soothing sores eyes (also see Herrera et al. 2010: 102).

- **Emp** or **Omp** (P), **Oahp** (Mw), **Oip** (Pi), **Hue lauligi** (Nu), **Huuwe** (Ka)
  Ipomoea littoralis

Description: *Ipomoea littoralis* is a thin-stemmed, creeping or climbing non-woody vine in the morning glory family It has simple, alternate, heart-shaped leaves, 2–10 cm (1–4 in) long with smooth surfaces. The pale purple or lavender, funnel-formed flowers, up to 7 cm (~3 in) long, are shallowly 5-lobed, funnel-formed, and are borne at the leaf axils. The fruits are small, papery capsules containing four seeds. Flowers and fruits are reportedly produced throughout the year.

Distribution: The indigenous morning glory plant *I. littoralis* is a wide-ranging native coastal vine that occurs in the strand vegetation of many islands from Madagascar to Hawai'i, including reported collections from a large number of high and low islands in Micronesia. *I. littoralis* can be found from the shore to some inland areas with relatively low elevation, usually in open areas where it climbs over low-lying vegetation. According to Fosberg and Sachet (1977), this vine species is particularly common in open or disturbed places and along sea coasts, but is less common at middle elevations: "In Caroline atolls [it is] found on edges of taro pits, around villages, and generally in interior in openings and thin places in woods and coconut groves" and it is most likely "impossible to determine where this plant was native and where [it was] accidentally introduced by man."

General uses: According to Austin (1991: 253), "there is a long record of human usage of *I. littoralis*," and these uses fall into two categories: medicine or food. On some Micronesian islands, this native morning glory vine has historically been a famine food source as well as a source of traditional medicine. Nevertheless, this species can be become an environmental threat since it can develop "dense floating mats of intertwined stems over water surfaces, shading out native submersed plants and competing with native emergents."

Uses in Pohnpei: Known as **Emp** or **Omp** in Pohnpeian, *I. littoralis* is commonly found in the strand or low vegetation on rocky and sandy shores. *I. littoralis* leaves are used to feed pigs on Pohnpei. Metaphorically, *I. littoralis* is associated with a Pohnpeian proverb, *Sahl en pwoapwoaud kin mwei, sahl en neitik soh*, meaning the bond of marriage may be easily broken, like the fragile stems of this morning glory vine species. In contrast, the bonds between mother and child or father and child are not easily broken. Traditionally on Pohnpei, a forlorn lover who desires to be reunited with his or her lover collects the tendrils of the *I. littoralis* vine. The morning glory flowers of *I. littoralis* are said to be combined to scare away the dwarf spirits (*sokolei*). *I. littoralis* leaves are also used on Pohnpei to treat boils as well as a number of other ailments; for example, "the leaves are used in relieving pains after childbirth and as an ingredient in soups, and the flowers are employed as a hemostatic in menstruation" (Fosberg and Sachet 1977, citing Glassman 1952.) More detail regarding uses of *I. littoralis* related to childbirth, menstruation, and other ailments on Pohnpei has been provided by Balick et al. (2009: 361–362):

> "To treat *soumwahu en lih* (woman's sickness)... To treat painful menstruation, where a woman suffers stomach pains and lethargy while menstruating... To treat *poahsoan mwekidida*, also known as *likond* (affecting women with stomach pain; the abdomen is hot to the touch and tender but not swollen, and the woman is feverish and urinates frequently)... To clean the uterus... To treat a displaced uterus after childbirth, which may be caused by lifting heavy objects or bending over too

much... To treat an unspecified baby's sickness... To treat *soumwahu en seri* (sickness in babies, such as fever and loss of appetite)... To treat ulcers... To treat sore eyes, four leaves are pounded, wrapped in cloth, and squeezed into a small basin of water; the face is placed in the water, and the eyes are opened underwater." (Also see Lee et al. 2010: 141).

Uses in Mwoakilloa: Skin rashes are reportedly treated traditionally on this atoll using the juice obtained by pounding the leaves of *I. littoralis*, which is known on Mwoakilloa Atoll as **Oahp**. Pounded and squeezed leaves of this vine are also said to be applied to the head of a person to relieve a headache on this atoll; this same leaf juice is also reportedly mixed with a few drops of appropriate oil and combed into your hair to "make your hair black, long, and strong," but it is best this mixture is applied to the hair "prior to sunrise or after sunset only" (www.comfsm.fm/~dleeling/angio/litto.html).

▸ **Likehdou**, **Likehmw** or **Likoahmw** (P)
Ipomoea mauritiana

Description: *Ipomoea mauritiana* is a perennial climbing herb that produces stems up to 10 m (~30 ft) long from a tuberous rootstock. The stems can be woody near the base, but they become herbaceous near the tips; as they grow from the rootstock, the stems spread out over the ground or twine into other plants for support. Leaves of *I. mauritiana* are simple, alternate, and palmately lobed; the distinct lobes are acute or acuminate at the apex. The flowers have fused petals, are reddish purple or pink, and are produced in many-flowered, axillary cymes. The fruits are egg-shaped capsules with a persistent calyx that produce four brownish to black, hairy seeds.

Distribution: The origins of *I. mauritiana* are uncertain, possibly native to tropical America, but the species is now widely naturalized in many tropical parts of the world, including some islands in the Pacific region. Fosberg and Sachet (1977) reported *I. mauritiana* as occurring in Palau, Yap, Chuuk, Pohnpei, and Kosrae. This *Ipomoea* species is often found in disturbed swamp forests as well as some lowland areas, "growing in thickets and edges of forest, climbing on rocks and on other plants."

Uses in Pohnpei: *I. mauritiana* is known in Pohnpeian as **Likehdou**, **Likehmw** or **Likoahmw**. Its leaves have been reportedly "used in relieving pains after childbirth" on Pohnpei (Fosberg and Sachet 1977, citing plant specimen notes by Glassman no. 2564). Specimen notes for #2744108 in the US National Herbarium, collected in 1968, refer to the use of this plant (*Ipomoea* digitata = *I. mauritiana*?) "for feeding pigs," but only "when the pig doesn't want to eat." Balick and his colleagues (2009: 362–363) provided a series of traditional medicinal uses of the introduced morning glory herb *I. mauritiana*; the therapeutic usage of this species in Pohnpei is particularly related to pregnancy and childbirth. These uses are partially quoted below:

> "To treat early or imminent labor... as a tonic for a pregnant woman... To ease labor... To ease delivery of a baby... To keep a pregnant woman's stomach smooth... [and] in unspecified women's medicine." (For more detail regarding traditional medicinal use of this morning glory species on Pohnpei, especially for its therapeutic use before, during, and after birth, see Balick et al. 2009: 362–363).

Also see Lee et al. (2010: 133, 140, 142) for additional information regarding regarding the "many application for women's health" of **Likehdou** (*I. mauritiana*).

▸ **Sonsol** (P), **Hue haangai biigi** (Nu), **Huuwe** (Ka)
Ipomoea pes-caprae

Description: *Ipomoea pes-caprae* is a very vigorous, creeping or scrambling, evergreen, perennial vine that develops from a large, thick root that can become 3 m (~10 ft) long and 5 cm (2 in) in diameter. From this hefty root, the vine produces prostrate, sometimes twining, slightly angular and hollow stems up to 30 m (~90–100 ft) long; these stems contain a milky juice, and it often produces roots at its nodes. The heart-shaped leaves are often 2–10 cm (~1–4 in), have smooth margins, and are alternately arranged along the stems; the leaves are often held erect to one side of the stem. The inflorescence of this vine is an axillary few-flowered cyme with rosy purple flowers that are darker in the center and tubular (funnel to campanulate) in form. Its seeds are densely covered with brownish hairs but are dull yellowish when the hairs are worn off.

Distribution: Two subspecies of *I. pes-caprae* have been identified: subsp. *pes-caprae*, which occurs in Arabia and tropical Asia, and subsp. *brasiliensis*, which occurs throughout the tropics. *I. pes-caprae* subsp. *brasiliensis* is found along the shores of many tropical islands, including those in Micronesia. It forms mats, sometimes dense on the ground, and is found in the strand vegetation on sandy or rocky shorelines. It is an important sand-binding plant. Some botanists have suggested that this beach morning glory vine (*I. pes-caprae*) is found more widely and in greater numbers than any other tropical Pacific beach plant. Indeed, the species *I. pes-caprae* is one of the most widely distributed beach plants throughout the world's tropical and subtropical areas. *I. pes-caprae* has been referred to as "an especially important pioneering species along tropical coastlines... [and its] ability to stabilize sand dunes is a first line of defense against damaging storms" (Devall 1992). *I. pes-caprae* plants creep over sand dunes, setting down adventitious roots, and eventually form large mats that prevent erosion.

General uses: As a relatively common sprawling coastal strand vine in Micronesia, parts of *I. pes-caprae* have been used as an emergency or famine food and as a cordage source, and it has other purposes such as traditional medicine. In terms of its therapeutic usage, this species has been found to have multiple medicinal applications:

> "The plant is mucilaginous and is considered astringent, tonic, alterative, diuretic and purgative. Poultices made from the leaves are commonly used to treat skin infections, ulcers, boils, swellings, stings and wounds. Decoctions are widely administered as anodyne to treat rheumatism, and seeds against stomach-ache." (Lemmens 2013)

Uses in Pohnpei: Known as **Sonsol** in Pohnpeian, the beach morning glory, *I. pes-caprae*, has been used for a variety of medicinal purposes. For example, the leaf buds are used in the baths of children to make them walk earlier, and the young flower buds and white spots on young leaves of this plant are also used to treat venereal disease in females. Other therapeutic uses of *I. pes-caprae* in Pohnpei include its use "... to increase breast milk... To cool the uterus... To soften or widen the uterus... To treat pregnant women with *soumwahu en naniak* (mangrove sickness)..." (Balick et al. 2009: 364). Furthermore, drinking the juice squeezed out of pounded leaves of this coastal plant is also reportedly used "to ease labor (birth) pains," and young leaves are also said to help heal boils (www.comfsm.fm/~dleeling/angio/ipomoea_pes-caprae.html).

▸ **Ketieu** (P), **Koassoau** (Mw), **Kaelaeseu** (Pi), **Gadiou** (Nu)
Ixora casei

Description: *Ixora* is a genus of flowering plants with more than 500 species belonging to the coffee family, Rubiaceae. One of these *Ixora* species is *Ixora casei*, an attractive, native shrub or small tree found on Pohnpei and other Micronesian islands. It grows up to about 2–3 m (6–9 ft) tall or more and produces variable, opposite, elliptic or slightly oval-shaped leaves that are shiny, up to about 25 cm (~10 in) long, and short-stemmed. Its handsome clusters of many, small, deep reddish, odorless flowers are about 7 cm (3 in) in diameter, and the calyx of each flower is tiny, with a corolla tube that is about 5 cm (2 in) long. Fruits, when they do develop, are berry-like and have two seeds typical of its family, Rubiaceae. Note: There is species identity confusion in some parts of Micronesia between *I. casei* Hance and *I. coccinea* L.; *I. casei* is native in Micronesia, and *I. coccinea* is probably native to South Asia, but now pantropical in cultivation as an ornamental plant.

In modern times, *I. coccinea* has been introduced into various islands in Micronesia. According to Fosberg et al. (1993: 87), "several varieties" of *I. coccinea* can be differentiated, "three of them cultivated in Micronesia as ornamentals." Today, the alien *Ixora* plants are often mistaken for the native *I. casei*.

Distribution: A variety of *Ixora* species is found across tropical America to Africa, India, and southern Asia. *I. casei* is apparently native to at least some of the islands in Micronesia. (Note: Most if not all of the local names for this species throughout Micronesia are cognate, i.e., are related linguistically). According to Fosberg et al. (1993: 84), *I. casei* "is native on at least all the high Carolines and introduced into cultivation in Australia long ago from [Kosrae], distributed from Sydney or Melbourne and now cultivated under various names throughout the tropics." As noted above, Fosberg and his colleagues also noted that *I. coccinea* is commonly confused with *I. casei*, and has been "introduced at least into the Marianas and northern Marshalls in Micronesia," but is now reportedly found in Pohnpei, where several horticultural forms and varieties with red, pink, or yellow flowers are sometimes cultivated. In addition, yet another alien *Ixora* species, *I. chinensis*, is also now found in Pohnpei and cultivated as an ornamental (Herrera et al. 2010: 171).

Uses in Pohnpei: *I. casei*, known on Pohnpei as **Ketieu**, has multiple uses. For example, Lessa (1977: 182, also see Glassman 1952) referred to the roots of **Ketieu** having medicinal uses including "as a hemostatic in menstruation," and in the past, "the very hard wood was used to make spears," and the stems of this woody species were also reportedly "cut into small pieces to use in a local Pohnpeian dance called *dokia*." More recently, Balick and his colleagues (2009: 383–384) provided the following impressive amount of information about the multiple uses of **Ketieu** on Pohnpei:

> "The wood is used as a pole on which to attach the breadfruit picker blade… The flowers are used in *mwaramwar* (head garland)… The plant is used to decorate houses… The stems are used for traditional dances that involve sticks… The wood of the stems is used to make outriggers [e.g., the struts connecting to the pontoon are typically made using **Ketieu** (*I. casei*) stems, e.g., see Brosi et al. 2007]… The stem is used to make *rahk* (part of the outrigger of canoes)… The wood is used to make *ketiuk* (upright sticks that are tied to the outrigger of canoes)… The stem is used as a fishing pole… and as a tool to husk coconuts… To use as a charm to protect a person from getting caught by a ghost while sleeping in the forest… To use as a charm to prevent romantic enchantments… To use as a charm to make a person leave his or her own land… The plant is used as a charm to make a person tell the truth about committing theft, for example, of *sakau*… To treat a baby who has lost its appetite… To treat *soumwahu en seri*, where an infant is crying with fever… To treat a woman with menorrhagia… To treat high fever in a baby… To treat sickness with boils on the head… To treat a cut on a person's body… To treat back pain… To treat *soumwahu en insensuwed* (depression)… To treat poison of crown-of-thorns starfish injury… To treat sea urchin punctures… To treat swelling from blue sea urchin poisoning…" (For use of *I. casei* planted as a land boundary marker, and its flowers for *mwaramwar* or garland decoration, see www.comfsm.fm/~dleeling/ethnobotany)

Uses in Mwoakilloa and Pingelap: Glassman (1953: 296) found this species to be cultivated, but uncommon on Mwoakilloa, while Lessa (1977: 182) saw it cultivated on Pingelap and noted it was for ornamental purposes.

Uses in Nukuoro: According to Drummond et al. (2016), I. casei is known on Nukuoro Atoll as **Gadiou**, and its attractive bunches of reddish flowers are "used to make flower crowns" or head garlands.

▸ Keleun Ahnd (P)
Kleinhovia hospita

Description: *Kleinhovia hospita*, known in English as the guest or puzzle tree, is an evergreen tree normally small to medium in height and bushy, with fibrous bark. Under optimum conditions, it can grow up to about 20 m (65 ft) tall, generally with a dense, rounded crown. Its leaves are ovate to cordate in shape with palmate venation. Its large flowers are freely branched with pinkish to yellowish petals that are produced in upright sprays (i.e., panicles up to 20–40 cm (8–16 in)

in length). Its 5-valved capsular fruits are inflated and papery, and each usually contains one globose white seed.

Distribution: The native range of *K. hospita* appears to be quite wide, stretching from tropical Africa and Asia through Malesia to the high islands in Samoa, Tonga, and the Society Islands of Polynesia, and in the eastern region of Micronesia. On the other hand, some have suggested that its native range is restricted to areas in Indonesia, Malaysia, and other parts of tropical Asia, but not Remote Oceania. Whether it is truly a native tree in Polynesia or Micronesia, or if it has been introduced by people, needs to be more fully resolved. Today, where it is found in some areas of Micronesia, including Pohnpei, *K. hospita* can either be relatively common or occur as a rare tree in disturbed, open areas of secondary vegetation and fallow agricultural plots. It can form into groves in disturbed areas. On some islands, it is reportedly occasionally found at the edges of forests and in low-lying areas, sometimes occurring in the vicinity of taro patches (e.g., for Yap, see Falanruw 2015: 222). In some areas where this small tree has been introduced, it is "sometimes cultivated as [a] street and landscape tree for its showy flowers and seedpods, and due to its tolerance to salt spray, is especially well-suited to seaside locations" (www.iplantz.com/plant/933/kleinhovia-hospita).

Uses in Pohnpei: *K. hospita*, known as **Keleun Ahnd** in Pohnpeian, is reportedly used to make canoes, with its trunk being "cut to length and width as needed and carved" (see Balick et al. 2009: 431); Rehg and Sohl (1979) reported **Keleun Ahnd** being "used for making outriggers."

▸ **Soumwehl** (P), **Nɛhnkoadkoad** (Mw), **Moougu** (Nu), **Walengaa dulii** (Ka)
Laportea ruderalis and *Laportea interrupta*

Laportea is a genus with about 20 or more annual or perennial herb species. Thes plants are similar to many species in the stinging nettle family Urticaceae; *Laportea* species have stinging hairs, and they may have both stinging and non-stinging hairs on the same plant. Botanists recognized two species of *Laportea* on Pohnpei: *L. ruderalis* (G. Forst.) Chew, which is probably indigenous on this island, and *L. interrupta* (L.) Chew, which is probably an alien species on Pohnpei that was introduced to this island before the arrival of Europeans. The latter has not been documented as occurring on any of the outer islands in Pohnpei State. Both of these *Laportea* species have been previously classified as *Fleurya interrupta* (L.) Chew. In the section that follows, we have combined our discussion of these two *Laportea* species on Pohnpei because of their similarity in appearance and use.

Description: *L. ruderalis* is a small, erect, sparsely branched, somewhat succulent, glabrous herb that grows up to about 40 cm (1.4 ft) tall with reddish stems. Its simple, ovate, slightly fleshy leaves are alternately arranged and have coarsely toothed margins. The small pink to whitish, individual male and female flowers are borne in open, axillary, paniculate clusters. The achene fruits are small and ovoid.

L. interrupta is a sparsely branched, slightly fleshy, erect herb, 20–80 cm (8–24 in) tall. Overall, *L. interrupta* can be distinguished by its irritant hairs. Its alternate, ovate-acuminate leaves are 6–9 cm (~2–4 in) long and 5–6 cm (~2 in) wide with coarsely toothed margins. The leaves have internally grooved petioles and are broader towards the base, narrowing near to the tip. Its small, light green flower buds are produced in long narrow panicles of inconspicuous clusters at intervals along the axis, and contain small, ovoid, seed-like fruits.

Distribution: *L. ruderalis* appears to be native from Malaysia to Micronesia, including Pohnpei, and in Polynesia, where it is most common on atolls. *L. ruderalis* can be found in sandy soils, even near the shore, but is also occasional in shady and moist disturbed sites in household gardens and coconut plantations in rural areas. On some atolls, *L. ruderalis* is common and locally dominant in open sites and in openings among the tree cover.

L. interrupta, on the other hand, is not reported as occurring on the outer islands, and no local names for this species have been recorded for those islands. *L. interrupta* is probably native to Southeast Asia, but now has a very wide distribution, and was an inadvertent aboriginal (pre-European contact) introduction as far eastward as Tahiti. Locally, it is uncommon to occasional in disturbed places, especially along roadsides and in gardens, fields, and plantations and around houses at up to over 500 m (1650 ft) elevation. It was once reported from Hawai'i, but probably no longer found there or in Tahiti. It should be noted that Fosberg et al. (1979: 60) listed *L. interrupta* as native to Micronesia, but that is a minority view.

Uses in Pohnpei: ***Both*** *L. interrupta* and *L. ruderalis* are known in Pohnpeian as **Soumwehl**. These closely related herbaceous species have been reportedly used ***traditionally*** in Pohnpei for medicinal and ritual usage (Herrera et al. 2010: 185).

L. interrupta has apparently been used in sexual and reproductive contexts in Pohnpei. For example, according to Balick et al. (2009: 507):

> " An ancient practice was to take the irritating leaves and rub them on the lips of the vagina, which would cause itchiness for about one or two hours (this was said to be favored by men during sexual intercourse, when the swollen lips can be tucked into the vagina)...[and was previously] used to lengthen a women clitoris before sexual intercourse for the first time, a leaf was rubbed onto the clitoris to cause itching and by pulling on the clitoris it could be made to lengthen (do this early before intercourse)..."

In the case of *L. ruderalis*, according to Balick et al. (2009: 507), this species of *Laportea* has been used to help discourage sexual intercourse or for treating skin rashes:

> " ... the plant is put inside a woman's vagina to keep a man from violating her because otherwise it will give him a rash/itch... To treat *kilin kidi* (literally, "dog skin," impetigo), many leaves are boiled and the body is bathed in it or showered with it until the sores go away..."

Uses in Pingelap: Glassman (1953: 301) noted that Fleurya ruderalis (a synonym of *L. ruderalis*) has reddish brown flowers and is common in coconut groves, but did not record any uses on Pingelap. He noted that its Pingelapese name is "Ne-kirrir-ir," but we could not confirm or correct it during research for this book.

▸ **Dangandangan** (P) (Mw)
Leucaena leucocephala

Description: *Leucaena leucocephala*, a member of the bean family, Fabaceae, is normally a relatively slender shrub or small tree growing to heights of 1.5 to 5.0 m (5–25 ft) or more. *L. leucocephala* is facultatively deciduous, which means it can prematurely shed its leaflets in response to environmental stress. A key botanical characteristic affecting the persistence of this species is its ability to develop especially deep energy storage roots; it normally develops a long taproot up to 5 m (~15 ft) in length that is quite robust. In shallow soils, roots may branch and grow laterally to about 30 cm in length due to clay layers (e.g., see Brewbaker 1987). *L. leucocephala*'s bipinnately compound leaves, with 12–20 small leaflets per pinna (a leaflet or primary division of a pinnate leaf or frond), are alternately arranged on the slender but dense stems and branches. The pale green to white, spherical (globose) flower heads, when fertilized, produce many clusters of flat, brown, strap-shaped pods 15–20 cm (6–8 in) long, each containing about 15–25 small, glossy brown seeds.

Distribution: *L. leucocephala*, a thornless woody plant native to southern Mexico and northern Central America (Belize and Guatemala), was introduced into the Pacific region before World War I. It is now naturalized and widespread as a very invasive species in a number of tropical Pacific islands, especially in the dryland regions of the Hawaiian and Marquesan Islands and in the limestone areas of the Mariana Islands, e.g., on Guam. It also occurs in severely disturbed wet areas, but not as a dominant species. Although *L. leucocephala* normally favors limestone soils, it appears to do well on some volcanic soils and in especially disturbed areas, and it is naturalized in dry to mesic habitats up to 700 m (2300 ft) on some islands in the tropical Pacific. It is said to have been broadcast (seeded) deliberately on a number of battle-eroded islands after World War II (e.g., in the Mariana Islands). *L. leucocephala* can be useful, but frequently grows where it is not wanted, forming dense thickets that may exclude other plant species. Due largely to its relatively deep roots, *L. leucocephala* is quite difficult to remove from an area where it has become well established and mature. "It resprouts from cut stumps, grows readily from both seeds and cuttings, and tends to produce pure stands" (Falanruw 2015: 116). In Micronesia, *L. leucocephala* can be seen growing in large stands along some roads, especially in the Mariana Islands as well as other islands such as Peleliu in the Republic of Palau.

Uses in Pohnpei: Known as **Dangandangan**, or the "lead tree," on Pohnpei (e.g., see Balick et al. 2009: 389), *L. leucocephala* is reportedly naturalized and invasive on this island, but is also listed as a source of animal forage and fuel, as well as contributing "to an environmental improvement or remediation, i.e., replenishes malnourished soil or used for erosion control of land" (see Herrera et al. 2010: 7, 120).

- **Delahrik** or **Deleurik** (P), **Kiri** (Mw), **Kuhri** (Pi), **Helii lodo bela** (Nu), **Geinga lodo wai** (Ka)
  Ludwigia hyssopifolia and *Ludwigia octovalvis*

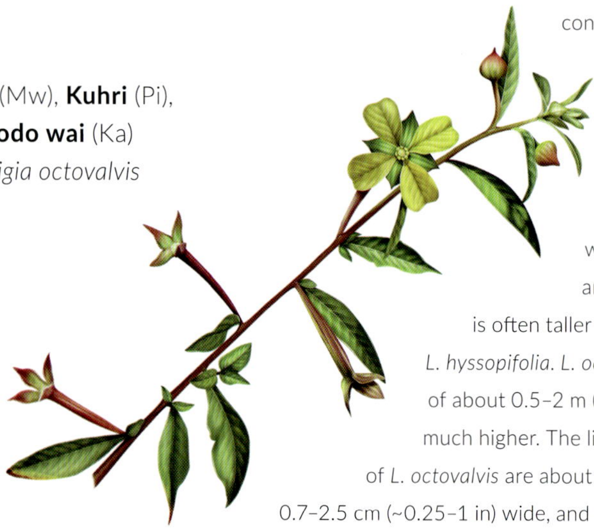

Ludwigia is a genus of about 82 species of aquatic or wetland plants with a widespread but mainly tropical distribution. These herbs are characterized by stems which are "creeping, floating, or emergent and ascending to erect, rooting at nodes" (Hoch et al. 2015). In recent times, there has been considerable debate among botanists and plant taxonomists about the classification of many *Ludwigia* species (e.g., see Wagner et al. 2007: 83). The two closely related herbaceous *Ludwigia* species now naturalized on Pohnpei are *L. hyssopifolia* and *L. octovalvis*.

Description: *L. hyssopifolia* is an erect, branched, annual herb, normally smaller than *L. octovalvis* (see below), ranging from about 15 to 150 cm (0.5–5 ft) in height, with 3–4 angled stems. *L. hyssopifolia* produces simple leaves with entire margins. The leaves are arranged alternately along the stems and are linear to lanceolate in shape with pointed tips. The solitary, yellow flowers of *L. hyssopifolia* have four more or less round-tipped petals and are produced at the stem axils. The finely haired, nearly cylindrical fruit capsules of *L. hyssopifolia* are 12–25 mm (~0.5 to 1 in) long. *L. hyssopifolia* can reportedly produce as many as 200,000 seeds per plant, which are gradually released and are able to float in water for two weeks or more and germinate in their aquatic habitat before sinking; the seeds apparently do not germinate under submerged conditions or when buried in soil.

L. octovalvis is an erect, perennial herb with grooved stems that are sometimes somewhat woody at the base. The annual species *L. octovalvis* is often taller than the perennial species *L. hyssopifolia*. *L. octovalvis* reaches heights of about 0.5–2 m (20–80 in) or sometimes much higher. The linear to lanceolate leaves of *L. octovalvis* are about 5–7 cm (~2–3 in) long, 0.7–2.5 cm (~0.25–1 in) wide, and arranged alternately. The yellow flowers have four petals, each 10–16 mm long with a notch at the tip. The fruits are ribbed capsules 2.5–5 cm (1–2 in) long and produce a multitude of very small seeds.

Distribution: *L. hyssopifolia* and *L. octovalvis* are closely related herbaceous species somewhat similar in appearance, especially in terms of their bright yellow flowers. In addition, both of these weedy species have been introduced in many tropical regions of the world, including on a number of Micronesian

Islands, including Pohnpei, where they have become naturalized plants. This means they are not native species, but reproduce on their own in new areas such as islands where they have been introduced by people. The natural habitat of these two *Ludwigia* species is in wet places, including shallow freshwater pools and ditches, along margins of canals and in humid lowland crop fields. For example, *L. octovalvis* is a common weedy species found in moist locations or wet places in or near taro patches in various tropical Pacific islands (Holm et al. 1997: 448). Some botanists believe that *L. octovalvis* was an aboriginal (pre-European contact) introduction on some if not most of these islands, perhaps including those in Micronesia. If so, *L. octovalvis* was probably brought to these islands accidentally by early voyagers when tiny seeds attached to transported cuttings of taro plants, such as sweet taro (*Colocasia esculenta*) and giant swamp taro (*Cyrtosperma merkusii*). "It is well known that Austronesian colonizers and later migrants to tropical and subtropical Pacific islands introduced many plants of economic and agricultural importance [but these ancient and more recent immigrants] sometimes introduced plants of no apparent economic importance, presumably inadvertently" (Athens and Stevenson 2012: 35). *L. octovalvis* is an example of a small-seeded "weedy" species that early human colonizers may have introduced accidentally to Pohnpei (also see Athens 1997: 269; Fall 2010: 264).

Uses in Pohnpei: Balick et al. (2009: 456) provided medicinal information regarding the use of the leaves of *L. hyssopifolia*, which, according to these authors, is known as **Delahrik** on Pohnpei:

> "To treat fever, gather as many leaves as will fit in a piece of cloth, squeeze into a bucket of water, and use this to bathe. This should relieve fever and is to be used for four days or more if needed, making a new bucketful of leaf extract each day..."

Balick et al. (2009: 457) provided information regarding the use of the leaves of *L. octovalvis* in traditional medicine:

> "To use as a charm so a parent will not punish a child having left and returned home without permission, four leaves are placed in the child's pocket on the way home... [and to] treat *ingin* (blood blister), a handful of leaves is pounded, wrapped in a piece of cloth, squeezed in a basin of water, and showered over the blister..." (Balick et al. 2009: 457)

▸ **Sellap** or **Sallap** (P)
Luffa cylindrica

Description: *Luffa cylindrica* is a vigorous, fast-growing, trailing, or climbing vine. Its square stems and tendrils can grow up to 10 cm (4 in) long. The large leaves of *L. cylindrica* have a rough surface and may be up to about 25 cm × 25 cm (10 in × 10 in) in length and width. Its deep yellow to golden-yellow flowers are up to 10 cm (4 in) in diameter. The fruit has ten furrows on the surface and can reach up to about 30–60 cm (24 in) in length and 8–12 cm (3–5 in) in width. The fruits turn brown when mature, are elliptical to cylindrical in shape, and enclose hundreds of flat, smooth, black seeds. As an annual herbaceous vine, *L. cylindrica* has a general growth period of 70–80 days from transplanting the germinated seed to development of immature fruit, and 100–120 days to production of its mature fruit.

Distribution: The genus *Luffa* includes seven species. Four of these are native to the Old World tropics, and three more distantly related species are apparently native to South America. One of these species, *L. cylindrica*, is now widespread in the tropics and subtropics, both as a cultivated and as a naturalized (growing wild on its own) species. *L. cylindrica* is apparently an ancient cultivated plant and therefore it is hard to determine whether this plant originated in Africa or Asia, but it is not native to Pohnpei:

> "The plant occurs wild in West Africa, but this is often believed to be a result of escape from cultivation, as the plant is known

as "white people's sponge" in several communities in the region. However, evidence of an Asian origin is rare. How the plant has spread over the entire tropical zone is also not clear… Some argue for ocean currents as a dispersal mechanism, but more often human dispersal is raised as the probable cause of the wide distribution of *Luffa cylindrica*." (Mather 2012)

General uses: *L. cylindrica* has general uses as a source of sponges, food, and medicine. In a number of areas in the tropics, the fruits of cultivated varieties of *L. cylindrica* are eaten fresh or cooked as a vegetable and are used in soups. The fibrous network of the fruit serves as a sponge, and it is also used as a filter. In addition, table mats, insoles, sandals, shock-absorbing material, and gloves are made from the fruit's fibrous network, and an edible oil is extracted from the seeds. *L. cylindrica* hybridizes with other species of the genus *Luffa*, but in most cases hybrids show a great reduction in fertility or even sterility. Hybrids of *L. cylindrica* and *L. acutangula* are found in cultivation. These hybrids produce bitter and inedible fruits but are suitable for the production of sponges. The multipurpose vine species *L. cylindrica* is also used therapeutically, indicating its medicinal value. Partap et al. (2004: 127) reviewed the widespread occurrence of *L. cylindrica*, which is found in many areas of the world and is incorporated into traditional systems of medicine for the treatment of several human diseases. They reported that "phytoconstituents present in it mainly belong to the category of flavonoids," which are antioxidant compounds that help to eliminate cell-damaging free radicals and metallic ions in the human body. Among the many health benefits attributed to flavonoids are reduced risk of cancer, heart disease, asthma, and stroke. Additionally, flavonoids may also play a special role in protecting the brain. Different part of *L. cylindrica* vines are traditionally claimed to be used for the treatment of many different kinds of ailments, "including snake bites, convulsions, cramps, tetanus, emetic, cathartic, dropsy, nephritis, chronic bronchitis, asthma, sinusitis and fever to be list a few" (Partap et al. 2004: 127).

Uses in Pohnpei: According to the Pohnpeian–English Dictionary, *L. cylindrica* is known in Pohnpeian as **Sellap** and is reportedly "used medicinally." Balick et al. (2009: 367) provided the following ritualistic use of *L. cylindrica* on Pohnpei:

> " For use in an apology ritual when a person accidentally drops a *ngarangar* (a coconut shell cup that the pounded *sakau* root liquid is squeezed into) while the *Nahnmwarki* is sitting inside the *nahs*, 30 cm (1 ft) of the [*L. cylindrica*] vine is cut, the two ends are held together making a circle on the ground around the areas where the cup has been accidentally dropped, some spiritual words are spoken before turning the cup over again, and then the cup is brought up to the *Nahnmwarki*, whereupon forgiveness is requested. (Every person holding the *ngarangar* must make sure the cup is safe from falling)…"

Balick et al. (2009: 367) also provided the following therapeutic uses of *L. cylindrica* on Pohnpei:

> " To treat swollen testicles, which may be caused by a fall, many curly hairs of the fruit (unspecified maturity of fruit) are placed in a cloth, and then the testicles are massaged with the cloth… To treat muscle and joint pain, the vine is placed under a hot stone; then the person sits over it covered with a sheet and breathes in the vapor from the heated vine (*umwulap* treatment). This treatment is done daily for four or eight days… To increase breast milk, four or eight leaves are pounded and squeezed in 2 l (0.5 gal) of water and consumed by the nursing mother… To treat bruises, the leaves and stem are pounded, placed in a piece of cloth, and rubbed over the bruise…"

Referring to *L. cylindrica* as 'Selap' in Pohnpeian, Lee et al. (2010: 61) indicated that this "sponge gourd" species, which "has been widely cultivated and naturalized on Pohnpei and throughout the tropics," is traditionally used in this island to treat skin burns:

> " To prepare this treatment, gather many leaves [of *Luffa cylindrica*], pound together and squeeze the juice onto the burn. The leaves are placed over the burned area as a poultice. This treatment should be applied once daily until healed."

Lee et al. (2010: 82–83) also indicated that *L. cylindrica* is used on Pohnpei for treatment of fever:

> " Four large leaves are collected, pounded with a rock and wrapped into a piece of cloth. The juice is squeezed into a cup and consumed once daily until the fever has been reduced."

In addition, Lee et al. (2010: 135) referred to the use of *L. cylindrica* leaves "to increase milk production in a lactating woman." Leaves of this vine are "collected in any quantity, pounded together, wrapped in a piece of cloth and the juice squeezed into a cup of water." This should be consumed by the lactating female "once daily for four or eight days." However, Lee et al. (2010: 136) also pointed out that pregnant or lactating women should be cautious about ingesting the seeds of the *L. cylindrica* vine due to the potentially dangerous "proteins" they contain.

It should also be pointed out here that Englberger (2009) referred to "*L. aegyptiaca*" (a synonym for *L. cylindrica*, i.e., **Sellap**) as an invasive weed in Pohnpei:

> This fast-growing vine, with its smothering growth, can be very aggressive. The plant is often found on fences and overgrows native vegetation. It is usually found in open, disturbed areas like roadsides."

Furthermore, for comparison, it should be noted here that *L. cylindrica* has also been identified as an invasive species on Kosrae Island, especially in disturbed areas in the lowland regions (e.g., see Mather 2012).

▸ **Weingal** (P), **Waingal** (Mw)
Lumnitzera littorea

Description: *Lumnitzera littorea* is a native mangrove tree in Pohnpei. In marginal areas, it may only grow to heights of a shrub or small tree, but in favorable sites it can grow up to 25 m (80 ft) tall. It produces thin, reddish, knee-shaped pneumatophores (breathing roots) that are exposed to the air at low tide. The smooth, glossy, semi-succulent, spoon-shaped (spatulate) leaves of this mangrove are 8–12 cm (3–5 in) long, flat, and much longer than they are wide, but the emarginate tips are wider than the base. The leaves are arranged alternately at the branch tips. The bright red flowers of this tree are about 2.5 cm (1 in) long and grow in clusters beyond the leaves at the branch tips. The ellipsoid fruit are about 2 cm long (~1 in), vase-like in shape, and ribbed; they turn black as they mature.

Distribution: The genus *Lumnitzera* includes two closely related mangrove species which are found in tropical coastal regions from eastern Africa, on islands in the Indian Ocean, along coastal India and much of Southeast Asia, and out to the tropical islands of the Western Pacific. It has two closely related species: *Lumnitzera racemosa*, which dominates in the western part of the range, and *L. littorea*, which dominates in the east. As native mangrove trees, these *Lumnitzera* plants are found in the middle and landward areas of the mangrove vegetation and swampy depressions. Birds are said to be very fond of the fruit. In Micronesia, *L. littorea* is found on many islands, from Palau all the way to Kiribati and Nauru. Although *L. littorea* is widespread and common, by 2010 it was "threatened by the loss of mangrove habitat throughout its range, primarily due to extraction and coastal development, and there has been an estimated 22% decline in mangrove area within this species range since 1980" (Ellison et al. 2010a).

General uses: *L. littorea* produces hard, strong wood that can be used for poles and house posts. This tree tends to have a very straight trunk and can be sawed into boards. *L. littorea* wood is used on tropical Pacific islands to make the boom poles that connect outriggers to canoe hulls. The attractive bright red flowers can be used for decorative purposes, and it is cultivated for ornamental purposes in some areas of Southeast Asia, such as in parts of Thailand and Singapore.

Uses in Pohnpei: The very strong mangrove wood extracted from *L. littorea*, which is known in Pohnpeian as **Weingal**, has been used traditionally in local house construction, including its utilization as house posts. *L. littorea* tends to have a very straight trunk and is thus preferred for sawing into boards. *L. littorea* also has medicinal uses in Pohnpei:

> To treat diarrhea, the bark is stripped and pounded, placed in a piece of cloth, squeezed into a cup of water, and consumed frequently until the diarrhea stops... Alternatively, to treat diarrhea, an unspecified amount of bark is scraped off, pounded, placed in a piece of cloth, squeezed in a glass of water, and consumed once daily for four or eight days... or until a person gets better..." (Balick et al. 2009: 355; also see Lee et al. 2010: 34)

▸ **Kidi en mal** (P), **Suhkae krismas** (Pi), **Dala dalaa moa** (Nu)
Lycopodiella cernua

Description: *Lycopodiella cernua* (formerly called by many biologists as *Lycopodium cernuum*) is an ancient, non-flowering, herbaceous plant in the clubmoss family Lycopodiaceae. *L. cernua* is commonly known in English as the "staghorn clubmoss" but also has other names. It develops its above-ground parts from rhizomes that produce erect slender stems up to about 1 m (~3 ft) high. These stems are generally multi-branched in the upper parts, with nodding, cone-like tips. The stems are densely covered with short, bristle-like leaves.

Distribution: This lowland species is a clubmoss (lycopod). Clubmosses are very ancient plants that do not produce true flowers. *L. cernua* has one of the widest biogeographical distributions within its family, Lycopodiaceae, and is found in many tropical regions. *L. cernua* is a common pioneer species in non-swampy open areas. (Pioneer plants are those that are first to naturally grow up into an area after the vegetation has been cut, burned or otherwise disturbed.) This species of small clubmoss also grows as an epiphyte or "air plant," sometimes out of the base of a large bird's nest fern (*Asplenium nidus*, known as **Tehnlik** in Pohnpeian). *L. cernua* also occurs within patches of mosses on the trunks of trees in some of the wet forests of Micronesia.

General characteristics: Some of the local Micronesian names for this clubmoss species, *L. cernua*, syn. *Lycopodium cernuum*, refer to a cat or a cat's tail, probably due to its superficial resemblance to the tail of the cat. In several Southeast Asian areas and some tropical Pacific islands, this clubmoss species (*L. cernua*) has traditional medicinal use.

Uses in Pohnpei: *L. cernua*, known as **Kidi en mal** in Pohnpeian, is commonly found in open grassland areas with paddle grass (*Ischaemum polystachyum*, known as **Rehpadil** in Pohnpeian) and other herbaceous species. It also grows as an epiphyte or "air plant," sometimes out of patches of mosses on tree trunks in some of the wet forests of Pohnpei. **Kidi en mal** is reportedly used in homes as a cockroach repellant on Pohnpei, and as a decoration on Chuuk and Kosrae as well as Pohnpei (see www.comfsm.fm/~dleeling/svp/lycopodo.html). Balick et al. (2009: 239–240) provided extensive reference to a wide variety of decorative, hunting, magical, and medicinal uses for this common clubmoss on Pohnpei. For example, they refer to the contemporary use of **Kidi en mal** as a Christmas decoration and as a deer bait (deer are not native to Pohnpei, but were introduced by people and are now established as a wild population in some upland areas of Pohnpei). **Kidi en mal** is also used on Pohnpei as a magical charm and as a medicinal source for many ailments, such as headache, back pain, muscle pain, and joint pain, as well as for stomach ache, arthritis, and influenza (flu).

▸ **Apwid** (P) (Mw)
Macaranga carolinensis

Description: *Macaranga carolinensis* is a common, small to medium-sized tree that grows up to 12 m (40 ft) in height. It has large, simple, alternately arranged leaves that are peltate (attached to the leaf stalk inside or away from the leaf edge). The leaves, with three main veins radiating from the point of attachment to the stem, are about 10–20 cm (4–8 in) long, and their margins are shallowly toothed and somewhat wavy. *M. carolinensis* is a dioecious species, i.e., its unisexual flowers, male or female, are produced on separate plants. The male trees bear tiny flowers on stalks; the female flowers are greenish-purple and the fruits are small, dehiscent capsules that are produced year round. This species is called **Apwid** in Pohnpeian, and according to Lee et al. (2010: 22), it is further differentiated by its flowers: the pink-flowered male (**Apwid mwahn**) is recognized as distinct from the purplish-green-flowered female.

Distribution: There are about 300 species in the genus *Macaranga*, with its center of diversity among the islands of Southeast Asia. *M. carolinensis* is regarded as endemic to some of the tropical Pacific islands. It is found in open, secondary forest formations in lowland areas of Micronesia. This successional forest tree species commonly occurs in cleared areas of tree gardens (agroforest), or other places disturbed by humans or natural events such as strong winds or landslides: "*Macaranga carolinensis* is a common secondary "bandage tree" that grows up in disturbed areas, rapidly forming a canopy and reducing soil erosion" (Falanruw 2015: 88; also see Stone 1967a: 112).

Uses in Pohnpei: The curved branches of **Apwid** are used for making fishnet frames. Its wood is also used for firewood as well as to start fires:

> " For use as a fire starter, the stem is cut in any length that is needed and carved flat on one side. A second stem, 15 cm (6 in) long and 2 cm (1 in) in diameter, is then cut and carved flat on one side. This second stem is placed on the flat side of the first and pushed [in an] up-and-down motion until smoke—and eventually a fire—starts." (Balick et al. 2009: 374–375)

Woody parts of this species are also used in Pohnpeian canoe production; specifically, **Apwid** wood is used to make the *kerek*, the upright part at both ends of a *wahrasapw* (the carved, decorated canoe used by the *Nanmwahrki*). Occasionally, **Apwid** is killed by girdling the trunk (cutting through the bark and cambium sheath) so that the dead trunk and branches can be used as a yam trellis. The large, flexible leaves of **Apwid** are still occasionally used for toilet tissue in emergency situations. Parts of this plant are also utilized for medicine; for example, the yellow leaf juice is reportedly used to prevent flu in babies, and its bark is used for ear infections as well as other illnesses such as diarrhea in children. Lee et al. (2010: 23) provided the following remedy for treating diarrhea:

> " ... four 7–12 cm long pieces of **Apwid** bark are scraped and pounded together. The pulp is wrapped in a piece of cloth and dipped in water. The juice is then squeezed into a small cup, roughly 50 ml in volume or the size of a coffee cup (6–8 oz). A new piece of bark is prepared each day by squeezing the pounded bark four times into a cup. This is consumed once daily at the onset of symptoms and continued for eight days or until healed (the number days are important to note)."

Lee et al. (2010: 138) also provided this remedy for therapeutic purposes:

> " **Apwid** is combined with **Mwehk** (*Glochidion ramiflorum* G. Forst.) and used for cleaning the uterus. Gather four young shoots of **Mwehk** and four young shoots of **Apwid**, clean, and rub them between the hands (or they can be pounded together); do not add water. The medicine is placed in the vagina for 8 hours and then removed. This should be applied once daily for four days."

For more detail regarding the various traditional therapeutic uses of **Apwid** on Pohnpei, see Balick et al. (2009: 375–377).

▸ **Luwehs** (P)
Mammea odorata

Description: *Mammea odorata* is a small, more or less erect tree with light-colored bark in the Mangosteen family. It has a wide-spreading crown up to 10 m (30 ft) tall, and its leaves are simple, oppositely arranged with smooth margins. The leaves are also leathery, obovate and rounded at the tip and narrow toward the base. *M. odorata* trees produce short, prominent, yellowish petioles with the lower side of the leaf lighter than upper side, which is more greenish and glossy. The fragrant, white-petaled flowers are produced below the leaves on branches and also on the trunk. The narrow-ovoid, long-pointed fruit are curved at the tip and average about 7.5 cm (3 in) in length. Each fruit contains one oblong seed when mature.

Distribution: *M. odorata* is indigenous to areas of Malaysia, Melanesia, and Micronesia, where it is mainly found in coastal forest regions among beach thickets, often upon limestone, as well as on the inner edges of mangrove swamps and in rocky places near the shore.

Uses in Micronesia: *M. odorata* produces hard wood that is used to make the handles of tools and other fabricated objects on a number of Micronesian islands, and the bark as well as other parts of this tree have had traditional medicinal and dye use in some areas of Micronesia: "Wood hard, light on outside of trunk with reddish heartwood that can be used to produce a dye; said to be used in Fiji to dye hair orange-brown" (Falaruw 2015: 74).

Uses in Pohnpei: *M. odorata* is known as **Luwehs** in Pohnpeian. Similar to other tropical Pacific islands where it is found, on Pohnpei it has had traditional use for its wood and some medicinal purposes:

> "The wood is used to carve knife handles... To treat *soumwahu en kau*... To treat *mehn warmwowi* (menorrhagia)... The bark of this species is part of a mixture described by Riesenberg (1968) to treat bruises. Okabe [1941c] noted that the stem of the tree was good lumber." (See Balick et al. 2009: 354–355 for more detail).

▸ **Kehngid** (P), **Koahngid** (Mw), **Kaehngid** (Pi), **Gaangidi** (Nu), **Geengidi** (Ka)
Mangifera indica

Description: Mango trees are large, evergreen plants 10–40 m (30–120 ft) tall with wide-spreading crowns. Their dense foliage is mainly composed of simple leaves that are smooth, leathery and oblong-lanceolate, 5–32 cm (2–12 in) long, 1.5–10 cm (0.6–4 in) wide, with normally undulate margins and a more or less acute tip. The four-to-five-parted flowers of mango trees are small, pinkish-white or yellowish, and are produced on axillary or terminal panicles. The fruits are relatively large asymmetrical drupes 5–15 cm (2–6 in) long and 6–8 cm (~2–4 in) wide. These desirable fruits are green with yellow spots or yellowish green to yellowish-orange at maturity and sometimes have a purple-to-red coloring; the thick, edible, juicy mesocarp is yellow-orange to dark orange, often stringy, and covers a large flattish, woody, adhering seed.

Warning: Scott and Thomas (2000: 108–109) provided this note of caution:

> "The leaves, sap, and fruit skin of the mango tree contain toxicodendrol, an oily resin closely related to the toxin in poison ivy. The yellow fruit inside the peeling is free of this toxin. As with poison ivy response, only some people are sensitive to this oil. Repeated exposures tend to produce worsening rashes. The skin of unripe fruit and tree sap are the most common cause of skin rashes. When blooming, the flowers may cause breathing difficulty or swell[ing] of the eyes and face in susceptible people."

Distribution: *M. indica* is apparently native to tropical Asia, possibly India (or Myanmar) as its epithet name (*indica*) suggests. Mango trees are cultivated in practically all tropical and some subtropical regions. Mango has also become naturalized throughout much of the tropics and subtropics. This wide-spreading naturalization has occurred in conjunction with the spread of human populations, and as such, the mango plays an important part in the diet and cuisine of many diverse cultures. In the Pacific region, all mangos were probably introduced from other parts of the world. They were early post-European contact introductions in most areas of the Pacific, but possibly pre-European contact introductions into some parts of this region. The earliest recorded introductions into Hawai'i were prior to 1825; however, most introductions to the Pacific Islands have occurred over the past 100 years.

General uses: Mango (*M. indica*) is a common garden tree throughout the tropics, including many of the tropical islands of the Pacific. Recent research indicates that medicinal use of the mango tree can have many therapeutic applications, including their use for "antidiabetic, antioxidant, antiviral, cardiotonic, hypotensive, and anti-inflammatory" purposes (e.g., see Shah et al. 2010). Essentially, most parts of the mango trees, such as its "fruit, pulp, peel, seed, leaves, flowers, and bark," have therapeutic value (Baumann 2016). For example, when ripe, the delicious dessert fruit of the mango tree is particularly high in vitamin A. The fruit is also eaten green, processed into pickles, pulps, jams, and chutneys, and is frozen or dried, and is an important source of sustenance for birds and insects, as well as bats and other mammals. Mango is now cultivated commercially throughout the tropical and subtropical world, and also as a garden tree and as a shade tree for stock. In addition to the fruits being eaten, the wood is also used in construction and building (e.g., see Bally 2006).

Uses in Pohnpei: Mango is known in Pohnpeian as **Kehngid**, a general term for several varieties or cultivars of this widespread and popular fruit tree (see p. 431). Balick et al. (2009: 324) provided some information about the use of mango trees on Pohnpei for food and medicine:

> "The fruits [of mango trees] are eaten as food… and the ripe fruit is a good source of vitamin A [see Englberger et al. 2003b/c]. To treat acute backache in both men and women radiating down the legs (neurological only, not for use with backache with hematuria—blood in the urine), the outer bark is scraped off and four pieces… of inner bark are pounded, put in a cloth squeezed into one cup of water, and consumed daily for four days."

- **Dapiohka** or **Kehp tuhke** (P), **Dapiohka** (Mw), **Dapioka** or **Maenioak** (Pi), **Dabiogaa** (Nu), **Tabiogaa** (Ka)
  Manihot esculenta

Description: *Manihot esculenta*, known in English as "cassava," "tapioca" or "manioc," is a semi-woody shrub or small tree. It can grow to about 1.3–5 m (4–15 ft) in height, with deeply palmately parted leaves that have long petioles and are arranged alternately. The main human use of *M. esculenta* is the cultivation of its fleshy, elongated, tuberous roots or rhizomes; these are more or less woody and only slightly thickened in wild varieties. Cultivation can produce very large tuberous roots up to about 2.5 m (7.5 ft) long and 10–15 cm (4–6 in) wide, weighing up to about 40 kg (~88 lbs), but averaging 4–7 kg (~9–16 lbs). Although this cultivated species rarely flowers in many areas where it grows, when it does, it produces small inflorescences in panicles, which are less than 1.2 cm (0.5 in) long. The seeds are produced in globular capsules about 1.2 cm (0.5 in) across, with six winged angles; each capsule produces three seeds. It is important to note that the cassava plant produces high levels of cyanogenic glycosides, which protect it against herbivory but can pose problems for people who consume the starch-rich tuberous roots without proper preparation and cooking (see the next page).

Distribution: *M. esculenta* can be cultivated in all well-drained tropical and subtropical areas of the world where there is a warm humid climate and requires a fairly well-spaced annual rainfall of about 150 cm (~80 in) for optimum growth. Although *M. esculenta* is believed to have originated in Brazil and Paraguay, today it is generally recognized as a cultigen (a fully domesticated plant) with no wild forms of this species having been documented. There are reportedly two centers of speciation, one in Central America from southern Mexico to Guatemala and another in northeastern Brazil. Archeological evidence indicates that cassava has been cultivated in one or more regions of the New World for thousands of years. This root crop was probably introduced into Micronesia at an early date during the era of Spanish colonization (e.g., Barrau 1961: 53):

> "The exact origins of cassava (*Manihot esculenta*) are not yet known, but it is thought that it arose in central Brazil. Cassava was likely domesticated between 7,000 and 9,000 years ago in the Amazon and is now grown throughout the tropical regions of the world. Root crops are an essential part of historic Pacific island diets, but taro, sweet potato, and yam are the traditional root crops of the region. Cassava was introduced to the Pacific sometime around 1800 during the early years of European contact. It has become an important dietary staple and in some entities is produced in [a] larger quantity than the traditional, preferred root crops. In the Pacific region, cassava is

generally not produced on a large scale. It is grown in subsistence and home gardens, and is available in local markets." (Also see Thaman and Thomas 1982, 1985).

General uses: *M. esculenta* is a very widely grown crop in the tropics, including insular areas in the Pacific region. It is cultivated for its enlarged, starch-filled tuberous roots or rhizomes, which, on average, contain approximately 30% starch, but very little protein. Its chief use for starch "has almost totally supplanted that extracted from tubers of arrowroot, *Tacca leontopetaloides*" on a number of tropical Pacific islands (see Barrau 1961: 53). The tuberous roots of *M. esculenta* are prepared for consumption much like white potatoes. Cassava roots are generally peeled and then boiled, baked, or fried. Consuming uncooked cassava roots is not recommended because of potentially dangerous concentrations of cyanogenic glucosides in these tuberous rhizomes; however, cooking can reduce these toxins to innocuous levels (e.g., see Kwok 2008).

Uses and cultivars in Pohnpei: Cassava is known in Pohnpeian as **Dapiohka** or **Kehp tuhke**. The tubers of cassava are cooked and consumed on Pohnpei as a starch-rich food. For example, "*pihlohlo en kehp tuhke*" is referred to by some as "a traditional baked pudding dish, made of grated cassava, ripe mashed bananas, and coconut cream" (Rehg and Sohl 1979). However, unlike the different types of taro, yams, bananas, and breadfruit, cassava is not part of the ancient agroforestry system of Pohnpei; indeed, it is believed to have been introduced into Pohnpei by the Japanese, and has become integrated into the farming system of this island in modern times (e.g., see Balick et al. 2009: 377). Recent research on Pohnpei surveyed nine varieties of cassava currently being cultivated on Pohnpei for their yields and tastes; overall, the goal of this research was to understand the potential and limitations of cassava varieties to aid cassava farmers and the general population on Pohnpei:

> "... to strengthen food security, fight the high incidence of diabetes and other lifestyle diseases, enhance economic development, and help Pohnpeians adapt to the challenges brought on by the impacts of climate change." (Pretrick and Tuivavalagi 2017)

These researchers found nine cultivars (see p. 431) and based on the 2016 and 2017 harvests identified

> "... four high yielders (***Tolonier***, ***Palau a***, ***Japan***, ***Palikir***), four low yielders (***Marsi***, ***Palau b***, ***Benga***, ***Navavalu***), and one (***Yap***) that gives "very high" yield in 2016 and "very low" yield in 2017. Tuber yields [were] generally higher in 2016 and this is attributed mainly to the wetter weather condition during the first six months of the growing period. Cassava has been considered a hardy crop that can withstand difficult conditions, but this study tends to indicate that even this hardy crop could be seriously impacted by climate change." (See reeis.usda.gov/web/crisprojectpages/1003410-assessing-cassava-varieties-for-food-security-and-economic-development-on-pohnpei.html, underlined for emphasis).

Pretrick and Tuivavalagi (2017) concluded that: (1) ***Tolonier*** variety is the "highest yielding variety," (2) ***Yap*** variety is possibly "the most sensitive to inadequate water during the initial part of the growing period," and (3) ***Benga*** variety is "most susceptible to lodging" (i.e., the displacement of stems or roots from their vertical and proper placement, which may cause lower yields and diminish nutrient density). In addition, this research also determined that ***Japan***, ***Navavalu***, and ***Yap*** cultivars "could be resistant to infestation with white scale insects."

▸ **Koahre** (P)
Manilkara hoshinoi

Description: The classification of the genus *Manilkara* is confusing. More research is need for *Manilkara hoshinoi*, which is a relatively large forest tree in Pohnpei, known in Pohnpeian as **Koahre**. Although *M. hoshinoi* is assumed by some botanists to be an endemic species in Pohnpei, Costion and Lawrence (2012: 88) briefly described the problems associated with the genus *Manilkara* in Micronesia, indicating that the genus *Manilkara* was under taxonomic revision, and early molecular evidence suggested that the genus *Manilkara* is not monophyletic, and that there are more than one species of *Manilkara*, such as the unique *M. hoshinoi* on Pohnpei (see www.theplantlist.org/tpl/record/kew-120171).

Distribution: *Manilkara* is a genus of tree species in the family Sapotaceae. Various *Manilkara* species are widespread in tropical and semitropical locations in Africa, Madagascar, Asia, Australia, and Latin America, as well as on various islands in the Caribbean and Pacific Oceans.

Uses in Pohnpei: *M. hoshinoi* is known as **Koahre** and has traditionally been associated with chiefs, and thus has much influence, especially in Kiti and Sokehs (see Merlin et al. 1992). It produces strong timber and can be found in both lower and upper forest areas of Pohnpei. "Glassman (1952) noted that the wood [of *M. hoshinoi*] was used to make furniture and walking canes" (Balick et al. 2009: 496).

▸ **Kisetikmei** or **Pisetikmei** (P)
Melastoma malabathricum

Description: The genus *Melastoma* has a confused taxonomy with difficulty in the classification of several species. For example, *M. malabathricum* and *M. candidum* are closely related species and have been misidentified by botanists in areas where these species grow as natives or alien plants (e.g., see Wagner et al. 1999). *M. malabathricum* is an erect to spreading melastome shrub that normally grows to 1.5–5 m (4–15 ft) in height, with scaly branches and petioles. Its leaves are rough on the upper surfaces, more or less elliptical, pointed, and about 5–15 cm (2–6 in) in length with 3–5 distinct veins stretching from base to tip. The flower petals are pinkish and about 2.5 cm (1 in) long, and its dark-colored, 5-celled, 1–1.5 cm (~0.5 in) long fruits (berries) are eaten by birds that scatter the seeds (e.g., see Wagner et al. 1999).

Distribution: *M. malabathricum* is reportedly one of 22 *Melastoma* species found in the Southeast Asian region and is considered native to tropical and temperate Asia and on some of the tropical Pacific islands. *M. malabathricum* is found in moderately wet to humid regions of Southeast Asia and parts of the tropical Pacific. This shrub is found in lower and upper forest tropical environments, usually in more open areas. *M. malabathricum* var. *mariannum* is believed to be native to the Marianas and Caroline Islands, including Pohnpei (PIER 2006).

General uses: *M. malabathricum* has a widespread herbal status and there is widespread folkloric belief in its usefulness within the medicinal traditions of India, China, Malaysia, Indonesia, and some tropical Pacific islands. In fact, the leaves, shoots, barks, seeds, and roots of *M. malabathricum* have reportedly been used to treat a variety of maladies, including "diarrhea, dysentery, hemorrhoids, cuts and wounds, toothache, and stomachache," and recent scientific research has "revealed the wide pharmacological actions of various parts of *M. malabathricum*, such as antinociceptive, anti-inflammatory, wound healing, antidiarrheal, cytotoxic, and antioxidant activities" (e.g., Joffry et al. 2012: 1).

Uses in Pohnpei: According to tradition, when people originally colonized Pohnpei, the small white fruits of *M. malabathricum* were eaten; indeed, the fruits of *M. malabathricum* are said to resemble small breadfruits. This probably explains the Pohnpeian names of this species: **Kisetikmei** or **Pisetikmei**. It is used today in Pohnpei for medicinal remedies. For example, its fruits are reportedly used in the treatment of shingles, a painful, viral, nerve disease that affects the skin, and the terminal buds of the plant are used to increase the appetite of young children. Local Pohnpeian medicinal remedies involving *M. malabathricum* and many other plant species have been provided in the "Micronesian Healing Plants Presentations" made by students in ethnobotany classes offered by the College of Micronesia (COM). For example, in September 2007, a Pohnpeian student's presentation referred to the use of **Pisetikmei** (*M. malabathricum*) and **Iol** (*Merremia peltata*) "to treat sprains." Several additional medicinal uses for *M. malabathricum* on Pohnpei are referred to by Balick et al. (2009: 437–438):

> “ To make a perfume commonly used at feast time... As a charm to protect a person from *soumwahu en sokolei*... To use as a protective charm against a spell known as *kau*

(which if left untreated can lead to death)... To use to protect land from sorcery causing your crops to grow poorly... Mix the plant with **Katar** (*Sphaeropteris* [*Cyathea*] *nigricans*, MB3994) for unspecified sorcery... To treat mouth sores... To treat *nan aumweredi* (thrush in a baby that causes a white cottony growth, preventing the baby from being able to eat or drink)... To treat influenza... To treat an itchy rash similar to shingles that starts small and grows and spreads around the hand... To treat *soumwahu en seri* (sickness in babies, such as fever and loss of appetite)... To treat ulcers... Alternatively, to treat ulcers or gastrointestinal acid [use as needed] when symptoms continue... To make tampons for menstruating women, four, six, or eight young shoots are pounded and shaped into an appropriate size and shape (in addition to absorbing blood, the preparation will remove unpleasant odors)... To treat damaged nerves..."

▸ **Iol** or **Salomp** (P), **Iohl** (Mw), **Iol** (Pi)
Merremia peltata

Description: *Merremia peltata*, a member of the morning glory family Convolvulaceae, is a stout, coarse, subwoody climbing vine with large tuberous roots and milky sap that produces white flowers and large, heart-shaped, peltate leaves 6–30 cm (2–12 in) long and 8–25 cm (3–10 in) wide. It grows very quickly, covering open areas in cultivated lands and the native forest. *M. peltata* produces clusters of white (or rarely yellow), funnel-shaped, shallowly 5-lobed flowers and dry, papery, oval-shaped capsules up to 2.5 cm (1 in) long containing 1–4 furry seeds. This vine grows to lengths of 20 m (65 ft) or more.

Distribution: This indigenous vine is very aggressive in areas that are disturbed naturally or through cultivation of crops. It grows very quickly, covering open places or food gardens, and even in native forest areas. If it is not removed soon enough, it may smother plants, including large trees such as breadfruit. The native range of this common vine is assumed to be in the tropical Pacific region, but those islands in which *M. peltata* is truly native and those where it has probably been introduced by people are still not completely understood. Indeed, there is considerable debate as to whether or not this creeping vine species is native to any or many Pacific islands:

> "The vine *Merremia peltata* is widespread in Micronesia and Polynesia. Information regarding *M. peltata*'s status throughout its range is often contradictory as on some islands the vine is considered to be native. For example, it has been considered both native and an introduced invasive on Kosrae Island, Federated States of Micronesia. On islands where it is particularly aggressive, *M. peltata* is capable of smothering trees or groups of trees. As well as affecting the productivity of the tree it climbs on, the vine affects the behavior of forest birds by preventing access through the canopy where they would normally build nests and forage for food. Its abundance in disturbed areas also causes a problem for forestry and plantations." (Pacific Invasives Initiative, www.issg.org/cii/PII/Biocontrol.htm; also see Space and Flynn 2002)

As noted above, *M. peltata* typically grows up and over shrubs and trees in forests and woody thickets, particularly dry lowland and mesic inland natural communities, but not so much in wet upland and cloud forests (Meyer 2000). It is common as a primary or secondary successional plant species in disturbed areas up to about 300 m (~1,000 ft) in elevation. In areas where it is especially "invasive," it can suffocate trees or groups of trees. However, on the positive side, *M. peltata* may provide rapid ground cover following land disturbance, which can reduce erosion and nutrient loss. There is debate over the extent to which external factors such as cyclones and land clearing drive the invasiveness of this species; indeed, *M. peltata* may be a successional component of regenerating forest in its native range. In summary: this vine is very aggressive in areas that are disturbed naturally or as a result of cultivation of crops; and if it is not removed soon enough, it may overwhelm other plants, including large trees such as breadfruit.

General uses: In Micronesia, this morning glory vine has some therapeutic use. For example, tips of twining stems of *M. peltata* are used for medicine. In addition, on some islands in the region, its flowers are used for decoration, and food is wrapped with the leaves.

Uses in Pohnpei: The large, lengthy and sprawling stems of **Iol** or **Salomp**, as *M. peltata* is known in Pohnpeian, are sometimes used as ropes for a variety of purposes, including as a type of living support (i.e., as a trellis) for yam crops. In this case, breadfruit trees, *Artocarpus* species (**Mahi**) serve as the main physical support of the trellis for growing yams, *Dioscorea* spp., known collectively in Pohnpeian as **Kehp**. The root crops of yams, especially *D. alata*, are a very important source of food and prestige on Pohnpei. Yams are planted at the base of breadfruit and then their vine stems are trained up into the canopy of these trees along ropes made out of the bark fiber of **Keleu** (*Hibiscus tiliaceus*) or the vine stems of **Iol** (*M. peltata*). In the case of **Iol**, Balick et al. (2009: 366) indicated that well-developed "vines are used as a yam trellis after roots are trimmed off the vines..." Balick et al (2009: 365) also referred to the use of the stems of the sprawling **Iol** vine to make ropes used "to pull rough-cut canoes out of the forest when carving trunks from the trees." Among the reported medicinal applications of **Iol** (*M. peltata*) on Pohnpei are use of its leaves and seeds to induce abortion (e.g., see Christian 1899; Glassman 1952: 100). Balick and his colleagues (2009: 366) also noted the use of **Iol** (*M. peltata*) for some medicinal purposes; for example, they reported that many "pieces of the youngest part of the vine are pounded, placed in a cloth, and squeezed into a coconut shell," after which a "limited amount of the liquid is consumed for as many days as needed to treat diarrhea," and a preparation of pounded **Iol** leaves is reportedly used to treat any *pwudong* (itchy rash)." Local Pohnpeian medicinal remedies involving the **Iol** vine and many other plant species have been provided in recent times via the "Micronesian Healing Plants Presentations" which have been made by students in ethnobotany classes offered by the College of Micronesia (COM) at the Pohnpei campus. For example, in September 2007, a Pohnpeian student's presentation referred to the medicinal use of **Pisetikmei** (*M. malabathricum*) and **Iol** (*M. peltata*) to treat sprains: "Four of the red growing tips of the **Iol** vine are mixed with four leaves of **Pisetikmei** and applied to the sprain" (www.comfsm.fm/~dleeling/ethnobotany/e73/healing.html).

▸ **Oahs** (P), **Oaj** (Mw), **Gala mau** (Nu), **Oodi** (Ka)
Metroxylon amicarum

Description: *Metroxylon amicarum* is a large palm tree. It produces a single, woody trunk about 90–140 cm (~3–5 ft) wide that grows to heights of about 10–33 m (35–110 ft) tall. Its erect crown develops large, pinnate, arching leaves (fronds) that are about 4–7 m (13–23 ft) long, with petioles about 1–3 m (3.3–10 ft) in length. Unlike other botanically related palms in the genus *Metroxylon*, *M. amicarum* palms flower regularly for a number of years rather than flowering once and dying. *M. amicarum* palms also differ from the other native four palms of Pohnpei in that their inflorescences are produced within, not under, the crown of palm fronds. The flowers that make up the inflorescence of *M. amicarum* are borne in crowded spikes, and its large, round, glossy, and variably shaped nuts (fruits) are about 10–15 cm (4–6 in) long. The fruits have a scaly skin or exocarp, a fibrous mesocarp, and a white "meat" interior or endosperm which covers the embryo that is more or less soft when young and hardens as it matures. Fallen fruits are eaten by pigs. Because the "meat" (endosperm) of the baseball-sized nut is white and hard when ripe, *M. amicarum* and its close relatives in the genus *Metroxylon* are known in English as "ivory nut palms" (for further information about this large palm tree see McClatchey et al. 2006).

Distribution: The native range of *M. amicarum* is still unresolved. According to McClatchey et al. (2006), "*M. amicarum* is native to the Caroline Islands (Federated States of Micronesia, states of Pohnpei and Chuuk), the Marshall Islands, and formerly in

Guam and Palau." It is possible that the fruits of this large palm or one of its ancestors could have been washed downslope in its native continental or tropical island habitat; eventually reaching the sea, the palm fruits could have then been dispersed widely by ocean currents, sometimes washing up on shores. Currently on Pohnpei, *M. amicarum* is reportedly found in "freshwater wetlands, either coastal or in moist upland rainforest." Stone (1970) referred to a few *M. amicarum* palm trees that were planted in Guam; however, because no local name for this species is known in Guam, the plant was probably introduced to that island by people, possibly in relatively recent times. This palm may have also been introduced to some other areas of Micronesia, including Chuuk, "either in the prehistoric period, or shortly after European contact began." In the case of Chuuk, it was most likely introduced from Pohnpei, where it is called **Oahs** and is considered to be an indigenous species. However, McClatchey et al. (2006) also suggested that "*M. amicarum* could have been introduced to Pohnpei by people in pre-European contact times from the Santa Cruz Islands as a cultivar of [another species of *Metroxylum*] *M. warburgii*."

> "One legend says that before Pohnpei was vegetated, people lived in houses of rock without any roofs. A lady with magical powers visited the island and took pity on the people there. She then sent them the **Oahs** plant [*M. amicarum*], which the people used for thatching and other significant purposes..." (see "Serehd" [the newsletter of the Conservation Society of Pohnpei, CSP], Volume 8, Issue 4, October 2006); could this Pohnpeian legend refer to the human introduction of the big, useful *M. amicarum* palm tree via its large, hardy fruits to Pohnpei?)

Whether or not **Oahs** is native to Pohnpei or was introduced by people, today, populations of *M. amicarum* are fragmented in their distribution on this island; *M. amicarum* is said to be found growing naturally only in some areas of the FSM, where it is, more or less, threatened by habitat change. Although today it is harvested for numerous purposes, *M. amicarum* may appear to be used sustainably; however, it currently has no formal conservation protection in the form of protected areas, and therefore, a more complete understanding of the size of the population and rates of harvest is needed (e.g., see Bachman and Chadburn 2015). For example, in recent decades, this impressive and useful palm species has been progressively threatened in Pohnpei as it is being removed to make room for additional **Sakau** cultivation (see Merlin and Raynor 2004, 2005). As a result, it was listed on the IUCN redlist as a vulnerable species (see "Serehd" 2006, 8: 4).

Uses in Pohnpei: The **Oahs** palm tree is an occasional component of the upland Pohnpeian rainforest, as well as occurring in lowland freshwater swamps. **Oahs** (*M. amicarum*) is the tallest of the five native palm species on Pohnpei. The traditional use of the long fronds of *M. amicarum* for roof thatching on Pohnpei is important (e.g., see Merlin et al. 1992: 31). In fact, the word for "roof" in Pohnpeian is *oahs*, which is also the name for the native ivory nut palm, *M. amicarum*. Indeed, this palm is both a main source of roof thatch on Pohnpei as well as providing wood in its trunk, which is used to make floorboards for local huts and also for support of a hut's roof. In terms of Pohnpeian roof thatching, two types have been reported:

> "... *doakoahs en Ruk* and *doakoahs en Pohnpei* or simply *doakoahs*. *Doak* means "to pierce" with a needle-like object. **Oahs** is the Pohnpeian name for *Metroxylon amicarum* [and its] frondlets are folded so as to create a one-third, two-thirds split in relation to the length of the frondlet. The base is one-third, the apical end is two-thirds... The minor rafters over which the frondlets are folded are made of either *Phragmites karka*. (Chuukese: **Woowo** or **Niwo**, Kosraen: **Loa**, Pohnpeian: **Lirau**) or *Saccharum sponaneum* (Kosraen: **Ac**, Pohnpeian: **Ahlek**). The minor rafters are called *rahu* in Pohnpeian..." (www.comfsm.fm/~dleeling/ethnobotany/text/ethnobotany.html).

It has also been reported that the frondlets of **Parem** (the nipa palm, *Nypa fruticans*) are not wide enough to use as "proper *doakoahs en Pohnpei*" (thatch-making material), which are typically made on Pohnpei from the frondlets of **Oahs** (*M. amicarum*):

> "*Doakoahs en Pohnpei* differs [from *doakoahs en Ruk*] in being laid on a 45 degree diagonal. Note that traditionally Pohnpeian thatch was not covered by a net to protect the roof from wind damage. On the flat, outer island atolls in mid-ocean, the wind is a more significant source of damage. On Pohnpei rain and subsequent thatch rot are apparently more problematic. Pohnpei thatch is not netted down. The thatch is designed to catch the wind and "fluff" slightly, aiding in drying of the thatch and slowing rot... The *nohk* is the back half of the midrib. Removing the *nohk* is necessary when using **Oahs** due to the thick midrib. The **Oahs** frondlet cracks and breaks unless this is done. This is not apparently strictly necessary with **Parem** [*Nypa fruticans*]." (www.comfsm.fm/~dleeling/ethnobotany/e83/thatching.html)

There are other uses of **Oahs** on Pohnpei besides it use for thatching material. For example, use of the *nohk* to make a whisk broom (the *nohk* is the broken off, back half of the **Oahs** palm frond midrib): "Teeth are used to crack the midrib and split the *nohk* [which] can also be removed from [the] nipa [palm], but with more difficulty and a smaller resulting [broom]" (www.comfsm.fm/~dleeling/ethnobotany/text/ethnobotany.html). The floral bracts of **Oahs** have also been used traditionally to carry **Sakau** and for making toys. Moreover, the "white meat" or endosperm of the **Oahs** fruit is fed to pigs and shrimps. Furthermore, **Oahs** has traditional value as a source of medicine; for example, the young leaves, corky parts of the stem, and roots of this palm are reportedly used as important components of local therapeutic remedies. According to Balick et al. (2009: 265), parts of **Oahs** have been used: "To treat *soumwahu en ohl* (men's sickness)... To treat *soumwahu en rerrer* (or *soumwahu en lusulus*) (high fever associated with the eyes going back in the head showing the whites of the eyes, shaking the body, and a need to lie down) [and]... To treat *lusulus* (seizures)..." Lee et al. (2010: 120) provided this description of the use of **Oahs** roots to treat high fever symptoms:

> "For shaking sickness due to high fever, the roots are cut into eight 5 cm long pieces, cleaned and pounded. The pulp is wrapped in cloth and squeezed four time into a cup of water. This drink is consumed until the shaking stops."

Overall, **Oahs** is a large attractive palm that is desirable as an ornamental plant, with useful fronds and other parts; and its hard "nut" is a vegetable source for ivory-like material and may become a valuable export item in the future. Indeed, the hard white "nut" of the **Oahs** has been a good provider of raw material from which craft artifacts have been fabricated by people living on Pohnpei, and *M. amicarum* has been referred to as the "world's finest source of vegetable ivory," with individual trees producing "up to two tons of seeds per year," and these "nuts" have been used to make buttons for commercial export for many years (see "Serehd" 2006, 8: 4).

Uses in Kapingamarangi: *M. amicarum* is certainly not a native palm on Kapingamarangi, but as a major driftwood resource it has been reportedly utilized for canoe construction in this remote atoll. Although the tree trunk of choice for making the hull of new canoe on Kapingamarangi was normally that of breadfruit tree (*Artocarpus* sp.), other parts of the boat could be put together with the relatively rare but desirable wood from the ivory nut palm, *M. amicarum*—from its floating trunk that periodically washed up on the shores of the atoll:

> "The canoes of Kapingamarangi were made from materials that could be found on their atoll. Occasionally this supply was augmented by drift logs. These drift logs were often preferred due to their seasoned state. The wood of choice found on the atoll was breadfruit. It should be noted that this is a different type of tree than the breadfruit found on Pohnpei. The breadfruit of Kapingamarangi has a different leaf shape, the fruit has seeds and the wood is thought to be of superior quality [this may refer to the Mariana Islands seeded breadfruit, *Artocarpus mariannensis*]. Other wood used in the construction was from the ivory nut palm. The trunk of this tree is soft in the center and very hard and sinewy along the outer ring. The wood, though small in cross section, is quite strong and used where minimum size and maximum strength is important (i.e., the platform (*hata*) and the float [stanchions] (*hatoto*)." (Roberts 1997)

The people of Kapingamarangi descent who reside on Pohnpei, notably in Pohnrakied village, are known for their handicraft skills. They produce some of the finest carvings, ornaments, and souvenirs in all of Micronesia. They use a variety of plant resources in their work, with the ivory nut from *M. amicarum* being a very popular material for small sculptures and necklace pendants.

▸ **Peipeisapw** (P)
Microlepia speluncae

Distribution: *M. speluncae* is indigenous to the high islands in the FSM, including those in Chuuk lagoon and in Pohnpei, where it grows in semi-exposed locations along the edges of forests, roads, banks, etc.

Uses in Pohnpei: *M. speluncae*, known as **Peipeisapw** in Pohnpeian, is used "as a charm to help **Sakau** (*Piper methysticum*) grow well, and for this purpose, the planting stick [stem cutting] of **Sakau** is rubbed on the base of the fern when planting." The **Peipeisapw** fern (*M. speluncae*) is also reportedly used to treat "cuts or wounds from a machete," in which case, "the unrolled fiddleheads are put onto the cut or wound" (Balick et al. 2009: 233).

▸ **Sapeleng** (P), **Jɛplɛng** (Mw)
Miscanthus floridulus

Description: *Miscanthus floridulus* is a very tall, coarse, reed-like grass that grows up to about 3 m (10 ft) tall. Its leaves are about 60 cm (24 in) long or more and about 2.5 cm (1 in) wide. The inflorescence of *M. floridulus* is a plume-like panicle (cluster of flowers with pedicels on a common stalk) that is up to about 30 cm (12 in) long.

Distribution: *M. floridulus* is an indigenous or introduced large grass species commonly found in many tropical Pacific islands, especially in open or disturbed and fernland areas. It is believed to be native to Japan, the Ryukyu Islands, Taiwan, and other islands in the western part of the Pacific Ocean. *M. floridulus* has become an invasive pest in Guam and other Pacific islands outside its native range. Today, it is reportedly common in the Mariana Islands, Chuuk, and Pohnpei.

Uses in Pohnpei: The stems of this large grass, known as **Sapeleng** in Pohnpei, were used in a local game known as *pei sihr*, especially before European contact (*pei* means "game"; and *sihr* means "stick"). Now, only children, if anyone, play this stick-throwing game, in which the objective is to sling your stick as far as possible by bouncing it at a low angle off the ground. **Ahlek** (wild cane, *Saccharum spontaneum*) is also used for this game, but **Sapeleng** is preferred.

▸ **Weipwul** or **Wempwul** (P), **Wɛhnpwul** (Mw), **Uhpwul** (Pi), **Bugaliaa** (Nu), **Nonu** (Ka)
Morinda citrifolia

Description: There are approximately 50–80 species in the genus *Morinda*; most are found in the tropical areas of the eastern hemisphere. One of these species is *Morinda citrifolia*, a useful, upright, small evergreen shrub or tree known in English as the "Indian mulberry" or more recently as "noni." Its underground rooting system is similar to citrus crops and coffee, "with an extensive lateral root system and a deep taproot" (Janick and Paull 2008). The lateral roots allow the plant to absorb light rains that only penetrate the upper surface of the soil or substrate, while the deep tap root lets the plant take up water regularly from far below during dry periods. Above ground, *M. citrifolia* can reach heights of 3–10 m (10–33 ft), with heavy, hard, yellowish wood. Occasionally, *M. citrifolia* will support itself on other plants or sprawl over them as it can develop as a kind of woody vine or liana. Overall, there is much variation in the lifeform, fruit size, leaf size, tastiness, aroma of the ripe fruit, and amount of seeds per fruit of *M. citrifolia*. Its shiny, dark to pale green, stout leaves are broadly oval in shape and are 20–45 cm (8–18 in) long and 7–25 cm (3–10 in) wide. The flowers are tubular, perfect, small, white, and clustered with each having five lobes. Its unusual, fleshy, edible, greenish- to yellowish-white, compound (syncarp) fruit is approximately 3–10 cm (1–4 in) long, spherical, and multi-segmented. When ripe, the fruit becomes soft and putrid, but it has been a useful famine food plant for people and for domesticated pigs (see Nelson 2006 for a very thorough description and discussion of this versatile species).

Distribution: *M. citrifolia* is assumed to be native to tropical and subtropical Asia and Australia, though it is now distributed across the tropical regions of the world. *M. citrifolia* occurs roughly between the latitudes 20°N and 20°S, including the Pacific Ocean region, where it commonly occurs throughout much of Melanesia, Polynesia, and Micronesia. "The species has also become naturalized on the open shores of Central and South America (from Mexico to Panama, Venezuela and Surinam) and on many islands of the West Indies, the Bahamas, Bermuda, the Florida Keys and parts of Africa" (Janick and Paull 2008). A study of genetic relationships involving nucleotide sequence data by Razafimandimbison et al. (2010) suggests that *M. citrifolia*'s biogeographical origin in Micronesia precedes human dispersal influences; in other words, these scientists' genetic analyses imply that the "large-fruited *M. citrifolia* var. *citrifolia* might well have been present in the Pacific before the [human] arrival of the Micronesian and Polynesian ancestors from Southeast Asia," and that the "wide distribution of this form of *M. citrifolia* var. *citrifolia* is attributed partly to the trans-oceanic dispersal of its buoyant seeds, self-pollination and its ability to produce flowers and fruits year-round." It is also possible the early Pacific Islanders brought *M. citrifolia* fruits with them as part of the transported landscape of cultivated plants and animals and planted them on the islands they colonized.

General uses: *M. citrifolia* ("noni") can be found growing widely in the tropical Pacific islands. *M. citrifolia* is a multipurpose plant offering people many natural resources. In addition to its traditional use as a famine food, dye, and firewood, it also provides the wood from which tools such as digging sticks, coconut husking sticks, and poles are made. Furthermore, one of its most significant traditional uses is as a source of traditional medicine; Nelson (2006) provides this comprehensive summary of therapeutic uses:

> "Leaves, fruits, stems, and roots are used in various medicinal preparations, healing protocols, and treatment methods throughout the Pacific region. Medicinal uses (traditional) include: treatment for malaria, general febrifuge, and analgesic (leaf tea); laxative (all parts of the plant); jaundice (decoctions of stem bark); hypertension (extract of leaves, fruit, or bark); boils and carbuncles (fruit poultice); stomach ulcers (oils from the fruit); scalp insecticide (seed oil); tuberculosis, sprains, deep bruising, rheumatism (leaf or fruit poultices); sore throat (gargling a mash of the ripe fruit); body or intestinal worms (whole fresh fruits); laxative (seeds); fever (leaf poultice); cuts and wounds, abscesses, mouth and gum infections, toothaches (fruit); sties (flowers or vapor from broken leaves); stomach ache, fractures, diabetes, loss of appetite, urinary tract ailments, abdominal swelling, hernias, stings from stonefish, and human vitamin A deficiency (leaves). The leaves are also used as a medicinal poultice or body wrap (e.g., Micronesia)."

Also summarizing the traditional and contemporary medicinal use of noni, Krishnapillai (2010: 185) indicated the importance of this shrub or small tree in many Pacific societies, especially in Micronesia:

> "Noni is the most commonly used medicinal plant in Micronesia. Since its journey from South East Asia along with early human migration over 2000 years ago as a canoe plant for survival, it found a place of pride in the annals of traditional healers in the Western Pacific region. Although modern science is now widely spread, traditional medicinal practices involving [the] noni plant still exist in all islands of this region. Traditional healers employ every part of [the] noni plant including leaves, roots, bark, seed, flowers and fruits for various ailments, externally and internally. Some of its uses as tonic, demulcent, emollient, stimulant, analgesic and diuretic are exemplary."

Noni has, in fact, become an important economic product across the world over the past few decades, both because several different cosmetic products are produced from its leaves and fruits and because of its presumed therapeutic powers. Nevertheless, many scientists and others remain skeptical of the almost "cure all" status of noni: "The medicinal applications, both traditional and modern, span a vast array of conditions and illnesses, although most of these have yet to be scientifically supported" (Nelson 2006). More recently, Assi et al. (2017) published a lengthy "comprehensive" review regarding industrial uses, pharmacological activities, and clinical trials of *M. citrifolia* (noni). Their review cautiously concluded that much more scientific study is needed to validate the efficacy and safety of the remarkably variable uses and expected effectiveness of *M. citrifolia* as a dietary supplement, a food functional ingredient, or as a natural health enhancer:

> “*M. citrifolia* has been used as a medicine for the overall maintenance of a good health as well as the prevention of some diseases including skin, brain, GIT, heart, liver and cancer. [As of 2015], the only information available for daily recommended oral dose of *M. citrifolia* is 2 g; however, there is no information on recommended doses range for topical preparations. It is important to understand that *M. citrifolia* properties require further extensive studies in terms of identification of the active compounds, their specific mechanism of action and safety. In addition, more in vivo and clinical studies are required with a significant number of subjects and a wider range of concentrations especially in the cancer field. Since further studies are recommended on *M. citrifolia* and its products, thus they should be used with caution particularly for customers with concomitant illness and medications. Companies which are producing *M. citrifolia* products should provide relevant information regarding the bioactive components of *M. citrifolia* and any extra nutritional elements added to the products, along with their concentrations in order to arouse a sense of awareness among consumers for the safe purchase.” (Assi et al. 2017: 703)

Another critical study of the varied medicinal use of *M. citrifolia*, involving the review of 92 research studies, focused on use of the roots, leaves, and seeds of this species and came to the following conclusions: “... although *M. citrifolia* is widely and successfully used for the treatment or prevention of various diseases, it should be consumed carefully, and only after exhaustive studies into its chemical constituents and mechanisms of action, both in in vitro and in vivo models, as well as clinical trials” (Torres et al. 2017: 971).

Uses in Pohnpei: *M. citrifolia* is known as **Weipwul** or **Wempwul** in Pohnpeian. It is traditionally a multipurpose plant of significant and varied usage throughout most of Micronesia, including Pohnpei and the outer islands. For example, it is used to a produce a red dye, and its heavy, hard, yellowish wood is used to make digging sticks, coconut husking sticks, and poles for other uses. The dense wood is also used to fabricate the *peirar*, a tool for pounding baked taro and cooked breadfruit. In addition, the vines of some cultivated yam species (*Dioscorea* spp.) in Pohnpei, especially the smaller ones, are trained to grow up onto a living **Weipwul** plant, which can serve as a yam trellis. The leaves, fruit, and other parts of this plant have a variety of therapeutic uses on Pohnpei, such as treating asthma (*mwop*), as well as serving as a general health tonic. Glassman (1952: 94–95) referred to the medicinal use of various parts of *Weipwul* on Pohnpei:

> “The leaves of this species are used in conjunction with other plants as [a] cure for rheumatism; the stipules are employed in treatment of wounds caused by scorpion fish; the terminal buds, on abscesses; the inner bark and roots, as a hemostatic in menstruation; the young flowers, to relieve pains after childbirth; and the young fruit, to alleviate pain in heart attacks.” (Also see Lessa 1977: 183).

More recently, Balick and his colleagues (2009: 487–489) provided the following variety of therapeutic applications of **Weipwul** in Pohnpei:

> “To use as a charm to cast off *soumwahu en eni* (ghost-sickness) in children... To use as a charm to remove an *eni* (ghost spirit) from a person... To treat *soumwahu en eni* in babies that are crying at night... To use to protect against harm from *enihn nansed* (the evil spirit from the sea)... To get people out of a trance or to “cure” *soumwahu suwed* (possession by an evil spirit)... To treat urinary tract infection and cramps... To treat *kilin kidi* (impetigo)... To treat wounds... To treat open cuts... To use to remove a foreign body from the eye... To treat asthma... To treat type II diabetes, six remedies are reported... To treat kidney stones... To treat toothaches... To treat earache... To treat abdominal pain... To treat stomachache... To treat *medek* (stomach pain in a woman after she has delivered a baby)... To treat stomach pain in a man, such as caused by hunger... To treat stomachache... To treat sore eyes... To treat *soumwahu en saleng medek* (earache)... To use as a vermifuge... To treat a baby with *karakar* (fever)... To treat *karakar* (fever when a person feels cold and is shaking and shivering but feels warm to the touch... To cleanse the uterus,

as when menstruation is too long… To make *wini en neitik* (medicine for delivery)… To treat impotency in men around fifty years old… To use to make *wini en dokala* to treat the poisonous sting of **Nehu** (stonefish) when someone steps on the spines and their foot swells… To treat stonefish stings… To treat pain… To use for pain and therapeutic massage… To treat *koakon medek* (joint pain common in older men and women associated with cold temperatures)… To treat joint or back pain from carrying heavy loads… To use as an antisyphilitic… To use as a daily treatment for heart disease, type II diabetes, kidney problems, and skin infections…"

Lee et al. (2010: 34) reported another traditional medicinal remedy in Pohnpei involving the use of young fruit with flowers of **Weipwul** to treat diarrhea:

> " Two fruits are gathered, pounded and placed in a piece of cloth. The juice is squeezed into a cup of water. One cup of this is consumed daily until symptoms are alleviated."

Additional local Pohnpeian medicinal remedies involving **Weipwul** have been provided in the "Micronesian Healing Plants Presentations" which were made by students in ethnobotany classes offered by the College of Micronesia (COM). For example, in Spring 2005, according to one student, **Weipwul** has been used on Pohnpei to treat high blood pressure (hypertension):

> " Obtain a root of [**Weipwul**] about a centimeter in diameter. Clean the root. Remove the skin. Boil in water, drink the tea that forms to reduce blood pressure."

In the ethnobotany class at COM during Spring 2006, another college student referred to therapeutic uses of flowers, fruit, and other parts of **Weipwul** on Pohnpei: "The flowers are used to cure pink eye, the fruit for stomach ache and diabetes, the *salongen* (ears) on the petiole are used for stomach and head pains" (www.comfsm.fm/~dleeling/ethnobotany/healing_plants_presentations.html). In September 2007, a student presentation in the COM ethnobotany class also dealt with the multiple uses of **Weipwul**; this presentation referred to "a unique treatment for a [stye] in the eye," and other therapeutic purposes:

> " The central vein of a mature leaf is brought near the eye, the top of the leaf down. The leaf is snapped near the petiole right in front of the eye, thus spraying the eye with juice from the central vein. This procedure is repeated until the tip of the leaf is reached. Puncture wounds from rusty nails can be treated with the juice of the outer skin of this tree. The roots can be boiled and the liquid then consumed to build up one's blood supply. And the ripe fruit treats a particular type of *soumwahu en ohl* (men's sickness)." (www.comfsm.fm/~dleeling/ethnobotany/e73/healing.html)

Uses in Mwoakilloa: Glassman (1953: 296) observed *M. citrifolia* (known as **Wɛhnpwul** on Mwoakilloa) growing "as a tree along the [coastal] strand" on this atoll. The Mwoakilloa Dictionary (Harrison, and Albert 1977) defines **Wɛhnpwul** simply as a "kind of tree with hard wood."

Uses in Pingelap: *M. citrifolia* is called **Uhpwul** in Pingelapese language. Lessa (1977: 183) noted that "the natives use the fruit as an edible part and as a medicine." Damas (1994: 239) described this species as being found "away from the village and apparently not native" on Pingelap Atoll. According to Damas, *M. citrifolia* has served as a food source or could be "used as a medicine" on Pingelap. More recently, Hattori (2012: 278) found it planted near houses on Pingelap and used on this atoll as a source of medicine for wounds, as well as "for general health [and] it is said that eating the fruit will protect you from old-age blindness." A Pingelapese student report in the healing plant presentations of a COM ethnobotany class (2007) also dealt with the multiple uses of **Uhpwul**, including remedies to "heal men's illnesses" (www.comfsm.fm/~dleeling/ethnobotany/e73/healing.html).

Uses in Nukuoro: According to Drummond (et al. 2016), *M. citrifolia* is called **Bugaliaa** in Nukuoro and its "cheese fruit" *hua bugaliaa* has been "used to treat diabetes, high blood pressure, and heart problems" (Drummond et al. 2016).

Uses in Kapingamarangi: *M. citrifolia* is known among the people associated with this atoll as **Nonu**. As part of the complex preparation of preserved breadfruit on Kapingamarangi, Hiroa (1950: 37) observed the glossy leaves of **Nonu** being laid out on woven mats in the sunlight, where they are traditionally covered with prepared and cooked breadfruit so it can dry before being rolled into large parcels that are carefully covered with pandanus leaves for long-term storage and intermittent use of the hardened breadfruit paste (see Hiroa for a lengthy description of this traditional food preparation). In addition to the use of **Nonu** leaves in the breadfruit preservation process, the people on Kapingamarangi Atoll reportedly use the fermented fruit juice as a traditional therapeutic "tonic" drink (see www.comfsm.fm/~dleeling/ethnobotany/e73/healing.html).

▸ BANANAS

Musa, the genus that contains banana species, is the largest and most economically important in the monocotyledon plant family Musaceae. The genus includes at least 35–50 species that are native to regions in Africa, Asia, and Australia. The taxonomic status or classification of common bananas and plantains has had a complicated history. Bananas (or "dessert" bananas) are typically eaten raw without the peel and are usually smaller, rounder, and sweeter than plantains. Plantains (or "cooking" bananas) include less sweet cultivated varieties and are typically eaten cooked, sometimes along with their leaves and fibers, and are usually larger, more angular, and starchier than bananas. Most edible bananas are seedless and therefore sterile, and consequently they are propagated vegetatively (planted asexually using cuttings). Presently, there is no formal scientific distinction between plantains and bananas, which are all hybrids classified collectively as *Musa × paradisiaca*.

Musa × paradisiaca includes cultivars that are crosses between two species of *Musa*: (1) *M. acuminata*, a species of banana that is probably native to Southeast Asia; and (2) *M. balbisiana*, a species of wild banana that is probably native to eastern South Asia, northern Southeast Asia, and southern China. Most cultivated bananas and plantains are triploid cultivars, either of the *Musa × paradisiaca* hybrid types or of *M. acuminata* alone (triploid = three copies of every chromosome instead of the normal two). However, a recent study in the journal Frontiers in Plant Science has revealed that bananas, "it turns out, are not what we thought they were." This new research study indicates that "the banana family tree" overall is more varied and even more "mysterious" than previously assumed. This recent study uncovered domesticated bananas with genetic markers that connect them to "three types of wild bananas that have not yet been found." If, in fact, scientists and farmers can find these putative wild ancestors of our widely cultivated, domesticated banana varieties, which are vulnerable to crippling infection, it could "provide potential ways to strengthen the crops against disease" (Sardos et al. 2022).

Based on the recent research by Sardos et al., Whang (2022) also reported on how the use of genetic markers recovered from samples of different banana "varieties" has identified more genetic complexity in bananas than known previously:

> " Today, it is possible to use genetic markers to trace these bananas back to their ancestors by simulating breeding patterns in a computer program. This procedure can reveal what kind of trading routes and agricultural practices were established in different communities."

In the future, with the use of advanced genomics and computer modeling, we may better understand how banana fruits were transformed from being full of large hard seeds to those that are seedless, fleshy, and edible and which today represent the fourth most important staple food crop in the world. With approximately 100 million tons of bananas and plantains consumed each year, more than 500 million people depend on them as their staple food source. Unfortunately, only a few of the varieties dominate the world markets, and thus are susceptible to widespread infection:

> " There are thousands of banana [and plantain] cultivars; naturally occurring mutants selected for vigour, yield, hardiness, fruit quality and, importantly, seedlessness. Since banana cultivars are seedless they cannot reproduce without human help, so once in cultivation they must be multiplied by vegetative propagation. Evidence from

> Papua New Guinea indicates banana cultivation may be up to 10,000 years old. Despite all the cultivars grown, only two nineteenth-century cultivars have found favour in international trade, the thick-skinned 'Gros Michel' and 'Cavendish'. The paucity of useful cultivars, combined with vegetative propagation, makes commercial banana production particularly vulnerable to disease outbreaks. Since the 1950s, commercial production of 'Gros Michel' has been uneconomic because of Panama disease. Today, subsistence and commercial production of 'Cavendish' bananas is also threatened, by a new strain of Panama disease. Other than prevention, cultivar replacement is the only viable option to maintain productivity. The future of bananas is supermarkets may be less bleak than imagined; within bananas as a whole there is plenty of fungal resistance, the trick is to get this resistance into commercial bananas." (Harris 2020)

Following this general discussion of bananas and plantains, we will consider *Musa* × *paradisiaca* in more detail, and then move onto two additional important species of the same genus: *Musa textilis*, a fiber resource banana plant known widely by its English name, "Manila Hemp," and in Pohnpei as **Utinwel** or **Tikahp**; and then second, on another type of banana, *Musa troglodytarum*, known widely by its Polynesian name of **Fe'i**, and in Pohnpei as **Karat**.

▸ **Uht** (P), **Wus** (Mw), **Wis** (Pi), **Gahudi** (Nu), **Hudi** (Ka) *Musa* × *paradisiaca*

Description: Bananas and plantains are variable, large, tree-like, perennial herbs, lumped together botanically as a group of hybrids that are all classified as *Musa* × *paradisiaca*. They may grow up to 2–10 m (6–30 ft) in height, with a single non-woody "trunk" (pseudostem) which is actually a fleshy, watery stem made up of large, spirally arranged, overlapping leaf bases called sheaths. The pseudostem propagates from an underground corm and can grow to be as much as 30 cm (12 in) in diameter. The large leaves of these herbs are generally 1–4 m (3–12 ft) long. The flowers develop as a "cluster," which in fact is an elongated, plump, purple to green "bud" (sometimes called the "bell" or "heart"). This inflorescence, containing about 2–20 flowers per cluster, develops from the center of the crown, and the edible, generally seedless fruit are variable in color, shape, and size, depending upon habitat and variety. Each banana plant bears fruit only once, with the average weight of a cluster about 10–12 kg (~25 lb). As a group, the flowering parts and fruit are referred to as the "bunch," and the individual clusters of fruits are known as "hands," while individual fruits are known as "fingers" (Nelson et al. 2006).

Distribution: The actual geographical origin of edible *Musa* × *paradisiaca* hybrids is not known, but the generally accepted primary center of origin is in Malesia, a biogeographical region including the Malay Peninsula, Indonesia, the Philippines, and sometimes New Guinea, while India is a secondary center. From these putative centers of origin, edible banana and plantain clones probably spread out of Southeast Asia and/or western Melanesia entirely due to conscious human transport:

> " An examination of linguistic terms for "banana" within Island Southeast Asia and Melanesia sheds light on the history of *Musa* spp. domestication. Linguistic investigations suggest a westward dispersal of banana from New Guinea, mixing with a Philippine variety, *Musa balbisiana* (or at least a sphere of cultural usage), then westward again to mainland Southeast Asia, and (as far as can be linguistically inferred) onward to the western edge of South Asia. The linguistically derived interpretation accords generally with the archaeobotanical evidence and botanical

models for the dispersal of banana cultivars (Donohue and Denham 2009: 293; also see Kennedy 2008, 2009, and De Langhe et al. 2009 for discussions about the complex domestication histories of bananas which appear to have been first domesticated in New Guinea thousands of years ago)."

Edible banana plants in the *Musa* genus are thus assumed to be native to parts of Melanesia and Southeast Asia and so are not indigenous in the islands of Remote Oceania (i.e., eastern Melanesia, Micronesia, and Polynesia), where people have introduced them for their valuable food and other products. Today, the distribution of *Musa* banana cultivars is pantropical (worldwide in the tropics); they can even be cultivated in some temperate, relatively frost-free climates such as parts of California, but generally do not fruit in those areas because of the cooler temperatures. Some varieties of bananas are said to have been introduced in most tropical Pacific islands very early in pre-contact times, while other varieties were introduced during the post-European contact period, mostly prior to World War II. Today, the common bananas of commerce are grown on virtually all high tropical Pacific islands and on some atolls which have adequate rainfall and soil. These cultivated bananas occur "as an intercrop in shifting agricultural areas, in small contiguous stands in garden and fallow areas, and around villages," and are commonly grown in home gardens, where they are "often planted along borders or in lines at intervals among other ground crops" (Clarke and Thaman 1993).

General uses: Bananas are an important fruit crop with other uses besides food. Generally, bananas are a well-adapted subsistence crop in many areas of Micronesia, typically grown in back yards for family consumption. The seedless *Musa* banana cultivars or clones are sterile domesticated plants that must be reproduced vegetatively by suckers arising from the mother plant's base, thus their distribution is dependent upon humans who must reproduce the plants by these rooted basal shoots. An advantage of cultivating bananas is that the crop has to be planted only once and then the root suckers from this first main plant can be used to produce the next plant. Nelson et al. (2006) emphasized the ethnobotanical significance of *Musa* species within the high islands of the tropical Pacific, where bananas fit in very well into traditional Pacific island agroforestry, either interplanted within diversified farming systems or in plantation-style growth under full sun conditions. The multipurpose use of bananas as more than a food source has played a significant role in their traditional status among tropical Pacific islanders:

" *Musa* species attained a position of central importance within Pacific societies: the plant is a source of food, beverages, fermentable sugars, medicines, flavorings, cooked foods, silage, fragrance, rope, cordage, garlands, shelter, clothing, smoking material, and numerous ceremonial and religious uses." (Also see Kepler and Rust 2011).

Furthermore, banana plants have also long been used medicinally and can be a source of shade when grown near settlements. Moreover, the large banana leaves are also used in Micronesia in breadfruit preservation as pit liners and covering material to help protect the cooked and prepared breadfruit. As an edible crop, bananas are consumed cooked or raw in various stages of maturity. Therefore, they are one of the most significant sources of calories for the human diet worldwide, containing relatively large amounts of potassium, as well as being a valuable famine food in the tropical Pacific:

" All banana plants have starchy corms and pseudostem bases that were and still are used in times of famine on Pacific islands (mostly SW islands and New Guinea). This was evidently the original use of banana for human food." (De Langhe 1995; also see Nelson et al. 2006)

Lois Englberger (2003: 2), a Pohnpei resident for many years, pointed out that bananas were one of the staple foods in various islands of the FSM. However, in these and other Micronesian islands since the 1970s, "there has been a great decrease in the consumption of bananas and other traditional locally grown foods, along with an increased consumption of imported foods, such as rice, wheat flour, sugar and fatty meat" along with a shift from bananas with a deep yellow to orange colored flesh to introduced varieties with light-colored flesh. In her conclusion, Englberger (2003: 4–5) underscored the significant health benefits of consuming traditional banana cultivars in Micronesia, especially what she referred to as notable cultivars, including those grouped under the name **Karat**:

" This study has shown that some cultivars contain high levels of provitamin A and total carotenoids, in addition to important levels of vitamin C, fiber and potassium (Dignan et al. 1994). Epidemiological research suggests that carotenoid-rich foods protect against certain chronic disease including diabetes, heart disease and cancer. Thus, these Micronesian carotenoid-rich bananas may also protect against those diseases, which are now serious problems in FSM. Further research in FSM and elsewhere is needed to identify further carotenoid-rich cultivars that may affect health improvement and that may be acceptable for increased production and consumption." (Underlined for emphasis).

The use of the Pohnpeian word **Karat** can be confusing because it refers to cultivars in two different banana species: *Musa × paradisiaca*, discussed here, and also *Musa troglodytarum*, a separate species discussed below. **Karat** bananas that are *Musa × paradisiaca* are of "Maoli-Popoulu type" and are locally called ***Karat en Iap*** ("**Karat** from Yap"). These bananas are slightly yellow, not orange, inside and have none of the *M. troglodytarum* traits. It may be that ***Karat en Iap*** bananas are classified as **Karat** because they have "fat fingers" (individual fruits).

In any case, many scientific studies indicate that consuming foods rich in carotenoids can reduce the risk of age-related diseases by reducing oxidative stress by interrupting the propagation of free radicals (unstable atoms that reportedly damage cells, causing illness and aging; e.g., see Kaulmann and Bohn 2014). It is therefore hoped that more people will become aware of the health benefits of eating more fresh fruits and vegetables in general, and that bananas will be one of the favorite choices of Micronesians. In 2011, the year that Lois Englberger unfortunately passed away, a co-authored publication, of which she was the lead author, summarized the "Approaches and Lessons Learned for Promoting Dietary Improvement in Pohnpei, Micronesia," with special reference to consumption of bananas in Pohnpei, which also applies to many of the Micronesian communities in modern times:

" The island state of Pohnpei, Micronesia, has experienced much change in diet and lifestyle since the 1970s. Serious problems of micronutrient deficiencies and non-communicable disease such as diabetes, heart disease and cancer have emerged, following the neglect of traditional local foods and the shift to rice and imported processed foods. An awareness campaign on the benefits of local food, especially carotenoid-rich bananas for nutrition is the subject of this [discussion]. A community, inter-agency, participatory [program has been] implemented focused on raising awareness on island food production and consumption. Messages were shared on horticulture, cooking, food processing and conservation through mass media, posters, print materials, photography, national postal stamps, workshops, displays, youth clubs, farmers' fairs, competitions, e-mail and slogans: "Go Yellow" and "Let's Go Local." Research was undertaken in food analysis, gene bank collections and community case studies as part of a global health project. As yellow-fleshed carotenoid-rich foods (banana, taro, pandanus and breadfruit varieties) were identified and promoted, banana and taro consumption increased as did the number of the varieties consumed. Carotenoid-rich banana varieties not previously marketed such as ***Karat***, ***Utin Iap***, and ***Daiwang*** became popular. Foods ready for consumption using local banana and taro varieties appeared in the markets where these had not been sold previously. The awareness campaign stimulated great interest in Pohnpei and throughout the region with invitations to present at international meetings and "Let's Go Local" workshops outside Pohnpei. It was proclaimed as an awareness success and consideration should be given to applying this approach to other Pacific islands." (Engleberger et al. 2011: 224, underlined for emphasis).

Uses in Pohnpei: Edible bananas and plantains have long been cultivated as a crop in Pohnpei, while some varieties only appear to have been cultivated and consumed in Pohnpei and the outer islands since European contact began, i.e., during the "historic period" for which we have written records. According to Levin (2015a), a combination of archaeobotanical evidence (e.g., ancient plant remains, e.g., phytoliths) recovered from confirmed archaeological contexts indicates that banana leaves were used in breadfruit cooking in the "historic period." Thus we may assume that bananas were being grown on Pohnpei sometime after European contact in Pohnpei began, if not before. Commentary on the limited social status of bananas in Pohnpei is noted by Levin (2015a: 17–18), which may be indicative of the limited antiquity of the presence and use of bananas in Pohnpei:

> "Bananas, also a staple starch, have comparatively lower status (Fischer and Fischer 1957; Englberger 2003), so there are not as many particulars surrounding their cultivation, but conversely, no notable secrecy. Bananas are typically integrated into agroforests, and although they are botanically large herbs rather than trees, they perform an arboreal function in the Pohnpeian agricultural system. They can be prepared in a number of ways, including mashing (especially to feed to infants), baking, and frying, or some varieties are eaten plain. Despite their high nutritional value, there is a certain stigma attached to eating bananas for some modern Pohnpeians because of their relatively lower status (Englberger et al. 2009); it is not clear if this is a recent development."

Regardless of the reported lowly status of edible bananas on Pohnpei, there are many names for *Musa* × *paradisiaca* varieties on this island, including those eaten raw (i.e., bananas), as well as those eaten after being cooked (i.e., plantains). The numerous names for edible bananas are evidence of the wide and perhaps deep contemporary interest in bananas in Pohnpei, if not the great cultural significance of this general type of fruit.

Balick et al. (2009: 286) offered additional information about the consumption of the edible fruits of *Musa* × *paradisiaca* bananas (**Uht**), and reported that the leaves of these large herbs are used to treat bruises and stomach aches, as well as to cover the stone earth oven (*uhmw*):

> "The fruits of this group are eaten as fresh fruit. To treat sprains and external and internal bruising, the meristem or "heart" of a young plant is mashed and put on the sprain (which gives a cool feeling and lessens the swelling and bruising), while liquid from the mass of the stem is consumed (the more the better). Alternatively, the mashed meristem is put in a tub and the meristem is applied to the sprain..."

Cultivars in Pohnpei: The general term for any banana or plantain in Pohnpei is **Uht**, but there is a large number of distinct cultivars. For example, Lessa (1977: 147) referred to a multitude of banana varieties and their uses on Pohnpei:

> "41 native varieties of bananas and plantains that were recorded, most being planted while others grow wild; almost all are edible, being eaten raw, baked, roasted, boiled, or fried, depending on variety, but a fruitless variety [actually a separate species, *Musa textilis*, which is discussed below] is used for its fibers [and] the leaves of some of the [many banana] varieties are used in the treatment of bruises and stomachaches."

Raynor (1991) recorded 55 names referring to "*Musa* spp." on Pohnpei, 42 of which were found and photographed by Englberger and Lorens (2004). They created an unprecedented photo collection of Pohnpei's banana diversity and noted Pohnpeian names for an additional six cultivars they could not locate. In addition, Jeff Daniells, a world-recognized banana specialist working with local experts in Pohnpei, surveyed a large number of banana varieties in Pohnpei and produced a report on identification training, characterization of banana varieties, and development of the local banana industry (Daniells et al. 2004). Cultivar names referred to in that guide book for Pohnpei banana varieties have been correctly identified according to international equivalent taxonomic terms, as well as their Pacific banana relatives, genome, flesh color, etc. (see www.musalit.org/seeMore.php?id=10797). We have used the above references and the Pohnpeian–English dictionary to list of banana cultivars on Pohnpei for this book (see pp. 417–418). That list includes known cultivar names of all *Musa* species on the island.

Uses in Mwoakilloa: In the middle of the 20th century, Murphy (1950) noted that "the banana is the premier medicinal plant" on Mwoakilloa and that it is "used in a number of food combinations." Murphy pointed out that although edible banana plants do not "thrive as well as they do in the neighboring high island groups such as Pohnpei and Kosrae, banana fruits provided "a valuable addition to the human diet of people on [Mwoakilloa] atoll," and that dishes containing boiled bananas were common in several mixtures. Throughout this atoll, Murphy observed planted bananas in places where the soil was richer and moister than the average for the atoll, and he noted that some banana plants invade the edges of the large wet taro gardens. Two popular banana dishes on Mwoakilloa are *pihlohlo* (pudding cake of mashed bananas or taro or tapioca) and *komluj* (pounded boiled banana balls, sweetened and rolled in grated coconut) (see www.comfsm.fm/~dleeling/ethnobotany/63food_presentations.html).

Cultivars in Mwoakilloa: The general term for bananas in Mwoakilloa is **Wus**. Murphy (1950: 73–74) referred to seven "cooking" varieties of banana on Mwoakilloa and to seven others that were said to be consumed raw or without cooking. Our perusal of the Mokilese–English dictionary revealed 12 cultivars, which are listed on p. 418.

Uses in Pingelap: St. John (1948) referred to all bananas on Pingelap as belonging botanically to "*Musa paradisiaca*" (= *Musa* × *paradisiaca*) and observed them cultivated in the Pingelap village and infrequently along with sweet taro (*Colocasia esculenta*), sugar cane (*Saccharum officinarum*), and giant swamp taro plants (*Cyrtosperma merkusii*) around the swampy depression in the center of the island. Damas (1994: 55–56) referred to bananas as one of the "lesser foods" on Pingelap, "planted as a shoot cut from the tree [actually a large herb] after the fruit is picked, each plant bearing one large bunch which is harvested after about six months." Damas also reported that "an area beginning about 150 yards south of the taro patch" had served the community as a "banana plantation where large numbers were formerly grown." However, by the time Damas made his visits to Pingelap (1975–1983), "bananas, papayas, and the one mango tree grew only in close proximity to dwellings," to protect their fruit from being taken "by children, adolescents, and young men." The Pingelapese–English dictionary by Hattori (2012) lists a number of Pingelapese banana dishes: *doluwis*, *kaesipaeraepaer*, *pali koikoi*, *pihlohlo*, and others.

Cultivars in Pingelap: The Pingelapese general term for bananas is **Wis**, but, as St. John (1948) noted, there are many names for distinct varieties of bananas on the atoll (see p. 419).

Uses and Cultivars in Kapingamarangi and Nukuoro: Hiroa (1950: 8) reported very little about bananas on Kapingamarangi, indicating that they were "grown among [giant swamp taro] in the cultivations." Stone (1966a: 131–132) referred to 'Tamo' as the Kapingamarangi name and 'Kahuti' as the Nukuoro name for *Musa* × *paradisiaca*. Accurate names in contemporary spellings are **Hudi** and **Gahudi** in Kapingamarangi and Nukuoro, respectively. Both appear to be cognates (linguistically related) to **Uht**, the general name for bananas on Pohnpei, from where they may have been introduced to Nukuoro and Kapingamarangi. Stone's '*tamo*' corresponds to the Kapingamarangi word *daamoo*, which refers only to banana fruits, not the plant. Several cultivars are grown on both atolls (see p. 419).

▸ Utinwel, Utisel, Tikahp (P)
Musa textilis

Description: *M. textilis* is referred to in Pohnpeian as **Tikahp**, **Utinwel**, and **Utisel**. In English it is known widely as "Manila hemp," or **Abacá** from Spanish (written **Abaka** in the Philippines). It is a large, herbaceous, perennial species that is a noted source of valuable fiber, but produces inedible fruit. It grows erect up to about 1.5–6 m (5–20 ft) tall when mature with a normal stem diameter of up to 30 cm (12 in) at the base. *M. textilis* produces a fleshy, watery stem made up of large, spirally arranged, overlapping leaf bases called sheaths. Leaves are 1–2 m (3–6 ft) long and up to about 60 cm (24 in) wide. The buds are shiny and apple-green and the fruits may be 10–25 cm (4–10 in) long, green, yellow, or brown, cylindrical or angled, and either seeded or seedless. Normally, it grows for about 1.5–3 years before the first harvest and has an economic viability for about 6–15 years. The inedible fiber species is commonly grown from seed.

Distribution: *M. textilis* is adapted to the hot humid tropics with an average relative humidity of about 80%. It appears to be native to the Philippines and the Moluccas, and indeed most Manila hemp fiber comes from that general region, although *M. textilis* has also been introduced elsewhere, such as areas of Central America. In the Philippines, it is successfully produced in a zone of cultivation within the latitudinal range of 15°N–5°S, where it is grown mostly in regions under 500 m elevation. However, *M. textilis* can be cultivated in the wet tropics at altitudes between sea level and about 1100 m, although it does very poorly in shady and windy habitats, preferring moist soil.

General uses: *M. textilis* is produced commercially for the high-quality, tough, flexible, and elastic fiber that is extracted from its petioles. This fiber is a prominent global fiber crop and is favored for making ropes because of its buoyancy, resistance to saltwater damage, and its mechanical strength. As one of the sturdiest of all natural fibers in the world, it is cultivated today as a commercial crop in the Philippines, from where it originates. On average, many millions of metric tons are produced annually, which form a very large percentage of the world's total production. This species was the main source of fibers in the Philippines for centuries, and over time this fiber also came to be named "Manila hemp" in English. *M. textilis* has been cultivated in the Philippines since the 1500s and became available more or less worldwide in the 1800s. The other major producing country for this significant crop is Ecuador. Although Manila hemp has now been largely supplanted by synthetic materials, such as polypropylene, its fibers may still be woven into a cloth used to make bags, hats, twine, and other goods. Manila hemp fibers have the following qualities: they are very robust and durable; they are resistant to saltwater decay; they can be made into many tough, long-lasting products; and they are comparatively inexpensive to produce. Generally, the harvesting of *M. textilis* fibers occurs about 2–4 times a year, when the trunk is chopped down above the roots to extract the fibers; subsequently, fresh sprouts from the roots will grow into trunks with leaves relatively quickly after cutting. Once the trunk is harvested, the leaf sheaths are stripped free and the pulp is scraped off to obtain the fiber strands, which are twisted together to make cordage. The strands are mainly composed of cellulose, pectin, and lignin (see Kepler and Rust 2011 for a world-class study of bananas, including discussion of the species used as a fiber source).

Uses in Micronesia: *M. textilis* has a lengthy and rather unique history as a cultivated fiber crop in some of the islands of Micronesia. In these islands, *M. textilis* was the exclusive source of fiber used through the cultural process of backstrap weaving to make elaborate belts, sarongs, and breechcloths. The cloth made with the natural pale straw-colored fibers of *M. textilis* has a high luster and a supple elasticity. It is important to note that this form of loom weaving, which involves a horizontal tension backstrap loom, is the only loom-weaving form in the Pacific and has been restricted to islands in what today is the FSM, and a few of the more remote atolls on the northern margins of Melanesia:

> "It is absent from the rest of Micronesia and Melanesia and, eastward, from all of Polynesia proper. But westward it occurs in virtually all of Indonesia. The Pacific distribution of weaving technique and loom type indicates that the basic instruments of the craft spread from a southeast Asiatic source through Indonesia into the Carolines and thence into northern Melanesia." (Riesenberg and Gayton 1952: 342)

Loom weaving, with traditional use of the banana fibers extracted from *M. textilis*, does appear to have spread out of Southeast Asia and into some areas of the Central Pacific; the hypothetical cultural diffusion of loom weaving, presumably along with the *M. textilis* plant, has been summarized:

> "We may suppose that ring-warp weaving was introduced from Indonesia into the Western Carolines first, that it by-passed Palau and spread eastwards. The loom with cylindrical beams diffused to the end of the Caroline chain. Later, open-warp weaving was introduced on Yap, but did not pass beyond this northwesterly Caroline island." (Riesenberg and Gayton 1952)

Traditional knowledge and use of *M. textilis* plants have greatly dwindled in some areas where this banana fiber plant had an important ethnobotanical relationship and specialty craft use, such as in parts of Micronesia. On the other hand, demand for the fiber as a pulp product, e.g., to be used in paper production, has been increasing in recent times in some regions.

Uses in Pohnpei: Although bananas locally called **Tikahp**, **Utinwel**, and **Utisel** are strikingly similar in appearance to other banana and plantains trees, the major difference between them is edibility. The fruits of *M. textilis* are reportedly inedible because they are generally small, very seedy, and unpalatable (e.g., see Kepler and Rust 2011: 416).

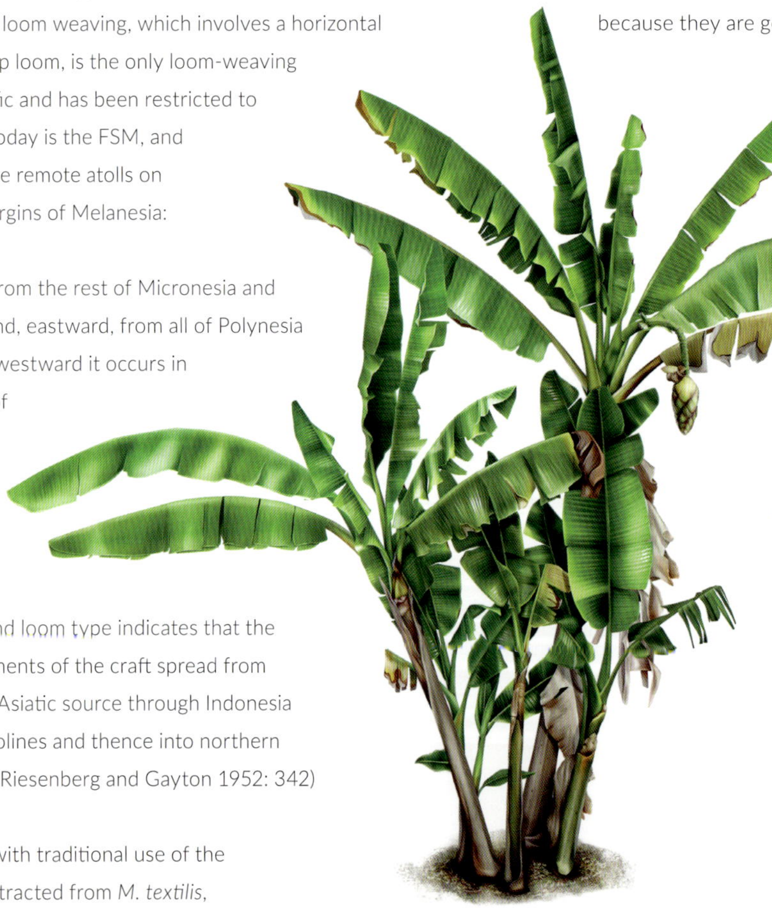

According to Glassman (1952), while the fruit of *M. textilis* "is inedible, the succulent petioles are employed as a source of water" on Pohnpei (Balick et al. 2009: 285). However, the Pohnpeian–English Dictionary refers to **Tikahp** and **Utisel** as "mostly eaten cooked," and **Utinwel** as "usually eaten cooked." In any case, the main if not exclusive use of *M. textilis* in Pohnpei has been for production of fiber for weaving cloth. A century ago, during the Japanese occupation period, Kanehira (1915) "observed that [*M. textilis*] was growing widely in Pohnpei as a naturalized plant and that it was used to make an "excellent cloth." (Balick et al. 2009: 285). Indeed, the "warping bench" for weaving fibers into cloth has been described as a special cultural development peculiar to the islands of Pohnpei, Kosrae, and Kapingamarangi. (Note: Mark Merlin observed use of such a "warping bench" on Ulithi Atoll, Yap State, FSM, in the 1990s). It has been stated that "the elaborate elegantly designed fabrics" woven using *M. textilis* fibers on Pohnpei and Kosrae were based on local weaving techniques developed over time which enhanced progress of the "basic elements common to the Micronesian–Melanesian weaving area, namely, the knotted yarn, the ring warp, and the warp-face weave" (Riesenberg and Gayton 1952: 371).

Uses in Kapingamarangi and Nukuoro: On these Polynesian outlier atolls, loom weaving was a traditional culture trait (e.g., see Reisenberg and Gayton 1952). According to Hiroa (1950: 148), this kind of weaving was basically a craft developed and essentially confined to Micronesia, where it was documented historically "by the German Expedition" as being present in Nukuoro and Kapingamarangi. However, on these atolls it was the inner bark fiber extracted from "wild hibiscus" (*Hibiscus tiliaceus*) that provided the weaving material. On the other hand, Hiroa also noted that men traditionally did the weaving on Kapingamarangi, while on Nukuoro Atoll, "where a sufficient growth of banana plants provided an alternate fiber for weaving, the women operated the looms." Hiroa pointed out that banana plants also grew in the moist areas where giant swamp taro (*Cyrtosperma merkusii*) was being cultivated on Kapingamarangi; however, the supply of banana fiber on that atoll "was evidently too small to lead to its fiber being used for weaving." On the other hand, for over a century now, migrant populations from Nukuoro and Kapingamarangi atolls have been living on Pohnpei, where *M. textilis* is more accessible; in that context, it would be beneficial to study the status and the possible future growth and use of the *M. textilis* banana on Pohnpei.

▸ **Karat** (P), **Wus karas** (Mw), **Hudi karati** (Ka)
Musa troglodytarum

Description: *Musa troglodytarum* is the kind of banana that Pohnpeians call **Karat**. In English, it is better known as the **Fe'i** banana after its Polynesian name. Some taxonomists do not consider it to be a separate species and prefer to call it a group of cultivars within *Musa × paradisiaca*. In any case, like all banana species in the genus *Musa*, these plants are tall herbs. Typically, banana plants can grow to be about 10 m (33 ft) or higher in height. Although they appear tree-like, the "trunk" of any plant belonging to these banana species is actually a false trunk; in other words, it is technically a "pseudostem," formed from the tightly wrapped sheath bases of the leaves. When a banana plant reaches maturity, each pseudostem produces a single flowering stem that grows up inside it and eventually emerges from the open upper part of the stem. Then, as the inflorescence elongates, female flowers appear and progressively form fruit (the bananas). Eventually, male flowers are produced.

The main difference between **Karat** banana plants (*M. troglodytarum*) and other more familiar bananas (*Musa* × *paradisiaca*) is the upward-pointing growth of the **Karat** fruit bunches, as opposed to "regular" bananas, which hang downwards. Typically, **Karat** bananas can be recognized by their "erect bunches, pink-red to purple sap and deep yellow or orange colored fruit pulp" (Nelson et al. 2006: 4; also see Ploetz et al. 2007; Kepler and Rust 2011).

Distribution: Although **Fe'i** banana plants are unique and distinct from other more familiar banana species or their hybrids, their precise geographical and genomic (ecological and evolutionary) origins are poorly understood. If *M. troglodytarum* is a "true species," it is probably derived from *Musa maclayi*, a seeded banana species native to Papua New Guinea and possibly the Solomon Islands. As the ancestors of Pacific islanders spread out over Remote Oceania, they carried their traditionally useful "canoe plants" with them, consciously transporting cuttings (clones) of **Fe'i** bananas; on the other hand, some biogeographers and other scientists believe that bananas (with seeds) may have also been dispersed, at least to some areas, by native pigeons and parrots (e.g., see Sharrock 2001).

General uses: Nutritionally, the fruits of some clones of *M. troglodytarum* are exceptionally high in beta carotene (precursor of vitamin A), an antioxidant that is believed to have a very important role in human health. Beta carotene is involved in producing the red, yellow, and orange coloration of some fruits and vegetables; indeed, the name beta carotene is derived from the Latin word for carrot. (For more information regarding the relationships between beta carotene, banana species, and human health, see the various publications by Lois Englberger et al.). **Fe'i** banana plants grow slowly but are generally vigorous, tolerating most diseases and pests, and require little care once established (e.g., see Kepler and Rust 2011; Ploetz et al. 2007: 3). **Fe'i** banana cultivars are traditional infant food and can be ingested raw, though they are generally eaten after being cooked. They have been described as "delicious and nutritious when baked or boiled, especially if the slices are swathed in fresh coconut cream" (Ploetz et al. 2007: 3). Banana saps have long been used in many cultures as a dye source; in fact, the sap from any banana plant "will stain clothes indelibly black [and] the bright magenta sap of **Fe'i** bananas still finds uses in SW Pacific cultures" (Nelson et al. 2006: 25).

Uses in Pohnpei: **Karat** has been used traditionally in Pohnpei as a food and medicinal resource, and its other uses include as a prestige offering; in fact, it is one of the few kinds of banana that can be used in ceremonial presentations in Pohnpei. Some varieties of **Fe'i** bananas may be unpleasantly astringent unless cooked; however, some cultivars are less so and are eaten raw when fully ripe. Generally, **Fe'i** bananas have a soft texture and a sweet taste. Therefore, they have been a traditional weaning food for babies in Pohnpei and consumed generally by Micronesians for hundreds of years. However, there has been a shift away from eating traditional foods towards eating imported foods, and so bananas with whiter flesh are now often preferred over traditional banana varieties with deeply colored flesh. One issue with **Fe'i** bananas is that eating them causes the production of lime-green urine, caused by the excretion of excess riboflavin present in the fruit. This effect led people to believe that **Fe'i** bananas might not be safe to eat, particularly for children (Englberger and Lorens 2004). As a result of the decline in eating traditional foods, including **Karat** bananas with their deeper colored flesh that contain high levels of beta carotene, vitamin A deficiency has apparently increased. A 1999 promotional campaign in Pohnpei that aimed at encouraging the consumption of **Karat** cultivars had some success in increasing the purchase and consumption of these nutritious bananas (Englberger 2003). **Karat** cultivars reportedly have the highest provitamin A carotenoid content among all the bananas of the world. Englberger (2003: 3–4) referred to the traditional use of this type of banana as an infant food with healthy benefits:

> "It is quite distinctive for its very fat fingers and deep yellow-orange flesh. Although this banana was first identified simply as **Karat**, further work in Pohnpei shows that there are at least two types: ***Karat pwehu***, which has a fat oblong finger (often 200 grams in weight), and ***Karat pako***, which has an even larger fat oblong finger (often 400 grams in weight). Informants explained that **Karat** is prepared in a special way for young children. Because of its thick skin, the unpeeled finger can be pressed and the flesh softened without the banana skin rupturing. The top of the finger is removed and children enjoy sucking out the contents, somewhat like eating an ice cream cone. It also has an important cultural significance and is one of the few banana cultivars in Pohnpei that can be used in ceremonial presentations. ***Utin lap*** is much rarer than **Karat** and has a distinctive cone-like shaped bunch, small fingers and deep orange flesh. **Karat** became rare due to neglect, similar to all other local foods, and also because newly introduced bananas were easier to grow. It was first analyzed in 1998 and found to contain high levels of carotenoids."

A mother and son posing with bananas and other delicious local foods for sale at a store in Kolonia.

Karat bananas have a rich yellow color and are highly valued in Pohnpei.

A cart loaded with freshly harvested green bananas.

Local Pohnpeian dishes, *uht idihd* (cooked, grated green bananas dressed with coconut cream) surrounded by *uht mwehl* (boiled banana with coconut milk).

Uht idihd, wrapped in banana leaves and baked in an *uhmw*.

▸ **Karara** (P) (Mw)
Myristica hypargyraea

Description: *Myristica* is a genus of over 150 tree species in the family Myristicaceae. Trees in Myristicaceae often develop well-formed and straight trunks that can grow to heights of about 18–27 m (60–90 ft), with diameters of about 50–120 cm (20–48 in), sometimes supported by buttresses along the lower parts of the trunk. The most economically important member of this family is the commercially cultivated nutmeg tree, *M. fragrans*, which is the main source of the valuable nutmeg spice. *M. hypargyraea*, which is closely related to it, is a large tree in the lower forest of Pohnpei and is assumed to be native to the island. It produces relatively large and somewhat rough-surfaced, oblong or lanceolate leaves about 20–35 cm (8–15 in) long and 10–15 cm (4–6 in) wide. The flowers are axillary, and the large fruits are globose-ovoid in shape with very small hairs on the surface and enclosed in a deeply laciniated mace (the reddish seed covering or aril of the seed inside the nutmeg fruit). The trunk tapers off at the base and is supported by roots branching on each side. Note: *M. hypargyraea* (or *M. hypargyraea* subsp. *insularis*) is currently an "unresolved" scientific binomial taxon; a synonym for this species is *Palala hypargyraea*; however, here we refer to this Micronesian nutmeg species as *M. hypargyraea*.

Distribution: The various species in the *Myristica* genus are reportedly distributed in the Indo-Malayan region, Indonesia, the Philippines, and Western Pacific islands. *M. hypargyraea* is reportedly a successional tree in the rainforests of tropical Pacific islands, where it occurs naturally and is at least partially dispersed by large native pigeons wherever these birds still exist in remote island ecosystems.

Uses in Pohnpei: *M. hypargyraea* is called **Karara** in the Pohnpeian language. It is used as a source of traditional medicine and wood for construction and was formerly used in the making of Pohnpeian canoes and for other purposes including medicine. Balick and his colleagues (2009: 449–450) reported many different uses of **Karara** in Pohnpei, including several for traditional therapeutic relief of various ailments:

> "The hard wood is used for furniture or construction… and to make canoe paddles and flooring… The sticky inside of the fruits can [be] used to seal a crack in a canoe to keep the water out (it will not hold the split surfaces together)… To make a canoe, a tree that is 120 cm (4 ft) in diameter and 30 m (ca. 100 ft) tall is cut down and carved into a canoe. The left over branches are used for firewood… To treat *soumwahu en naniak* (mangrove sickness), cut four pieces of bark, clean and pound them, squeeze into 2 lit (0.5 gal) bottle of water, and drink the entire amount every day for an eight-day period… To treat [heavy menstrual bleeding], four pieces of the bark are pounded, squeezed in 1 lit of water, and consumed until better… To treat high fever… To treat open cuts chewed bark is placed on the wound. [For high blood pressure] use bark pounded and squeezed into to a bottle of water and drunk three times daily for four or eight days… To treat painful diarrhea, the bark is pounded, squeezed into a cup of water and all of it consumed… To treat toothache, use inner bark that is chewed…" (Also see Herrera et al. 2010: 154).

Lee et al. (2010: 146) reported that the traditional treatment for back pain on Pohnpei involves boiling scrapings of the "skin of the stem" in water, which, after sufficient cooking, is consumed "until the symptoms are alleviated." Lee et al. (2010: 118) also reported the traditional treatment for "shaking due to high fever" in which "one piece of skin from the stem of both **Karara** and **Karer** (*Citrus limon* [lemon])" are combined, boiled in water for five minutes, then the liquid is "consumed when fever is high and shaking starts."

Warning: The seeds of this native nutmeg tree are considered to be toxic and therefore dangerous to consumed in any manner.

▸ **Rehdil** (P), **Pwoa** (Mw), **Pwe** (Pi), **Lluhe** (Nu), **Logo daha** (Ka)
Nephrolepis acutifolia, *Nephrolepis hirsutula*, and *Nephrolepis obliterata*

Three or more species in the fern genus *Nephrolepis* (family Nephrolepidaceae) are closely related botanically and somewhat similar in general physical appearance, so we will review them together. They are classified by botanists as *Nephrolepis acutifolia* (Desv.) Christ; *Nephrolepis hirsutula* (G. Forst.) C. Presl; and *Nephrolepis obliterata* (R. Br.) J. Sm. All three are referred to in English as "sword ferns" and in Pohnpeian as **Rehdil**. In Pohnpeian, *reh* means "grass" and *dil* means "through," thus **Rehdil** is a fern that grows through open grassland or fernland areas.

Description: (1) *N. acutifolia* is an indigenous, epiphytic fern that typically grows on other plants, producing long pendulous fronds 1–6 m (3–20 ft) in length. Hanging from the crotches of trees in damp, shady habitats, this large fern is rather attractive with the "weeping habit" of its long fronds. Each stalk (stipe) of its fern fronds develops from buds at the tips or nodes of its prostrate stems (stolons), located at or just below the surface of the area where the fern grows. The basal portions of its stipes are densely covered with lanceolate, rusty-brown scales. These scales sometimes have marginal cilia. The pinnae of the fronds are leathery (coriaceous) and the sori (clusters of the spore-producing sporangia) are located along the margins of the pinnae and are linear and continuous or interrupted. (2) *N. hirsutula* is a terrestrial (land-based), bipinnate fern that grows up to 1 m (3 ft) tall with thick, rusty-colored hairs along the midribs of its fronds. Its sori are usually located on the lower surface of fern leaves, lined up separately at the pinnae margins. It is sometimes identified as *N. biserrata*, yet another fern species that is also called **Rehdil** in Pohnpeian. (3) *N. obliterata* is a large, attractive, ground-dwelling or terrestrial fern which grows upright and arching to heights of 60–90 cm (24–36 in), spreading slowly by rhizomes (below-ground stems) as well as by stolons (above-ground, lateral stems). The leaf stems or petioles of *N. obliterata* ferns have sparse, red-brown hair-like scales with pale margins and a few longer hairs. Its evergreen fronds are long and feathery, and the leaflets have margins which are lightly scalloped. This sword fern reproduces either by spores or is divided to establish new plants. *N. obliterata* is regarded as among the most beautiful of fern species.

Distribution: *Nephrolepis* is a pantropical and subtropical genus of ferns comprised of about 20 species. The range of distribution of each species is as follows: (1) *N. acutifolia* is indigenous in the Paleotropics (tropical regions of Asia and Africa) and found in many islands of Polynesia and Micronesia. (2) *N. hirsutula* is also native in many tropical areas of the Old World, where it occurs often on trunks of palms with persistent leaf bases and as a ground fern in some areas such as on roadsides and in open fields, or in gardens as an ornamental. Indeed, *N. hirsutula* is widespread in tropical Asia, especially in Southeast Asia, and it can also be found throughout much of the tropical Pacific. In Micronesia, it has been reported from a large number of islands. (3) *N. obliterata* is adapted to rainforest conditions as a typically terrestrial fern, often occurring in soil near streams or water bodies, or in crevices of rock outcrops, sometimes as an epiphytic plant. It is apparently native in some of the rainforests in northeastern Australia, New Guinea, and some other areas of the tropical Pacific, including many islands in Micronesia.

Uses in Pohnpei: *N. acutifolia* has therapeutic and ritualistic uses on Pohnpei:

> "To treat *loalohla* (or *lesehkida*) (homesickness), medicine for forgetting a person you care about that has left, eight leaf tips (taken from the unfolding leaves [fiddleheads], symbolic of a person taking the treatment getting away from someone) are chewed and swallowed repeatedly until the person forgets or feels better... The plant is used along with **Konok** (*Piper ponapense* C. DC.) and **Omp** (*Ipomoea littoralis* Blume) in sorcery to put a spell on a person who has put a spell on you, to make them feel sick as well..." (Balick et al. 2009: 235–236; also see Herrera et al. 2010: 20)

N. hirsutula leaves and "buds" have reportedly been used on Pohnpei as medicine for babies, which is believed to help them grow healthier, and it has been used as a medicine for emotional problems (Glassman 1952). More recently, the "fiddleheads" from *N. hirsutula* have been reportedly used for therapeutic purposes and to repel alien cockroaches (see www.comfsm.fm/~dleeling/ethnobotany/text/flora.html). Finally, *N. hirsutula* is said to have another name on Pohnpei: **Limenkasar**. This also refers to two other ferns, *Blechnum orientale* and *Pteris vittata*. At least one of these ferns is said to be used to forget a "lost lover."

N. obliterata is used on Pohnpei for a number of purposes as documented by Balick et al. (2009: 238):

> "This fern is used in *mwaramwar* (head garland)... The plant is used to decorate a church, party, or funeral when mixed with flowers... To treat depression four curled apices are chewed before sunrise... Alternatively, to treat depression, four shoots of the young leaves are eaten daily for four or eight days... Also, to treat depression, four shoots of the young leaves are crushed in the mouth and swallowed before the sunrise daily for four or eight days... To induce vomiting, the tip of the fiddlehead with brown scales is chewed along with **Marasau** and **Peinuhp** and swallowed (if vomiting is not induced within twenty minutes, then the person has been hexed)... To treat long labor (lasting four or five days), the young fiddleheads are chewed and swallowed and the baby will be born within an hour or so... To treat children's sickness, the stem is pounded, wrapped in cloth, squeezed in a cup of water, and consumed daily for four or eight days..."

Uses in Mwoakilloa and Pingelap: *N. acutifolia* is also used medicinally on Mwoakilloa Atoll to treat fever by rubbing leaves of this fern together and then rubbing "the crumpled leaves all over the body. No oil is used, just the leaves" (www.comfsm.fm/~dleeling/ethnobotany/healing_plants_presentations.html). On Pingelap, people use one or more *Nephrolepis* species "mixed with grass and used as fertilizer for taro" (Damas 1994: 231).

▸ **Tipaker** (P), **Dipaker** (Mw), **Dabaage** (Nu), **Baagee** (Ka)
Nicotiana tabacum

Description: Tobacco (*Nicotiana tabacum*) is a coarse, erect, annual herb with large oblong-ovate to lanceolate leaves 30–60 cm (12–24 in) long by 10–20 cm (4–8 in) wide. It is believed to be a hybrid of *Nicotiana sylvestris*, *Nicotiana tomentosiformis*, and possibly *Nicotiana otophora*. The big, odoriferous leaves of tobacco plants have pubescent surfaces, as do the inflorescences, which produce pink flowers (e.g., see Stone 1970: 519).

Distribution: Tobacco as a hybrid series of plants is probably native to tropical areas of western South America. Where it has been introduced on tropical islands, including areas of Micronesia, it can become weedy, sometimes "escaping into areas along roadsides, fence rows, fallow fields, and other disturbed habitats" (Wiggins and Porter 1971: 475).

Uses in Pohnpei: Balick and his colleagues (2009: 502), referring to other authors, reported that the use and importance of tobacco on Pohnpei is similar to what happened on other islands in Micronesia after tobacco was introduced by Europeans:

> "This species has a long history of use on Pohnpei, introduced in trading times during the early contact period (1830–1886) (Ragone et al. 2001). It was one of the most important items of exchange of that time, with a stick of tobacco being payment for a day's labor (Hanlon 1988). Today, it is one of the plants mixed with betel nut and chewed." (Also see Hezel 1984, A Brief Economic History of Micronesia; and Marshall 2013, Drinking Smoke: The Tobacco Syndemic in Oceania).

Problems associated with tobacco use: Tobacco use—whether through smoking, chewing, or other forms of consumption—poses serious health risks worldwide, and Micronesia is no exception. Both government and non-government organizations in the region recognize it as a deadly public health crisis. The remainder of this text contains statistical information and a number of excerpts, all taken from the 6th edition of the Tobacco Atlas (see www.tobaccoatlas.org). Together, they provide startling insight into the contemporary problems associated with tobacco consumption in the FSM, including Pohnpei.

> " Tobacco harms the health, the treasury, and the spirit of the Federated States of Micronesia. Every year, more than 156 of its people are killed by tobacco-caused disease. Still, more than 190 children (10–14 years old) and 9,490 adults (15+ years old) continue to use tobacco each day. Complacency in the face of the tobacco epidemic insulates the tobacco industry in the Federated States of Micronesia and ensures that tobacco's death toll will grow every year. Tobacco control advocates must reach out to other communities and resources to strengthen their efforts and create change."

In 2015, 20.8% of adults (15+ years old) were using tobacco daily in the FSM. Although fewer men smoke on average in the Federated States of Micronesia than in medium-HDI countries, more than 7,200 men still smoke cigarettes each day, making it an ongoing and dire public health threat. According to the same report, 1.92% of children were using tobacco daily in the FSM, indicating that more boys smoke in the Federated States of Micronesia than on average in medium-HDI countries. This is largely due to the fact that:

> " The tobacco industry markets its products aggressively to lower-income populations and youth in Federated States of Micronesia.

In addition, about 12.9% of people in the FSM were using smokeless tobacco daily, translating to about 9,100 persons, highlighting an ongoing public health challenge, including heightened levels of oral cancers.

There are societal harms associated with tobacco use in the FSM:

> " Although economic costs of smoking in the Federated States of Micronesia are not known, the total economic cost of smoking globally amounts to 2 trillion dollars, when adjusted for 2016 purchasing power parity (PPP). This includes direct costs related to healthcare expenditures and indirect costs related to lost productivity due to early mortality and morbidity."

Tobacco use also harms development in the FSM:

> " Buying tobacco robs families of the resources they may need to rise out of poverty. A smoker in the Federated States of Micronesia would have to spend 17.47% of their average income (measured by per capita GDP) to purchase 10 of the most popular cigarettes to smoke daily each year!"

Tobacco use also harms the environment worldwide, and in the FSM:

> " Cigarette butts are the most commonly discarded pieces of waste worldwide. It is estimated that 2.6 million tonnes of butts and packs wind up as toxic trash each year. This is roughly equivalent to the weight of 500,000 endangered African elephants."

Tobacco use also has a negative impact on noncommunicable diseases such as heart disease, cancer, chronic respiratory disease, and diabetes in the FSM.

Tobacco use also harms industry in the FSM:

> " The combined revenues of the world's 6 largest tobacco companies in 2016 was more than USD 346 billion, 89,950% larger than the Gross National Income of the Federated States of Micronesia. The industry is a powerful force that does not fear the actions of nation-states because of their extensive resources and global market power."

▸ **Parem** (P), **Param** (Mw)
Nypa fruticans

Description: The nipa palm, *Nypa fruticans*, is adapted to brackish marshes or swampy areas. The form of *N. fruticans* is unfamiliar for a palm plant, having long, upright leaves and underground, branching stems. Its rhizomatous stems, emerging from the mud of a stream or mangrove areas, grow out level to its substrate, except for its ascending leaves. Nevertheless, the nipa palm can attain a natural height of about 6 m (20 ft) or more along the inner edge of the mangrove forest, where the water is less salty. The erect fronds, 3–7 m (10–22 ft) long, have lengthy leaflets, consisting of about 42 pairs of segments or pinnae, that are mostly alternate, linear, acute, or attenuate. The uniquely structured flowers are borne near the ground (female flowers in a golden cylindrical spike and male flowers in a globe-shaped head). The fruits are syncarps, i.e., many individual small fruits within a common receptacle that are produced in a globular mass about 12–30 cm (5–12 in) wide.

Distribution: The native range of *N. fruticans* includes areas such as India, Sri Lanka, Southeast Asia, Indonesia, Malaysia, Philippines, Australia (Northern Territory, Queensland), Japan (Ryukyu Islands), the Solomon Islands, the Mariana Islands, and Micronesia from Palau to Kosrae (e.g., see Ellison et al. 2010c). *N. fruticans* is found along the landward edge of the mangrove forest, and it often forms dense groves at river mouths and other places where fresh water flows into swampy coastal areas. However, *N. fruticans* is not a "true" mangrove since it does not exploit truly littoral environments, nor can it endure inundation with undiluted saltwater for protracted periods. The nipa palm occurs most commonly where there is brackish water near the mouths of streams and rivers, and populations of these palms can extend far upstream into areas of permanent freshwater where tidal-influenced water-level fluctuations are able to disperse the ripe seeds. The seeds separate from the globular syncarp and float away on the tide, occasionally sprouting while still in the moving water. The nipa palm's trunk grows lying down, usually underwater. New plants grow vegetatively from each branch of it rhizomatous stems, often creating dense, monospecific stands. This palm species can be invasive in some regions, where it then becomes naturalized.

General uses: The nipa palm's long, feathery leaves are used as roof material for thatched houses or dwellings in areas where it grows. The leaves are also traditionally used in many types of basketry and thatching, and the seeds in the fruits are "edible when still young" (Stone 1970: 142). In fact, *N. fruticans* is a multipurpose plant with a great variety of uses throughout its large range, as the diversity of applications below indicates:

"Large stems are used to train swimming in Burma as it has buoyancy. On the islands of Roti and Savu, nipa palm sap is fed to pigs during the dry season. This is said to impart a sweet [flavor] to the meat. The young leaves are used to wrap tobacco for smoking. In the Philippines and Malaysia, the flower cluster (inflorescence) can be tapped before it blooms to yield a sweet, edible sap collected to produce a local alcoholic beverage called tuba, bahal or tuak. Tubas can be stored in *tapayan* (balloon vases) for several weeks to make a kind of vinegar known as *sukang paombong* in the Philippines and *cuka nipah* in Malaysia. Tubas can also be distilled to make arrack, locally known as *lambanog* in Tagalog and *arak* in Indonesian. Young shoots are also edible and the flower petals can be infused to make an aromatic tisane. Immature fruits—sweet, translucent, gelatinous balls [are] used as a dessert ingredient in Thailand, Malaysia, and Singapore. The nipa palm has a very high sugar-rich sap yield. Fermented into ethanol or butanol, the palm's large amount of sap may allow for the production of 6,480–15,600 liters (per year) of fuel per hectare. Sugarcane yields 5,000–8,000 liters per hectare (per year) and an equivalent area planted in corn would produce just 2,000 liters (per year) per hectare. The flower stalk of *Nypa fruticans* can be tapped, yielding a sweet sap. This sap can be drunk as is, boiled to produce a brown sugar or fermented to form alcohol or vinegar. The young fruits are often eaten as a desert in Singapore. *Nypa fruticans*, known as *rola*, is a dreaming or totemic plant for some Tiwi people. Large mud mussels (*Geloina coaxans*) that grow around the base of the stems in deep mud are eaten after roasting." (www.palmpedia.net/wiki/Nypa_fruticans)

Uses in Pohnpei: *N. fruticans* is called **Parem** in the Pohnpeian language. Unlike in many other islands where it is found, the nipa palm is not important for thatch on Pohnpei because the fronds of the large **Oahs** (ivory palm, *Metroxylon amiracum*) are preferred. This plant is said to be named after Parem, a small volcanic island in Pohnpei's lagoon, where it was first put on land when people brought it from across the sea. At that time, it may have been used in the sails of voyaging canoes. The fruits of **Parem** are used for medicine on Pohnpei; for example, according to Balick et al. (2009: 267), "To treat menorrhagia, a fruit [of the nipa palm, **Parem**] is pounded, wrapped in cloth, squeezed in 2 lit (0.5 gal) of water, and consumed daily for four or eight days..."

▸ **Kidi** (P), **Kajpwar** (Mw), **Kaesaepwar** (Pi), **Gahao** (Nu), **Ganniu** (Ka)
Ochrosia oppositifolia

Description: *Ochrosia oppositifolia* is a medium-sized tree that can sometimes reach heights of 15 m (50 ft) or more. Its bark is grayish and smooth. The leaves are obovate to oblanceolate in shape with 1–7 cm (0.5–3 in) long petioles. They have smooth margins, are arranged opposite one another, and are produced in whorled groups of 3–5. The tips of the leaves are rounded to broadly acuminate, and their upper surfaces are dark, glossy green, glabrous, and convex, while the lower sides are lighter in color. The leaves have prominent veins, including a distinct intra-marginal vein. The inflorescence of *O. oppositifolia* is a cyme up to 20 cm (8 in) long which produces many white, fragrant flowers with five petals and yellow centers. The elliptically shaped fruits, produced in pairs, are approximately 5–8 cm (2–3 in) long. The fruits turn yellow when mature and then "after falling to the ground, the fruit pulp decays, leaving a fibrous network" containing a single large seed (Falanruw 2015: 40, also see Kitalong et al. 2008: 121). Like other members of the dogbane family, Apocynaceae, parts of this tree produce a milky latex.

Distribution: The thick, fibrous mesocarp of the fruits of *O. oppositifolia* allows them to float and, consequently, they can be spread far and wide by ocean currents. Therefore, this widespread woody plant is found in many tropical island coastal forests from islands in the Indian Ocean to the Western Pacific, including all the major island groups in Micronesia except Kiribati (where it may have become extinct after humans arrived). Where stands of this species are still intact, the dense canopy of soft wood species creates heavy shade under which few other species can survive. Today, these sparse forest fragments are found only in the relatively moist islands, such as in some of the atolls of the Marshall Islands, where there are remnants of almost pure stands of tall, clear-trunked *O. oppositifolia*. This native coastal tree was probably much more common in Micronesia in the past, especially prior to the arrival of humans. Indeed, it was an important component of inland forests of the high islands of Micronesia, certainly in moist habitats, especially in limestone areas with sufficient rainfall. Although this tree is grows among many other species, it also occurs in almost pure stands on some islets of isolated atolls. These single-species forests are probably the result of a lack of competition from the few tree species in the area and of frequently stressful environmental conditions (e.g., recurring high winds, seismic wave destruction), which do not favor "weaker" species. In these nearly pure stands, *O. oppositifolia* trees are tall with clear, relatively wide trunks and dense, rounded crowns of large dark green leaves. The ground under this tree's single-stand forests is commonly covered by a dense stand of seedlings of the same species, about 10–20 cm (4–8 in) tall, and apparently growing slowly, if at all.

Uses in Pohnpei: *O. oppositifolia* is known in Pohnpeian as **Kidi**. According to a local male informant, on Pohnpei, the wood of *O. oppositifolia* "is used to make paddles for canoes... and for firewood" (see Balick et al. 2009: 331; these authors refer to **Kidi** as *Neiosperma oppositifolium*, a synonym of *O. oppositifolia*).

▸ **Kadireng** or **Kadiring** (P), **Kadring** (Mw), **Dehko** (Pi), **Aloagi** (Nu) (Ka)
Ocimum tenuiflorum

Description: *Ocimum tenuiflorum* (also known as *O. sanctum*) is an aromatic, branching herb in the mint family that grows erect, with woody lower branches. It is known in English as "sacred basil." Its stems are 4-angled in cross-section. The leaves are simple, pointed, arranged opposite each other, and produce an agreeable odor. Its flowers are small and commonly clustered in racemes at branch tips or leaf axils. The fruits contain four one-seeded nutlets.

Distribution: The genus *Ocimum* is a member of the Lamiaceae or mint family that includes at least 150 species. *Ocimum* species are widely distributed, with the greatest number of species in Africa. The best-known species are the strongly aromatic herb *O. basilicum* (Thai basil) and *O. gratissimum* (African basil), as well as the medicinal herb *O. tenuiflorum*, also known as *O. sanctum* (sacred basil). *O. tenuiflorum* is an annual or short-lived perennial plant that has been used as a culinary and medicinal herb for centuries. However, *O. sanctum* is considered by many to be an agricultural and environmental weed. *O. tenuiflorum* is an introduced plant in Micronesia but is native to tropical Asia and is also now common in many tropical regions. It is often planted in house gardens and occasionally escapes from cultivation. As an early European introduction to Micronesia, this strongly scented member of the mint family is now found in the Marianas, Caroline, and Marshall Islands.

General uses: *Ocimum* species (the basils) are frequently referred to as the "king of the herbs," especially in South Asia, where the common use in daily ceremonies of this plant known as **Tulsi** is a manifestation of traditional Ayurvedic medicinal wisdom, providing what is reportedly "an example of ancient knowledge offering solutions to modern problems." Therefore, of all the plants used in this traditional Hindu therapeutic system, "none has a status comparable to **Tulsi** or holy basil (*Ocimum tenuiflorum*)" (Cohen 2014: 251). Indeed, this revered species is linked with wholesomeness and is associated with purification of the mind, emotions, and the physical body:

> " Many Hindu households have a **Tulsi** plant growing in a brick column either in front or in back of their home. Every morning the eldest female in the family lights a lamp, waters the **Tulsi**, and then offers a prayer while walking in a circle around the **Tulsi** plant. Although the **Tulsi** is edible, Hindus will not eat **Tulsi** as the plant is considered sacred. **Tulsi** is present at weddings and funerals and is believed to bring blessings upon the household. The plant is a protector of the family. Elsewhere in southeast Asia, **Tulsi** is used as a medicinal plant, thus the plant has both spiritual and healing uses." (www.comfsm.fm/~dleeling/ethnobotany/text/#chapter10)

Pattanayak et al. (2010: 95) summarized the therapeutic value of sacred basil (*O. tenuiflorum*) and the role of ethnobotany in the search for and confirmation of traditional medicinal knowledge:

> " In traditional systems of medicine, different parts (leaves, stem, flower, root, seeds and even whole plant) of *Ocimum tenuiflorum* Linn. have been recommended for the treatment of bronchitis, malaria, diarrhea, dysentery, skin disease, arthritis, eye diseases, insect bites and so on. The *O. tenuiflorum* L. has also been suggested to possess anti-fertility, anticancer, antidiabetic, antifungal, antimicrobial, cardioprotective, analgesic, antispasmodic and adaptogenic actions. Eugenol (1-hydroxy-2-methoxy-4-allylbenzene), the active constituents present in *O. tenuiflorum* L., have been found to be largely responsible for the therapeutic potentials. The pharmacological studies reported in the present review confirm the therapeutic value of *O. tenuiflorum* L. The results of the above studies support the use of this plant for human and animal disease therapy

and reinforce the importance of the ethno-botanical approach as a potential source of bioactive substances."

Uses in Pohnpei: On Pohnpei, *O. tenuiflorum* (syn. *O. sanctum*) is known as **Kadireng** or **Kadiring**. This introduced herb species is reportedly planted on either side of a path or driveway and it is believed by some that it will prevent evil spirits from passing. Therefore, **Kadireng** may be cultivated as a pair of plants in front of some Pohnpeian homes. Sacred basil is reportedly used in Pohnpei as a "seasoning, steam bath for congestion, and by some for protection," e.g., "evil cannot pass between two plants planted on either side of a path" (www.comfsm.fm/~dleeling/ethnobotany/text/flora.html). A student at the College of Micronesia on Pohnpei reported the following uses of this widely known herb species:

> "**Kadiring** is useful in many ways... the leaves are used for cooking. The leaves are also used for curing headaches and colds [through] *umwulap*, steam bath... [and] the [leaves] are edible." (www.comfsm.fm/~dleeling/botany/1999/vhp/ocimum_sanctum.html)

Uses in Nukuoro: The bright green leaves of *O. tenuiflorum* are reportedly collected and cooked with turtle on Nukuoro Atoll (Drummond et al. 2016).

▸ Kalak *(P)*
Palaquium karrak

Description: *Palaquium* is a genus that includes about 120 species of trees in the Sapotaceae family, which is characterized by producing white sap and, in the case of some species, yielding edible fruits. The leaves of *Palaquium* species are typically spirally arranged and often clustered near twig ends. Their white flowers are mostly bisexual, though some unisexual flowers have been reported. Greenish fruits are one or two-seeded with rare instances of several seeds. *Palaquium karrak*, known in Pohnpeian as **Kalak**, is a large endemic tree in the rainforests of Pohnpei (Costion and Lorence 2012: 88). This canopy tree can reach up to 15 m (~50 ft) in height with a maximum trunk width of about 70 cm (30 in). Its bark is soft, brown, and flaky, producing a milky sap. The leaves are generally yellowish green, and the flowers are fragrant, with a green calyx and yellowish-white corolla. The fruits are green and ellipsoid.

Distribution: The geographical range of the many *Palaquium* species reaches from India across Southeast Asia, New Guinea, and Australia to the Western Pacific islands. Within their general range, *Palaquium* species are mostly found in the Philippines and Borneo. Many species are recorded in the Malaysian states of Sabah and Sarawak. The most extensive remaining native rainforests in the FSM exist on Pohnpei, where this forest type occurs throughout the uplands approximately 1000–2000 ft in elevation. Dominant species in this Eastern Caroline, Pohnpeian Island forest type include *Campnosperma brevipetiolata* and *Elaeocarpus carolinensis*, with *Parkia korom*, *Palaquium karrak*, *Myristica hypargyraea*, *Cinnamomum carolinense*, *Ficus tinctoria*, *Barringtonia racemosa*, *Terminalia carolinensis*, and *Cynometra ramiflora*.

Uses in Pohnpei: **Kalak** produces delicate flowers which drop off quickly. The woody parts of the **Kalak** tree have many purposes on Pohnpei, and its flowers are also used for adornment:

> "The wood is used for lumber; the hardwood is carved into wooden plates and to make knife handles... The wood is also carved into rice scoops... The plant is used in house construction... The wood is used for firewood, flooring, and timber... The plant is used in *mwaramwar* (head garlands)..." (Balick et al. 2009: 498; also see US National Herbarium specimen 3558426 for original collection notes, 2 Nov 2001, R. Gallen 115)

▸ PANDANUS SPECIES

The genus *Pandanus* includes numerous species, many of which have significant use to people in tropical Asia, Australia, and the Pacific Islands. In Pohnpei State and more broadly in Micronesia, people have a long history of living with and using plants in this genus. This is especially true for *P. tectorius* (**Kipar** in Pohnpeian), which is widely distributed throughout the region and occurs in an impressive range of cultivated and wild varieties. In the past, some of these were described as separate species but are now understood as cultivars of *P. tectorius*. Only a few species—*P. conoideus*, *P. dubius*, and *P. patina*—are indeed distinct and we will examine them first in alphabetical order, though our focus remains on *P. tectorius* and its stunning diversity of cultivars.

▸ **Mwatal** (P), **Mwasal** (Mw)
Pandanus conoideus

Description: This short *Pandanus* species develops many prop roots with prickles, along with short, sharp spikes on the truck and its branches. *P. conoideus* [syn. *P. cominsii*] trees may produce up to 8 or 10 main branches and grow to heights of about 5 m (15 ft) tall. The long leaves are 1–2 m (3–6 ft) long and about 10 cm (4 in) wide with thorns along their margins. The oppositely arranged leaves are twisted, producing a spiral appearance. The hard, rough-surfaced (lumpy or spiky) compound fruit (syncarps) contain numerous, bright red (infrequently yellow) drupes and can weigh as much as 8 kg; these grow at the ends of the branches between the leaves and at maturity are up to about 1 m (3 ft) in length with three straight leaf-like bracts along the edges.

Distribution: *P. conoideus* is found in parts of Southeast Asia, such as the Maluku Islands in Indonesia, but is semi-domesticated in the highlands of New Guinea, where it may be native or introduced by people. According to Stone (1982), *P. conoideus* is endemic to the island of New Guinea and nearby islands. It reportedly grows throughout much of Papua New Guinea from sea level up to about 2000 m above sea level (Jebb 1992), becoming relatively common above 500 m. In Micronesia, *P. conoideus* is found in the lowlands or swamp forests on some of the high islands. In such environments, it provides the overall greatest understory canopy cover of all *Pandanus* species found in the forests of the FSM (see Donnegan et al. 2011: 37).

General uses: In the highlands of New Guinea, *P. conoideus*, with its many cultivars, is one of the most important cultivated tree species in plantations and near homes; its large, conical-shaped, compound fruits are often sold along roadsides and walking tracks, as well as in urban markets. In this Melanesian region, the fruits of *P. conoideus* are generally cored of the soft interior material and then steamed in an earthen oven, boiled, or roasted. Then, after cooking, they are eaten by directly sucking off the edible mesocarp, or they are consumed in the form of a pleasant-tasting, oily, vitamin-A-rich, ketchup-like sauce that is eaten with greens, pumpkin, or bananas. In addition, the oil extracted from the fruits is also "used as hair and body oil, polish for arrow shafts, and for paints and dyes [and the] leaves sometimes used for thatching" (Clarke and Thaman 1993). The large fruit of *P. conoideus* is also eaten in areas of Indonesia such as the Moluccas, Seram, and Halmahera. In these areas of Indonesia and New Guinea, the useful, red-fruited *P. conoideus* trees are normally planted from suckers or cuttings, particularly cuttings extracted from the ends of branches where the adventitious roots or "suckers" are produced faster (see Cook 2016: 66–70, for an exemplary, detailed description of the cultivation and utilization of *P. conoideus* by a highland New Guinea people, the Amungme of the Tsinga Valley).

Uses in Pohnpei: The Pohnpeian word **Mwatal** refers to both *Freycinetia ponapensis* and *Pandanus conoideus*, as well as to the dried leaves of *P. conoideus* "commonly used for making sleeping mats" (Rehg and Sohl 1979). To prepare the leaves of *P. conoideus* for weaving purposes on Pohnpei, older long leaves are specially selected and then softened over a fire for about an hour until they become whitish; following a day of waiting, these fire-dried leaves are then smoothed flat and soft by rubbing them across a tree sapling. "The leaves are thus prepared to weave into mats used on the ground for sleeping or for guests of higher rank to sit on as a sign of respect..." (Balick et al. 2009: 291, see under *P. cominsii* Hemsl. var. *micronesicus* B.C. Stone). The dried **Mwatal** leaves are also reportedly used to make skirts in Pohnpei. In addition, this species is reportedly used as a source of medicine.

▸ **Bagua** (Nu), **Baehoo** (Ka)
Pandanus dubius

Description: *Pandanus dubius* is described as "a very distinctive species, individuals of which are usually trees up to 10 or 15 m tall, and bear relatively broad leaves of a thick leathery texture, ending in a rather abruptly incurved, shortly caudate tip. The fruits of this species are also very characteristic, ripening to a purplish green color. The individual keys contain one to five fairly large white seeds near the base, enclosed in a relatively thin woody tissue. These seeds are rich in endosperm, edible, and have a pleasant flavor much like that of young coconut meat" (Stone 1966b: 431–432). *P. dubius* can be differentiated from the more common *P. tectorius* by the following: its long peduncle, a stalk supporting an inflorescence, which is 60–80 cm (24–32 in) in length; its large, white, edible, tasty and rounded seeds about 1–1.5 cm (0.4–0.6 in) in diameter, which are situated at the base of the phalanges (individual fruits); its trunk, which is covered with sharp prickles; and its roots, which are covered in vertical rows of prickles (e.g., see Thompson et al. 2006).

Distribution: *P. dubius* is reportedly native to insular Southeast Asia, New Guinea, the tropical Western Pacific islands of Melanesia and Micronesia, and possibly also to the Andaman and Nicobar Islands. *P. dubius* occurs "in coastal forests, including beaches and rocky shores in Southeast Asia and the Western Pacific region" (Thompson et al. 2006). In Pohnpei State, this species commonly occurs only in the two southernmost atolls, Nukuoro and Kapingamarangi.

Uses in Nukuoro and Kapingamarangi: Stone (1966a: 132) noted this species on Nukuoro and wrote that it is "very similar to *P. compressus*, found in Kapingamarangi." At present, *P. compressus* is considered to be a synonym for *P. dubius*, so, until proven otherwise, only two *Pandanus* tree species are believed to exist in Nukuoro and Kapingamarangi: *P. dubius* (**Bagua** in Nukuoro and **Baehoo** in Kapingamarangi) and *P. tectorius* in its many varieties (collectively known as **Hala** in Nukuoro and **Binu** in Kapingamarangi). *P. dubius* is said to be cultivated occasionally for its edible seeds, but more so for its thick leaves that are used to make baskets and floor mats. According to Stone (1966a: 132), although the seeds of *P. dubius* are eaten on Kapingamarangi, it is probably "most highly valued for the leaves, which are preferred for mats because of their thick texture, which makes a slightly springy mat, and their waxy surfaces, which tend to shed water."

Stone (1966b: 431–432) provided some additional detail about the similarity of *P. dubius* and *P. compressus* (before they were understood to be a single species) and their unusual occurrence in both Nukuoro and Kapingamarangi, but not significantly anywhere else in the islands of Pohnpei State:

> "[*P. compressus*] is found elsewhere in Micronesia in Palau, and also in the Solomon Islands, New Hebrides, and Shepherd Islands. It is very similar to *Pandanus dubius* Sprengel, a plant common in the Marianas Islands, where the seeds are frequently eaten, as was pointed out long ago by Chamisso [1836]. Curiously, it is *Pandanus dubius* which is found in Nukuoro Atoll, just to the north of Kapingamarangi, where it plays a similar role ecologically, is used occasionally for its edible seeds and frequently for its leaves [for weaving mats], and is called 'Pakóa', a name clearly cognate of the Kayangel (in Palau) name 'Pōk', the Chamorro (in Mariana Islands) name 'Pahong'. [as well as the Kapingamarangi name 'Pahó; Paehó', see Stone 1966a: 131]

▸ **Piht** (P)
Pandanus patina

Description: *Pandanus patina* is an endemic native tree in the upper wet forest of Pohnpei, with relatively thick prop roots, seldom exceeding 6 m (~20 feet) in height, but normally about 4–5 m (~12–15 ft) tall when mature. *P. patina* usually dominates summit areas located around the tallest peaks of Pohnpei, where it grows in the high forest areas (Stone 1967b: 109), and it reportedly never occurs in lowlands on Pohnpei (Kanehira 1935: 51). Although its leafy crowns cover up to 98% of the upper story foliage in its high mountainous habitat, this special *Pandanus* species (*P. patina*) needs more botanical and ecological study.

Distribution: *P. patina* is found only in Pohnpei, where its local name **Piht** is also used to refer to pandanus leaves in general. According to Stone (1960: 409), this relative of **Kipar** (*P. tectorius*) is a "beautiful species" and "one of the most perplexing of the [Pohnpeian] flora if the question of origin is raised." *P. patina* is only found in several very wet upland summit areas of Pohnpei, usually in pure stands with relatively few understory plants or epiphytes (air plants) growing on it (except some mosses). The fallen leaves of *P. patina* usually cover much of the floor of the upland forest underneath these endemic trees.

Uses in Pohnpei: **Piht** leaves are used to weave long-lasting, high prestige mats called *lohsalap* (or *pereklap* in honorific speech). These **Piht** mats require a relatively long time to make and become the distinctive mats upon which a high-ranking person will sit during special occasions. According to Kanehira (1935: 51), "the seeds [of *P. patina*] are edible." Stone (1960: 409) also referred to the edible seeds contained in the magenta-colored, "one-carpellate drupes" (fruits) of *P. patina*, which he noted are "comparable in flavor to coconut, and sometimes important to hunters or travelers."

▸ **Kipar** (P) (Mw) (Pi), **Hala** (Nu), **Binu** (Ka)
Pandanus tectorius

Description: *Pandanus tectorius* is a wild-growing or cultivated tree, in English referred to simply as pandanus after the genus name, or "screwpine" because of the twisted form in which the leaves are produced on a branch, and the gross resemblance of its many-sectioned syncarp (i.e., compound fruit) to a large pineapple-like fruit. *P. tectorius* can be described as a wide-branched shrub to relatively small tree that grows up to heights of about 6–9 m (~20–30 ft) or more. An individual tree can live for about 60 years. The stout trunk has branches that grow at wide angles to the trunk and produce many aerial prop roots. Its long, strap-like leaves are spirally arranged towards the ends of the branches and leave a coiled pattern on the trunk when they fall.

> "As [*P. tectorius* trees] age, they produce smaller leaves and infructescences [a group of fruit arranged on a stem in a characteristic pattern] and prop roots grow out from the terminal leaf clusters. If these branches should break off, perhaps during strong winds, the entire terminal cluster of leaves and roots can form a new plant… However, the resulting plant may have a different genetic makeup and therefore different leaf or fruit qualities than the mother plant." (Gallaher 2014: 98)

The leaves of *P. tectorius* can grow up to 3 m (~10 ft) in length with serrated margins (thorny leaf edges). Most varieties have spines along the edges and on the midribs of the leaves, but spineless and variegated forms also occur. *P. tectorius* is a dioecious species, and thus has male and female reproductive organs in separate individuals. Male (staminate) trees produce large clusters of tiny, fragrant flowers surrounded by white- to cream-colored bracts. These flowering clusters are about 30 cm (~1 ft) long. Female (pistillate) trees produce the large, compound fruit as an aggregate of many angular, wedge-shaped sections or keys (drupes or individual fruits, also known as phalanges). The individual fruits turn orange as they mature. A large single fruit or key may be 7.5–10 cm (~3–4 in) long, while in width, 4–5 cm (1.5-2 in) at the outer end, and 2–2.5 cm (0.75–1 in) at the inner end near the core. The external surface of an individual large fruit may be green or yellow when ripe, but the surface of the inner edible end is always yellow-orange in color. The inedible varieties have fruits which, though smaller, look just like the edible types; however, they are extremely astringent and bitter. The main palatable fruit part of those *P. tectorius* cultivars is its fleshy base. The fruits' reddish-brown seeds are obovoid, ellipsoid or oblong and are 6–20 mm (0.25–0.8 in) in length (e.g., Miller et al. 1956; Murai et al. 1958; Thomson et al. 2006).

Distribution: *P. tectorius* is the most widespread of more than a thousand species in the family Pandanaceae. This extremely variable species can also be considered a group of species, or a "species complex," which, according to Gallaher et al. (2015: 20) "originated in Eastern Queensland within the past six million years and has spread to encompass nearly the entire geographic extent of the family from Africa through Polynesia." It is most likely that the extensive dispersal of *P. tectorius*, in its many forms, has occurred over time via long-distance ocean current movements and aided by a combination of traits such as asexual reproduction, in which seeds are produced from unfertilized ovules (agamospermy), wind pollination (anemophily), and the multitude of seeds produced in buoyant fruits which allow establishment of colonizing populations in remote locations. Therefore, much of the widespread distribution of *P. tectorious* in the tropical Pacific, including Micronesia, is apparently natural. This highly variable species complex occurs primarily and spontaneously in semi-natural vegetation in more or less coastal habitats throughout the tropical and subtropical Pacific, to a large degree because these woody plants can withstand drought, strong winds, and salt spray. However, its presence on different islands can be attributed to the conscious transport and introductions of this valuable, multipurpose tree by humans. People certainly carried cultivars with desirable traits to many remote islands over hundreds if not thousands of years. Indeed, *P. tectorious* is commonly cultivated or "protected in garden areas and homegardens, sometimes in monocultural patches or as border plantings," along with other useful species, including, for example, breadfruit trees (*Artocarpus altilis*) and sweet taro (*Colocasia esculenta*) (Thomson et al. 2006: 7). On the other hand, *P. tectorius* is a typical native tree along tropical shorelines of wet coastal areas in a wide range of geographical areas where it is an early colonizing species. In fact, *P. tectorius* is often among the first plants to colonize the shores of tropical Pacific islands. For example, in Southeast Asia, *P. tectorius* was already well established on Rakata Island just 14 years after the huge 1883 eruption of Krakatoa volcano in Indonesia, which "sterilized" the nearby island of Rakata; and *P. tectorius* was recorded on Anak Krakatau in 1934, only 3 years after the initial emergence of that younger volcanic island (Whittaker et al. 1989). Although *P. tectorius* commonly grows in moist coastal locations, it can also be found in open grasslands on valley slopes up to an elevation of about 600 m (~2,000 ft), as well as in scrub secondary vegetation on rugged, fossilized, coralline raised reef known widely as "*makatea*" (e.g., see Wagner et al. 1999, Franklin and Merlin 1992, Merlin 1991).

Although this important tree species is regarded as native in Micronesia, early Pacific islanders probably brought useful, artificially selected varieties (cultivars) of *P. tectorius* to these islands originally from one or more areas of Southeast Asia or Near Oceania. Over many generations of cultivation and exchange of clones beyond as well as within Micronesia, selection of preferred types for cultivation has produced many different cultivars of *P. tectorius*. Therefore, this is especially so on atolls, where *P. tectorius* is a dominant species along many sandy shorelines. As the Marshallese proverb points out: *Liok tūtbok*, "Pandanus roots grab deep through sand to hold firmly." Although it grows readily from seed, *P. tectorius* is also widely propagated from branch cuttings by local people in agricultural areas and also in home gardens. *P. tectorius* trees are grown from cuttings for a variety of natural products such as edible fruit and broadly useful long pliable leaves; however, these cultivated *P. tectorius* trees rarely produce seed, while those that sprout naturally from seed without human help and do develop into mature plants are not usually useful for food. This is a strong indication that people have been reproducing *P. tectorius* for many generations by planting cuttings (asexual reproduction), and in some cases from seed, to encourage certain preferred resources provided by these cultivated trees. Within the limited number of native plants found in the atoll and reef island environments, *P. tectorius* is one of only a relatively few important species that may produce edible plant material. *P. tectorius* is found as an assumed indigenous species on Pacific islands in Polynesia, Micronesia, Melanesia, and as far west as northern Australia.

General uses:

> "The genus *Pandanus* includes many species that are useful to people around the world, and humans in the remote oceanic islands such as those in Micronesia have a long history living near and using plants in this genus, especially *P. tectorius*. In addition to the leaves, which were and continue to be the most widely used natural fiber material in Oceania, nearly every part of the plant has important uses." (Gallaher 2014: 100, underlined for emphasis).

Among peoples living on tropical Pacific islands, the pandanus tree (*P. tectorius*) has long had multipurpose importance, second only perhaps to coconut (*Cocos nucifera*) for its cultural, health, and economic significance; this is especially, but not exclusively, the case on atolls.

> "In Kiribati, pandanus is called the "tree of life" as it provides food, shelter and medicine. In the Marshall Islands, it is called the "divine tree," like coconut, because of its important role in everyday life. Pandanus is also an important staple food in the Federated States of Micronesia..." (Secretariat of the Pacific Community 2006, underlined for emphasis).

Indeed, the large, nutritious compound fruits of this pandanus species have been a staple food, providing as much as 50% of the daily energy consumption in some eastern islands of Micronesia, including the FSM, Marshall Islands, and Kiribati (see Miller et al. 1956, Englberger et al. 2003c, 2003f, Thomson et al. 2006: 16). The chief edible part of this pandanus is the fruit, especially the fleshy base of the keys (botanically, also known as the phalanges or individual fruits in the compound syncarp). Although pandanus does grow from seed, the fruit rarely produces fully grown trees. In addition, as noted above, those *P. tectorius* trees that sprout naturally from seed without human help and eventually develop into mature plants do not usually produce edible fruits. For example, these self-sown pandanus fruits contain irritating oxalates, although normally at low levels, but enough to be extremely astringent and bitter (Miller et al. 1956: 3). In order to maintain a population of *P. tectorius* trees that produce edible compound fruits, Pacific islanders will cut the terminal clusters of leaves off older, mature pandanus trees that yield edible fruits because they produce prop roots, and commonly these cuttings are used to propagate new trees vegetatively: "This method of propagation ensures that the resulting tree will have properties similar to the parent tree" (Gallaher 2014: 98). In general, vegetatively reproduced *P. tectorius* trees produce fruit about 2–4 years after a cutting is planted.

Some Micronesian varieties of pandanus fruits are eaten raw, some cooked, and some may be consumed either raw or cooked. Besides serving as a food, when eaten raw, the fibrous fruit of the edible varieties help clean the teeth, acting as a natural floss. Raw or cooked pandanus fruits contains 14–18 percent carbohydrate with no fat, along with important vitamins and minerals, and therefore they provide highly nutritious and healthy food:

> "Cooked and processed, pandanus fruits of *P. tectorius* can be preserved as a paste or a powder; and in a preserved form it can last for years. Kayser (1936) pointed out that food made from processed *P. tectorius* could be stored in houses on Nauru Island for up to six years after complex cooking and drying of the ripe fruits. Thus, cooked pandanus fruits can be a good emergency food source, and indeed it has been taken on board canoes in the past for use as sea rations. Traditionally, the starchy fruit pulp of *P. tectorius* is preserved for future use in different ways in Micronesia. For example, the fruits or keys may be boiled and baked, after which the pulp is extracted, processed and dried into an edible paste for later consumption. The texture and appearance of pandanus paste has been likened to the fruits of date palms; and even though pandanus cultivars with large quantities of pulp are normally favored, taste or flavors differ between cultivars." (Thomson et al. 2006: 16)

In recent decades, the consumption of edible *P. tectorius* fruits has decreased in some areas of its use in Micronesia, for example, on some of the high volcanic islands, and in Nauru, where the fruits of this species had previously been a main traditional staple food. However, *P. tectorius* fruit still remains an important food source in other regions of Micronesia where "adults may commonly consume 20 fresh keys or about 1 kg (2.2 lb) of fruit per day" (Thomson et al. 2006: 16). Those varieties of *P. tectorius* with edible fruits provide a valuable source of vitamin A and contain about the same amounts of calcium, phosphorus, and iron as peaches. Edible fruits of *P. tectorius* also have small amounts of thiamine, riboflavin, and vitamin C (e.g., see Miller et al. 1956; Murai et al. 1958). More recent research by Englberger et al. (2009c) sampled the nutritional value of 11 pandanus cultivars collected from Mwoakilloa and Kapingamarangi atolls in Pohnpei State; Englberger and et al. also sampled the dried pandanus paste "packaged traditionally in dried pandanus leaves and tied tightly with hand-woven coconut rope" from the Marshall Islands, which was the first time that this paste preparation (known as *mokwaņ* in Ratak chain and *jāānkun* in Rālik chain) had been assessed for its nutritional content. The analysis by Englberger et al. confirmed that the selected cultivar fruits they tested contained substantial concentrations of provitamin A carotenoids (110–370 mg b-carotene/100 g) and total carotenoids (990–5200 mg/100 g), while the Marshallese pandanus paste contained 1400 mg b-carotene/100 g, 5620 mg total carotenoids/100 g, and ten vitamins (including 10.8 mg/100 g vitamin C). Note: Englberger et al. (2009b) suggested strongly that "these [pandanus] cultivars and foods should be promoted in Micronesia and possibly elsewhere in the Pacific in order to reduce vitamin A deficiency and provide further health benefits and enjoyment."

In addition to use as food and medicine, other uses of this *Pandanus* species have and continue to be very important more or less throughout Micronesia; for example, its trunk and stems are used in building construction, and leaves are used to weave hats, mats, and walls of thatch as well as to make rope and aid in the construction of canoes. Barrau (1961: 57) also pointed out that "the heart and extreme tips of the young aerial roots of *Pandanus tectorius* are utilized as food." The region-wide traditional importance of this plant as a "keystone species" was summarized by Gallaher (2014: 94):

> " [*Pandanus tectorius* has] had a profound effect on the people of Oceania. Their use of [sails made from its leaves] enabled them to move beyond the reef and become an ocean-voyaging people capable of exploring the vastness of the Pacific. In addition, every part of this tree has uses that have been critical to the cultural heritage and survival of the people of this region."

Uses and cultivars in Pohnpei: The longtime College of Micronesia instructor and student of Pohnpeian flora and its traditional and modern use Dana Lee Ling offered this overview of the importance of *P. tectorius* in Micronesia in general, and Pohnpei in particular:

> " *Pandanus tectorius* is one of the most useful trees in Micronesia. The leaves, which are the most useful part of the plant, are woven or plaited into mats, thatch, sails, baskets, hats, local fans, marmars [garlands], handicrafts, and many other items. Tips of prop roots are eaten and are often used as native medicine in Pohnpei and maybe in other Micronesian islands as well. The fruit is a major source of food in Micronesia especially on the atolls. They can be eaten raw or cooked. Either way, it is very healthy. Besides serving as a food, its fibrous helps clean the teeth, acting as a natural dental floss." (www.comfsm.fm/~dleeling/angio/pandanus.html)

Pandanus tectorius is known as **Kipar** in Pohnpei, as well as in Mwoakilloa and Pingelap. Besides serving as an occasional food and medicinal resource, parts of **Kipar** are reportedly used as a dye source (e.g., see Balick et al. 2009). Furthermore, on Pohnpei the dried leaves of **Kipar** are traditionally woven into hats, baskets, clothing, and other items such as sleeping mats (*pitenlos*, e.g., *pitenlos pwetepwet kohsang ni piht mour*), with the "white strips of a sleeping mat [being extracted from] green pandanus leaves" (see Rehg and Sohl 1979).

The main fruiting season of **Kipar** pandanus on Pohnpei is reportedly from November to January. Traditionally, pandanus fruits have served as one of three main "seasonal *nohpwei*" (plant offerings), an obligatory tribute of a first fruit or *mwowe* presented to the chiefs (Anson and Raynor 1993: 138). Although the consumption of edible parts of certain *Pandanus* cultivars on Pohnpei does occur, Stone (1966b: 430–431) argued that such use of **Kipar** cultivars (see p. 423) as food on Pohnpei (as well as on the volcanic islands of Kosrae and Chuuk) "could in virtually all instances be traced to immigrant populations residing on those islands which retain a good deal of the culture of the atolls from which they had come." Stone pointed out, for example, in Pohnpei alone, there were and still are:

> " ... resident immigrant populations representing Kapingamarangi, Nukuoro, Mokil [Mwoakilloa], Pingelap, Ngatik [Sapwuahfik], and Mortlock Islands people [and the] presence of their "own" cultivars had led, however, to the erroneous suppositions that the [Pohnpeians] themselves made considerable use

of edible pandanus, and that these atoll cultivars represented wild [Pohnpeian] species of *Pandanus*."

Uses in Mwoakilloa: Murphy (1950: 71) referred to the "numerous" uses of **Kipar** on Mwoakilloa with "first place" being the use of its leaves as prime thatching material, with its second most important use (of some varieties) as a food source; Murphy also noted that the "leaves are used in making mats, hats, and various other handicraft, and trunks of pandanus trees serve as crosspieces in all boat houses since they have the virtue of not sagging." Glassman (1953: 298) also provided some details and referred to the **Kipar** tree as a very important plant on this atoll, indicating that the "Nuts [seeds] of some of the varieties are edible, and leaves of most of the varieties are used in thatching."

In a recent college class report, a student from Mwoakilloa presented the following extensive comments regarding past and contemporary use and the cultural significance of *P. tectorius* on Mwoakilloa Atoll:

> "The pandanus tree is a very important tree to the Mwokillese people. To this day, the pandanus tree still holds special meaning to them. Because of this tree, people gather together for feast, activities, or just to enjoy the day. They use this plant to build houses, and to weave mats, hats, even masks for the warrior dances. They also use this plant as a means of food and, in the old days, dental floss. All the parts of the tree [have their] own unique use. The bark, for example, was used in the old days to build houses for the people of Mwokil. They thought that this was the strongest tree that could support their houses. It was used as the foundation to the house. Then they completed it by using the leaves to make thatched roofs. But then again that is not the only use for the leaf. The people had a lot of warrior dances that needed masks to be used. Since there were no cardboard boxes to make masks with and no stores to buy them at, they had to make the masks locally. These, the masks, are held dear to the old people who are still alive today and their kids. The mats that they slept on and the hats that they wore to block the heat of the sun were also made from the leaves. The fruit also comes in handy. Back then, they had a season for the pandanus tree. Whenever the fruits started getting ripe. They would harvest the crop and gather together to make whatever kinds of recipes they could. They would boil it and add grated coconut to it, or make a special plate from it and all sorts of things that we, now-a-days, don't do anymore. In these days, the Mwokillese on Pohnpei would look forward to going down to the Mwokil Atoll. The Mwokillese on this island don't weave mats because they are too caught up with the western technologies on this island. If they don't have mats they buy them. This is why it never stops to excite them thinking about all those mats that they know would be waiting for them. Not only are they long-lasting than the ones in the stores, but they are also FREE. Each lady would be very proud of her mat if she got one. They would brag about how many they have or how new it looks or just because they have one. The Mwokil Atoll uses these mats as their only means of gifts to everyone and anyone who visits the atoll. The fruit attracts a lot of visitors to where it grows. When it is ripe, all family members gather together to enjoy the fruits and, of course, the day. These are only just some of the reason why the pandanus tree is so special to us." (S.L. from Mwoakilloa, www.comfsm.fm/~dleeling/angio/pandanus.html)

The Mokilese–English dictionary mentions a lot of pandanus-based dishes, such as *pehru* (boiled pandanus juice with coconut milk), *rodma* (cake from taro and pandanus), and *mahdailik* (pandanus baked with taro). The latter is explained in the following food recipe from Mwoakilloa Atoll published by the Secretariat of the Pacific Community (2006: 5):

> "Recipe for pandanus fruit mixed with giant swamp taro [dish called *mahdailik* from Mwoakilloa Atoll]. The ingredients: Fresh ripe pandanus fruit, Coconut cream, Syrup prepared from sweet toddy or imported, refined sugar, Giant swamp taro (*Cyrtosperma*), boiled or baked. Recipe instructions: 1. Boil pandanus keys; 2. Scrape out the pulp; 3. Add coconut cream; 4. Add sweet toddy syrup or sugar and mix well; 5. Grate the cooked giant swamp taro; 6. Form the grated taro into balls and put into a baking tin; 7. Add the pandanus mixture; 8. Bake and serve."

When traditional sailing canoes were still made and used on Mwoakilloa Atoll, the use of **Kipar** leaves for producing durable and effective mat sails was of great practical and customary significance:

> "The sailing canoe, which is a single outrigger canoe equipped with a triangular sail, is traditionally the only water craft for fishing, transporting, and voyaging on Mwoakilloa, a low coral atoll in the Eastern Caroline Islands. Before wooden boats became prevalent in the 1960s, canoes (*war*) held vital importance, and every

Softening de-spined pandanus leaves is a crucial step in the meticulous preparation of weaving material. Traditionally, the leaves are pounded with a stone pestle, but here they are rubbed against a metal post instead.

extended family (*peneinei*) owned at least one sailing canoe and one paddling canoe." (Nagaoka 2014: 306)

Glassman referred to the earlier observations of Christian (1899), a European visitor to Mwoakilloa Atoll, who reported that leaves of pandanus were still being used to make canoe sails at the end of the 19th century by the people of this atoll. More recently, Nagaoka (2014: 316–317), reporting about the traditional production of sailing canoes on this atoll, pointed out that "Mwoakillese people traditionally use a ruler made from a pandanus leaf (*pis*), called a *pisen war* (literally, "pandanus leaf of canoe") for more detailed measurements of canoe parts." Nagaoka described the making and use of this device:

> "... a pandanus leaf is cut to an arbitrary length, such as one-fourth or one-eighth of the hull length, and is halved and re-halved... The *pisen war* was used for measuring different parts of canoe: the slant of the keel line, the *ere* [local design] design, and the curved lines of the triangular sail."

Nagaoka (2014: 315) also referred to the special efforts of canoe makers to keep others from replicating the design of their own sails: "To avoid others copying their sail design, some builders used to eat the pandanus leaf ruler to keep a secret of the measuring formula and would use a Western-style ruler, converting their pandanus leaf ruler system into inches."

In addition, Nagaoka (2014: 318) also commented about the men of Mwoakilloa traditionally needing "to guard (*poaldi*) themselves and their fishing equipment from the influence of women," and that violation of this customary taboo was thought to bring about "bad luck" during fishing forays; furthermore, Nagaoka referred to the gender roles involved in traditional canoe construction on Mwoakilloa, and specifically about who wove the pandanus leaf mat sails:

> "Therefore, women were excluded from riding on sailing canoes. Indeed, the work of sailing canoe building is largely done by men while women are only in charge of weaving pandanus leaf mat sails, making coconut sennit twine and rope..., and preparing food for helpers." (Underlined for emphasis).

Cultivars in Mwoakilloa: On some Micronesian islands, especially the drier atolls, more than a thousand years of artificial selection by local inhabitants has helped produce cultivars with extremely large, edible, aggregate fruits. Many varietal names for different types of *P. tectorius* have been reported from the atolls of Pohnpei State (e.g., see Stone 1963; Stone, B.C. 1966b; Englberger et al. 2005a, b, c). In Mwoakilloa, Murphy (1950: 71) commented on the varieties and distribution of *P. tectorius* during the middle of the 20th century:

> "Pandanus, like the coconut, is primarily an atoll tree, and there are 15 or more varieties in Mokil [Mwoakilloa]. Some varieties are planted or at least encouraged; others are part of the natural forest. In one variety or another the tree is fairly evenly distributed throughout the coconut land of the three islands though nowhere does it constitute any very large proportion of the forest."

A few years later, Glassman (1953) made the first list of cultivar names in Mwoakilloa. More recently, Englberger et al. studied cultivar recognition by community members (2005c), documented *Pandanus* cultivars consumed by the Mokilese diaspora on the main island of Pohnpei (2009b), and studied in detail their nutrition content, finding ***Juaipwehpw*** to be especially high in beta carotene (2009c; see table 2). See pp. 423–424 for the comprehensive list of *Pandanus* cultivars in Mwoakilloa.

Uses in Pingelap: In the middle of the 20th century, according to St. John (1948: 106), pandanus trees were "important to the inhabitants [of Pingelap Atoll] furnishing edible fruit, timber from the trunks, and thatch from the leaves." More recently, Damas (1994: 231) reported that pandanus trees are used on Pingelap to make sleeping mats that are woven out of their leaves which "are cut, spines removed from the margins, put on a fire to heat dry or dried in the sun, folded into rolls when dry, and then split into widths for weaving" (also see Balick field notes for specimen MB4286 in the herbarium of the New York Botanical Museum). In addition, Damas referred to the use of pandanus leaves to make hats as well as thatching for roofs, and he noted that the trunks of pandanus are "used for rafters when split into sixths for subsidiary support for thatched roofs." Balick et al. (2009: 192) also referred to parts of *P. tectorius* that are used in traditional house construction on Pingelap: "The stilt roots are split and used as crossbars in traditional houses, thatch is woven on them, and roots are sometimes used for the beams and posts (although this wood is weaker than coconut, *Cocos nucifera*)." Balick et al. (2009: 292) also reported that, besides being planted, **Kipar** trees are allowed to grow close to the shorelines to control erosion. Specifically referring to Pingelap Atoll, Damas (1994: 55) noted that "the female, or fruit-bearing, pandanus trees occur mainly along the shorelines of the three islands [islets of Pingelap], while the male, or staminate form is found in the interior."

Damas (1994: 231) acknowledged that many **Kipar** varieties are recognized on Pingelap, but asserted that pandanus fruits did "not form a very important part of the diet, the fruit usually being eaten raw and serving at something of a confection." Damas referred to the enjoyable "sweet flavor" of pandanus fruits, but commented that "the stringy fibers clog the spaces between one's teeth, thus making pandanus a questionable treat to outsiders." Nevertheless, for local inhabitants, the stringy fibers can serve as a kind of organic floss and play a role in dental or oral health. Balick et al. (2009: 292) referred to the fruits of *P. tectorius* as providing "the best food on Pingelap" where the basal portion of the individual fruits of edible varieties are eaten raw, boiled in water, or ground up with true taro (*Colocasia esculenta*) and cooked in an earth oven. Damas (1994: 231) did refer to the "sap or juice of pandanus roots squeezed for juice [that] is used for stomach or other internal ailments."

More recently, Hattori (2012) defined several Pingelapese words or terms associated with pandanus plants and included notes which reveal some traditional uses of this tree and its fruits. For example, *kipar mwisikel* is the male pandanus tree that does not bear fruit but does bear "white flowers that look like leaves, [and that] the flowers have some powders on it, and smell great" and are used locally for head garlands; *aehki paes* is the game in which "one team hides the pandanus fruit leftovers and the other team searches for it"; *daeipw*, an "unripe pandanus fruit"; *iahkipis*, which refers to stripping off "the pricky edges off from the pandanus leaf"; *kaesiwaraewar*, which is a "rolled-up leaf of the pandanus tree [with which] children played"; *naekae* means the rolling of "dried green pandanus leaves"; *pis* being "brown and fallen pandanus leaves used for mats"; and the now sadly archaic term, *saeraekin mahng*, which is a canoe sail made from strips of dried green pandanus leaves.

Cultivars in Pingelap: St. John (1948) referred to pandanus trees of "good size" that were common both on the beaches and in the interior of Pingelap and recorded the names of 14 cultivars, specifically stating that more exist but his informant could not remember. He noted similarities between the names and appearances of some varieties on Pingelap and those he had observed on atolls of the Marshall Islands, suggesting long-term contact between those places. Subsequent researchers commented on pandanus cultivars on Pingelap and provided more names. See p. 424 for the comprehensive list.

Uses in Nukuoro and Kapingamarangi: Stone (1966b) referred to the traditional and ongoing importance of the commonly cultivated *P. tectorius* trees on both Nukuoro (where it is known as **Hala**) and Kapingamarangi (where it is known as **Binu**), as well as many other atolls in Micronesia, noting that:

> "The large number of cultivars named is indicative of the significant role which these common atoll trees [*P. tectorius*], with their swordlike spiny leaves, fibrous trunks, and large heads of fruit, play in the ecology and economy of atoll-dwelling peoples."

During the centuries of human settlement on Kapingamarangi and Nukuoro atolls, major change in land vegetation occurred due to human impact. As a result, land on the larger islets came to be dominated by coconut, breadfruit, pandanus, and swamp taro, and on the smaller islets by coconut. Commenting on pandanus (*P. tectorius*) on Kapingamarangi, Niering (1956: 5) suggested that it has two sources of origin: one natural, i.e., the ancestral colonizer of the truly wild trees, and the other, a series of cultivars brought in via human introduction as "canoe plants" in the period before European contact. Niering described their distribution on the atoll:

> "... common either as scattered mature trees or as immature localized stands. Numerous native varieties of it occur. Although seedlings are often abundant under some trees, most of them are propagated by burying the cut end of a branch from a known productive variety."

Based on information gathered from local informants, Niering (1956: 22, 29) also referred to several changes during the century before his research visits to Kapingamarangi, including the greater abundance of breadfruit and pandanus in the past on the atoll. The decline of traditionally important breadfruit and pandanus appears to have resulted from vegetation removal during the "Japanese occupation so that more coconut palms could be cultivated for the copra trade." However, following the end of World War II, the number of pandanus trees began to increase after the American government took over because the demand for mats and other handicrafts, made at least in part with pandanus leaves, was growing:

> "Although [people on Kapingamarangi play] the dominant role in regulating the abundance of coconut, breadfruit, and [swamp taro], pandanus reproduction in the plantations is of considerable importance. The natives claim that such trees produce only useful leaves and that the fruit is of inferior quality."

The research publications by Te Rangi Hiroa (Sir Peter H. Buck) provide us with much information about the traditional material culture and ethnobotany of the Polynesian societies, including the presence and use of *P. tectorius*, on Kapingamarangi:

> " Pandanus grow luxuriantly on all the islands of Kapingamarangi [atoll] and retains the Polynesian name of 'Hara' [**Hala**]. The hara almost equals the coconut in the variety of its uses. On Touhou [an islet of Kapingamarangi], the plants are scattered among the dwellings but each tree is privately owned. The female trees are not very tall, but they all bear fruit of an enormous size and each head of leaves appears to bear fruit." (Hiroa 1950: 28)

The female trees of *P. tectorius* have long been an important seasonal food source in Kapingamarangi. Hiroa (1950: 28–30) provided much information regarding preparation and consumption of edible material derived from the compound fruits (syncarp) of this species on Kapingamarangi. Hiroa (1950: 28) first describes how the ripe fruit or keys (*'hara reu'* [*hala leu*]) are separated and the inner fleshy part is eaten raw (chewed to loosen the edible material) (also see Stone 1966a: 131); Hiroa continued his discussion of how the fruits in "a stage greener" than the ripe *hala leu* fruits are cooked and then eaten, and how pandanus flour (*'paku harahara'*) is produced through a lengthy procedure lasting 2–3 days. This process involves cooking a mass of the inner fleshy parts of very ripe individual fruits (*'kihara'*) in an earth oven, followed by the cooked mass being pounded in a wooden bowl until it forms into a paste, from which handfuls are then formed into cakes that are eventually cooked on a grill over heated coral pieces. This is followed by flipping of the cakes until both sides are hardened. The hard cakes are subsequently sun dried the following day, after which the cakes are cooked again. The resulting quite dry, brittle cakes are then pounded and chopped in large wooden bowls until the flour has been produced. This relatively long-lasting flour can be mixed with water and eaten as a pleasant-tasting porridge (see Hiroa 1950: 29–30 and figures 16 and 17):

> " The most common way of eating pandanus is to chew on the inner soft end. In the raw state the edible portion is a juicy liquid which is sweet and pungent. In the cooked state the starch causes the juice to thicken and the edible end becomes soft and pulpy, which as stated by Murai (1958: 71, figure 20) resembles mashed potatoes. According to Buck (1950: 286), in Kapingamarangi the keys are cooked in an earth oven and eaten by hand. The dried eaten keys are stored and used for fuel. This might have some implications for fuel on voyaging canoes." [underlining added for further study] (Schattenberg 1976: 37)

Hiroa (1950: 28) described the taller male trees on the outer islets of Kapingamarangi Atoll as being encouraged to grow tall "by cutting off the side branches to procure the green leaves for plaiting," and these branches are cut off "a short distance from the trunk in order to provide a series of steps by which the trunk can be readily climbed to remove the higher branches as the full leaf heads develop." The trunks of some of these tall trees are used in the construction of houses on the atoll.

One way to appreciate the importance of pandanus in Kapingamarangi, as well as in Nukuoro, is to explore some of the rich local vocabulary connected with this plant. The Kapingamarangi Lexicon (Lieber and Dikepa 1974) offers many examples: *dahida* and *higu*, terms for the opposite edges of a thatch sheet: the thick edge made of the doubled-back butt end of pandanus leaves and the leafy tip edge; *tau heleheu* is the pandanus season; *e ihi ni lau hala* is the action of peeling of thorns from a pandanus leaf; *ga baehoo* is bast to make string from the pandanus tree root; *hada bagu* is the grill for pandanus flour cakes; *magana* is the paste made from pandanus fruit flesh when extracted and boiled; *hau hingala* is a head garland of pandanus flowers. There are also instruments such as *hugu tawe* made of a flying fish tail and used for pandanus leaf-splitting; *niha mala* made of barracuda teeth and used for cutting thatch; *lou heleheu* and *lou lau hala* respectively used for picking pandanus fruit and leaves; and *dugidugi tugi lau hala*, the mallet used for pounding and softening pandanus leaves.

The same dictionary also lists many names associated with parts of the pandanus plants: *lau hala* (pandanus leaf), *manga hala* (pandanus tree branch), *binu* (the whole tree); pandanus fruits in different stages maturity: *lilingaa* (immature), *hala modo* (green), *hala tongo* (partially ripe), *heleheu* (ripe), *heleheu gu lengelenge* (fully ripe), *bagu heleheu* (dried), *heleheu leu madaali* (poorly ripening); and individual pandanus keys: *hala gu kala* (almost ripe key), *huaahala* (ripe key). Some of these terms can be matched to terms recorded by Hiroa (1950: 28).

Similarly, Carroll and Soulik (1973), in their Nukuoro Lexicon, and more recently Drummond et al. (2016) list many interesting examples of pandanus-related terms from Nukuoro: *baabaa tugi lau* and *hadu tugi lau* are the log and stone on which pandanus leaves are pounded until softened; *lau* are the leaves thus softened; *biningaa lau* are softened leaves packaged into rolls; *dugidugi tugi lau* is the local tool used to unstiffen the pandanus leaf; *goobai lau* is a pandanus leaf hat; *gaso* are roof thatch rafters made of pandanus slats; *hagahala* is a sleeping mat woven from pandanus leaves; *hala dudu* describes the loud crackling noise similar to that of burning pandanus; *hui hala* are ripe pandanus compound fruits; *laho* are aerial roots of pandanus tree; *singano* is the flower of a pandanus tree; and *ulunga hala* is a pillow produced using a pandanus log, an object which can have a number of "magical uses."

Among all these pandanus-derived objects, probably the most typical and widespread in all of Micronesia are the mats and other items woven from pandanus leaves. In Kapingamarangi, Hiroa (1950) described the use of *P. tectorius* leaves and preparation of mat-making in significant detail with intricate line drawings. His words and illustrations provide us with excellent documentation of a traditional and important craft involving plait weaving using the long leaves of this widespread pandanus species to produce as wide variety of items:

> "The objects made from pandanus leaf were enumerated to me as roof sheets ('*kaurama rouhara*' [*gau lama lau hala*]), mats ('*kahara*' [*gahala*]), sails ('*ra*' [*laa*]), baskets ('*kata*' [*gada*]), fisherman's hats ('*kipai*' [*goobai*]) and a "square" ball ('*pare*' [*bale lau hala*], formerly called '*tipitipi*'). The roof sheets have been described under [Hiroa's description of Kapingamarangi] houses, the baskets in rectangular shape are modern and made for trade, mats are made in fair quantity for local use, trade and presents, the pandanus sail has been abandoned, the fisherman's hat is still in use, but the pandanus ball was not seen in any native game." (Hiroa 1950: 105)

He named and described in detail the pandanus mats that were common on the atoll when he was there (middle of the 20th century). The sleeping mats, '*kahara moeng*' [*gahala moenge*], averaged about 75 in (~2 m) long and 34 in (~1 m) wide. These mats were being used in pairs on cold nights, with one underneath as a mattress spread over the coarse coral gravel of the shelter floor, and the other placed above as a cover known as the '*kahu*' [*gahu*]. Hiroa also referred to a more elaborate sleeping mat known as '*wawa*' [*waawa*], made extra-long so one end can be rolled as a pillow or head rest with finer texture than common sleeping mats. Coarser mats utilized for covering bare flooring when needed were referred to as '*kara pata*' [*gahala pada*] or '*kahara te*' [also, *gahala holo hale*]. Hiroa describes these sleeping mats as well as other kinds of woven pandanus leaf mats and their construction in additional detail. The '*kahara turi*' [*gahala dulii*] mats are the small-sized traditional mats worn by women as skirts when working in areas where giant swamp taro (*Cyrtosperma merkusii*) and true taro (*Colocasia esculenta*) are cultivated. Men reportedly wore traditional fine pandanus mats called '*kahara tangata*' [*gahala dangada*] (Hiroa 1950: 106).

According to Stone (1966b: 433–434), leaf material from the two *Pandanus* species found on Kapingamarangi is sometimes combined for a special effect:

> "The leaves of 'Ti paho' [**Baehoo**] (*Pandanus* compressus [= *P. dubius*]) are preferred for mats, while those of 'Ti pinu torongahai' [***Binu dolongohai***] (*Pandanus fischerianus* [= *P. tectorius*]) are preferred for thatching. It is of interest to note that coarser, "cheap" mats may be made of leaves of [***Binu dolongohai***], and these often have a few darker strips of [**Baehoo**] leaves interwoven to produce various patterns."

Hiroa (1950: 116–117) also described the traditional use of pandanus leaves is sail making, even though imported canvas was the sole material being used to make sails by time Hiroa visited the atoll. Only two "matting sails" still existed on the atoll when he did his research there, and fortunately he was able "to procure one of the complete strips of matting used in manufacture of sails" (Hiroa 1950, specifically p. 116, fig. 73 and p. 182, fig. 108) that was subsequently deposited in the Bishop Museum in Honolulu (C.10083). This 34.5 ft (10.5 m) long, 9.2 in (23 cm) wide specimen was essentially "identical with that of the pandanus mats" he described earlier in his research publication, but with some modifications, and as Hiroa pointed out, these sails "were made of [many] long narrow panels of plaited pandanus leaf, 9 to 10 in wide, and of the same technique as the strip already described" (Hiroa 1950, specifically p. 199, fig. 120). The strips were combined by overlapping the plaited bands one over the other by about an inch for each and stitched together using cordage to form the complete traditional triangular pandanus sails for the Kapingamarangi canoes (also see darkwing.uoregon.edu/~mspp/kapingamarangi/k09-mate.htm).

Another traditional use of woven *P. tectorius* leaves in Kapingamarangi has been for the production of the conical-shaped fisherman's hat. These have been commonly used here during the day while fishing or canoeing, obviously for some protection from the intense solar radiation; however, its local Polynesian name, *goobai waga llama*, refers to a cover or hat associated with canoes and torches:

> "... it was [also] used at night on canoes while torching for fish, usually flying fish. The torch, made of dry coconut leaves, gave off a goodly number of sparks, and the fisherman, who held the torch in one hand and used the scoop net with the other, wore the hat to protect his head (see Hiroa 1950: 118–122 for a very detailed description of the construction of these woven hats)."

Hiroa also discussed the puzzling origin, spread, and distribution of these specialized hats, as they are "not present in any part of Polynesia," though unplaited pandanus leaves have been traditionally sewn together in "Guam and the Carolines" of Micronesia. Hiroa (1950: 123) concluded that "conical hats probably spread into Micronesia from the west and diffused to some of the Polynesian outliers where changes in technique were developed. The subject requires further investigation through the collection of specimens throughout the individual groups of Micronesia."

Another simpler use of the long flexible leaves of pandanus on Kapingamarangi is reported by Hiroa (1950: 36) in his detailed description of the preparation of preserved breadfruit or *'pakukura'* [*bagu gulu*].

> "The women peel the green mature breadfruit with shell scrapers or use their teeth if it is very ripe. The fruit is cored and torn into lumps and packaged in breadfruit leaves and tied with dry pandanus leaves. These leaf packages are placed in an earth oven for two hours. The next day the packages are opened and emptied into a bowl. The women mash the contents into a yellow paste. This mass is spread out by hand in the sun on mats of green woven coconut leaves. The women go over the paste carefully removing any lumps. After several hours the paste is turned to dry on the other side. At this point the men take over and roll the bread fruit paste into packages of pandanus leaves tied around with 2 ply cord. The packages when finished, as recorded by Buck (1950: 38), are 19 in long and 3.8 in in diameter. This preserved breadfruit is said to keep a long time and is used as a reserve food and a suitable provision for a long voyage at sea. As needed, one end of the roll is opened and the desired portion is cut off." (Bruce 1976: 35–36, citing Hiroa)

Other traditional non-food uses of pandanus trees in Kapingamarangi include the utilization of the long trunks of the male trees that can reach up to more than 7 m (over 20 ft), which are integrated into the woodworking of houses (Hiroa 1950: 28). In addition to the many other uses of *P. tectorius* in Kapingamarangi are the medicinal applications, e.g., "The tip of a descending aerial root [aga binu], the meristem, is cleaned, pounded, and squeezed into a cut to staunch the bleeding" on Kapingamarangi Atoll (www.comfsm.fm/~dleeling/ethnobotany/healing_plants_presentations.html).

Cultivars in Kapingamarangi: Stone (1966b) contemplated the impressive diversity of pandanus cultivars on Kapingamarangi Atoll and concluded that all, or at least the great majority, are derived from artificial selection of *P. tectorius* over many generations:

> "During an intensive study of Marshallese pandans it soon became obvious that these "species" [previously believed by some to belong to different *Pandanus* species] were merely cultivars, usually clones, valued for their fruit. I now believe that all the Marshallese forms can be reduced to one common species [*P. tectorius*]." (Stone 1967b)

Stone concluded that all but one of those cultivars were clonally propagated strains of one species: what today is recognized botanically as *P. tectorius*. Stone noted that "the majority of these *P. tectorius* clones were being cultivated for the edible parts of their fruits." Stone also pointed out that at least one cultivar of *P. tectorius*, and plants belonging to a second species, of what today is recognized botanically as *P. dubius* "are used for sources of leaf material for weaving or thatching, and the seeds of the latter species are also on occasion eaten" (see the next section).

Stone also suggested that most of the *Pandanus* cultivars "appear to have an indigenous character and may be termed Kapingamarangi clones, but a few have been introduced from the Marshall Islands, a major source of cultivars to the east, and these few are accorded names reflecting their introduced status." Furthermore, Stone indicated that "a few Kapingamarangi cultivars have been introduced in other islands, including adjacent Nukuoro Atoll to the north, and around the Kapingamarangi immigrant village in [Pohnpei]." Finally, Stone, with reference to the nutritional value of the pandanus fruits (wsee Miller et al. 1956), suggested more than a half century ago that "the importance of these edible cultivars will diminish as the amenities of foreign civilization preempt a place in the economy of the atoll, [however] they will not disappear entirely, and it may be added that it would be very desirable to ensure their maintenance (1966b: 435–436). It is not known how many cultivars that Stone observed are preserved to this day, but the list of cultivar names in the Kapingamarangi–English dictionary (Lieber and Dikepa 1974) bears witness to an impressive variety for a single atoll (see p. 425).

▸ **Rawahn** or **Duhrien** (P), **Duhrion** (Mw), **Duhriaen** (Pi), **Duuliono** (Nu), **Dulion** (Ka)
Pangium edule

Description: *Pangium edule* is a medium to large tree. It grows to heights of as much as 25 m (80 ft). Its large, shiny, pointed leaves have smooth margins, but the leaves on the saplings are 3-lobed. Overall, the leaves are generally cordate or subcordate to ovate in shape and about 15–25 cm (6–10 in) long, with the petioles nearly as long as the leaf blades. The leaves of *P. edule* are clustered at the tips of branches. As it is a dioecious species, the flowers are unisexual and individual trees are either male or female. The male flowers, each about 5 cm (2 in) wide, are produced in bunches with 2–3 calyx lobes, 5–6 petals, and many stamens. The solitary female flowers are similar to the male flowers but lack functional stamens. The large fruits are about 15–30 cm (6–12 in) long and approximately half as wide. The rough, brown-skinned fruits become yellowish when mature and produce many compressed, ovate, grayish seeds about 5 cm (2 in) long. The seeds are surrounded by creamy white to yellow-fleshed pulp (e.g., see Stone 1970: 427, Falanruw 2015: 134).

Warning: The fruits and seeds contain hydrogen cyanide and can be deadly if consumed without prior preparation.

Distribution: *P. edule* is assumed to be native to insular Southeast Asia and is indigenous from Indochina possibly to Vanuatu in Melanesia. It was brought to Micronesia after European contact and now is reportedly found in Palau, Yap, the Mariana Islands, Chuuk, Pohnpei, and Kosrae. It is sometimes common near human settlements and can normally be found growing in village agroforests, where it can be mixed with secondary forest. This tree needs only about 6 months to reach full maturity. In English, it is sometimes called the "pangi fruit" or "football tree."

Uses in Micronesia: *P. edule* fruit is also occasionally referred to as "durian," perhaps because it tastes mildly like the fruit of the durian (*Durio zibethinus* in the Malvaceae family), but the true durian fruit is spiny and much stronger smelling. The yellow-fleshed fruits of *P. edule* ("false durian") are sometimes eaten and reportedly contain valuable levels of provitamin A and other carotenoids (Englberger et al. 2003b). "The pulp of the ripe fruit is delicious; however, the seeds, unripe fruit, and other parts contain prussic (hydrocyanic) acid, which is quite poisonous." (Falanruw 2015: 134, underlined for emphasis). The fruit reportedly becomes edible after it is soaked in water for a sufficiently long period of time, and the seeds, which are rich in oil, also become edible after adequate pounding and washing with water. Furthermore, the leaves, resin, and bark of this tree are also poisonous and are sometimes reportedly used as a fish poison. The seeds are also said to be useful as an "insecticide," which can be used to preserve fish for future consumption (e.g., see Kanehira 1933, Etoh 2001).

Uses in Pohnpei: *P. edule* is known in Pohnpeian as **Rawahn** or **Duhrien**. Today, this introduced fruit tree is widely naturalized on Pohnpei; as a common tree in secondary forest regrowth, *P. edule* is difficult to kill by girdling the trunk. According to Sproat (1968: 99), the ripe fruits are consumed by people and are fed to some animals on Pohnpei:

> "The yellowish, sweet, aromatic pulp surrounding the seeds is quite tasty. The mature fruits should be picked without bruising and kept till they soften before cutting and eating. The seeds contain prussic acid [hydrogen cyanide] in quantity in the oils and may be dangerous for eating unless properly treated and prepared by an experienced person."

The large fruit are also reportedly used as pig food, and Balick et al. (2009: 319) also commented on the use of the poisonous *P. edule* seeds as bait to kill rats as well as to trap and poison other animals; for example, "as bait for wild chickens, a trap is made from plants and the seeds of this plant are put in it to poison the wild chickens that enter the trap..."

▸ **Kurumw** (P)
Parkia korom

Description: *Parkia korom* is a big tree in the bean family Fabaceae that can grow up to 20 m (~65 ft) tall with a trunk as much as 1 m (3 ft) in diameter. Its flowers are yellow with a combined spherical shape and are thought to be pollinated by bats. Like other species in Fabaceae (the bean or pea family), the fruits of *P. korom* are pods, in this case about 25–32 cm (10–13 in) long, including the stem, and about 3.5–4.5 cm (1.5–2.5 in) in diameter. Each pod contains at least 20 seeds. Typically, fine red-brown hairs are found on the pods and young twigs, while the small leaflets characteristically produce very small whitish hairs.

Distribution: *P. korom* is found only in Pohnpei. Its scientific name derives from its Pohnpeian name, **Kurumw**. The genus *Parkia* is named after Mungo Park, one of the first Western explorers in Africa. Some botanists believe there are two woody species of *Parkia* in Micronesia and that each is a single-island endemic with relatively small population size: (1) *P. korom* (found naturally only on Pohnpei, where it is rare today, e.g., Costion and Lorence 2012: 77–78), and (2) *P. parvifoliola* (found naturally only in Palau, and also very rare). **Kurumw** has been reported to be more common in more humid areas of the upland Pohnpei forests (see "Serehd" 2006, 8: 4). Unfortunately, this unique Pohnpeian tree is thought to be in danger of becoming extinct. It is listed as "vulnerable" to possible disappearance on the "red list" of the International Union for the Conservation of Nature (IUCN). In recent years, the CSP has conducted surveys on this species' current distribution and found only a small number of living young trees, none of them apparently bearing flowers or fruits.

Uses in Pohnpei: **Kurumw** is reportedly used as a medicine to treat unspecified skin disorders, and its wood is utilized to make flooring for houses (e.g., see Merlin et al. 1992). It has also been referred to as an important source of wood used traditionally in the construction of canoes and "has a history of being in Pohnpei long before Nan Madol" (Glassman 1952; also see "Serehd" 2006, 8: 4, and Balick et al. 2009: 392). *P. korom* trees were reportedly harvested significantly during the period of Japanese occupation during the first half of the 20th century.

▸ **Pwompwomw** (P), **Pwoamwpwoamw** (Mw) (Pi), **Hua dolo** (Nu)
Passiflora foetida

Description: *Passiflora foetida* is a perennial, sprawling vine with hispid stems that have tendrils. Its 3-lobed leaves are about 5–10 cm (2–4 in) long and equally wide. Its leaf petioles are 2–5 cm (1–2 in) long. Its flowers are solitary, axillary, bisexual, bracteate (gland tipped) with five white to purple petals. Its hairy, dark-yellow to orange berry fruits are thin, leathery skinned, globose to subglobose, glabrous, and surrounded by pinnately branching bracts. The scanty but sweet and tart pulp of the fruit is edible and has many seeds. "The very finely laciniate sepals enclosing the fruit probably gave rise to the curious common name" (i.e., "love in a mist," Stone 1970: 430). Another of its common names, "stinking passion flower" (*foetida* means stinking in Latin), is probably derived from odor emitted when its leaves are crushed (www.comfsm.fm/~dleeling/ethnobotany/text/flora.html).

Distribution: *P. foetida* is native to tropical America and is now an invasive species widespread on a number of Pacific islands including some in Micronesia, such as Pohnpei (see Englberger 2009). It is found in disturbed areas, pastures, roadsides, secondary thickets, waste areas, landscaped areas, and greenhouses, and it often grows over low vegetation or forms dense ground cover which prevents or delays the establishment of other species. Its seeds are commonly eaten and spread by birds and mammals. However, although this perennial, herbaceous vine is a common weed, its fruit is sometimes consumed by people.

Uses in Pohnpei: *P. foetida* is known in Pohnpeian as **Pwompwomw**, which comes from a common English name for this species: "pompom." When the fruits are orange, they are sometimes eaten on Pohnpei and have a taste "like a tart passion fruit." Furthermore, it has been reported that parts of the plant are also used medicinally on Pohnpei, for example as a remedy for boils and to treat babies with a fever and skin diseases, as well as for "an unspecified children's sickness" (Balick et al. 2009: 461).

▸ **Ngih** (P), **Kaingi** (Mw), **Kaingae** (Pi), **Nngie** (Nu) **Laagau hai goo** (Ka)
Pemphis acidula

Description: *Pemphis acidula* is a coastal woody shrub or small tree with very heavy, dense wood and wiry twigs. It can grow as tall as 5 m (15 ft) or more, producing small, fleshy, oblong leaves, and small white, wrinkled solitary flowers. It is sometimes called "ironwood" in English because of its heavy, hard wood. However, this name causes confusion because another notable hardwood tree, *Casuarina equisetifolia*, also found widely in the Pacific, but only recently introduced to some areas of Micronesia such as the Marshall Islands, is also known as "ironwood."

Distribution: This often common, dense hardwood, coastal plant is indigenous on many tropical islands in both the Indian and Pacific Oceans, including those in Micronesia. *P. acidula* is usually found in strand vegetation close to the high tide line along sandy shores of coral islands, and along the landward edge of mangrove areas. On atolls, it frequently occurs on coastal limestone rocks, cliffs, and on limestone bedrock outcrops on atolls. A striking description of the harsh habitat adaptation of the densely wooded *P. acidula* plant in the exposed coral reef environment of Ebon Atoll in the Marshall Islands was provided by Wiens (1957: 6):

> "In the extreme northeast part of the lagoon, an amazingly flat limestone reef bare at low water is almost smooth enough for roller skating. The jagged remnants above the reef flat support a good growth of *Pemphis acidula*, with roots anchored in erosional pockets containing some sand and holding fresh water after rains. These remnants rise 4.5 feet [~1.5 m] above the surface of the reef flat. These erosional remnants are similar to eroded bedrock in other areas, in the sharp and jagged character of the surface, which type of surface appears to occur most often where rain water [which is slightly acidic] dissolution rather than wave action is the disintegrating factor involved."

Although *P. acidula* was once common along many low-lying sandy or coral reef limestone regions in its natural tropical range, the International Union for the Conservation of Nature (IUCN) referred to its possible overuse and the loss of its habitat in its native range as it is relatively slow growing:

> "If sea-level rise is a continued trend over this century, then there will be continued mortality and re-establishment of species zones. However, species that are easily dispersed and fast growing/fast producing will cope better than those which are slower growing and slower to reproduce." (www.iucnredlist.org)

General uses: The local name of *P. acidula* in much of Malaysia, Indonesia, and parts of the Philippines is **Mentigi**, **Sentigi**, **Cantigi**, or **Bantigi** and its various cognate names in many different Pacific islands of Micronesia are also quite similar, e.g., **Ngis**, **Gangiy**, **Hangi**, **Gaingi**, **Engi**, **Engiy**, **Eengi**, **Ngih**, etc. The whole *P. acidula* tree is often left where it grows naturally near the shores on tropical Pacific islands to protect inland areas from salty sea spray. In addition, it is a preferred fuel wood because it produces a very hot flame. Moreover, its very dense, hard wood has a very wide range of traditional uses among Pacific Islanders, as indicated by Thaman (1992b: 152):

> "[Its] extremely hard wood [is favored] for carved objects such as house frames, canoe parts, keels, connecting pegs, and paddles, digging sticks, clam knives, tool handles, thatching needles, pipes, back scratchers, fish hooks, fish-net frames, fishing poles, shuttles and meshing needles, fish, eel, and rat traps, spears, fish clubs, war clubs, darts, food containers, mortars and pestles, pounders, pump drill pieces, coconut huskers, combs, drums, tops, throwing sticks and other toys, etc." (Also see Clarke and Thaman 1993).

One popular use of *P. acidula* in some tropical areas is as an ornamental bonsai plant; however, extraction (or theft) of these trees in their youthful state may be limiting the reproduction and growth of these trees in their natural environments (e.g.,see IUCN Red List of Threatened Species at www.iucnredlist.org, which currently lists it as a native species of "least concern" but does refer to its major current threat, "collection for use in the ornamental bonsai trade.")

Uses in Pohnpei: *P. acidula* is known in Pohnpeian as **Ngih**. Its wood can be used to make planting sticks, coconut husking stakes and firewood (e.g., Merlin et al. 1992: 34). Wood from this plant was also formerly used to make war spears and for many other purposes:

> "It is used for house posts and is considered the strongest tree on the island... It is used to make utensils such as long- or short-handled spoons... It is also used for carving into spear handles, shafts, and other carvings... and to make knife handles, ax handles, and sticks to husk coconuts... To use as a fish attractant, leaves and branches are carried near the reef where fish are hiding, and then when the fish congregate they are speared by people wearing goggles and swimming underwater... To use as a charm to protect a person from sorcerers when going into a big crowd, such as at funeral feasts or ceremonies, the bark of the stem is chewed and swallowed... [and it is also used medicinally on Pohnpei to treat dysentery and ulcers]." (Balick et al. 2009: 422)

Uses in Mwoakilloa and Pingelap: This tree is called **Kaingi** in Mwoakilloa and **Kaingae** in Pingelap. Glassman (1953: 293) noted that it is a "common [coastal] strand plant" on Mwoakilloa but did not document any uses.

Working on Pingelap, St. John (1948: 11) indicated that this small tree could be found in the lagoon beach environment and that its wood was "used for handles, pipes, etc." According to the informant of Damas (1994: 238), *P. acidula* has the "hardest wood in the Pacific," and the Japanese, during their occupation and administration on Pingelap, used its dense wood to make smoking pipes. Hattori (2012: 228) noted that **Kaingae** is known locally to be "very strong," and used to make the *daroawa*, the wooden connectors between the hull and the outrigger of a traditional Pingelapese canoe.

▸ **Lirau** (P) (Mw)
Phragmites karka

Description: *Phragmites* is a genus of common reeds or large perennial grasses found in wetlands throughout most temperate and tropical regions of the world. Many botanists divide the genus *Phragmites* into three or four species, although *P. australis* is sometimes regarded as the sole species (e.g., Kiviat 2013; Lansdown 2013). Based largely on the fact that *P. australis* is generally a temperate species, while *P. karka* is tropical in its distribution, here we recognize *P. karka* as one of four species of *Phragmites*. A tall, erect reed, *P. karka* grows to heights of up to 5 m (15 ft) and has large stems, up to 2 cm (1 inch) in diameter at the base. *P. karka* produces widespread colonies by means of underground rhizomes, ground-creeping stolons, and aerial shoots (known among reeds and grasses as "culms"). Its coarse, sword-shaped leaves have broad bases and acute tips up to 50 cm (20 in) long and about 2 cm (1.5 in) wide. *P. karka* is strongly tufted, producing bushy, silvery white clusters of tiny flowers at the tops of the stems, which are quite noticeable. These many-branched panicles of flowers (inflorescences) are 30–70 cm (12–28 in) long and have abundant silky hairs.

Distribution: Native from tropical Africa through parts of southern Asia, Australia, and on some Western Pacific islands, *P. karka* is a common large reed or grass in Micronesia. Large areas of *P. karka* occur on Saipan, Guam, and Chuuk (e.g., see Mueller-Dombois and Fosberg 1998: 221). *P. karka* also invades abandoned wetland taro patches. In fact, it occurs in a variety of marshy habitats, including shallow water at the margins of lakes, ponds, in and along streams, and ditches. In irrigation canals, it has become highly invasive in some tropical area of India and southern China, where it is in need of control. The use of large "ornamental" grasses or reeds in landscaping has become popular in recent years. Although these types of *Phragmites* reeds can be found in commercial nurseries, some species are invasive (developing into troublesome weeds), including at least two species of common *Phragmites* reeds with variegated forms: *P. australis* and *P. karka*, sometimes referred to as the "candy stripe reed."

General uses: *Phragmites* species have an ancient history of use by marsh peoples in South Asia, and their many uses have been relied on by several cultures living within their natural range. The young shoots of *Phragmites*, especially *P. australis*, can be cooked and eaten like asparagus or bamboo sprouts. On the other hand, *P. karka* has been used to form brooms, and the stems have been used as building material, fuel, or to weave useful items such as hats and mats. The stems are also sometimes used to form beds for water purification, and this reed is also planted by water to protect the land from erosion because of its dense root system. Indeed, the marshy *P. karka* species is sometimes used as a reliable indicator of fresh water, and in the Philippines, leaves of *P. karka* are reportedly used for "pond-field fertilizer" (e.g., Bodner and Gereau 1988). *P. karka* can be artificially propagated through the use of rhizome cuttings.

Uses in Pohnpei: *P. karka* is known in Pohnpeian as **Lirau**. It occurs in fresh or brackish water marshes, often along with *Saccharum spontaneum*, another large grass or reed known in Pohnpeian as **Ahlek**. The split stems of **Lirau** are used to make attractive materials for house walls. Its stems, along with the stems of its relative **Ahlek**, are used as *rahu* or "minor rafters" (www.comfsm.fm/~dleeling/ethnobotany/e83/thatching.html). In Pohnpei, *P. karka* stems have also been used to make traditional flutes and as a source of traditional medicine. For example, Balick et al. (2009: 304) provided information regarding the medicinal uses of *P. karka* on Pohnpei:

> "To treat *koaloalamwahu* (sadness), four shoots of young leaves 4 in long are pounded, wrapped in a piece of cloth, squeezed in 2 lit (0.5 gal) of water, and consumed. This treatment is repeated daily until the person becomes well again... To treat open cuts, the soft part of the young leaf is chewed and squeezed onto the wound... To treat *kau* (a sickness caused by eating poisonous food and drink) that might have been caused by sorcerers, four shoots of the young leaves are pounded, squeezed into a cup of water, and consumed, whereupon the person will vomit to get the poison out..."

- **Limemeirpwong** (P), **Limoahmoair in pwo** (Mw), **Limaeimaeirpwong** (Pi), **Bua mai moni** (Nu), **Laagau kiikii** (Ka)
Phyllanthus amarus

Description: *Phyllanthus amarus* is a small, semi-prostrate to erect, diffusely branching herb that grows to heights of 10–60 cm (4–24 in). It is somewhat woody at its base and produces very small, simple, alternately arranged, oblanceolate to oblong leaves that are rounded at both ends. Its flowers are tiny with pale green to white calyx lobes produced under the leaves at the nodes. The small, green, round fruits or capsules split into six sections with 5–7 ribbed seeds.

Distribution: Native to tropical America, *P. amarus* is a weedy plant now widespread as an alien herbaceous species naturalized in many other tropical areas, including the high and low islands of Micronesia. It occurs in dry, generally lowland, disturbed places and is common to abundant in lawns and roadsides and as a weed in villages, gardens, and cultivated fields.

Uses in Pohnpei: According to Balick et al. (2009: 464), the small weedy *P. amarus* herb is known in Pohnpeian as **Limemeirpwong** and has been used medicinally on Pohnpei:

> "To treat diarrhea, leaves are boiled, chewed, and swallowed... To treat joint pain... To treat back pain... To treat congestion and heart failure of older people... To treat asthma, congestion, or heart failure of older people (respiratory distress), one plant is boiled for ninety minutes in water and consumed like hot coffee (This treatment is also good for infants with colic)... To treat diabetic sores on a person's leg... To treat amoebic dysentery... Glassman (1952) noted that the entire plant was boiled and used to treat gonorrhea." (Also see Lee et al. 2010: 29).

Uses in Mwoakilloa: *P. amarus* is known as **Limoahmoair in pwo** on Mwoakilloa, where it is said to be used to treat "dysentery—diarrhea with blood plus headache and fever" by boiling the leaves of this plant in water and then drinking the water (www.comfsm.fm/~dleeling/ethnobotany/healing_plants_presentations.html).

Uses in Pingelap: *P. amarus* is known as **Limaeimaeirpwong** on Pingelap, and according to Damas (1994: 237), it is used on the atoll as "a fungus remedy."

- **Kideu** (P) (Pi), **Kamkam** (Mw), **Maaile** (Nu), **Gideu** (Ka)
  Phymatosorus grossus

Description: *Phymatosorus grossus* is a creeping fern with sturdy, dark-colored rhizomes (roots) that produce erect, glossy green fronds (leaves). The fronds are pinnately divided into 1–8 pairs of lobes up to 40 cm (24 in) long and are connected to a stipe (stem) of equal or greater length. The undersides of the fronds bear characteristic, circular, red spore clusters (sporangia), which are shallowly depressed, forming raised or bordered pits on the upper surface of the frond. *P. grossus* is a fern in Polypodiaceae, a common fern family. As noted above, this fern has had a rather confusing taxonomic (botanical classification) history. The Latin word *grossus*, the epithet or specific name, or second part of the taxonomic binomial of this species, refers to the fern as being "big, or coarse, apparently in reference to the larger fronds and sori as compared with other species in the genus." The fern's fragrant fronds are often lobed and somewhat resemble the large leaves of breadfruit trees.

Distribution: *P. grossus* is indigenous from tropical Africa to eastern Polynesia but has been introduced and subsequently naturalized in Hawai'i. *P. grossus* is a very common native fern in Micronesia, especially in disturbed areas or secondary forests. In fact, it is found in a wide variety of habitats including in coral rock crevices and scrub vegetation near sea level on both atolls and volcanic islands as well as on tree trunks up in the rainforests at the higher elevations of over 1,000 m (~3,000 ft) on the high islands. *P. grossus* probably arrived on the atolls and high islands of Micronesia well before humans. As a common ground cover in the coastal forests of atolls, this creeping fern also become epiphytic (grows on other plants), especially around the base of cultivated coconut palms and breadfruit trees.

General uses: *P. grossus* is used as an ornament and for medicine on many tropical Pacific islands, including those in Micronesia. In Polynesia, it is apparently a very important source of traditional medicine, being a component in many medicinal treatments (e.g., see www.tetiaroasociety.org/island/plants/microsorum-grossum for medicinal use in French Polynesia under the synonym of *Microsorium scolopendria*).

Uses in Pohnpei: On Pohnpei, this fern is called **Kideu** and has had many ancient and modern uses as a "charm" due to its reported magical or spiritual powers. It has been used for several cultural reasons, as well as for numerous medicinal purposes (see Balick et al. 2009: 244–245). For example, is said to be "used by young dancers as a *mwaramwar*, the erect crown of fronds acts as a "fence" against ghosts being brought on the dancers," and thus helps protect "dancers from anyone trying to cast magic onto them" (www.comfsm.fm/~dleeling/botany/1998/vhp/gob.html).

Uses in Mwoakilloa: *P. grossus* is known on Mwoakilloa as **Kamkam**, but the Marshallese name **Kino** is also used as a loanword to refer only to the immature leaves without spores, used to make head garlands. This fern also has medicinal applications; for example, the "tip of a non-fertile" frond of the *P. grossus* fern is reportedly used by the people on Mwoakilloa to help heal cuts (www.comfsm.fm/~dleeling/ethnobotany).

Uses in Pingelap: According to Damas (194: 231), these ferns are "chewed or pounded to a liquid and used for [treating] open wounds."

▸ **Kapwoi** (P) (Mw) (Pi), **Gabui** (Nu), **Lau gai buu** (Ka) *Piper betle*

Description: There are probably well over 1,000 species within the genus *Piper* in the tropical regions of the world. One of these species, *Piper betle*, is a highly variable, high-climbing vine that is widely important in the lives of millions of humans due to its widespread use in the cultural practice of betel nut chewing (see pp. 147–150). *P. betle* is a perennial, semi-woody climber that is dioecious; normally, the small male and female flowers are produced in narrow spikes on separate plants. Male *P. betle* vines are cultivated, and its stems may grow robustly up to a height of about 20 m (65 ft) with diameters of 15–20 cm (6–8 in). The leaves of this vine are oval or somewhat heart-shaped, normally about 7–10 cm (3–4 in) long and unequally rounded at the base with 5–9 (usually 7) prominent veins. When crushed, the leaves are "aromatic and somewhat acrid to taste" (Fosberg and Sachet 1975: 17). Female spikes, when mature, are red, fleshy, more or less oblong, 3–8 cm (~1–3 in) in length and 0.5–1 cm (~0.4 in) thick. The fruits are coalescing, fully embedded in the pulp, and fused with the rachis. The tiny, smooth seeds are oblong to globose-obovoid.

Distribution: Assumed to be native to South or Southeast Asia, *P. betle* is a climbing vine commonly cultivated in India, Sri Lanka, Malaysia, Indonesia, East Africa, and the Philippines as well as in several island areas of the tropical Western Pacific. The *P. betle* vine, like many other cultivated plants, was brought to western Micronesia by early migrants in the ancient times as a "canoe plant." It thrives under humid forest conditions with high relative humidity and ample soil moisture. It flourishes in areas with 2,250–4,750 mm annual rainfall and is cultivated at altitudes up to 900 m. It prefers shade and needs protection from wind. *P. betle* vines prefer deep, well-drained, friable loamy and clayey soils, rich in organic matter and with a pH of about 7–7.5. Zumbrich provided (2007–2008: 94–96) the following descriptive aspects of this widespread cultivated species:

> *Piper betle* L. ("betel pepper") is a climbing plant typically propagated asexually from stem cuttings rather than from seeds. The leaves of *P. betle* L. have a relative[ly] high content of phenolic compounds which not only taste refreshingly but also exert a range of pharmacological activities. The broad heart shape of the leaf makes it ideally suited for assembling *Areca* [betel nut] and lime paste on its surface and to be folded into a betel quid."

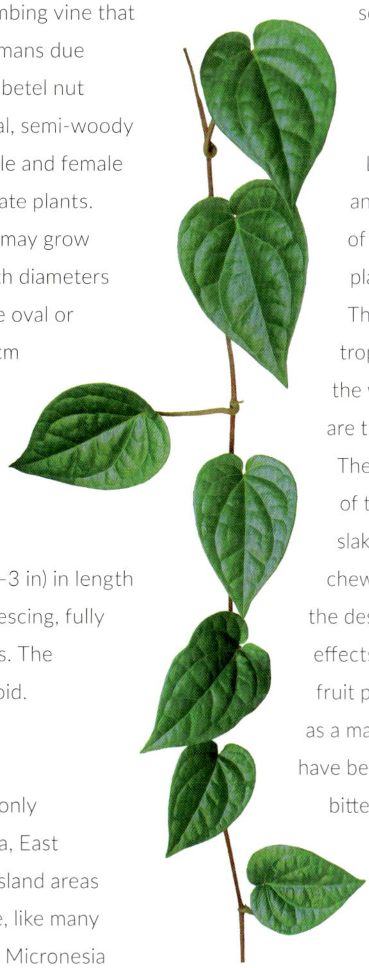

General uses: Although two species of *Piper* (*P. ponapense* and *P. fragile*) are endemic to Micronesia and do have some therapeutic ethnobotanical significance, today, the introduced *P. betle* is more or less "inseparable from daily life" throughout much of the tropical Western Pacific (Costion and Lorence 2012). Within Micronesia, Fosberg and Sachet (1975: 17) described the distribution of this vine as "usually planted or persisting from plantings, around old dwelling sites and gardens." The parts of *P. betle* traditionally used in various tropical regions including Micronesia (especially in the western islands of Palau, Yap, and the Marianas) are the vine itself, plus the leaves, roots, and fruits. These parts are used in conjunction with the fruit of the betelnut palm, *Areca catechu*, along with slaked lime, and are wrapped up in a quid that is chewed, or more correctly sucked on, to produce the desired psychoactive and other physiological effects. Although the endosperm of the betel palm fruit provides the stimulating psychoactive alkaloids, as a masticatory (chewable mass), the leaves of *P. betle* have been described as warm-tasting, pungent, and bitter. Traditionally, the use of betel quids has been a cultural trait of a large percentage of people in an area stretching from East Africa to islands in the Western Pacific. In fact, this amounts to "between a quarter and a tenth of the world's population, which makes it one of the most popular of all psychoactive substances" (Bissa et al. 2007: 26).

Uses in Pohnpei: The more or less recently introduced *P. betle* vine is known as **Kapwoi** in Pohnpei and is associated with use of the betel nut palm fruit. The "chewing" of the betel palm fruit ("nut") only has only become common in Pohnpei during the past few decades. This relatively recent rise in betel nut use is associated with increasing contact between the western and eastern Caroline Islands, and especially with the formation of the FSM. This newly formed political entity includes Yap State, where betel nut palm trees and their use have ancient traditional importance, and the other three states, Chuuk, Pohnpei, and Kosrae, where the use of betel nut was absent until recent times. In Pohnpei, the "Tahitian chestnut tree" (*Inocarpus fagifer*), another introduced crop tree known in Pohnpei as **Mworopw** or **Mwoaroapw** (Pauku 2006: 417) provides good support for cultivating **Kapwoi**. As in many other areas, in Pohnpei, the preparation and use of the betel quid is more or less similar: the pepper betel vine leaf is wrapped around a betel nut from the *Areca catechu* palm and slaked lime (calcium hydroxide), and in some cases tobacco, is added. This mixture is then chewed (Balick et al. 2009: 467). Recent preliminary research among 7th and 8th graders in the schools of Yap and Pohnpei indicates that most adolescents in these islands have used the betel nut wrapped in *P. betle* leaves, and, furthermore, many have included tobacco in their betel quid, which has its own dangers of dependency (Milgrom et al. 2013). These researchers have called for more extensive and intensive research on the use of these substances in Micronesia. Note: For extensive details on the past and modern use of the betel nut quid in Micronesia, see the species section on **Pwuh** (*Areca catechu*), p. 147.

▸ **Sakau** (P), **Jakau** (Mw), **Saekau** (Pi), **Sagau** (Nu), **Tagaao** (Ka)
Piper methysticum

> "From its divine origins and revered status in Pohnpeian traditional culture to its modern-day use in sacred ritual and recreational activity, **Sakau** is without doubt the most important plant on Pohnpei. It has influenced the way Pohnpeians view each other, their community, and the world ever since it arrival, presumably from Vanuatu via innumerable voyages, trading, and back migrations." (Balick et al. 2009: 199)

Description: *Piper methysticum* (kava, known in Pohnpeian as **Sakau**) is a member of the pepper family, Piperaceae. It is a shrubby plant with a root stock or stump but no true rhizomes. All cultivated *P. methysticum* plants belong to a series of sterile clones with no floral reproduction that are all assumed to be derived hypothetically from *Piper* wichmannii, a truly wild endemic species found in some areas of Melanesia. An alternative taxonomic perspective, based on a study involving observational botany and genetic research, concluded that the accepted taxon (scientific name) for kava that embraces all the truly sterile clones, including those in Pohnpei, should be *P. methysticum* var. *wichmannii* (see Applegate and Lebot 2006). Whatever the valid evolutionary origins of this culturally important species may be, all cultivated kava shrubs today must be propagated via vegetative reproduction (e.g., planting viable stem cuttings). Kava plants produce heart-shaped leaves up to about 30 cm (12 in) long with 10–13 veins emerging from the leaf base. Their stems have swollen nodes and can reach up to about 3 m (~10 ft) in length. The tiny male flowers are arranged in solitary, greenish white spikes up to about 6 cm (2–3 in) long. Each flowering spike ascends from a stem axil opposite a leaf. Female flowers are either unknown or rarely recognized due to their very small size and infertile nature. This sexual infertility is a clear indication that artificial selection by people has been a long and effective domestication process of vegetatively reproducing traits desirable to humans, and as a result has eliminated reproduction via fertilized seeds.

Distribution: Lebot and Simeoni (2004) offered an explanation for the relationship between the "quality" of kava (*P. methysticum*) and the tropical Pacific geographical distribution of its use for psychoactive and other physiological effects:

> "We argue that kava (*Piper methysticum* Forst. f.) is a Pacific domesticate that originated in Melanesia. We provide botanical, chemical, genetic and cultural evidence to suggest that farmers in the northern part of Vanuatu were the first to select the species as an asexually reproduced root crop. From Vanuatu, cultivars were carried eastward into Polynesia and westward into areas of New Guinea and [northwestward into] Micronesia. Using herbarium data, isozyme and AFLP markers, we correlate the information gained from field surveys to HPLC analyses and attempt to demonstrate that chemotypes result from a selection process that is still active. The selection of particular mutants by farmers must have been, and still is, a rational process to preserve new characters when they appeared. Growers have selected cultivars to improve the chemical composition responsible for the physiological effects. Field experiments demonstrate that the

chemotype [e.g., the presence or absence of specific kavalactones, which are the physiologically and psychologically active ingredients] is genetically controlled although the kavalactones content is determined by both genetics and environmental factors. The control and improvement of quality is therefore a cultural approach that aims at the identification of locations suitable for the cultivation of particular kava varieties. The appreciation of quality appears to reflect the different cultures within Melanesia and between Micronesian, Polynesian and Melanesian consumers. Different ways of benefiting from the psychoactive properties of the plant explain the use of particular chemotypes and therefore the selection operated to preserve them. Clearly, the word kava refers to different beverages that produce different physiological effects according to what consumers desire."

Cultivated *P. methysticum* is generally adapted to grow in relatively open areas of year-round wet tropical forest with well-drained soils which allow adequate air to reach the roots. It, therefore, thrives in many hillside locations, usually as an understory crop because too much sunlight can be harmful, especially in its early growth. On Pohnpei, it is usually grown in more or less shady, damp areas at the edge of forested areas. Since kava cultivars are all sterile and cloned from its wild ancestor *P. wichmanii*, it is almost exclusively propagated vegetatively and "only grows wild in abandoned garden sites where branches bowing off some abandoned shrub occasionally sprout new plants from internodes" (Lindstrom 2007). Today, kava is grown across much of the tropical Pacific, commonly for production of a relaxing therapeutic and mentally stimulating psychoactive beverage. Each of its cultivars not only has different requirements for successful cultivation, but also displays unique characteristics both in terms of its appearance, and in terms of its psychoactive properties (Lebot et al. 1997).

The traditional cultural patterns of kava consumption throughout most of Polynesia indicate possible cultural connections with Pohnpei, or an even more ancient history of kava use in Remote Oceania; in other words, since the theoretical origin of the domesticated kava plant and its initial dispersal point is thought to be in Vanuatu, well beyond Polynesia and Micronesia, it is clear that this keystone cultural plant must have been brought to other parts of the Pacific hundreds or maybe thousands of years ago. Local chiefs in the many islands of Polynesia, especially in western Polynesia (e.g., Tonga and Samoa) closely controlled the consumption of kava and related rituals and ceremonies associated with this cultural keystone plant. Based on language associations, Crowley (1994) suggested that kava use was introduced to some areas of Micronesia through contact with Polynesians, which may indicate direct exchange and influence between these areas well before European contact. In Micronesia, only Pohnpei and Kosrae have traditional histories of using kava. On Pohnpei, ethnographic sources and oral histories tell us that kava use follows a strict protocol controlled by elites and that it is used in prescribed social circumstances. Although its use was reserved traditionally for high-ranking individuals on Pohnpei, today its use in Pohnpei is widespread. It is believed by some scholars that the **Sakau** plant was introduced to Pohnpei directly from Kosrae, where it is known as **Suhka** or **Seka**, but it is little used today on that island.

General uses: Traditional consumption of kava, involving a psychoactive beverage that includes rootstock material pounded and strained from the *P. methysticum* plant, has long been and in cases remains a significant cultural trait generating communal acceptance. The physiological effects of this kava beverage are analgesic, anesthetic, and anticonvulsant and are generally those of a narcotic rather than of a stimulant. The immediate effects on drinkers are mainly relaxing, rendering those who sit through a long session of kava consumption calm and sleepy (e.g., see Pollock 1995: 12). Traditionally, kava is prepared by either chewing, grinding, or pounding the plant's rootstock. In Pohnpei, kava rootstock is pounded on a large stone slab using a round rock. After some cold water is added to the ground-up rootstock, the resultant liquid is consumed at the discretion of traditional rules, or nowadays sometimes in non-traditional situations such as kava bars. Kava beverages are still consumed traditionally in a number of islands throughout Melanesia and western Polynesia, but in Micronesia they are now only consumed openly on Pohnpei, and more recently on some islands in the Kiribati. In fact, distribution and original ritual utilization of kava has been greatly reduced in several areas of its traditional use in the Pacific Islands; this generally occurred following the arrival of European missionaries in eastern Polynesia and American missionaries on Kosrae in Micronesia (Brunton 1989; Lebot 1991; Lebot et al. 1997).

Prior to the arrival of missionaries and other foreigners, kava-drinking in Micronesia was well established in Kosrae and Pohnpei (Petersen 1977, 1995). However, in Kosrae, the kava beverage was banned by the Christian missionaries in the 19th century starting in 1852 (McGrath 1973: 64; also see Riesenberg 1950), and in recent times its cultivation and usage is officially reported as having been completely abolished on that island (Lebot 1991: 174, Merlin et al. 1993; Marshall and Marshall 2002).

However, on Pohnpei, although missionaries denounced the use and cultural significance of kava, its use persisted:

> " The missionaries saw *sakau* "kava" as the ultimate symbol of the island's dark ways; its use separated heathen from Christian. Albert Sturges [1819–1887, an American Protestant missionary in Micronesia who helped set up the Congregational church on Pohnpei] wrote: "To this people, kava is the only means of communication with their spirits; they hold a cup of drink, always in their hands, when addressing the object of prayer… Kava is here what the cross is to the Christian; it fell from heaven and is the only means of obtaining a hearing here." The missionary was quite correct in one aspect of his assessment; *sakau* did indeed constitute the sacrament of the island. Pohnpeians employed *sakau*, a gift from the god Luhk en Leng, at all ceremonial functions." (Hanlon 1988: 114)

Consequently, Pohnpeians today are the only Micronesians to have openly maintained the traditional cultivation and drinking of kava in this region. On the other hand, in some areas today, the use of kava has increased, especially as a recreational activity within a variety of different cultural contexts.

Since the capital of the FSM is located in Pohnpei, many government officials from other islands that live and work in Pohnpei have become exposed to kava consumption; and after indulging in the beverage, some of these non-Pohnpeians have continued consuming it. In addition, some residents of the Republic of Kiribati, the eastern Micronesian nation consisting of 32 atolls and 1 raised coral island, have now introduced kava to some places, where it has become a relatively popular drink (e.g., see kava.com/kiribati-and-kava/ or southpacificspecialist.org/nightlife-entertainment).

However, in Kiribati and other places where kava is consumed by Pacific islanders who do not cultivate it, kava is entirely separated from its ancestral spiritual and religious significance. Without its "traditional" associations, and when it is connected to commerce involving money exchange, kava becomes merely another commodity, as it has in several regions in the Pacific, including some areas of Micronesia (e.g., see Lebot et al. 1997). There has been an ongoing controversy concerning the positive or negative aspects of kava-drinking that has had legal, social, economic, and other cultural and medicinal impacts in various areas of the Pacific, and well beyond. In Kiribati, many residents and respected leaders have objected to incorporation of kava-drinking, which is not traditional in Kiribati, and decried its use generally. Kiribati's former president, Sir Ieremia Tabai (2017), called on the Kiribati government to deal with what he characterized as "the excessive consumption of kava in Kiribati, saying people are getting sick and lazy because of kava," and warned that "there was no development in the country when men were sleeping all day" (www.rnz.co.nz/international/pacific-news/337777/kiribati-people-cursed-by-kava).

In contrast, Aporosa (2019a) argued for a "de-mythologizing and re-branding of kava as the new 'world drug' of choice":

"It is now the turn of kava to be re-evaluated, to see whether there are properties in this plant that might be readily substituted for more conventional and harmful drugs, for instance tobacco and alcohol. However, as highlighted by Norton and Ruze (1994), kava like cannabis, has an enduring reputation that still makes it difficult for many to accept. Kava has been mythologized as an illicit alcohol, highly addictive, and causing physical harm. When examining the history of kava use in traditional contexts and considering the evidence now available, it is possible to demythologize this characterization. Looking at the potential benefits, it is time to re-brand kava, not only on the grounds as a relaxant, but in possessing life enhancing medicinal properties and as an alternative to alcohol, understanding that will be beneficial to policy makers, doctors and pharmacists." (Also see Aporosa 2017, 2019b).

In this new context, Aporosa et al. (2020) have also addressed the modern "problem" or dangers presumably associated with drinking large amounts of traditional kava and subsequently driving a motor vehicle:

"Kava drinking is a tradition among Pacific island people, although growing in popularity with other ethnicities. However, drinking substantial quantities of kava has raised concerns regarding physical manifestations of slow response and lack of precision in bodily control. These impairments can have significant consequences when after consuming large volumes of kava an individual makes a choice to drive."

In their attempt to identify potential risks for related to "kava drink-drivers," Aporosa et al. (2020) measured a set of "cognitive functions following high traditionally consumed kava volumes (greater than 2,000 mg of kavalactones)." Their study results revealed no statistically significant differences "between control and active participants at any measurement point over a 6-hr testing period regardless that the movements and speech of the active participants were observed to slow as the test session and kava consumption progressed." In conclusion, these researchers referred to discrepancies "between test results and observations during testing and by road policing officers" and argued urgently for more research about this kava-drinking and driving safety issue, and emphasized the urgency for additional study in many areas where kava is consumed large or small doses, including in Pohnpei. It should be acknowledged that we need additional scientific and cross-cultural investigation of the associated risks of kava use, especially non-traditional consumption of psychoactive kava beverages and other preparations, wherever these take place in modern societies in which motor vehicles and kava consumption are common. More specific discussion regarding the safety of kava use is presented below.

Uses in Pohnpei:

"Kava occupies a central place in the lives of the people of Pohnpei... In few—if any—societies is it of more importance. Pohnpeian kava's mana—its spiritual value and power—crowns a complex hierarchy of social rank. It invests every activity with solemnity, purpose, and meaning. It commands greater and more consistent respect than any leader or other element of the social order. Kava retains ascendant symbolic value, even for those who do not drink it." (Petersen 1995: 34)

It is hard to imagine a sound more emblematic of Pohnpei than the ringing of water-worn stones being pounded against large flat, specially chosen basalt slabs (*peitehl*). Three or four men sit surrounding the **Sakau** pounding rock, rhythmically

crashing their hand-held stones upon cut pieces of the rootstock. The pounding releases the psychoactive substances (kavalactones) from the rootstock. After sufficient pounding, the mashed **Sakau** is moistened with water and laid out on the fibrous bark of **Keleu** (*Hibiscus tiliaceus*). The slimy bark is then rolled around the mashed rootstock and strained over the stone or into buckets.

This process has been happening on Pohnpei for centuries, going back perhaps to the dawn of human presence on the island. In her archaeological doctoral dissertation, Levin (2015a) referred to the early human occupation of Pohnpei, and its high island neighbor Kosrae, as beginning 2,000 years ago or more by Pacific island voyagers. These early human colonizers probably came north from the ancestral Lapita culture in Remote Oceania. Eventually, they arrived in these Micronesian islands, where they established sustainable farming and fishing systems. The agriculture began with slash and burn farming, and over time developed into intensive cropping systems focused on agroforests and cultivated swamps. The archaeological evidence for these ancient Micronesian settlement patterns and environmental adaptations included "the presence of storage pits [for crops] and **Sakau** (*Piper methysticum*) pounding stones" (Levin 2015a: 115). Using this and other forms of evidence to reconstruct ancient Pohnpeian "lifeways" by studying the material and non-material traditions of modern Pohnpeian society, Levin suggested that the early islanders eventually began growing "non-utilitarian crops... for the purpose of ceremonial feasting and gaining status on Pohnpei." Levin's work focused largely on the discovery and documentation of ancient plant remains which contained long-lasting phytoliths ("plant stones"), which are microscopic structures made of silica that develop in some plant tissues, and which persist after the plant has decayed. This microscopic evidence can be used to identify the ancient plant species. According to Levin (2015a: 111), among the plant families found in tropical Pacific islands that have been observed to produce phytoliths "with some taxonomic value" is the pepper family, Piperaceae, which "suggests that phytolith analysis will be useful in the study of **Sakau** (*Piper methysticum*)." However, while some traditionally important plant species do produce phytoliths, not all of these species do so in a part of the plant that is useful for food or beverages, or other culturally significant products, and this situation seems to apply to the kava plant:

> "Notably, while the Piperaceae family (including **Sakau**) produces taxonomically useful phytoliths in the leaves, in Pohnpeian archaeological contexts, the most heavily used part of the plant is the roots, <u>which do not contain diagnostic phytoliths</u>. Thus, Piperaceae phytoliths are most likely to be useful in understanding landscape use [and not necessarily direct use of a kava beverage in ancient times]. It should be noted that it is ethnographically known that the stems and leaves [of **Sakau**] are transported into the *nahs* [i.e., feast house, community meeting house, or place for worshiping during feasting, see Balick and Lee 2009], so they may provide some information on **Sakau** processing and consumption, but not as much as would be produced if the heavily processed roots contained diagnostic phytoliths." (Levin 2015a: 123; also see Levin 2019; Piperno 2006, underlined for emphasis).

Legends and proverbs:

> "Uitanngar [Udenngar] came with Luhk from an unknown place. They made an offering from which came the creation of **Sakau**. They collected **Sakau en Na** and another small grass, pounded these, and then Uitanngar mixed skin from his heel and pounded that with the grasses; and finally they drank. That is why **Sakau** smells bad like a festering wound. Then they came down to a place in Uh and made another offering to the gods, especially the god of thunder [Luhk Nahnsapwe]. They planted the mixture and followed the growing stem up to the heavens. They made another offering there. The pounded **Sakau** fell to earth and landed in various *nandiwi* [fountain or spring of water] here and there. People saw rats eat the **Sakau** and grow dizzy and sluggish. Following this observation people tried the **Sakau**, and thus it came to be used by humans."

It is of interest to note that another drug plant (in this case, a stimulant) is said to have also been discovered by observing an animal consuming part of the plant; in this case, it is said that people noticed goats eating the beans of a certain tree in the highlands of eastern Africa and becoming more active afterwards. These people tried the beans themselves and thus discovered coffee. Such legends illustrate the traditional and scientific practices of observation and experiment—noticing an effect on animals and trying, or experimenting, to see if there would be a similar effect on humans. Scientists often use rats in this way now. The legends also address another point of interest to scientists: the origin of plants. Have these plants developed, or evolved as species in the area they are now found through a process known as natural selection, or have they been brought in from another place by people?

Takain wel ("Stone of the bush") is a traditional Pohnpeian proverb (*lepin kahs*) that refers to *sakau* use. The word "stone" in the proverb refers to the kava pounding stone or flat rock platform. One interpretation of this proverb is that in "the old days, commoners could not drink kava by themselves, so they sometimes went into the bush to do it secretly" (Reisenberg and Fischer 1955: 16):

> "After the drink was prepared, each participant offered the filled cup to the chiefs and guests, turning away his head in so doing. This traditional gesture dates back to ancient times, when natives without rank were forbidden to look at kings or chiefs. Before contact with white men, [*sakau*] could be prepared only by noblemen. During the German period (1899 to 1914) this restriction was removed, and since that time anyone may make the drink. During [World War II], the Japanese limited [*sakau*] drinking to once a week and to authorized ceremonial occasions (Civil Affairs handbook, 9). The restriction was enforced because [the Japanese believed] indulgence made the [Pohnpeians] lazy and because it was believed that the drink had certain toxic effects on the body." (Glassman 1952)

Environmental impact: **Sakau** is planted from stem cuttings, often in places near the homes of the farmers so that the plants can be watched and thus kept from being harvested by others without their permission. As noted above, **Sakau** is also cultivated in various other areas in the *nansapw*, and some places in the *nanwel* which are kept secret by farmers. Areas where **Sakau** is cultivated in the forest are known as *diwi* (spring); some are special clan-cultivated areas (*kahpw*). The suitability of a *diwi* is indicated by the presence of the **Kehrari** / **Koahrari** plant (*Boehmeria celebica*). Traditionally, the *sakau* beverage was reserved for high-ranking individuals on Pohnpei, but in recent decades, because of the growing use of kava-based beverages and their commercialization, **Sakau** cultivation and consumption have become very common. **Sakau** is now a significant cash crop in Pohnpei, and as a result, the indiscriminate clearing of land (often hidden in forested areas in the uplands and in steep forest areas where slope instability is a problem) to grow **Sakau** became excessive and damaging. Opening the forest, in many areas, causes severe harm to mountain slopes and causes shallow landslips, higher risks of landslides, and increased erosion in the important watershed areas of the forest. This ecological damage also affects marine environments on Pohnpei as runoff into the ocean increases with land clearing. By the turn of the millennium, it became sufficiently clear that education and conservation measures had to be undertaken to curtail commercial cultivation of **Sakau** not just hidden in the uplands but also in the open. In an integrated approach to watershed conservation (including the "Grow Low" campaign), farmers in Pohnpei have been encouraged to grow **Sakau** in lowland areas, and not to clear native forest, especially in the uplands, to cultivate **Sakau** (e.g., see Merlin and Raynor 2004, 2005).

Medicinal uses: According to Balick et al. (2009: 468), the **Sakau** plant "has an effect similar to taking a laxative," and can be used to treat stingray stings, open cuts and "*soumwahu en insensuwed* (sadness or depression)." Regarding the treatment of open wounds, leaves of **Sakau** can be used by persons on Pohnpei who cut themselves accidentally while working in the forest; the stoppage of bleeding can be accomplished by pounding or squeezing kava leaves until liquid seeps out, after which this juice should be dripped on the bleeding cut, and then the wound should be covered with a **Sakau** leaf that can be tied in place to protect the injured area (see www.comfsm.fm/~dleeling/botany/1998/vhp/sakau.html). Consuming a lot of *sakau* has a specific effect on the person's skin. *Mwelengen sakau* is a Pohnpeian phrase referring to dried, scaly skin that people get from drinking too much *sakau*. Although this dermatological ailment may be distressing to observe, it is reversible if *sakau* consumption is curtailed.

For much more information on **Sakau**, including its many ceremonial uses and cultural significance on Pohnpei, see Balick et al. 2009; Lebot et al. 1997; Pollock 2009; and www.comfsm.fm/~dleeling/ethnobotany/text/ethnobotany.html.

Cultivars in Pohnpei: Two basic cultivars of **Sakau** are grown in Pohnpei. ***Sakau Rahmedel*** (or ***Rahmoadoal***) has smooth, light-colored stems with long internodes. ***Sakau Rahmwanger*** has darker, non-smooth stems with pronounced purplish or blackish spots. Each has additional names or sub-types (see p. 431). When ***Sakau Rahmedel*** and ***Sakau Rahmwanger*** grow together, they are called ***Malahd***. There is the possibility of new cultivars being imported to Pohnpei (see danaleeling.wordpress.com/2011/12/17/hawaiian-kava).

Toxic aspects and potential side effects—fact or fiction: Based on its latest (2016) comprehensive review of kava safety, the Food and Agriculture Organization (FAO) and the World Health Organization (WHO) made a balanced statement about the potential benefits and dangers of kava use and abuse. These international organization acknowledged in their review that kava beverages have a lengthy history of consumption in the tropical Pacific region areas where they have been traditionally used, especially in "community ceremonies." They also noted that in more recent times, kava use has largely transitioned from traditional patterns of usage to consumption for recreational purposes, both within Pacific islander societies as well as beyond in the more expansive international community, and generally where it is used, people believe, based on their often long-term experiences, that kava consumption is both safe and pleasurable:

> "There is little documented evidence of adverse health effects associated with traditional moderate levels of consumption of kava beverage, with only anecdotal reports of general symptoms of lethargy and headaches. Whether this reflects genuine low incidence or an under-reporting of adverse health effects is unclear. Clinical trials examining the efficacy of aqueous extracts of kava in treating anxiety, although limited, have also not identified adverse health effects. On the other hand, there is strong evidence that high levels of consumption of kava beverage can result in scaly skin rash, weight loss, nausea, loss of appetite and indigestion. These adverse health effects, while significant, are considered to be reversible upon cessation of kava use. Other possible effects include sore red eyes, laziness, loss of sex drive and general poor health. No effect on cognition, which might be associated with the pharmacological activity of kava, has been identified. No information is available on the potential for kava beverage consumption to impact on the incidence of chronic disease. Moderate to high kava beverage consumption also produces a reversible increase in the liver enzyme gamma glutamyl transferase (GGT), which may be an early indicator of cholestasis. Clinical surveys in Aboriginal communities in northern Australia with a history of heavy kava use have not revealed any evidence of long-term liver damage associated with consumption of kava beverage." (FAO and WHO 2016; also see Teschke et al. 2011; also see Balick and Lee 2002, 2009)

Lee Ling offered the following sobering comments about past and present use of **Sakau** on Pohnpei in his discussion "Drugs: For good or for evil?":

> "For over two millennia the peoples of Pohnpei have used *sakau* as a central part of their custom and culture. The use of *sakau* did not prevent Pohnpeians from thriving and growing as a civilization. Yet [modern] overuse of recreational *sakau* may be altering and damaging the extended family structure. Some children are effectively "*sakau* orphans" due to parental "almost every night" *sakau* habits. *Sakau* used non-traditionally is damaging to the transmission of cultural memes [ideas]. This is true of many plant-based psychoactive compounds. Used in their traditional setting the drug is [an] integral part of the custom and culture. Used in non-traditional ways, especially when over-used or abused, then the compound will damage the culture, the custom, and in some cases the health of the people..." (www.comfsm.fm/~dleeling/ethnobotany/text/ethnobotany.html#sumner2008)

Today, the use of *sakau* on Pohnpei is extremely widespread, with men and women of various ages and socioeconomic and political standing partaking informally. A survey conducted in Pohnpei some years ago revealed that 83.6% of males and 53.2% of females drink *sakau*, with males reporting that on average they consume the beverage for 5.4 hours per week (Balick and Lee 2009: 189–190). Typically, at the end of the work day in Kolonia, the island's capital and urbanized center, several *sakau* bars open for business. In many of them, one can hear the sounds of pounding, but some bars now use mechanical grinders to crush the rootstock. Some doctors on the island prescribe *sakau* as a cure for hypertension and high blood pressure. Today, more *sakau* is probably being consumed on Pohnpei than ever before.

> "There is still much to be learned about contemporary cultural use of **Sakau**, as well as the science of how its effects influence human physiology. Potentially, this plant could play a much more important role in global health care, but only if questions about the benefits as well as the adverse effects of some of its preparations are answered via research in basic science and clinical trials. We suspect that **Sakau** will continue to "speak" on Pohnpei for many generations to come, even as globalization exerts such a dramatic change on local lifestyle." (Balick et al. 2009: 199).

Wengid sakau: squeezing the liquid from pounded **Sakau** root wrapped in **Keleu** bast (inner bark of wild hibiscus).

Pasapas en sakau: plants placed on platforms side by side as an offering at an important cultural event.

Sap sakau: cutting and preparing plants to be cleaned and pounded.

Sukusuk sakau: the traditional pounding of roots on a *takain sukusuk*, the designated pounding stone.

Consuming sakau from a *delen sakau*, traditional cup made from half of a coconut shell.

Family and friends enjoying a *pohn takai*, a *sakau* "party on the rock."

▸ **Konok** or **Konoak** (P)
Piper ponapense

Description: A member of the Piperaceae (pepper family), *Piper ponapense* is a dioecious, evergreen climbing vine that is thin-stemmed and perennial. The chartaceous (paper-like textured) leaves of this vine are 3–5 cm (1–2 in) wide and 8–15 cm (3–6 in) long, and they are arranged alternately with an oblong-elliptic shape with a distinct palmate venation pattern. The inflorescences of *P. ponapense* are mostly solitary, producing reddish male flowering spikes. Note: *P. fragile* Benth., another native pepper vine found in some islands in Micronesia, is only referred to in the context of the following discussion that focuses mainly on *P. ponapense*.

Distribution: Indigenous *Piper* vines in Micronesia, specifically *P. ponapense* and *P. fragile*, are often found in wet-forested areas of the volcanic islands of Micronesia. Glassman (1952: 55) referred to *P. ponapense* as being present in Pohnpei and Chuuk. These vines also exist on some low coralline islands; for example, see Stone (1959: 90, 1960a) and Fosberg (1960: 6) for reference to the presence of native *Piper* vines on the ground and trees in humid, breadfruit-dominated forests on atolls in the region.

Uses in Pohnpei: *P. ponapense* (and *P. fragile*) are reportedly used for a number of medicinal purposes.

Glassman (1952: 55) indicated that parts of *P. ponapense*, known in Pohnpeian as **Konok** or **Konoak**, are used to alleviate stomach aches. Regh and Sohl (1979: 42) noted that this species "is used in the treatment of toothache." Balick et al. (2009: 473) referred to the personal adornment and ornamental use of this plant, and to a wide variety of medicinal applications:

> "The vines with leaves are used in *mwaramwar* (head garland), especially for men during a *kamadipw* (feast)... To treat *soumwahu en moang medek* (headache) caused by *soumwahu suwed* (possession of an evil spirit)... To treat joint pain... To treat back pain... To treat *ngih medek* (toothache)... To treat *soumwahu en ohl* (men's sickness)... To treat *tihnsou medek* (backache from heavy lifting or staying somewhere cool and wet)... To use to stop urination... To treat chills... To treat *pwinan* (pinkeye)... To treat sore eye... To treat boils on the buttocks, under the testicles, or underarms... To treat depression... To treat headache... To treat toothache... To stop bleeding of cuts... Alternatively, to use to stop bleeding... To treat a cut, leaves are chewed and put on the cut to stop the bleeding and promote healing... To use as a treatment for syphilis for males... To treat gonorrhea... To treat *soumwahu en lih* (women's sickness)... To treat babies that fall down... To clean the *sakau*... This plant is used as an ornamental..."

It has also been reported that if one blows into the end of the stem of **Konok**, "a liquid will come out the other end of the stem that relieves the symptoms of pink eye" (see danaleeling.blogspot.com/2016/09/plants-that-heal.html). The medicinal efficacy of *P. ponapense* may be, at least partly, due to the presence of tetrahydropyridine alkaloid, piperidine, produced in the stems and leaves of this species (e.g., see Patel et al. 1997).

▸ **Kehrari** or **Koahrari** (P), **Oarmo** (Mw), **Oaraema** (Pi), **Olongaa** (Nu), **Walengaa** (Ka)
Pipturus argenteus

Description: *Pipturus argenteus* is a native shrub or small tree that grows to heights of 5–7 m (15–20 ft) or more, with dark reddish brown branches typically bearing noticeable leaf scars. Its broad, soft, alternately arranged leaves are oval in shape, about 12 cm (~5 in) long and half as wide; in addition, the leaves are acuminate and have distinct 3-nerved venation, coarse serrations on the margins, and white, woolly growth underneath. *P. argenteus* is a dioecious species, which means the male and female reproductive organs are produced on separate plants. The male flowers have four free stamens normally producing abundant pollen. The female flowers are greenish white, lacking petals, and are produced in small head-like clusters; the fruits are small, fleshy, whitish false-berries. *P. argenteus* belongs to Urticaceae, the nettle family, which includes several species with stinging hairs. Some contain mild poison, which can cause blisters and pain that may last for several days. This native species of *Pipturus* is not one of the poisonous nettle plants. Indeed, *P. argenteus* is the host plant for several butterflies and other insects; for example, its flowers are used by adult butterflies as a source of nectar.

Distribution: *P. argenteus* is a widespread species that is reportedly indigenous from some islands in the Indian Ocean to insular southeast Asia, as well as throughout much of Melanesia to areas in eastern Polynesia, and as far east as the Marshall Islands and Kiribati in Micronesia. Locally, *P. argenteus* is a relatively common shrub or small tree in disturbed habitats, especially in the eastern range of its distribution; it is found mostly at elevations near sea level, especially on limestone and on atolls, but may be found up to about 1,250 m (~4,000 ft) on some high islands closer to the equator. As a colonizer of disturbed areas, this species is most often found in the lowlands, especially on the coral islands of Micronesia, near the ocean on rocky cliffs, in coastal thickets, as well as in dry or open forest or on the outskirts of forested areas and in clearings. It is also an important early successional species (pioneer plant) in abandoned gardens or formerly cultivated areas.

General uses: Strong cordage extracted from the bast fiber of *P. argenteus* is "used for fishing lines and nets throughout the Pacific" Clarke and Thaman 1993), and in some Pacific islands, *P. argenteus* bast stem fibers are reportedly "used in fine mats for ceremonial exchange" (Thaman 1994: 169). The raw fruit of *P. argenteus*, which is small, whitish, and mulberry-like in appearance, contains many tiny seeds and it is eaten in some areas as wild food, though some do not find it particularly tasty. The seeds are also eaten traditionally in some cultures by pregnant women and newborn babies, and the cooked leaves are consumed as a vegetable, often cooked in coconut milk (e.g., Clarke and Thaman 1993). There are also many medicinal uses for *P. argenteus* in Southeast Asia and New Guinea (e.g., see www.proseanet.org, World Health Organization 2009: 191). Generally, the therapeutic uses of *P. argenteus* include aiding in birth as an abortifacient, as well as helping cure sores, boils, aching teeth, dysentery, cold, cough, asthma, stomachache, tuberculosis, malaria, and headache (e.g., see Nick et al. 1995).

Uses in Pohnpei: *P. argenteus* is known in Pohnpeian as **Kehrari** or **Koahrari**. Balick and his colleagues (2009: 510) briefly referred to some of its uses: "The wood is used for firewood... To use to make a fishing line... The plant is used for fiber... The leaves were once used as toilet paper because they are strong, fuzzy, and soft... The leaves are used as a fertilizer for swamp taro..."

Uses in Mwoakilloa and Pingelap: This species in known as **Oarmo** in Mwoakilloa and **Oaraema** in Pingelap. Glassman (1953: 295) reported it growing to be a tree "20 feet high, in coconut grove[s]" where it could be found frequently on Mwoakilloa. St. John (1948: 110) found it in the inland forest areas in Pingelap, growing to "several meters in height." He also noted that bast fiber was extracted from this woody species and "used for making fish nets." More recently, Balick and his team collected this species on Pingelap in 2002 (specimen 4301 deposited in NYBG herbarium).

Uses in Kapingamarangi: Hiroa (1950: 123, 127–129) referred in some detail to the now-abandoned traditional use of bast fiber from a "nettle" species occurring on this atoll, specifically the former use of the inner bark of the 'Warenga' (*P. argenteus*). Indeed, Hiroa described the production of traditional cordage on Kapingamarangi Atoll, where "fiber obtained from 'Warenga' bast is recognized as being much stronger than **Hau** [*Hibiscus tiliaceus*] or breadfruit, but owing to the labor of preparing enough from the scanty bast, it has gone completely out of use." Indeed, use of bast fiber from this plant was abandoned by the time Hiroa did his extensive field work on that atoll in the late 1940s. Hiroa (1950: 131–133) noted that the disappearance of 'Warenga' bast fiber production and its use had "been accelerated by the easily acquired trade cordage." However, Hiroa meticulously described how to prepare the traditional cordage made using the fiber of *P. argenteus* (see figure 78 in Hiroa 1950: 129 for an image of a high-ranking male extracting fiber from this stems of *P. argenteus* in Kapingamarangi). The cordage made from this plant's fibers was "called *roahi* as are hibiscus and breadfruit cords," and in past times it was "used for fishing lines and some forms of net." The major techniques used to catch fish, the main source of protein for the people on this atoll, involve "the use of nets, hooks and line, and traps," and in the past, previous to "trade contacts," Hiroa (1950: 210) tells us that much time must have been expended in preparing breadfruit, hibiscus, and 'Warenga' bast and "twisting the cord to make the various forms of nets and the long lines for deep-sea fishing." Hiroa (1950: 132) also referred to what a local informant told him about the Kapingamarangi name for *P. argenteus*: "the old name for the plant was 'Oronga' [which is still the pronunciation used in Nukuoro Atoll] but that it changed to 'Warenga' because of the sticky sap (ware) which clings to the fingers during scraping." For clarification, according to the standard Kapingamarangi orthography, the contemporary spelling of 'Warenga' is **Walengaa**, and the sticky sap is *walee*, which is also the term used to describe slimy and sticky things.

▸ **Kamal** (P)
Pittosporum ferrugineum

Description: *Pittosporum ferrugineum* is an evergreen woody shrub or small tree species that ranges in height from about 1.5 m (4–5 ft) to 3–15 m (~10–45 ft), but sometimes even reaching 22 m (~70 ft). The trunk is usually no more than ~30 cm (12 in) wide, and like many others in its botanical family, Pittosporaceae, *P. ferrugineum* produces yellow sapwood, which is fragrant when freshly cut. Its leaves are spirally arranged and clustered at the twig tips, and the small whitish flowers, which have a honey-like scent, are formed in many-flowered terminal panicles. Its bright red seeds are formed in capsules that have a sticky pulp.

Distribution: *P. ferrugineum* occurs naturally from Southeast Asia through areas of New Guinea and northern Australia to the Solomon Islands, with an altitudinal range from sea level to about 600 m (2,000 ft). It grows in seasonal and dry forests but is most commonly found in beach forests and along streams. In Micronesia, *P. ferrugineum* is indigenous to the high volcanic islands in Chuuk lagoon, including Weno, Tonowas, Tol, and Fefan, as well as on Pohnpei and Kosrae, and on some atolls including Satawan and Nomwin.

General uses: This small tree is used in horticulture as an ornamental and is noted for its ease of cultivation, producing masses of small white flowers and yellow fruits. The bruised leaves and fruits of this species have been used as a fish poison. The bark is used as an antidote to poison. It is shredded, baked in bamboo, and then eaten with traditional ash salt and green vegetables, which is said to enable poisoned individuals to vomit the toxic materials they have ingested. The root bark is pressed into the cavity of an aching tooth to bring temporary relief, while in the treatment of malaria, a mixture of the chewed bark and traditional ash salt is dripped into the nose to treat an enlarged spleen caused by malaria and to treat stomach ache. Parts of this plant are also used generally to reduce a fever (e.g., see Quattrocchi 2012: 2968; World Health Organization 2009; Burkill 1966).

Uses in Pohnpei: *P. ferrugineum* is a native tree known as **Kamal** in Pohnpeian, and its bark is used to make a coloring dye. It is also said to be used as a magical medicine to chase away dwarf spirits (*sokolei*). According to Balick and his colleagues (2008: 473), **Kamal** is used traditionally "to make *wimi en sokolei* to treat painful joints (this sickness starts with pain in the leg joints, then spreads up the body)..." In the legend of **Kamal**, it is connected to **Kempenial** (*Psychotria hombroniana*) via a legend recounted below); both of these native plant species are found in the *nanwel* (upland or native forest), but grow primarily in the *nansapw* (cultivated or human disturbed areas) of Pohnpei.

The Legend of Kamal: A man-like being, named Taimuan, that was part-spirit and part-human had a sister, Litapinmalekelek. Taimuan caught a human, put him in a cave, used a magic word to shut the cave, and called his ghost friends for a (cannibalistic) supper party. Taimuan's sister heard the man crying and helped him escape. When they fled, she asked every plant not to tell her brother the direction in which she went. Taimuan followed them, using his sense of smell. He asked all the plants where they had gone. The escaping man had urinated on the **Kamal**.

Now the **Kamal** has a bad smell. The woman had forgotten to ask the **Kempenial** plant not to tell, and it eventually told Taimuan which way they had gone. A lady approached and hid Taimuan's sister and the escaping man, who had something wrong with his eyes. The lady and girl put medicines and soil in water streaming into his eyes and that was the end of him.

▸ **Pwohmaria** (P) (Mw) (Pi), **Bua maliaa** (Nu), **Laagau hai akai** (Ka)
Plumeria rubra and *Plumeria obtusa*

Description: *Plumeria rubra* and *P. obtusa* are two closely related species and look similar in many ways. As ornamental shrubs or small trees, they are known in English as either *Plumeria* or frangipani. These widely used decorative species produce broad, somewhat umbrella-like shaped crowns about 4–6 m (15–20 ft) tall and thick succulent branches that bear five-parted, fragrant flowers, expanding into a "pinwheel." Each flower is about 5 cm (2 in) in diameter and may be white, red, yellow, pink, or multiple colors. The leaves are about 20–30 cm (8–12 in) long and 5–10 cm (2–4 in) wide. They are usually glossy green, though may be dull green, and are generally ovate in shape, though may be blunt-tipped (*P. obtusa*) or pointed (*P. rubra* var. *acuminata* or var. *acutifolia*). *P. obtusa* and its hybrids tend to retain their leaves all year round, while the deciduous types tend to lose them seasonally, with new leaves emerging during or following the flowering period. The fruit pods are long, oval-shaped, oppositely arranged, and are produced infrequently. *Plumeria* trees produce a milky sap, typical of members of its family, Apocynaceae, which is known in English as the periwinkle or dogbane family. *Plumeria* sap contains a little resinous rubber, and although it has medicinal properties, this sap can be poisonous in large doses.

Distribution: *Plumeria* trees can be cultivated rather easily from branch cuttings. They are native to tropical America from Mexico to Panama and were introduced widely to several Micronesian islands after European contact. In most cases, the first introductions of *P. rubra* plants occurred before WWI, while they were possibly later for *P. obtusa* plants. These *Plumeria* trees are now common ornamentals on many islands, normally in house gardens and particularly in urbanized places on tropical islands, including those in Micronesia.

General uses: *Plumeria* trees have widespread use as ornamental plants throughout much of the tropical world today. This includes their relatively recent introduction to many Micronesian islands, where their colorful and fragrant flowers are often used in making garlands and flower arrangements for local residents, and in some areas also for tourists and other visitors. Medicinal usage has also developed among some groups of Micronesian islanders (e.g., see Clarke and Thaman 1993).

Uses in Pohnpei: *Plumeria* (**Pwohmaria** in Pohnpeian) trees were reportedly introduced into Pohnpei by the Japanese for use as an ornamental (Ragone et al. 2001). Kanehira (1916) observed local Pohnpeian use of **Pwohmaria** flowers in *mwaramwar* (head garlands), with the flowers strung together with thread. Raynor and Fownes (1993: 47) listed the use of **Pwohmaria** for their attractive flowers and as an ornamental plant, but they also referred to its use as a trellis to encourage cultivated vine growth (e.g., for yams).

Uses in Mwoakilloa and Pingelap: Glassman (1953: 295) recorded *Plumeria rubra* on Mwoakilloa and Damas (1994: 238) listed it as "*Plumeria acutifolia*" (= *P. rubra*) on Pingelap. Both authors noted that it is cultivated as an ornamental tree.

▸ **Kausekesek** (P), **Kakine** (Mw), **Kawisaekaesaek** (Pi), **Manu lauligi** (Nu), **Laagau pogo ligiligi** (Ka)
Polyscias guilfoylei

Description and distribution: *Polyscias guilfoylei* is an erect, evergreen shrub or small tree species, considered to be native to the Old and New World Tropics. It has no confirmed truly indigenous range, though according to Whistler (2000: 376), it is possibly "somewhere in Melanesia." *P. guilfoylei* can grow to about 7–8 m (23–26 ft) tall. Its relatively few branches grow in an erect

manner and eventually assume a columnar shape with slender, spindly bunches of vertical stems. The long compound leaves have variably shaped pinnate leaflets that are oppositely arranged and commonly variegated, sometimes with white or pale yellow, and often with serrated margins. The small flowers and drupe fruits are borne in terminal panicles.

General uses: This panax species (*P. guilfoylei*), often referred to as the "geranium aralia," is widely cultivated in tropical lowland areas to serve as an ornamental hedge, windbreak, or boundary plant. It is valued particularly for its attractive foliage: "Two kinds of plants are mainly cultivated, the typical pinnate type and, less commonly, the more dissected types" (e.g., see Whistler 2000: 377). *P. guilfoylei* also reportedly has uses for medicine and food.

Uses in Pohnpei: *P. guilfoylei*, known as **Kausekesek** in Pohnpeian, has been used in this island in the making of *mwaramwar* (head garlands), and also as an aid in ironing garments by dipping the branches in a tub of starch water produced from processed **Mwekimwek** (arrowroot, *Tacca leontopetaloides*), and then shaking them over the clothes being ironed (Balick et al. 2009: 334).

▸ Kedei (P)
Ponapea ledermanniana and *Ponapea hosinoi*

Description: Despite its relatively small size, Pohnpei's diverse native palm flora includes five palm species, three of which are indigenous (native in Pohnpei, but also native elsewhere). These indigenous palms include: (1) *Ponapea ledermanniana* (known in Pohnpeian as **Kedei**); (2) *Nypa fruticans* (known in Pohnpeian as **Parem**); and (3) *Metroxylon amicarum* (known in Pohnpeian as **Oahs**). The other two native palm species occurring on Pohnpei are endemic (found nowhere else). These include: (1) *Ponapea hosinoi* (known in Pohnpeian as **Kedei**, same as its close, indigenous relative, *P. ledermanniana* (see above); and (2) *Clinostigma ponapense* (known in Pohnpeian as **Kotop** or **Koatop**). Note: Coconut and betel nut palms are, of course, also found on Pohnpei, but these two useful palms are not native on Pohnpei (i.e., they were introduced into Pohnpei by people).

Here, we focus on the two native *Ponapea* species in Pohnpei, known locally as **Kedei**: (1) *P. ledermanniana* is a fast-growing, tall, slender indigenous palm with edible fruit. It grows up to 12 m (~40 ft) tall and rarely extends beyond the surrounding tree canopy. Unlike its close relative *P. hosinoi* (see below), it has bright white inflorescences. The fruits of *P. ledermanniana* are about 5–7 cm (~2–3 in) long and tapered at both ends in an egg-like shape. (2) *P. hosinoi* is an endemic species of palm in the genus *Ponapea* that is only found on Pohnpei and is closely related to the indigenous *P. ledermanniana*. *P. hosinoi* grows to about 12 m (~40 ft) tall and bears flowers year round below its crown of leaves. *P. hosinoi* can be differentiated from *P. ledermanniana* by its much broader leaves, its smaller male flowers, and the dark brown hairs on the inflorescence branches (Glassman 1952: 111).

Distribution: *P. ledermanniana* is native (indigenous) to both Pohnpei and Kosrae Islands, and although it is common on Kosrae, it is reportedly rare on Pohnpei, where it is said to occur as isolated individuals, or in small populations scattered in low-elevation forests. The endemic *P. hosinoi* palm is found mainly in native Pohnpeian forest areas.

Uses in Pohnpei: *P. ledermanniana* palms produce woody material in their core, which has been used traditionally to make construction pegs, parts of local houses, canoes, and spears, and its nuts are reportedly "edible" (e.g., see Glassman 1952: 111): "The most popular weapon used by traditional Pohnpeian warriors was the spear, commonly made from **Ketieu** (*Ixora* sp.) and **Kedei** (*P. ledermanniana*) [and it] was only after the machete was introduced that the local weapons lost their uses" (Anson and Raynor 1993: 14).

▸ **Oahr** or **Topwuk** (P), **Supwuk** (Mw), **Sopwuk** (Pi), **Valovalo** (Nu), **Wolowolo** (Ka)
Premna serratifolia

Description: *Premna serratifolia* is a native (indigenous) shrub or small tree throughout Micronesia, reaching heights up to 7 m (20 ft). Its simple leaves are oval to elliptical and 5–18 cm (2–8 in) long. Numerous, dense, multi-branched inflorescences (corymbose panicled cymes) are produced at the tips of the leaf stems. The greenish white flowers are bisexual, and the many small drupe fruits are fleshy and dark blue to black when mature. All parts of the tree have a strong odor. Reportedly, the two Pohnpeian names for *P. serratifolia*, **Oahr** and **Topwuk**, refer to two clearly different types of this species that are "easily distinguished" locally on Pohnpei:

> " **Oahr** has distinctly smaller leaves that are more densely packed. Where **Topwuk** has [a] gangly, unkempt look, **Oahr** is neat, compact, and orderly. This is not a soil difference—**Oahr** and **Topwuk** planted next to each other maintain their distinctness." (www.comfsm.fm/~dleeling/angio/premna_obtusifolia.html)

Similarly, according to Lee et al. (2010: 115, citing Smith 1991), these Pohnpeian names are related to locally recognized differences among the individual plants of *P. serratifolia*:

> " On Pohnpei, the local varieties—**Topwuk** and **Oahr**—are recognized as two distinct species, however, taxonomically they are *Premna serratifolia*. **Oahr** has smaller, denser leaves and is considered the female and **Topwuk** has larger leaves with a more open flower head and is considered the male."

Distribution: *P. serratifolia* is a native (indigenous) shrub or small tree that occurs throughout Micronesia. It is found near seashores, in some lowland native forests, and in agricultural or other open areas, mainly in rocky places as a colonizer of disturbed areas, but also in limestone areas. It is not only widespread and indigenous throughout Micronesia, but is also found in many Pacific islands and in areas of tropical Asia. Its fruits are eaten and dispersed by birds and bats. Although it is rarely cultivated, it is generally not removed by farmers when it is found growing wild in areas where crops are planted.

General uses: The wood of *P. serratifolia* has traditionally been used along with that of *Hibiscus tiliaceus*, *Morinda citrifolia*, and other species to make fire; see under the uses on different islands described below. *P. serratifolia* is extensively used in the traditional medicine of India and other areas in tropical Asia and the Pacific for the treatment of a variety of afflictions. Recent research indicates that the root wood, root bark, and other parts of this small tree produce biological components with significant therapeutic value that helps explain its long-term historical and potential future medicinal use (e.g., see Bose et al. 2013; Habtemariam and Varghese 2015; Mali 2015, 2016).

Uses in Pohnpei: *P. serratifolia* (**Oahr** or **Topwuk**) is said to be one of three plant species used traditionally on Pohnpei when starting a fire by rubbing sticks together; the other species are **Keleu** (*Hibiscus tiliaceus*) and **Nihn** (*Morinda citrifolia*) (e.g., see Glassman 1952). *P. serratifolia* is also used for firewood on Pohnpei, and its leaves and crushed fruit have medicinal applications, including treatment for various skin diseases. This woody plant is rarely planted on Pohnpei since it usually sprouts spontaneously and then is allowed to remain where it grows in *nansapw*, lowland agroforest or cultivated areas. The trunks of larger individuals of *P. serratifolia* are normally found further inland on Pohnpei and are used to make the local drums, which were the only means of long-distance communication before more modern times (see *aip*, a Pohnpeian drum made from wood of **Topwuk**, on the picture to the right).

Balick et al. (2009: 411) provided much additional information about a number of traditional medicinal and other uses of *P. serratifolia* on Pohnpei:

> The timber can be used to build a house… The seeds are used in a game by children to shoot through a hollowed papaya petiole (*Carica papaya* L.) as a peashooter… The sticks are used in *pahi* (a game)… [It is also used] [t]o chase mosquitoes away… To use as a charm to break spells… To use as a charm to improve luck while fishing… For a charm to make females want a young man, the man speaks to a tree early in the morning and then shakes it to spill the early morning dew onto them… To use as a charm for protection from *silasil en madau* (ocean spirits)… To treat a person under the influence of a spell called *karotongehla* (cast by another person, which causes sleepiness and aches and makes the person want to stay home and rest)… To treat asthma… To cleanse the uterine-vaginal area after childbirth… To cleanse the uterus… To treat uterine problems characterized by stomach pain, uterine cramps, and female problems… To treat menorrhagia… To cleanse a woman's system after labor… To treat menstrual cramps… To treat hepatitis… To treat difficult urination… To treat constipation… To treat diarrhea… To treat fungal infections… To treat *kihl* (a red itchy rash in babies that can be contagious if other children have not had it)… To treat a thick, red skin disease (to keep it from spreading)… To treat skin disorders of children… To treat boils… To treat *medek* (aches and pains)… To treat *mehn sahk* (headache with dizziness)… To treat damaged nerves… To treat asthma… To use to clear the lungs… To reduce a high *karakar* (fever)… To treat chicken pox…"

In addition, Lee et al. (2010: 15) referred to the healing preparation involving **Oahr** (*P. serratifolia*, female type) for the itching bites of sand flies (*Phlebotomus* sp., "no-see-ums," known in Pohnpeian as *meinlohsap*): "This treatment utilizes several pounded leaves of **Oahr** mixed with coconut oil and rubbed on the bites twice daily until healed." Lee et al. also noted the use of parts of *P. serratifolia* in Pohnpeian remedies for treating boils (2010: 51), bad coughs (2010: 83), bad headaches (2010: 94), "mangrove sickness" (2010: 127), and in combination with parts of other plants species for additional medicinal treatments (e.g., 2010: 147; for additional therapeutic uses of this *Premna* species on Pohnpei, also see www.comfsm.fm/~dleeling/angio/premna_obtusifolia.html and danaleeling.blogspot.com/2016/09/plants-that-heal.html).

Uses in Mwoakilloa and Pingelap: This species is called **Supwuk** and **Sopwuk** on Mwoakilloa and Pingelap, respectively. In 1949, during a short visit to Mwoakilloa Atoll, Glassman (1953: 297) noted it as a common woody plant on the atoll, found in the strand vegetation and up to 30 ft tall. Lee et al. (2010: 115) reported that the leaves of *P. serratifolia* are used on Mwoakilloa Atoll in a bathing preparation for people with hepatitis:

> The leaves are boiled in a pot of water. The patient showers in the infusion until the symptoms of hepatitis disappear. For men, the entire body is bathed including the testicles."

On Pingelap, St. John (1948: 112) identified this species as *Premna* integrifolia (= *P. serratifolia*), noted its name as 'Sokiik' and indicated that it grows in "moist woods."

Uses in Nukuoro: *P. serratifolia* is known as **Valovalo** on Nukuoro Atoll, where it has been a source of timber for various implements such as a wooden drum which was reportedly referred to as *'aisi'* or *'asi'* (see Eilers 1934: 297); the drum "was hourglass shaped and made from *Premna* wood" (Evans 1975). Hiroa (1950: 269) referred to the use of 'Werowero' (= **Valovalo**) wood in Nukuoro to make the "long single-pointed spear for fishing" and indicated that its local Nukuoro name is derived from the verb *'wero'* [valo], "to thrust." Note: Stone (1966a: 132) listed the Nukuoro name for *P. obtusifolia* (= *P. serratifolia*) as 'Varovaro' [**Valovalo**].

Uses in Kapingamarangi: Hiroa (1950) reported the multiple uses of the woody material provided by *P. integrifolia* (= *P. serratifolia*), a shrub or small tree known in Kapingamarangi as 'Woroworo' (= **Wolowolo**). According to Hiroa, 'Woroworo' was being used traditionally to make coconut husking tools, canoe stringer poles, thatch needles, spears, and other wooden implements. Weins (1956) indicated that *P. obtusifolia* (= *P. serratifolia*) was a common plant in the understory environments on the islets of Kapingamarangi Atoll. A translation of part of a Kapingamarangi legend refers to this plant (Elbert 1948; Emory 1965): *Te mataku e au i toku aroho i ti koe fi rau woroworo tera. I reperep ira ke tapa.* (The *Premna* tree leaf there trembles like flashing lightning. Then you know that it is me.)

▸ **Pwaked** (P)
Procris pedunculata

Description: *Procris pedunculata* is a rather coarse, fleshy, succulent, indigenous herb with green stems that partially sprawl over the ground (decumbent) or grow erect to 30–60 cm (12–24 in) tall. *P. pedunculata* is sometimes woody at the base. Its leaves are alternately arranged and dimorphic, i.e., one of a pair is normally quite small, while the other is oblong-lanceolate, about

15 cm (6 in) long. The leaves are also oblique, pinnately veined, and have an acute base and acute-acuminate tip. The leaf margins are entire or obscurely sinuate near the tip. *P. pedunculata* produces axillary stipules and short petioles up to about 1 cm (~0.5 in) long. Its male flowers are borne in short white cymes, the female flowers are produced in the leaf axils as immature green, dense, hemispheric sessile heads that ripen to a more or less whitish, orange, or reddish soft fleshy fruit up to 2 cm (~1 in) in diameter. Note: This species normally produces whitish fruit, but a red-fruited "variety" can be found in some areas.

Distribution: *P. pedunculata* is a native species ranging from tropical Asia to the Marquesas and Tuamotus in eastern Polynesia. In Micronesia, it is found in shady, terrestrial coral limestone environments, and it is usually found growing among limestone rocks, sometimes as an epiphyte, or on dead logs. *P. pedunculata* is one of few epiphytes occurring in the Micronesian littoral vegetation and agroforests.

General uses: Although records of traditional or medicinal uses of *P. pedunculata* in Micronesia are meager, there are reports of its traditional use from some Pacific tropical islands beyond Micronesia. For example, *P. pedunculata* is said to be used to help facilitate human birth in Vanuatu (Jolly 2002: 158). On the Polynesian island of Rotuma in Fiji, the ripe, fleshy fruits of "*Procris pedunculata* (J.R. and G. Forster) Wedd. var. *pedunculata* J.W. Parham" (known locally on Rotuma as hat 'en sina) are reportedly "edible and eaten by children and often fed to pigs" (Rigamoto 2001: 184). In addition, McClatchey (1996: 153) indicated that the leaves of *P. pedunculata* are used on Rotuma in treatment of dysmenorrhea (painful menstrual spasms), "associated with abdominal cramps." It is worth noting that at least some native tropical Pacific island fruit doves consume the fruits of *P. pedunculata*. For example, the fruit of *P. pedunculata* has been observed as the favorite food of *Ptilinopus insularis*, which is a fruit dove species endemic to the exceptionally remote South Pacific island of Henderson (relatively near Pitcairn Island); these birds may have played a role in the long-distance spread of *P. pedunculata* into various islands in Remote Oceania.

Uses in Pohnpei: *P. pedunculata* is an indigenous, non-woody plant on Pohnpei, where it is known as **Pwaked**. Although *P. pedunculata* on Pohnpei normally produces whitish-colored fruit, a red-fruited type found in some areas is reportedly used therapeutically. According to Balick et al. (2009: 510–511), P. pendunculata is used medicinally:

> "To treat boils occurring under arms, on buttocks, or under testicles... To treat boils associated with fever... To treat stomach pain after delivery and menstrual cramps... To treat swollen testicles... To treat an upset uterus... To treat "uterine cancer" (an unspecified condition)... To treat *kilsarawi*... To treat herpes..."

Lee et al. (2010: 57) pointed out that **Pwaked** is found on Pohnpei in "agroforested areas, rocky coasts, cliff, in open forest, and growing on trees and rocks," and that the leaves and/or fruits have been used traditionally to treat rashes, some with quite large blisters. Lee et al. (2010: 152) also reported that the juice extracted from the "whole stem" of this plant is said to be specially prepared to relieve back pain on Pohnpei.

▸ **Kuwahpa** (P), **Kuahpa** (Mw), **Kuapa** (Pi), **Guaafaa** (Nu), **Kuabaa** (Ka)
Psidium guajava

Description: *Psidium guajava*, the "true guava," is a shrub or small tree that grows up to 10 m (~35 ft) in height. Its hard, durable stems are finely pubescent, and its light reddish brown bark is smooth with some thin, flaking irregularly shaped scales. Its leaves, arranged opposite each other, are ovate-elliptic or oblong in shape and about 4–10 cm (2–5 in) long with prominant midrib and side veins. Its numerous axillary flowers are showy with multiple white fragrant stamens. The mature globose to ovoid fleshy fruits are about 5 cm (2 in) long with a persistent calyx at the end. The edible fruits produce a yellow rind with pinkish, sweet, and seedy pulp.

Distribution: *P. guajava* is native to tropical America but is now found throughout much of the world's tropical regions as a cultivated plant, and often as an invasive alien species. This introduced fruit species has frequently escaped from gardens and naturalized. In fact, the true guava invades disturbed sites and can form dense thickets. In Micronesia, it was apparently introduced early to Guam by the Spanish, and from there it was taken to the Philippines and eventually throughout tropical Asia as well as other islands in Micronesia (Stone 1970: 454–455). Its presence on other tropical Pacific islands was also recorded early, e.g., by the early 1800s it was reportedly becoming common in Hawai'i, where it occurs in dry and wet disturbed areas such as pastures, waste places, and scrub forest at up to about 1200 m (~3800 ft) elevation. This potentially serious weed is spread by humans, birds, a variety of introduced hoofed mammals, and other animals.

General uses: *P. guajava* was originally introduced in many Pacific islands for its edible fruit. However, its leaves are often boiled into a tea to treat diarrhea in many of these islands including some in Micronesia (e.g., see Balick et al. 2007). Guava is very high in vitamin C and also contains large amounts of vitamin A. The fruit can be eaten raw, but in some areas uncooked guavas are normally sliced and used in salads or desserts.

Uses in Pohnpei: The true guava, *P. guajava*, is known in Pohnpeian as **Kuwahpa**. Its hard wood is used on Pohnpei to make tool handles. Medicinally, the bark, leaves, and buds are used in a variety of remedies to treat diarrhea, and its fruit, a good source of vitamin C, is eaten both green and ripe. *P. guajava* is also used in Pohnpei in the treatment of "an upset uterus," for which "as many leaves as needed are boiled, poured in a large basin, and the person sits in the bath..." (Balick et al. 2009: 451–452, also see Lee et al. 2010: 28).

▸ **Kempenial** (P)
Psychotria hombroniana

Description: *Psychotria hombroniana* is a small native shrub in the coffee family that grows to an average height of about 3 m (6 ft). It is woody with relatively thin stems. Its leaves, which are arranged opposite in pairs and whose leaf blades have smooth edges, and are about 15 cm (6 in) long and 10 cm (4 in) wide. P. hombroniana produces flowers with white petals about 1–2 cm (~0.5–1) in length and width. Four stamens per flower protrude beyond the petal and are about 4–5cm (~2 in) long. Like its relatives in the genus *Psychotria*, *P. hombroniana* produces round, fleshy, yellow fruits with two seeds per fruit.

Distribution: *P. hombroniana* is a common, indigenous tropical rainforest species found in both lowland and upland forest areas. It is a relatively small, successional woody plant that grows best in disturbed places, such as along foot paths and in forests on limestone that has been significantly elevated above sea level. These limestone forests represent one of the most important vegetation types in Micronesia. According to Fosberg and M.-H. Sachet (1980a: 28), there are "several varieties [of this species] throughout the Caroline and Marianas high islands," while "atolls do not, apparently, provide suitable habitats for these plants." In fact, more than one native species of *Psychotria* has been recorded from Pohnpei (see Herrera et al. 2010: 172).

Uses in Pohnpei: Lee Ling indicated that *P. hombroniana* in Pohnpei is called **Kempenial** (or **Koampoanial** in Kitti) and listed the use by some children who chew and swallow the sweet pulp of the fruit, "but spit out the black seed," and that use of the wood of this plant is preferable "when Pohnpeians do the cultural stick dance (*dokia*) because of the nice sound it [produces]" (www.comfsm.fm/~dleeling/botany/2000/vhp/psychotria_hombroniana.html). The rough translation of the Pohnpeian name for this plant is the "plant along the path," which is where it can be found, since it is an early successional plant which grows best in relatively open places which humans have disturbed. Balick et al. (2009: 489–490) referred to a number of its traditional uses and symbolic meanings in Pohnpei:

> "To use in traditional dancing... make *wini en silasil* to protect a person from harm by sorcery such as love potions or poisons... help a child walk earlier, five methods are reported... help a child talk earlier... treat *soumwahu en insensuwed* (child's depression)... treat frightened or shocked babies with advanced *pwirimwei* that cry every night and cannot sleep and also have high fever during the nighttime only... treat a frightened boy with early stage *pwirimwei*... treat children who have a protein and vitamin deficiency... treat *kilipwet* (a thick white skin disease)... This treatment is used only on male children... This plant [also] plays a role in the legend of Siningamwahu, one of the Saudeleurs' [ancient rulers of Pohnpei] daughters..." (see Balick et al. 2009: 489–490 for more detail on the above uses of this plant)

Note: Christian (1899: 339) reported that a *Psychotria* species on Pohnpei, known as 'Muerk' was a "bush-tree," from which a "Decoction of bark [was] used for curing aphthae or thrush in children." However, Christian may have misidentified 'Muerk' as a species of *Psychotria*, and perhaps the "bush-tree" he referred to was *Glochidion ramiflorum*, known in Pohnpeian as **Mwehk**.

▸ **Peipei eni** (P) (Mw), **Peipei aeni** (Pi)
Pteris tripartita

Description: *Pteris tripartita* is a fern that ranges in height from 50 cm–2 m (~20–78 in). Indeed, the rachis (stalk of a frond), which is green in color but flushed with red or purplish red areas, is usually about 1.5 m (~4.5 ft) long, though occasionally longer. The reproductive parts (the brown-colored sori) are continuous along each side of the frond lobes, but only on the lower two-thirds, not reaching the apex of the lobes (see sweetgum.nybg.org/science and www.nparks.gov.sg). The genus name *Pteris* is derived from the Greek word *pteruc*, meaning fern. The species name *tripartita* means "divided into three segments" and refers to the pinnae form, the primary division of a pinnate leaf, especially for fern fronds.

Distribution: *P. tripartita* is a fern species belonging taxonomically to Pteridaceae, the bracken or brake fern family. *P. tripartita* is widespread in the Old World Tropics, including some islands in the Pacific region, and is assumed to be indigenous on Pohnpei (e.g., see Herrera et al. 2010: 28). It is reportedly found on Pohnpei in cleared areas "where soil is disturbed and around houses," and it is naturalized (i.e., an alien species growing wild on its own) in some tropical areas in a number of countries such as USA (in Florida) and Puerto Rico, where it has escaped from cultivation.

Uses in Pohnpei: This fern's local name in Pohnpei is **Peipei eni**, which is shared with other ferns such as *Diplazium ponapense* and *Asplenium pellucidum*. The word *eni* refers to ghosts, figuratively placing these species in a spiritual domain. In striking contrast, at least two other fern species (*Macrothelypteris torresiana* and *Histiopteris incisa*) have the opposite name, **Peipei aramas**, where *aramas* stands for human beings. Some spiritual significance of **Peipei eni** can be inferred from the following use described by Balick et al. (2009: 247):

> "To use during funeral feasts, on the fourth day of the feast a leaf is placed on the ground where the dead person is buried and the people bringing fish will go to the grave where the person is buried to show honor by placing the fish on the leaf, and then people sitting nearby remove the fish right away..."

Balick et al. (2009: 247) also list this medicinal use for **Peipei eni**: "to treat arthritis, leaves are pounded, wrapped in cloth, squeezed into 1 gal of water and consumed (1 gal daily) for eight days..."

▸ RED MANGROVES

Collectively known as **Ahk** in Pohnpeian, *Rhizophora* stilt or prop-rooted mangrove species are found on many Pacific islands, as well as beyond the Pacific region. These "red mangroves" include *Rhizophora apiculata* Blume, *Rhizophora mucronata* Lam., and *Rhizophora stylosa* Griff. However, it has been suggested *R. stylosa* could be eliminated by being grouped with the verified *R. mucronata* species. For example, Falanruw (2015: 184) added a cautionary statement: "*Rhizophora mucronata* and *R. stylosa* are quite similar, and *R. stylosa* may someday be changed back to *R. mucronata* var. *stylosa* (Griff.) Schimper" (also see Yan et al. 2016). In any case, these "red mangrove" species need further study and resolution of their species classifications, as well as of their varieties and suggested hybrids. Below we include discussion about two of these species, *R. apiculata* and *R. mucronata*. They are found in wetland regions of high islands, and also in some protected swampy areas on atolls. Their combined ranges include much of the coastal areas of insular Southeast Asia and west to parts of the tropical east African coast, and seem to have natural distributions that extend east into various islands in the tropical Western Pacific. Much of the following general discussion of these ecologically and ethnobotanically important *Rhizophora* mangrove species is based on research by Norman Duke (e.g., see Duke 2006, 2010, 2011).

- **Ecological services of Rhizophora mangroves:** Probably the greatest overall value of stilt or prop-rooted mangroves involves their role in "supporting marine food chains, protecting coastal areas, and improving water quality." Among the environmental advantages of protecting and maintaining the native (or even introduced mangroves in some cases) include the following: stabilization of the soil, and provision of natural windbreaks, woodlot reserves, wildlife habitat for indigenous species, and food for native animals including birds, forage for bee colonies, and environmental refuges for fish and other biota in the marine food webs (see Duke 2006: 13–15). In the case of soil stabilization, *Rhizophora* red mangrove forests hold wet soils in place with their durable, overlapping prop roots that reduce water movement and promote sediment build-up in areas that normally would be eroded. Preservation of indigenous *Rhizophora* mangrove vegetation acts generally as a windbreak along the margins of the coastal areas, while also offering wildlife places of refuge during strong storms such as typhoons, habitat for indigenous species, and food for native animals including birds.

Cultural uses of Rhizophora mangroves: Mangrove forests have been generally well-appreciated by Micronesians since the initial colonization by people in prehistoric times, both for their importance as nursery areas for fish and other marine species and for their role in coastal protection and sustainable use of their wood resources. However, during the Japanese occupation (1914–1945), mangrove forests were logged extensively for construction wood and tannin production. Then, following the end of WWII in 1945, most cutting activities stopped, and since then mangrove forests have generally been used for subsistence purposes only. This allowed for mangrove forest regeneration in most areas of the region (see Haws 2006: 106–107). Although there are many traditional ethnobotanical and more modern economic uses of stilt or prop-rooted red mangroves such *Rhizophora* spp., overall these mangroves are most significant as a source of durable wood. In fact, the Pohnpeian name, **Ahk**, is identical to the name for durable digging and husking sticks, *ahk*, which are often made of the strong and tough *Rhizophora* wood. The mangroves' woody parts are used for various other purposes, including carving of handicrafts and construction of homes and canoes. Their use as cooking fuel is widespread. Note: The extensive and intensive use of mangrove species, such as the *Rhizophora* trees described below, for their wood has led to problems of overcutting in more accessible areas (e.g., on Kosrae Island), so all islanders should aim to use the species sustainably when pursuing their traditional or modernized lifestyles. Other widespread Micronesian uses for these *Rhizophora* mangrove trees include as medicine, dye or tannin, and an ecotourism attraction.

As noted above, here we deal with the general environmental and cultural importance of two *Rhizophora* species (*R. apiculata* and *R. mucronata*), which are recognized as one or more species by people living in Micronesia where these mangroves are found. In addition to the following descriptions of the species and usage of *R. apiculata* and *R. mucronata* mangrove trees in Pohnpei, we also draw attention to the current and potential climatic and related sea-level rise threats, and we should also acknowledge other dangers to the mangrove forests, such as roads altering the flow of freshwater, dredging operations, oil spills, and overharvesting for firewood. It is important to develop mangrove management programs and to protect adequate areas of mangroves in order to maintain their ecosystem services, which far outweigh their value simply as timber.

An aerial view of extensive mangrove swamps at Dewen Soundau, in Sokehs.

This home's "backyard" in the mangroves features traditional canoe and a raft, as well as a modern fiberglass boat.

Unfortunately, in some places around Pohnpei, the fragile mangrove ecosystem is marred by accumulated trash and scrap.

Small natural canals through the mangroves provide essential access routes from the shore to the reef.

Mangroves hold significant ecotourism potential, offering visitors to Pohnpei a unique and largely unfamiliar natural landscape to explore.

▸ **Akapah** (P), **Ak** (Mw) (Pi)
Rhizophora apiculata

Description: As noted above, the species identification of *Rhizophora* trees native to the Indo-Western Pacific region is challenging when based on morphological characters alone; for example, Duke pointed out the Indo-Western Pacific "*Rhizophora* species are difficult to distinguish based on morphological characters alone." That aside, *Rhizophora apiculata* (**Akapah**) is widely recognized as a common native mangrove species in many parts of Micronesia, and under favorable conditions it can grow to be 25 m (~80 ft) in height with its straight to arching prop roots, although it is often a much smaller tree. The leathery, elliptical dark green leaves are shiny above, arranged opposite each other, and 8–16 cm (3–6 in) long and 3–7 cm (1–3 in) wide. The yellowish flowers are paired or sometimes solitary on short thick stalks. The elongated, viviparous fruits (i.e., they germinate while still attached to the plant) may grow to be 35 cm (14 in) long before they drop off the plant.

Distribution: Locally, *R. apiculata* is often one of the more abundant components of the mangrove forests in Micronesia, where it is often found in riverine estuaries and in embayments, "the intertidal wetland zone, 0–6 m (0–20 ft) elevation between mean sea level and highest tides, with variable rainfall" (www.agroforestry.net/tti/Rhizophora-IWP.pdf).

Duke (2010) provided the following information about the species status for *R. apiculata*:

> "This species is widespread and common within its range. It is threatened by the loss of mangrove habitat throughout its range, primarily due to extraction and coastal development, and there has been an estimated 20% decline in mangrove area within this species range since 1980. Mangrove species are more at risk from coastal development and extraction at the extremes of their distribution, and are likely to be contracting in these areas more than in other areas. It is also likely that changes in climate due to global warming will further affect these parts of the range. Although there are overall range declines in many areas, they are not enough to reach any of the threatened category thresholds."

Nevertheless, a recent 2020 review in Science News focused on the potential threats to mangrove forests worldwide, including those in Micronesia, and specifically in Pohnpei. This global survey of past, present, and future interactions between global warming (due to continued and increasing release of greenhouse gases, such as carbon dioxide and methane) and the resulting climate change concluded that there may be dire consequences for mangrove forests. The scientists involved in the research concluded that rising of sea levels "could drown mangroves." The widespread drowning of mangroves, including the *Rhizophora* tilt or prop-rooted trees, will severely endanger many tropical island ecosystems and the resources they provide for the living things, including millions of humans, that depend upon them: "We need to get started yesterday to ensure these important [mangrove] ecosystems are around to protect us into the future" (Gramling 2020: 7; also see Saintilan et al. 2020, Jones et al. 2020, Lovelock 2020).

Uses in Pohnpei: Although individual *Rhizophora* stilt or prop-rooted mangrove trees struggle to grow back from a stump, they have long been partially harvested for a variety of uses:

> "Generally, if greater than 50% or more of the leaves are removed from a tree, they will die. There appears to be a variant of *R. apiculata* in the Philippines and in some islands of the Federated States of Micronesia (Pohnpei and Kosrae) that is used by local people for sustainable firewood collection. These trees regrow from regular cutting, which keeps them stunted." (Duke 2006: 10)

In addition to their use for fuelwood, their durable wood is also used to make planting sticks as well as house rafters and beams. According to Rehg and Sohl (1979: 4), *R. apiculata*, known in Pohnpeian as **Akapah**, is "commonly used in the building of feasthouses" as well as for dye, digging sticks for crop cultivation, house poles, and firewood. Balick et al. (2009: 476–477) referred to use of *R. apiculata* as a source of wood, and for other purposes, including medicinal:

> "The wood is used as construction material for framing a house... The wood is used to make handicrafts and a tool for husking coconuts after it has been sharpened... For protection against hexes... For protection against *enihn nansed* (evil spirits from the sea)... To treat nervousness... To treat diarrhea... To treat *mehn madau* (sea sickness)... To treat asthma... To treat high fever... To treat boils when in the unopened stage... To treat muscle pain..."

▸ **Akelel** (P), **Ak** (Mw) (Pi)
Rhizophora mucronata

Description: The red, prop-rooted *Rhizophora mucronata* is usually much shorter than its close relative *R. apiculata*. However, under favorable conditions, *R. mucronata* can grow to 20 m (70 ft) or more. The crown of *R. mucronata* is more or less rounded and multi-branched, and it produces aerial roots from its base, bearing numerous tangled roots, some arching outward into the mud, others tangled and stiltlike. Overall, these mangrove trees form almost impassable thickets. Its dark green, leathery leaves are about 5–15 cm (2–6 in), ending abruptly in a short, slender, pointed projection about 1 cm (~0.5 in) long. Note: The leaves of *R. mucronata* are a lighter shade of green on top than those of *R. apiculata*, with leaf margins often rolled under. The flowers of *R. mucronata* have four thin white or pale yellowish petals (which fall off shortly after opening) and four persistent yellow sepals. The inflorescences are produced in branching clusters containing 3–10 flowers. The viviparous brown fruits of *R. mucronata*, which are somewhat similar to those of *Bruguiera gymnorrhiza* (**Soumw**) mangroves, are about 2.5–3 cm (~1 in) long and germinate while still attached to the tree before dropping off the mother plant. These green radicles (hypocotyls: the stem of a germinating seedling) are very long on *R. mucronata* and can grow to 30–90 cm (l-3 ft)! They are slender at the base, becoming thicker at the apex so that when they fall off the tree they can become stuck in the mud with the appropriate end facing upward, thus the new tree seedling becomes planted. Indeed, this kind of fruit solves the problem of establishing new plants in a substrate that is underwater twice a day.

Distribution: *R. mucronata* often grows at or near the seaward edge of the mangroves along the coastal areas of those islands where it occurs in Micronesia; it may also be found widely along the sides of water channels, estuaries, and bays, as well as on reef flats from areas in tropical Asia to the Western Pacific. In fact, *R. mucronata* is one of the most widespread *Rhizophora* spp., growing well in soft mud, and, similar to other native mangrove trees, *R. mucronata* holds the muddy soil of its habitat in place with its numerous prop roots that droop down from the trunk to the mud. *R. mucronata* occurs mostly in areas subject to regular freshwater flows.

Uses in Pohnpei: *R. mucronata*'s wood is used as fuelwood and also to make poles for house rafters and beams. Medicinal uses reportedly include: "To treat headache, back pain, muscle pain, or joint pain... to cool the uterus..." (Balick et al. 2009: 478).

Uses in Pingelap: St. John (1948: 112) reported the occurrence and use of "*Rhizophora mucronata* Lam." on Pingelap Atoll. He described it as "8 m tall, by 3 dm in diameter, with proproots... growing in shallow salt water of [the] lagoon," and as a source of "good timber."

Mangrove wood is a principal material for the skilled Kapingamarangi carvers of Pohnrakied village.

A fisherman hauls a net through the mangroves, which provide numerous prime spots for local subsistence fishing.

Exquisite, lifelike carvings of Kapingamarangi carvers are celebrated as one of Pohnpei's most renowned handicrafts.

A mangrove tree growing in a sandy patch in U municipality, its arching roots submerged at high tide while its leaves remain above the water.

▸ **Sehu** (P), **Doa** (Mw), **Seu** (Pi), **Dolo** (Nu), **Mee ngau** (Ka)
Saccharum officinarum

Description: *Saccharum officinarum* (common sugar cane) is a very large, perennial grass species. It is represented by numerous forms known as the "noble canes," and is an important cultivated plant in many tropical areas of the world. The stems of *S. officinarum* vary in surface color from yellow or green to dark red or almost black, sometimes striped, and grow in clumps (or ratoons). The stems are strong, thick, jointed, fibrous stalks unbranched up to 4–6 m (~12–20 ft) tall and filled with solid, juicy pulp. The nodes of the stems are positioned close together at base of stems and more widely spaced distally (further upward along the stems); the short internodes are conspicuous. Its broad leaves are 30 cm (12 in) long or more, smooth except for the saw-toothed edges, and overlapping at the leaf base. The branched panicle or flowering stalk develops at the end of each stem. This inflorescence is dense, long, erect or drooping and produces silvery white to pinkish flowers, sometimes described as a feathery, light-colored tassels, 30–60 cm (1–2 ft) long. Although small seeds do develop from its flowers, sugar cane is usually grown from cuttings (i.e., it is normally cloned).

Distribution: Sugar cane is probably native to some tropical regions of New Guinea, where it has been cultivated for many thousands of years (e.g., see Benson 2012: 98–102). It was brought to Micronesia by early Pacific voyagers in ancient times, and in some islands it has remained a valuable plant. In other parts of the tropical Pacific, including some islands in Micronesia, sugar cane (or at least some of its varieties) did not arrive until it was introduced by foreigners such as Europeans or Japanese. Sugar cane is commonly cultivated in rural garden areas of many tropical Pacific islands today, often in adjoining stands, as an intercrop or border planting, and in home gardens in both rural and some urban areas. Although it is cultivated to some extent on many atolls, sugar cane is still not very common in these types of island environments, including those in Micronesia. Sugar cane can become naturalized, but only occasionally because it is a mainly a domesticated crop.

General uses: "Sugar cane has been utilized from early times in both Polynesia and Micronesia" (Barrau 1961: 65). It was originally grown for the sole purpose of chewing in tropical Southeastern Asia and the Pacific due to its high sugar and juice content, and it is still commonly used this way on some of the more remote tropical Pacific islands. This process involves removing stem rind and sucking on or chewing the internal tissues. Today, it is a common supplementary food plant throughout the tropical Pacific, with many named cultivars on most islands, including some in Micronesia, where the juicy stems are chewed as a source of refreshment, energy, and in some cases for medicinal purposes.

Uses and cultivars in Pohnpei: Sugarcane, called **Sehu** in Pohnpeian, has multiple reported medicinal uses. For example, according to Balick et al. (2009: 307), it is used to "treat depression and anxiety" by skinning the stem of *S. officinarum*, which is then "pounded, squeezed into a cup of water with other unspecified ingredients, and consumed daily..." In this case, the sweetness of the sugar cane juice would dampen the distaste or bitterness of the other therapeutic components in a medicinal potion. Some cultivars have specific cultural uses. ***Sehu pwet*** (or ***Seupwet***) is the local name for a white cultivar of sugar cane that has long been used as a royal invitation to a feast or a social function for the *sohpeidi* (title holders in the *Nahnmwarki* line). ***Sehu weita*** (or ***Seuweita***) is a red cultivar of sugar cane, used to ask for a wife (marriage offering). As a bunch of planted **Sehu** grows upward, the long stems tend to fall over in several directions, unless they are held together by a rope tied around them; this is referred to by a Pohnpeian proverb: *Uhn sehu ehu pwoapwoaud* ("When so many sugar canes stand up together"). This suggests that if a marriage is not bound together, it may break up as it develops over time. In other words, a husband and wife may break up their marriage if it is not bound by mutual satisfaction. See p. 430 for other sugar cane cultivars identified in Pohnpei. One of them, ***Seuahlek***, is said to resemble **Ahlek** (*S. spontaneum*), a wild reed species related to sugar cane (see the next page).

Uses and cultivars in Mwoakilloa: Glassman (1953: 299) observed *S. officinarum* being cultivated on Mwoakilloa Atoll, and recorded its vernacular name as 'Tuh'. However, the name for sugar cane on this atoll today is **Doa** and is cultivated in six local varieties (see p. 430).

Uses and cultivars in Pingelap: *S. officinarum* is mostly cultivated on Pingelap Atoll in excavated swampy areas (*inipwel*) where *Cyrtosperma merkusii*, the giant swamp taro, is cultivated, as well as being planted close to homes. Sugar cane is known generally as **Seu** on Pingelap Atoll and has eight locally recognized varieties (see p. 430).

▸ **Ahlek** (P), **Alɛk** (Mw), **Ahlaek** (Pi) *Saccharum spontaneum*

Description: *Saccharum spontaneum*, wild cane, is a close botanical relative of common cultivated sugar cane, *S. officinarum* (**Sehu**). *S. spontaneum* is a highly variable, tall perennial grass species with hollow stems that reach heights of up to 2.5–4 m (~7–13 ft) tall and are about 2.5–5 cm (1–2 in) wide. This "wild cane" spreads by stout, often dense rhizomes, quite similar in general appearance to the cultivated sugar cane but more slender, with a hairy stalk supporting the inflorescence (a peduncle), which has a very fragile elongated axis (and extension of the petiole of a compound leaf that bears the leaflets, known botanically as a rachis). The inflorescence itself is a loose, plume-like, branching cluster (a panicle) 25–60 cm (10–24 in) long with silky hairs.

Distribution: *S. spontaneum* is a tall wild cane species with wind-dispersed seeds that is native to the "Old World Tropics." This naturally occurring cane species has been variously listed as native or introduced in the high islands in Micronesia, such as those in Palau, the Marianas, and the Federated States of Micronesia; however, more recently it has been reported as only native to some areas on the African and European Mediterranean fringe, as well as in some tropical or subtropical areas of South Asia (Mani 2013). Ecologically, *S. spontaneum* is a large grass that occurs near still and running water, especially in wet disturbed lowlands up to 500 m (~1500 ft) elevation; more widely, this wild cane occurs in "waste areas, fallow fields, marshes, banks of streams and ponds, on sand dunes, along railroads and highways, and in or around fields" (Holm et al. 1997: 695). *S. spontaneum* is considered to be an invasive species today in a variety of tropical and subtropical areas of the world, including some in Micronesia, and it is presently widespread throughout region. (For different references to "native or alien" varieties of this species in various high islands of Micronesia, see Fosberg et al. 1987: 57, Space et al. 2000, Fosberg and Sachet 1982: 88–90, Josekutty et al. 2002: 63–64, Herrera et al. 2010.)

General uses: *S. spontaneum* normally forms deeply rooted clumps and can serve as a useful soil binder, but it can also become a serious weed (invasive species) in irrigation systems. Nevertheless, wild cane (*S. spontaneum*) has and still is used for decoration, thatching, making ropes, as domestic animal fodder, and for numerous reported medicinal uses in some tropical areas.

Uses in Pohnpei: In the past, *S. spontaneum*, whose Pohnpeian name is **Ahlek**, was reportedly used to create low, thin walls of houses, and it was also traditionally used in Pohnpei as natural decoration in flooring, thatch, and walls, with the stems wrapped around thatch made out of the large leaves of **Oahs** (ivory nut palm) and **Nih** (coconut palm). **Ahlek** is also reportedly still used for medicine:

> "To stop a person from vomiting, four central leaves are crushed in the mouth and swallowed... This plant is used in association with **Aiau** (*Ficus prolixa* G. Frost.) to treat *soumwahu en lusulus*... To cure the shaking from a curse or high fever, gather a handful of leaves, pound them, put them into an *inipal* (coconut fiber), press, squeeze four drops into a cup of hot water, and drink one cup daily until cured..." (Balick et al. 2009: 308)

Additional medicinal use of **Ahlek** was provided by Lee et al. (2010: 16), who noted the use of the "small buds at the base of nodes along the stem of **Ahlek**" to treat dog bites "by pounding these buds and rubbing them onto the wound daily for four day or until the wound heals." Lee et al. (2010: 146) also reported the use of the leaves of this tall grass species for treating joint pain; "the treatment involves beating a number of leaves together, then wrapping them in some cloth and squeezing out juice into a cup and drinking the fluid daily for a series of days."

▸ **Remek** (P), **Roamoak** (Mw), **Raemaek** (Pi), **Manu gaba aasanga** (Nu), **Nau** (Ka)
Scaevola taccada

Description: The common coastal shrub *Scaevola taccada* is a succulent small to medium-sized shrub which can grow to heights of 3 m (15 ft) or more under some conditions; it has somewhat fleshy leaves that are 5–20 cm (2–8 in) long and crowded at the tips of the stems. It produces many white, five-petaled flowers 1–4 cm (0.5–2 in) long. The fruits are 8–15 mm (0.3–0.6 in) wide and green in color when immature, but turn white, roundish, and juicy when ripe.

Distribution: Widespread in the coastal strand vegetation of most tropical Pacific islands, *S. taccada* is among the most common plants on the atolls and other low coralline islands. Its fruits are buoyant and durable so they can float for long periods of time in salt water (for a year or more) and then germinate when washed ashore beyond the high-tide lint. *S. taccada* has long-distance dispersal ability and is an early colonizing species (one of the first to grow in open or disturbed places along tropical Pacific shore lines), often becoming the dominant shore plant on those islands that it colonizes. It occurs naturally in various areas of the vegetation, and in some cases it is cultivated. It is indigenous to Micronesia (i.e., it arrived and became established in Micronesia without human help), but it has not evolved into any unique species that would be endemic on any of the islands in Micronesia.

Uses in Pohnpei: *S. taccada* is called **Remek** in Pohnpeian and is regarded as an important medicinal plant because it is used to cure womb illness. The reported remedy uses pounded **Remek** leaves, which are squeezed through cloth and then drunk by the sick women. The attractive flowers are also used in making *mwaramwar*. In addition, the flowers, leaves, and buds of **Remek** are also mixed with coconut oil to make perfume. Balick et al. (2009: 402) provided the following information about the use of **Remek** on Pohnpei:

> "This plant is used as fencing for a traditional house... [It is also used for several purposes] as a charm to protect a house during the beginning of construction... To treat wounds... To treat *soumwahu en ohl* (men's sickness)... To treat eye infection where there is pus and blurred vision... To treat conjunctivitis (pinkeye)... To treat stomachache... To treat thick white skin disease... To reduce high fever... To treat asthma... To treat congestion of difficulty breathing... To use to clean the kidneys... To treat a gallbladder problem that results in a person having yellow stool[s]... To treat women during pregnancy... To treat type II diabetes... To treat an itchy rash that is bleeding from being scratched..." (Also see comments of Beauter Hainrick at: www.comfsm.fm/~dleeling/botany/2000/vhp/seasidemedicinal.html).

Uses in Mwoakilloa: *S. taccada* is known on Mwoakilloa as **Roamoak**. There are a number of medicinal uses of this native coastal shrub on Mwoakilloa Atoll. According to information provided by students in a biology course at the College of Micronesia (COM) on Pohnpei, these therapeutic applications include using the juice of the ripe fruits for "curing pink-eyes," pounded juice from green leaves for "curing lower abdominal and back aches," and along with massage, for hastening pregnancy, "especially for the newly wed." Additional medicinal preparations involving parts of *S. taccada* are also reportedly used on Mwoakilloa Atoll "for stopping discomforts after menstruation [as well as] prevention of uterus infection after delivery and miscarriages," and in addition, "ash or saps from its firewood is very effective in curing warts" (www.comfsm.fm/~dleeling/botany/2000/vhp/seasidemedicinal.html).

Students from the same biology course at COM also provided the following legend from Mwoakilloa cultural tradition, which refers to *S. taccada* as **Remek** (= **Roamoak**):

> "Once upon a time, there lived a couple in Likinreren, the eastern end of Mwandohn in Mokil [Mwoakilloa Atoll]. They had five sons and five daughters. The sons did not know of their fifth sister because their mother kept that a secret from them because she was delivered in an unusual way. She was delivered through the head and became a white flower in the 'Remek' without having petals. She named

her 'Raskinnad'. Everyday when the boys would go fishing on the reef, they would look back to the beach and [see] their four sisters playing with a fifth one whose beauty was beyond comparison. Everyday when they returned home and asked about what they saw, the sisters and their mother would deny their story. Then one day, they planned to hide and catch what they often saw. They hid themselves in the woods, while suddenly the four sisters came out to the beach and began playing. Suddenly, a fifth one jumped down from the 'Remek' (the flower became the most beautiful girl they [had] ever seen) and started to play with the others in the shallow water. The five boys jumped out of the woods and grabbed the beautiful girl and started to pull her around. Each of them wanted to have her for a wife. As a result, they tore the girl apart and then the girl disappeared into the 'Remek' shrub. Then after a long while, the 'Remek' had no flower. But soon another flower grew but it had five petals. The one that has no petals can still be found but very rarely. This one is called 'Raskinnad' and its story is still told on [Mwoakilloa] today." (www.comfsm.fm/~dleeling/botany/2000/vhp/seasidemedicinal.html)

Uses in Pingelap: *S. taccada* is known in Pingelapese as **Raemaek**. Damas (1994: 239) listed *S. fructescens* (a synonym of *S. taccada*?) as "cultivated" with an "edible fruit." According to information provided by Leo Cornelius in 2004, on Pingelap, *S. taccada* is used for a number of medicinal uses:

> "Leaves can be pounded and boiled. Drink the liquid to treat stomach ache in athletes. The juice of the fruit can be squeezed into the eye to treat pink eye (conjunctivitus). The leaves are used to treat gonorrhea in men. The leaves can be wilted over a fire and then placed on the breast of a young woman to enlarge her breasts. This is not for reasons of appearance but to enhance the potential milk capacity of the breast. Alternatively the leaf can be steamed and placed across the breasts." (www.comfsm.fm/~dleeling/botany/2000/vhp/seasidemedicinal.html).

▸ **Tuhkehn kilin waii** (P), **Kihdo** (Mw) (Pi), **Manu hai gili dabeo** (Nu), **Laagau hunu gili** (Ka) *Senna alata*

Description: *Senna alata* is a coarse, erect shrub or small tree in the bean family, Fabaceae. It can grow to about 3–4 m (~10–12 ft) tall with pinnately compound leaves up to 50–80 cm (20–30 in) long with 8–14 pairs of relatively large leaflets. It produces attractive clusters of bright yellow, crowded, and overlapping flowers, with its numerous seeds contained in black, papery, winged pods. This striking shrub, sometimes cultivated as an ornamental plant, is named for its flower buds, which grow in a column and resemble large yellow candles, each complete with a flame. During the nighttime, the leaves fold together.

Distribution: This alien woody species is a relatively recent introduction to several Pacific islands including Pohnpei. It is native to tropical America, but naturalized on many Pacific islands. Introduced to various tropical areas from the New World tropics as a decorative species, it is now often regarded as a weed. On Pohnpei, it can be found in gardens as an ornamental, and sometimes it becomes naturalized in poorly drained areas, along streams, and in other disturbed sites. Indeed, *S. alata* prefers open areas with abundant sunlight. It can form weedy thickets and is especially invasive in areas where there is a high water table. Thus, it is found along streams and rivers and on wet soils.

Uses in Pohnpei: *S. alata* is an introduced ornamental plant in Pohnpei that is sometimes planted around homes and along roadsides. It is known in Pohnpeian as **Tuhkehn kilin waii**. According to Balick et al. (2009: 394), parts of this plant have been used to "treat *Tinea versicolor*" and "ringworms" [*kilin waii*, which are common fungal infections causing rashes on the skin of some people (caused by fungi, not by worms!)].

Uses in Mwoakilloa: The name of *S. alata* on both Mwoakilloa and Pingelap is reportedly **Kihdo**. As on Pohnpei, the leaves of *S. alata* are used to treat a variety of fungal skin conditions. "The leaves are either rubbed vigorously into the affected skin, or a coconut oil is produced containing *Senna alata* and applied to the skin after bath" (see danaleeling.blogspot.com/2016/09/plants-that-heal.html).

▸ **Koiohnleng** (P)
Sida rhombifolia

Description: *Sida rhombifolia* is a small shrub or tough weedy plant in the hibiscus family, Malvaceae. It can grow to about 1 m (3 ft) tall and is distinguished by its small pale-orange to yellow-orange flowers. Its leaves are more or less ovate in shape and acute, with serrations on the margins near the leaf apex. They are glabrous above and pubescent beneath, about 1–2 cm (0.5–1 in) long on short petioles. Its flowers are produced alone or in pairs on the plant axils, and when fertilized they yield small black seeds (e.g., see Stone 1970: 415).

Distribution: Probably native to tropical America, this *Sida* species is alien and now naturalized on some islands in Micronesia. It was most likely introduced to these islands after European contact. However, a close relative, *S. fallax*, is native in dried areas on some islands in this region. On Pohnpei, *S. rhombifolia* and *S. acuta* (both alien species) are reportedly common in disturbed areas such as waste places, plantations, and along trails and roadsides, up to 1000 m (3,000 ft) or more.

Uses in Pohnpei: This small, tough-stemmed shrub is known as **Koiohnleng** in Pohnpei. It produces yellowish flowers with a pleasant fragrance, and it has been reported that the Saudeleur of old Pohnpei had **Koiohnleng** planted in his garden so he could smell it nearby and have it made into the *mwaramwar* he wore (e.g., Merlin et al. 1992). Balick et al. (2009: 433) referred to the use of *S. rhombifolia* in Pohnpei for making a broom by cutting off "as many branches as needed," stripping off the leaves, and tying these leafless stems together for use as a sweeping tool. Balick and his colleagues also referred to use of **Koiohnleng** leaves to treat new cuts, ear infections, and "*soumwahu en lih* (women's sickness)."

▸ **Koatoa** (P), **Koasoa** (Mw), **Koasa** (Pi)
Sonneratia alba

Description: *Sonneratia alba* is a large native mangrove tree of swampy areas. It may grow to 30 m (100 ft) tall with trunks that may measure as much as 1.5 m (5 ft) in diameter. *S. alba* often supports many forms of epiphytes ("air plants") such as ferns, orchids and other herbs attached to the trunk or growing out of crevices in the tree (epiphytes are more common on *S. alba* than *Rhizophora* spp. or *Bruguiera* mangrove trees). The most noticeable characteristic of *S. alba* trees themselves is their large breathing roots or pneumatophores that emerge from its muddy environment surrounding the trunk and are conical or spike-like in shape; they grow erect to about 20–60 cm (8–16 in) in height. When the ocean is at low tide, these stick-like structures, which are not always under the tree itself, are visible. *S. alba* also has the interesting ability to generate new sprouts from fallen branches that root in the soil and by setting aerial roots in the ground. Its leaves are more or less rounded, and its flowers are produced at the end of branches. The flowers are about 5–7.5 cm (2–3 in) long with many white stamens. They bloom in the evening time, and by morning, the stamens have fallen and thus leave a thick calyx enclosing the developing and more or less flattened fruit. *S. alba* produces very small seeds that are not easily noticed. These may sprout and develop into new plants. Individual plants of this species that are produced directly from seeds are more easily uprooted than those developing from cuttings or fallen branches.

Distribution: *S. alba* trees grow in mangrove swamps, usually (but not always) along the outer fringes of the mangrove forest. Indeed, *S. alba* grows in both sand and mud and can occur in the seaward, middle, and landward areas of the mangrove forest. With a wide natural range of distribution, *S. alba* can be found from Indo-Malaysia to the tropical Western Pacific, commonly in the mangrove forests of the high islands of Micronesia, and even as far east as the atolls of the Marshall Islands, where it may have been introduced by people for its valuable uses.

Uses in Pohnpei: *S. alba* is known in Pohnpeian as **Koatoa**. Its timber is very durable and has been used traditionally to make house posts in Pohnpei because it is resistant to rot. Christian (1899: 337) referred to the white-flowered *S. alba* mangrove species with "circular leaves and rounded fruits" as producing white, firm wood which was used for "cabinet-making." **Koatoa** can also be used as fuelwood, and people sometimes consume the sweet nectar of its flowers (see Merlin et al. 1992; also see Falanruw 2015: 216 for comment on sipping the nectar from the "calyx cup" of this tree's flowers in Yap). Balick et al. (2009: 422–423) provided additional information from Bascom (1947) about the uses of **Koatoa** on Pohnpei:

> "Bascom noted that the wood from this species was one of the best raw materials for house frames and general building purposes, as it was termite resistant. He observed that houses constructed two to three decades before were still standing. The trees are also known to protect the soil from erosion... Bascom also described a black dye made from the scraped bark of the sprout of this species, mixed with a type of black mud [and a number of remedies including parts of the *S. alba* tree have reportedly been used to] treat *soumwahu en naniak* (mangrove sickness)..."

▸ **Katar** (P), **Kasar** (Mw), **Kaesar** (Pi), **Manu haihale** (Nu)
Sphaeropteris nigricans

Description: *Sphaeropteris nigricans* is a native tree fern on Pohnpei that can grow to heights of 8–10 m (25–30 ft), with blackish, thick, fibrous trunks that have a soft inner core. However, the inner core is very longlasting and can remain in contact with the soil for as many as 20 years before rotting. *S. nigricans* produces large clustered fronds with petioles that are 1–2 m (3–6 ft) long, are divided twice, and are arranged at the furry, terminal ends of the stem (trunk), which is thick, dark, and fibrous but has a durable soft inner core. The midribs of the fronds are about 3–5 m (9–15 ft) long, blackish, and hairy (pubescent). The brown spores are confined in multiple round sporangia located on the undersides of fronds, where they are arranged oppositely on each of the sub-leaflets.

Distribution: The native distribution of the fern species in the genus *Sphaeropteris* reportedly ranges from southern China through eastern regions of tropical Asia to New Zealand, with a separate area in Central America and northwestern South America. Costion and Lorence (2012) referred to the different native tree fern species found in Micronesia belonging to the genus *Sphaeropteris* and provided a brief review of their distribution:

> "There are three native tree [*Sphaeropteris*] fern species in Micronesia, formerly placed in [the genus] *Cyathea*. Two of them are considered endemic. They are both locally abundant but *Sphaeropteris aramaganensis* is only known from three small islands in the N. Marianas on high volcanic slopes. *S. nigricans* occurs commonly in open savannas and in the forest understory of the Caroline Islands [in Chuuk, Pohnpei and Kosrae]. *S. lunulata* is less common, but also occurs in Southeast Asia." (Costion and Lorence 2012: 58)

Uses in Pohnpei: This large tree fern is known in Pohnpei as **Katar**. It is very common in some cultivated and native forest areas on the island. Indeed, it is the dominant plant in the lower canopy of the native rainforest and is also an important component of the montane cloud forest, where it is usually found in damp shady areas. The strong, durable inner trunk core of the **Katar** tree ferns are commonly used as house posts on Pohnpei and may last for up to 20 years. In addition, **Katar** wood is used to make fence posts and for numerous other purposes such as foot bridges on Pohnpei. The trunks of **Katar** have also been used as supports or planting sticks for commercial black pepper vines (*Piper nigrum*, **Pepa**). In addition, the large fronds of **Katar** are used as thatch for the roofs and walls of temporary shelters constructed in the forest. These short-term accommodations are useful when people are visiting the humid Pohnpeian rainforests, especially if it is cold and wet in the uplands. The fronds can also be piled up to serve as a mattress and blanket in these forest shelters.

Katar fronds and leaf buds are also sources of local medicine on Pohnpei, while the soft inner core of the **Katar** trunk is used as treatments for cuts and wounds. Traditionally, special uses have been connected with sorcery. According to Riesenberg (1948: 426), the leaves are reportedly pounded and squeezed, and the liquid is ingested for contraceptive purposes on Pohnpei. Lee et al. (2010: 131) reported that "**Katar** is used to alleviate excessive menstrual bleeding," and provided some detail regarding the treatment procedure (also see Balick et al. 2009 for an extensive description of traditional medicine and other uses of **Katar**).

Unfortunately, during the second half of the 20th century, the increasing number of black pepper farms on Pohnpei, along with overharvesting of the tree fern trunks for black pepper vine supports, resulted in the drastic reduction in the lowland population of **Katar** tree ferns. With newer roads now providing access to several inland locations, the upland populations of the **Katar** tree fern are also threatened (see Verma 2018 for a discussion regarding recent research focusing on "multiplication, sustainable, organic and climate-smart commercial cultivation of black pepper" on Kosrae Island where this tree fern is reportedly known as **Foa**; also see www.comfsm.fm/~dleeling/botany/1998/vhp/cyathea.html for several photos and a discussion regarding this tree fern listed on that website as *Cyathea nigricans*, a synonym for *Sphaeropteris nigricans*).

▸ **Moaresed** or **Ngkahu** (P)
Sphagneticola trilobata

Description: *Sphagneticola trilobata* was formerly classified botanically and known widely as *Wedelia trilobata*. This creeping plant is a somewhat woody invasive herb in the sunflower family (Asteraceae) that grows erect or sprawling to heights of about 1.5 m (5 ft). Slightly woody at its base, this bushy herb or trailing vine-like plant produces leaves 6–15 cm (2–6 in) long with prominent veins and toothed margins (edges). It produces bright yellow flowers and is sometimes referred to as the "beach sunflower," with its yellowish flowerheads forming clusters of tiny flowers characteristic of plants in the sunflower family.

Distribution: The native range of *S. trilobata* is reportedly found as a possible native species in areas of "Mexico, Central America, and the islands of the Caribbean, but now is grown almost worldwide in tropical and other warm places" (e.g., see Staples and Herbst 2005: 165). Although this yellow-flowered groundcover is widely used in open, disturbed lowland tropical areas, *S. trilobata* has been included on the "List of the world's 100 worst invasive species," which is compiled and posted by the International Union for Conservation of Nature (IUCN). This very aggressive weedy "beach sunflower" species is spread by people as an ornamental plant or groundcover in exposed places, and it is also planted in gardens, from where it spreads into surrounding areas by dumping of garden waste. It extends its range vegetatively (by clones), not by seed. It can quickly form a rather dense groundcover that blocks other plant species from redeveloping in those areas. It is considered to be a noxious weed in agricultural land, along roadsides, and in urban waste places and other disturbed or open sites. In addition, it is also invasive along streams, ditches, margins of mangrove swamps, and in coastal vegetation; in fact, it is now common as an alien, invasive plant on many tropical Pacific islands including some in Micronesia (e.g., in the Marianas, the FSM, and the Marshall Islands). It grows near the shore on high islands and in many places on atolls, especially along rocky shores, where it can be the dominant species. It can also become a very common weed on coconut plantations (e.g., see Weber 2003: 419).

Uses in Pohnpei: *S. trilobata* is reportedly known in Pohnpeian as **Moaresed** or **Ngkahu**. In his "Invasives hike plant checklist" COM, FSM Instructor Dana Lee Ling listed "Potential [plant] invasives on Pohnpei: Threats to the plants of Pohnpei." Among these potentially dangerous plant species, Lee Ling referred to the origins and consequences of the introduction of *S. trilobata* into Pohnpei about 50 years ago. Apparently, this aggressive member of the sunflower family, which was "brought from Hawai'i by women down at the Catholic mission around 1970 who thought it would look nice in their gardens," is all "over the island now" (www.comfsm.fm/~dleeling/ethnobotany). Although *S. trilobata* is reportedly very invasive and can take over areas where it is introduced, it is used on Pohnpei for medicinal and decorative purposes. For example, steam from the heated small fruits of this alien plant has been used in Pohnpei as a traditional medicine for illness of the groin. According to Balick et al. (2009: 339), the attractive yellow flowers of this plant "are sewed in *mwaramwar* head garlands." Note: Accurate botanical identification of *S. trilobata* is confusing to some, because another plant species that superficially resembles *S. trilobata* is *Wollastonia biflora* (also in the sunflower family). *W. biflora* has attractive yellow flowers and is known botanically by some as *Wedelia biflora*, and in Pohnpeian, *W. biflora* is also known as **Moaresed** or **Ngkahu** by some people. It should also be noted here that *W. biflora* has a number of its own medicinal uses on Pohnpei (see later in this text under "*Wollastonia biflora*," and in Balick et al. 2009: 338).

▸ **Kehnpap** (P)
Syzygium carolinense

Description: *Syzygium* is a genus of flowering plants that belongs to the myrtle family, Myrtaceae. The genus comprises about 1,100 species and has a native range extending from Africa and Madagascar through southern Asia and east through the Pacific. Most species of *Syzygium* are evergreen trees and shrubs. *Syzygium carolinense* is a native tree, known in Pohnpeian as **Kehnpap**. This description of *S. carolinense* is followed by descriptions of two introduced but useful other species in the genus *Syzygium*: *S. jambos* and *S. malaccense*.

Distribution: *Syzygium carolinense* is a forest shrub or tree native to the high islands of Chuuk, Pohnpei, and Kosrae. Another woody species in Pohnpei forests, *Syzygium stelechanthum*, known in Pohnpeian as **Kirekinwel**, is a native relative of *S. carolinense* and is endemic to the island, while another close relative, *Syzygium stelechanthoides*, is endemic to Kosrae.

Uses in Pohnpei: **Kehnpap** has been used traditionally on Pohnpei as a key source of wood for construction and also for firewood fuel and medicine (Herrera et al. 2010: 156). In addition, the fruits of *S. carolinense* are an important food for native birds. Balick et al. (2009: 453) noted the use of wood from this tree "in building houses to make every part of a house, both vertical and horizontal boards" and as such it is "the only tree that can be used like this." Balick et al. also note the medicinal use of this tree on Pohnpei "To treat *soumwahu en ohl* (men's sickness), after completing a treatment of *pwurin pohnpei* [and it is also used to] treat *tihnsou medek* (backache) [while to] treat arthritis, an unspecified amount of the inner bark is scraped off, boiled, and consumed as long as needed..." (For more therapeutic treatment details for *S. carolinense*, see Balick et al. 2009: 453).

▸ **Apel en wai** (P)
Syzygium jambos

Description: The rose apple, *Syzygium jambos*, is a small to medium tree that grows to heights of about 6–15 m (~20–45 ft) with grayish brown, smooth bark. Its light green leaves, which tend to hang and are thin with a leathery surface and are narrowly lanceolate in shape, are about 10–23 cm (~4–8 in) long. Its flowers are produced on the ends of stems, with four white to greenish petals and a multitude of creamy white stamens 1–5 cm (0.5 to 2 in) long. The edible, oval to bell-shaped fruit are normally red skinned, but slightly sour fruits have whitish yellow to pinkish yellow skins. The tastier fruit have crisp, white flesh that is slightly sweet and fragrant.

Distribution: *S. jambos* is indigenous to the Malay archipelago and has been introduced to southern Africa, Australia, and a number of Pacific islands, including some in Micronesia. In some of the islands where it is alien, it has become an invasive species threatening native flora. Generally, it can invade undisturbed native forest where it has been introduced. In fact, the tree thrives under a variety of soil conditions, and as a small to medium-sized tree it can form dense thickets which shade out native species. Most commonly, it is found in wet lowland tropical habitats up to about 500 m (~1,500 ft).

Uses in Pohnpei: This plant, whose Pohnpeian name, **Apel en wai**, means "foreign apple," was introduced from another tropical area, probably during the period of the German administration (1898–1914) for its pink, slightly sour fruit. Although it is known to be used for fuelwood and timber and is a nectar source for wild animals in some areas, on Pohnpei, its fruit is mainly consumed by children (e.g., see Merlin et al. 1992: 46; Herrera 2010: 156).

▸ **Apel en Pohnpei** (P), **Apɛl** (Mw), **Apol** (Pi), **Aabele** (Nu), **Aabele lauwehe** (Ka)
Syzygium malaccense

Description: *Syzygium malaccense*, known in Pohnpeian as **Apel en Pohnpei**, is a medium-sized tree with smooth, mottled gray bark that can grow up to about 15 m (45 ft) in height with dark green, oval, glossy, elliptic-oblong, opposite leaves. Its flowers have many deep red to red-pink stamens, and the fleshy, oblong-shaped, one-seeded fruit are glossy red or red and white on the exterior surface and are 3–7 cm (1.2–2.8 in) long and 3–7 cm (1.5–3 in) wide. The mildly sweet flesh of the fruit is white and surrounds a large, shiny seed.

Distribution: *S. malaccense*, or "Malay apple" as it is known in English, is now widespread in various parts of the tropical world, but it is considered to have originated in the Malesian region, perhaps Malaysia, where there are at least three *Syzygium* species which bear edible fruits (namely the "water apple" (*S. aqueum*), the "wax jambu" (*S. samarangense*), and Malay apple (*S. malaccense*)). The Malay or mountain apple is a useful tree that was introduced as a "canoe plant" throughout much of the tropical Pacific by ancient Pacific island voyagers for its edible fruit and wood. It is commonly found on many tropical Pacific islands in open forests, and is planted or protected in agricultural areas or near houses. Its combination of tree, flowers, and fruit has been praised as the most beautiful of the *Syzygium* genus (Morton 1987).

General uses: Whistler and Elevitch (2006d) noted the following uses:

> "It was once an important supplemental (and seasonal) fruit crop. Today, because other more prolifically fruiting and better tasting fruits are available (e.g., mango and papaya), cultivation of Malay apple has decreased. However, it is still a useful crop and is easily cared for in a homegarden. The wood is average but little utilized since other better woods were and are available. The ripe fruit is eaten raw. The skin is thin and the flesh is crisp and white. The slightly sweet taste is refreshing and can be a crunchy addition to a mixed fruit salad. Although not suited for jams or jellies, the half-ripened fruit can be pickled."

Uses in Pohnpei: Englberger et al. (2009b: 120) listed **Apel en Pohnpei** as a traditional food source on Pohnpei, and Balick et al. (2009: 454) referred to the medicinal use of parts of this tree species to treat diarrhea, skin disease or rashes, and *kilsarawi* (caused by the *Herpes zoster* virus).

▸ **Mwekimwek** (P), **Mwoakmwoak** (Mw), **Mwaekaemwaek** (Pi), **Bie** (Nu), **Mogumogu** (Ka)
Tacca leontopetaloides

Description: The large, perennial, stemless herb *Tacca leontopetaloides* (arrowroot or Polynesian arrowroot in English) grows from starchy tuberous underground stems. Although previously separated into its own family (Taccaceae), this species is now placed in the yam family (Dioscoreaceae). The deeply grooved, broad leaves, are 30–70 cm (12–28 in) long, as much as 120 cm (48 in) wide, and are divided into three parts; the leaves stand erect on long leaf stalks (petioles) 17–150 cm (6 in–4.5 ft) long. The striking greenish-purple flowers, with long hanging filaments (bracts) and spherical, smooth-skinned, yellowish fruit, are produced in clusters on top of a hollow stalk which reaches heights up to 1 m (3 ft) above the leaves:

> "The plant is usually dormant for part of the year and dies down to the ground. Later, new leaves will arise from the round underground tuber. The tubers are hard and potato-like, with a brown skin and white interior." (Wagner et al. 1990)

Distribution: *T. leontopetaloides* appears to be naturally distributed from western Africa through Southeast Asia to northern Australia, and possibly also to New Guinea and some Western Pacific islands. It was intentionally introduced into many tropical Pacific islands via early human migrations. Subsequently it was cultivated, and eventually it also became widely naturalized in many places. It is now found growing wild, or semi-wild, on some high islands as well as on all but the very driest atolls, often near the strand vegetation and in open areas. *T. leontopetaloides* is also sometimes found relatively far inland on some of the high tropical Pacific islands. It is a pyrophytic (fire-adapted) species and can regenerate from its tubers after natural or man-made fires.

General uses: The starch-rich tubers of Polynesian arrowroot (*T. leontopetaloides*) were an important food source for numerous Pacific island societies, including many in Micronesia. However, it often of greatest importance for the peoples living on the low islands and atolls, where this hardy perennial herb has had a long history of being an energy-rich starch source that was prepared into a flour to make a variety of puddings (e.g., see Whistler W.A. 1992b, Merlin et al. 1996, Etoh 2001). In historic and more recent times, *T. leontopetaloides* was cultivated commercially in sizable amounts for its starch content on some Pacific islands. For example, in Hawai'i during the middle part of the 19th century, its products were exported to Europe under the name "arrowroot" (Swanholm 1959, citing Coulter 1941). The starchy tubers are harvested when the aerial parts have died off, which usually takes about 8 months from the time of planting, but can sometimes take as much as 10–12 months before the tubers are sufficiently mature (e.g., see Kay 1987). The tubers are traditionally subjected to post-harvest treatments (labor-intensive processing) to eliminate the bitterness and toxicity. Thus, in order for the naturally poisonous tubers to yield edible starch, they have to be grated, squeezed, then allowed to soak in fresh water. Subsequently, the settled starch needs to be rinsed repeatedly to remove its bitterness, then cooked and dried. The flour of Polynesian arrowroot is traditionally mixed with a variety of other processed native foods, including mashed taro, breadfruit, or pandanus fruit extract, then often mixed with coconut cream, or even the sweet juice extracted from cooked ti plant (*Cordyline fruticosa*) roots to prepare puddings. Medicinally, tubers and crushed leaf stalks have been used in various treatments. For example, the tubers of *T. leontopetaloides* have been consumed to treat stomach ailments. In addition, the processed starch of this plant was used to stiffen cloth, while the stem fibers were woven into mats on some islands (Whistler 1992b, also see Spenneman 1994, Etoh 2001). Although its importance has declined, *T. leontopetaloides* is still cultivated for local use in many areas for food, local medicine, and ornamental purposes. Indeed, a recent study "identified alkaloids, saponins and tannins in the leaves of *T. leontopetaloides*, but only alkaloids in the tubers of this species" (Borokini and Ayodele 2012: 97). Most plants of *T. leontopetaloides* can produce numerous starchy tubers that are comparable in appearance to white potatoes, usually 10–15 cm (4–6 in) in diameter, but they can reach 30 cm (12 in) in nutrient-rich soils. The normal weight of individual tubers is 70–340 g, but they may weigh up to 1 kg (2.2 lbs). "Two distinctive forms" are reportedly found in some tropical Pacific islands: one that produces a single relatively large tuber and the other that yields a group of smaller tubers (e.g., see Kay 1987).

Uses in Pohnpei: *T. leontopetaloides*, known in Pohnpeian as **Mwekimwek**, was formerly "partially cultivated" for its tubers but now grows wild in relatively open coconut groves near the beach. Indeed, the nutritious tubers (when properly prepared and cooked) were an important traditional source of starchy food on Pohnpei until the introduction of **Kehp tuhke** (cassava or tapioca, *Manihot esculenta*). The bulbous, fleshy tubers of **Mwekimwek** were once more commonly collected, washed, dried, and pounded thoroughly into an edible flour, but with the advent of commercial supplies of flour, this has ceased. Balick et al. (2009: 278–279) referred to some of the medicinal uses of *T. leontopetaloides* on Pohnpei:

> "To treat an itchy, painful, spreading skin rash, leaves are pounded and rubbed on the rash… To treat *kilsarawi* (shingles), a few leaves are pounded and placed directly on the rash (nothing else should be added) two times daily for four days or more… To treat burns, the whole stem [leaf stalk] is pounded and squeezed onto the burned area so that when it heals it will not leave a scar…" (Also see Lee et al. 2010: 61).

Uses in Mwoakilloa: Murphy (1950: 74) reported in the middle of the 20th century that "arrowroot" (*T. leontopetaloides*) was scattered throughout the forested areas on the atoll's three sandy islets. He indicated that it was "especially abundant in southern Manton [= Mwandohn islet]," with most plants appearing to be "growing wild." However, Murphy also noted that some of these plants were being cultivated and the starch made from their tubers used as "a food by itself and… in various food mixtures." A few years after Murphy offered his account, Glassman (1953: 298) observed *T. leontopetaloides* as both "a cultivated and escaped plant in the coconut grove," and noted that the tubers were still "an important source of food." Similarly, working on Pingelap Atoll,

Damas (1994: 234) commented on earlier and contemporary use of this pre-European introduced crop species:

> "St. John (1948: 108) says both cultivated and spontaneous[ly occurring plants of this species are found on the atoll and are] used in plaiting hats as well as [as] a source of food. Albert [Damas' informant on Pingelap] says that the seed is dried in [the] sun and mixed with bananas, coconut milk, and sugar to make a [favorite dish] called *pilobo*."

In an article entitled "Food Security in Small Islands: Case Studies in the Federated States of Micronesia," Soto (2014: 30) emphasized the importance of maintaining traditional knowledge of edible and medicinally useful plants that may not be used in modern times:

> "On small islands and atolls, imported foods and medicines may not arrive for more than a month if a typhoon or an oil crisis [occurs]. However, a young generation has never eaten [these somewhat forgotten plants] and/or does not know how to cook them. It is necessary to document how to [prepare and] eat these plants for the next generation."

Uses in Kapingamarangi: In the middle of the 20th century, the botanist R.F. Fosberg observed *T. leontopetaloides* on Hare islet of Kapingamarangi, where he reported that it was locally common "on flat coral sand with considerabe humus; planted [with] coconuts and breadfruit, moist with rather little undergowth in most places." Stone (1966) listed the Kapingamarangi name for *T. leontopetaloides* as 'Pié' (here we report its Kapingamarangi name as **Mogumogu**). We know that the tubers of *T. leontopetaloides* have been processed and eaten on Kapingamarangi in the past (see specimen voucher 2431813, United States national Herbarium; notes by F.R. Fosberg, collected on Kapingamarangi, August 3, 1946). However, both Polynesian arrowroot (*T. leontopetaloides*) and turmeric (*Curcuma longa*), which were traditionally important crops in Polynesia and Micronesia, were not introduced to Kapingamarangi Atoll until the early 20th century (Emory 1965; also see Bayliss-Smith 2012: 16). If this is correct, then *T. leontopetaloides* has only a relatively recent history of presence and use on the remote Kapingamarangi Atoll.

▸ **Kehmah** (P)
Terminalia carolinensis

Description: The genus *Terminalia* includes about 100 species of large flowering plants in the family Combretaceae, which is found in tropical regions of the world. The name *Terminalia* is derived from the Latin word *terminus*, which refers to the fact that its leaves appear at the tips of its shoots. *Terminalia carolinensis* is a tall, straight, and traditionally valued tree. It can reach heights of about 35 m (~100 ft), and it has a strongly fluted trunk with buttresses at the base that may reach up several meters above the ground and can stretch out 10 m (~30 ft) or more from the base of the tree.

Distribution: *T. carolinensis* is native only to Pohnpei and Kosrae, and therefore considered endemic to these two islands. Although relatively rare on Pohnpei today, it is still fairly common on the island of Kosrae within forest wetland habitats (freshwater swamp), often as the dominant species. Typically, stands of this tree occur in depressions or on poorly drained soils, usually behind beach strand vegetation. On Pohnpei, *T. carolinensis* is now more limited due to widespread conversion of wetlands to agriculture (e.g., see Merlin et al. 1992; Paulson and Buden 2003). Where *T. carolinensis* is found in the lower forest of Pohnpei, and especially on Kosrae, it is usually found along river banks and in freshwater swampy areas. The buttresses at the base of the trunk help anchor the shallow-rooted tree, thus supporting its normally high crown. *T. carolinensis* often grows very straight, with distinctive lateral branches that make this native tree easily identifiable from a distance "with [its] distinctive pagoda-shaped crown and large buttresses" (Chimmer 2004; also see Chimner and Ewel 2004; Benstead et al. 2009: 455).

Uses in Pohnpei: *T. carolinensis* is known in Pohnpeian as **Kehmah**. It is used for timber, medicine, and as a source of edible nuts. Balick et al. (2009: 355) referred to two types of uses of this tree on Pohnpei: "The wood is used in construction for house posts, [and to] treat back pain, an unspecified amount of bark is scraped off, boiled in 4 lit (1 gal) of water, and consumed daily until better..." It should be noted that during the Japanese colonial period in Micronesia (1914–1945), *T. carolinensis* forests were heavily logged. After World War II, on Kosrae they recovered without much further significant exploitation (e.g., see Dahl 1993). Traditionally, **Kehmah** was used for canoe hull material, lumber, and medicine (Merlin et al. 1993). Since World War II, it has nearly disappeared from Pohnpei and is now common only in Kosrae.

▸ **Wahn dipwoapw** (P) (Pi), **Wehn doapwopw** (Mw), **Hua laagau** (Nu), **Gegebinu** (Ka)
Terminalia catappa

Description: *Terminalia catappa*, known in English as the "Indian almond," is probably an introduced species on Pohnpei. It is a small to large, fast-growing and spreading tree that is often noticeable because of its large, shiny, obovate leaves (with the widest part of the leaf toward the tip). The leaves turn red or yellow as they age and consequently become even more apparent. In addition, this tree typically develops a "pagoda shape" as it produces several horizontal branches from its central single trunk stem. This tree's many tiny white flowers are arranged on axillary racemes (unbranched, elongated inflorescence that bear stalked flowers). The green fruits are oval, almond-shaped, somewhat flattened, and have a narrow marginal ridge. As the green fruit matures and ripens, it turns red to purple, with smooth but tough outer skin covering an inner layer of corky fibers. This corkiness allows the fruits to float with their seeds, and eventually some become dispersed by ocean currents.

Distribution: *T. catappa* has a vast natural distribution in sandy areas near coastal shores of the Indian Ocean (perhaps why this tree is known in English as the Indian almond). Its native range continues through tropical Southeast Asia, extending to some islands in the Western Pacific Ocean. It is unclear whether the Indian almond is native to the islands in Pohnpei State. Indeed, the extent to which the geographical range of *T. catappa* has been increased through purposeful dispersal by humans is difficult to determine. Today, it is found on most of the islands of Polynesia and Micronesia; however, because it produces corky, saltwater-resistant fruits, it may disperse by means of ocean currents. Indeed, the Indian almond appears to be native even in places where it is known or believed to have been brought in by humans. For example, some assume that *T. catappa* was introduced by the Japanese to the Marshall Islands, where it is now common in some areas, especially in the littoral (coastal) forest. Here, we assume that *T. catappa* is an introduced (alien) species in Pohnpei. Regardless of its origins on many tropical Pacific islands, this coastal-adapted tree has a number of uses on various islands.

General uses: *T. catappa*'s edible nuts taste somewhat like almonds, although it is difficult to separate the edible nut from the rest of the hard fibrous fruit. Many years ago, the French ethnobotanist Barrau (1961) noted: "The nuts of the Indian almond... are eaten throughout Micronesia and Polynesia." However, there is more to this tree than its edible nuts. For example, Del Rosario and Esguerra (2003: 12) summarized this tree's multiple uses in tropical regions:

> "The species has traditionally been very important for coastal communities, providing a wide range of non-wood products and services. It has a spreading, fibrous root system and plays a vital role in coastline stabilization. It is widely planted throughout the tropics, especially along sandy seashores, for shade, ornamental purposes, and edible nuts. The timber makes a useful and decorative general purpose hardwood and is well suited for conversion into furniture and interior building timbers. Fruits are produced from about 3 years of age, and the nutritious, tasty seed kernels may be eaten immediately after extraction."

The Indian almond is also reportedly used widely as a source of medicine to treat sexually transmitted disease, as well as the tropical illness known as yaws (framboesia), and parts of this tree are used in some areas to treat heartburn or indigestion and cuts.

Uses in Pohnpei: *T. catappa* is known in Pohnpeian as **Wahn dipwoapw** or simply **Dipwoapw**. As it is an introduced species, its local name is borrowed from its indigenous relative *T. samoensis* (discussed separately below). On Pohnpei, *T. catappa* is a food and medicinal resource; for example, its seed kernel is eaten when ripe, and medicinally its bark is used in remedies to treat diarrhea and general dysentery. To treat these digestive problems, bark of *T. catappa* is scraped, pounded, wrapped in cloth, squeezed into a cup of water, and consumed daily until better. *T. catappa* is also used on Pohnpei to treat mouth rashes troubling babies or children (e.g., see Balick et al. 2009: 358; also see Lee et al. 2010: 25). Some people (e.g., the Mortlockese) use *T. catappa* therapeutically to "relieve the symptoms of pinkeye" by infusing in boiling water the large older leaves that have turned red and then, after the water has cooled, using the infusion to bathe the sore eyes (see danaleeling.blogspot.com/2016/09/plants-that-heal.html).

Uses in Pingelap: St. John (1948: 112) recorded that this tree species was "cultivated" on Pingelap Atoll in the middle of the 20th century.

▸ **Dipwoapw** (P), **Win** (Mw) (Pi), **Kini** (Nu), **Laagau daliha** (Ka)
Terminalia samoensis

Description: *Terminalia samoensis* is usually a small to medium-sized pioneering native coast strand tree that grows up to 2–15 m (6–50 ft) in height. Marshall (1957: 4) observed a "large" *T. samoensis* tree with a trunk "2 feet in diameter" on an atoll he visited in the middle of the 20th century. *T. samoensis* usually produces spirally arranged shiny leaves that are wider at the tip (obovate), alternately arranged, and clustered at the ends of the stems. The *T. samoensis* shrub or tree develops oval-shaped, almond-like single-seeded fruits that turn red when they mature. It is closely related botanically to two other local trees in the same genus: the much bigger native tree *T. carolinensis* and the introduced *T. catappa*.

Distribution: *T. samoensis* is an indigenous woody species that is native from coastal regions in Indonesia all the way to the low and high islands of Micronesia, and beyond to the Tuamotus in eastern Polynesia. This species is found in coastal strand forest vegetation, growing in coralline or sandy soils, especially on the islets of atolls, and it grows somewhat larger in open areas. It is rare on Pohnpei today, but it is still relatively common on the outer (atoll) islands.

General uses: Like its cousin, the tropical beach almond (*T. catappa*), the seeds of *T. samoensis* are edible, though due to their smaller size, they are not commonly consumed on most Micronesian islands where this species occurs as a native coastal woody species. The wood of this species is said to be used for traditional carvings throughout its tropical range.

Uses in Pohnpei: *T. samoensis* is known in Pohnpeian as **Dipwoapw**. In the past, *T. samoensis* was an important source of housebuilding timber in some areas, due to its durable wood. Its nut is sometimes eaten; however, today *T. samoensis* is reportedly very limited on Pohnpei.

Uses in Mwoakilloa and Pingelap: In contrast to its limited existence today on Pohnpei, Glassman (1953: 294) describes "*Terminalia littoralis*" (now classified as *T. samoensis*) as a common coastal stand tree about 20 ft tall on Mwoakilloa Atoll, and noted that its local name on this atoll is **Win**.

Lessa (1977: 170), who was probably referring to St. John's (1948) notes about this native coastal tree species, tells us that *T. samoensis* is also known as **Win** in Pingelapese, and on this atoll "the wood of [*T. samoensis*] is used for tool handles." More recently, Damas (1994: 238) referred to this tree as "*Terminalia littoralis* Seem" (a synonym for *T. samoensis*) and noted that its "fruit" is edible and its wood is "used for toolhandles..."

▸ **Oaloahd** (P), **Woaloahd** (Mw), **Waelahd** (Pi), **Helii a de honu** (Nu), **Miiamia** (Ka)
Thalassia hemprichii

Description: *Thalassia hemprichii*, known as "turtle grass" in English, is one of the "seagrasses" or saltwater herbaceous plant species commonly found in many marine wetlands of Micronesia. *T. hemprichii* is a dioecious seagrass, which means it has individual plants that only produce male flowers and other plants that only produce female flowers. As a macrophytic alga, it produces sickle-shaped leaves about 10–40 cm (4–16 in) long, and on these leaves are short black bars (tannin cells). It develops thick rhizomes with scars between the budding leaves (shoots).

Distribution: *T. hemprichii* has a broad geographical distribution. It is found along the coastal zones in southern Japan to Taiwan, extending to Hainan, China, the Philippines, throughout Southeast Asia, and in tropical regions of the Indian Ocean. It is also found in northern Australia and in the Western Pacific in saltwater environments of islands reaching as far as New Caledonia in Melanesia and the Marshall Islands in Micronesia. *T. hemprichii* grows submerged in shallow coastal bays with its roots embedded in sand or mud, or in pools of tidal flats, including those found in Pohnpei.

Uses in Pohnpei: This seagrass is called **Oaloahd** in Pohnpei, where it is found in shallow lagoon waters. Balick et al. (2009: 282) reported that *T. hemprichii* "is used as a kind of bait for turtles because turtles can be caught in the coastal lagoon beds where the turtles are foraging on this plant." *T. hemprichii* is also reportedly used in Pohnpei "for making rope and as a medicine for healing large wounds," and the large seeds of this plant are sometime eaten and "are considered a delicacy" (see Rehg and Sohl 1979).

Uses in Pingelap: According to Damas (1994: 231), *T. hemprichii* is known as **Waelahd** in Pingelapese. He described it as a "flowering weed," found in or close to the ocean, that was used mainly to "wave around to chase away evil spirits."

▸ **Pone** (P), **Pɛnɛ** (Mw), **Paenae** (Pi), **Milo** (Nu), **Mila** (Ka) *Thespesia populnea*

Description: *Thespesia populnea* is an attractive, small to medium-sized useful tree that grows to maximum heights of approximately 12 m (35 ft), but normally it is about 6–10 m (20–23 ft) at maturity. The relatively short trunk is frequently crooked, but it develops a broad, rather dense crown. Its leaves are broad, light green, pointed, glossy and about 7–30 cm (3–12 in) long. Its flowers, which are produced intermittently throughout the year in warm climates, are long, 5-petaled, and pale yellow with purple at the base. The buoyant, woody fruits, which have several seeds, are clustered, 2–3 cm (1 in) wide, and turn pale-grayish brown when mature. *T. populnea* somewhat resembles the coastal hibiscus (*Hibiscus tiliaceus*), but its small round nut does not split open at maturity, unlike those of the hibiscus.

Distribution: *T. populnea* is widespread from tropical Africa to Hawai'i. In the Pacific, it is found on all the high islands of Polynesia and Micronesia, but it was probably introduced in the eastern part of its present range, including Hawai'i. *T. populnea* can be propagated by both seeds and cuttings. It grows wild in the coastal forest and thickets. Although it is indigenous over a wide tropical area from eastern Africa and southern Asia through Malaysia to parts of Polynesia and Micronesia, it may not be native to all or most areas of Micronesia or Polynesia; future paleobotanical evidence may resolve its native range in Remote Oceania. Its buoyant woody seeds can be dispersed by ocean currents, at least over relatively short distances; however, we are still unsure about how long the seeds remain viable and able to germinate upon arrival in an appropriate coastal island forest.

General uses: Friday and Okano (2006c) described the usefulness of *T. populnea*, which they referred to as one of one of the most significant woody species among Pacific islanders:

> "The rich, dark wood is carved into beautiful bowls, tools, small canoes, and figures. Ropes are twisted from the bark. The trees provide protection against wind, salt spray, and the hot sun. The seeds, leaves, and bark provide medicine and food. In ancient times the trees were planted around temple sites. Today the tree is more rare than in the past because of overharvesting in some areas and increased urbanization in others. The tree is easy to grow and should be considered for reforestation and urban forestry projects in the Pacific where suitable sites are available." (Also see Elevitch 2006a).

Uses in Pohnpei: *T. populnea* is called **Pone** in Pohnpeian. It is a useful tree that was probably more common in the past but has been harvested extensively in Pohnpei and some other Micronesian islands because of its use as wood, including for craft products. In addition, it is used for medicine, and it was also possibly brought to Pohnpei by early Micronesian immigrants:

> "This wood is used for the construction of houses... The stem wood of this plant is used for making canoe paddles... This is the foremost wood [for] making canoe paddles because it is very strong, and also it is used to make the midrib for longer boats... To use for making the main part of a canoe... The wood is also used to make the wooden handle of the net used to catch flying fish... [and parts of this tree are used to] treat an infected wound..." (Balick 2007: 435)

Uses in Mwoakilloa and Pingelap: *T. populnea* is called **Pɛnɛ** in Mwoakilloa and **Paenae** in Pingelap. Glassman (1953: 294) observed that it is "a frequent tree along the strand" (coastal vegetation) on Mwoakilloa Atoll.

Working on Pingelap, Damas (1994: 237) noted that *T. populnea* produces a good-quality ("very hard") wood which is "used for breadfruit pole (*piled*), axe handles, and flying-fish net handles," while the trunk of the tree is used for lumber and bark for making nets.

▸ **Titin** (P), **Sisin** (Mw), **Sesen** (Pi), **Manu gili maau** (Nu), **Dogodogongo** (Ka)
Heliotropium arboreum

Description: *Heliotropium arboreum* (known in many sources by its older name *Tournefortia argentea)* is a small to medium-sized native tree in Micronesia. It belongs to the heliotrope family (Boraginaceae) and is often called "tree heliotrope" or "beach heliotrope" in English. This shrubby indigenous coastal tree grows to heights of 8 m (~25 ft) or more and has bark that is light gray or brown and deeply corrugated. It produces large, silvery-hairy, and somewhat fleshy leaves which are 10–20 cm (4–8 in) long in whorls at the branch tips. The many tiny flowers are arranged on long twisted stalks or panicles, and the small fruits, with 2–4 seeds enclosed in a corky tissue, mature from greenish to yellow and then eventually turn brown.

Distribution: *H. arboreum* is a very common native plant in the coastal strand vegetation of most sandy shores of tropical Pacific islands, but not in the Hawaiian Islands, where it is a species introduced by post-European contact. It is naturally restricted to the hot, salty, windswept, coastal stand environments, where it is an early colonizer (pioneer plant), often being replaced over time by larger, slower growing trees. Since this tree is native to many islands, including those in Micronesia, it can be considered indigenous in this region (but not endemic because it has not evolved into a unique species on any of these islands). The corky tissue enclosing the seeds within the fruits aids this plant in its oceanic dispersal via sea currents. Although infrequently considered to be pest species, *H. arboreum* has potential for invasiveness when introduced to new coastal environments; however, on some islands, the beach heliotrope is endangered due its use as an easily available fuelwood source (e.g., see Manner and Elevitch 2006b).

General uses: The tree heliotrope plays a significant role in the cultural adaptation and ritual activities of many atoll dwellers who live in coastal tropical environments of the Pacific region, including those in Micronesia. For example, the leaves of this plant are consumed by islanders as a vegetable and are sometimes used as fodder for pigs. Various parts of this plant are also widely used in traditional medicine and ceremonies, such as curing rashes on children, treating diarrhea, and alleviating fish poisoning on some islands, especially the atolls. The leaves of *H. arboreum* are also used as components in medicinal teas and steam baths, as well as to stop cuts bleeding and cover bruises to promote healing. Furthermore, the leaves are also used to cure diseases real or imagined that may be caused by the disrespecting "sea taboos." The wood of *H. arboreum* is used in house construction and to fuel earth ovens. It is also used to make canoe parts and handicraft items, and fabricate tool handles, frames for diving goggles, canoe bailers, and other items (Manner and Elevitch 2006b; also see Lessa 1977: 176; Thaman et al. 1994). The dead leaves and flowers of *H. arboreum* and *Guettarda speciosa* are preferred composting materials on many atolls. Manner and Elevitch (2006b) also referred to some of the environmental and ethnobotanical benefits of *H. arboreum* for tropical island coastal areas, as well as the need for more study of its potential for resource management in the Pacific region:

> "The species is important for its ecological benefits, particularly as a barrier to salt spray, a windbreak on exposed coasts, and for its presumed role in coastal stabilization. It is capable of growing near the ocean in saline conditions and in nutrient-poor sands and rocky soils. Tree heliotrope plays a significant role in the handicrafts, tools, canoe building, traditional medicine, and rituals of atoll island cultures. Unfortunately, it is not readily recognized by institutional agroforestry because it doesn't have nitrogen-fixing capabilities, grows slowly, and has limited use as a timber species. Little research has been conducted on this species in terms of its productivity, life cycle characteristics, and its potential as an agroforestry species."

Uses in Pohnpei: Balick et al. described several uses for *T. argentea*, which is known in Pohnpeian as **Titin**. These uses include for construction purposes as well as for traditional medicine, such as its use in medicinal steam baths and to cover bruises, and other purposes such as those noted below:

> "To use [**Titin**] as a meat tenderizer for octopus, the octopus is beaten with the leaves to make it softer right after spearing (when it is beaten sufficiently with leaves, the octopus will change color and the meat will be softer)... The wood is used for the main body of a canoe, including racing canoes... The fruits are used in a game by children as the 'peas' in a peashooter made from the petiole of **Memiap** (*Carica papaya*) leaf... The seeds are used as a toy when they are shot through the petiole of papaya by children... To [treat] *nan aumweredi* (thrush [mouth rash or ailment]) in children... To treat itchy skin disease... (*umwulap* treatment)... To treat an upset uterus... To treat menorraghia [menstrual periods with abnormally heavy or prolonged bleeding]... [Moreover this] plant is used to cleanse the womb of a woman whose baby died in labor. If this not done, it is believed that the next baby will die. The quantity of plants used depends on how sick the woman is. [Overall, **Titin**] is said to be a very useful plant..." (See Balick et al. 2010: 342–343; also Herrera et al. 2010: 95).

Uses in Mwoakilloa: *H. arboreum* is known in Mwoakilloa as **Sisin**. As on at least some of the other islands in Micronesia, its wood was sometimes used to make the frames for diving goggles, as well as for other purposes.

Uses in Pingelap: *H. arboreum* is known in Pingelapese as **Sesen**. It was listed by St. John (1948: 112) as *Messerschmidia argentea* (another synonym of *H. arboreum*) and noted that it was common on Pingelap when he visited that atoll in the mid-20th century. Damas (1994: 238) also listed this species as "*Messerschmidia argentea*," and noted that this plant is a "common" tree on Pingelap Atoll. As it is on Mwoakilloa, *H. arboreum* wood is sometimes used to make the frames for diving goggles as well as other purposes on the other atolls of Pohnpei State.

▸ **Wahpaili** or **Oangoahng** (P), **Doaudol** (Mw), **Saudul** (Pi), **Gaulimago** (Nu), **Dua gimoo** (Ka) *Vigna marina*

Description: The yellow-flowered vine, *Vigna marina*, is an herbaceous, prostrate or weakly climbing plant with a woody base. Known in English as the "beach bean" or "beach pea," this vine sprawls over rocky and sandy areas along the coast, or climbs over shrubs in some parts of the lowlands near the shore. Its trifoliate leaves, with three leaflets, are arranged alternately and are 4–12 cm (1.5–5 in) in length. The flowers are bright yellow, and its seeds are contained in dark-brown pods that are 5–8 cm (2–3 in) long. As a member of Fabaceae, the bean or pea family, *V. marina* fixes nitrogen in the soil and thus appears to make a contribution in sustaining soil fertility.

Distribution: *V. marina* is a very widespread indigenous plant species throughout the tropics, including the Pacific, and is very common in tropical coastal vegetation. Indeed, in Micronesia, *V. marina* is among "the most typical plants in sandy beaches, reducing evaporation and soil erosion" (e.g., see Manner 1990). *V. marina* "is probably second only to *Ipomoea pes-caprae* [**Sonsol** in Pohnpeian] in range and abundance on the beaches of the tropical Pacific" (Whistler 1992a).

Uses in Pohnpei: On Pohnpei, this coastal vine is called **Wahpaili** or **Oangoahng**. It is a common sprawling plant in coastal and lowland secondary vegetation and is reportedly used for medicine. According to Balick et al. (2009: 397–398), *V. marina* has a number of therapeutic or culturally motivated uses in Pohnpei:

> "As a charm to repel a hex placed on a person... To treat boils that do not open... To treat swelling boils on legs... To treat *mweipwet* (a skin fungus of white spots all over a baby)... To keep hair from falling out of the head... To treat *soumwahu en naniak* (mangrove sickness)... To help induce pregnancy... To treat breast pain before menstruation, but mostly after delivering a baby..."

According to Lee et al. (2010: 128), parts of **Oangoahng** are reportedly used to treat mangrove sickness:

"Pounded into a mash, the leaves (in any quantity) are soaked in a bucket of water and used to bathe with. The sick person is to shower with this water twice daily for four or eight days until symptoms are gone."

Uses in Mwoakilloa and Pingelap: *V. marina* is known on Mwoakilloa as **Doaudol**, and as on many islands elsewhere in Micronesia, it is common on along the coast of this atoll (e.g., see Glassman 1953: 295). Damas (1994: 237) wrote that, on Pingelap, "*V. marina* is known as 'Nimelitop', a compound name deriving from the words for "a woman who has recently given birth" and "having something to drink." Therefore, Damas believed that the local medicine "drunk after childbirth to heal the wounds of the womb" is prepared with parts of *V. marina*. However, Damas seems to have copied the name 'Nimelitop' from St. John (1948), although Glassman (1953) had previously noted that this was a mistaken association and had clarified that this name refers to another sprawling legume vine, *Canavalia cathartica*. Later, Hattori (2012) provided contemporary Pingelapese spelling as **Nihmaelidoap**.

- **Moaresed** or **Ngkahu** (P), **Moarjij** (Mw), **Kisiwael** (Pi), **Aigi mea** (Nu), **Dilidai** (Ka)
  Wollastonia biflora

Description: *Wollastonia biflora* has been classified by some botanists in recent times as *Melanthera biflora* (e.g., see Balick et al. 2009: 338; Herrera et al. 2010: 92), or as *Wedelia biflora* (e.g., see Glassman 1952: 98). Here, we recognize this plant species as *Wollastonia biflora*, a native, indigenous, somewhat woody herb in the sunflower family (Asteraceae). *W. biflora* is sometimes referred to as the "beach sunflower." It grows erect or sprawling to heights of about 1.5 m (5 ft), producing bright yellow flowers. This bushy herb or trailing, vine-like plant produces leaves 6–15 cm (2–6 in) long with prominent veins and toothed margins (edges). The yellowish "flowerheads" are actually clusters of tiny flowers, which is a key characteristic of species in the sunflower family.

Distribution: *W. biflora* is an indigenous species commonly found in coastal areas from tropical Asia to eastern Polynesia, including the Marianas, Caroline, and Marshall Islands in Micronesia. It grows near the coastline on high islands and in many places on atolls, especially along rocky shores, where it may be the dominant species. It may also become a very common weed in coconut plantations. On Pohnpei, Glassman (1952: 25) referred to this beach sunflower species botanically as *Wedelia biflora* (= *Wollastonia biflora*) and noted the normally poor development of strand vegetation on the main island of Pohnpei, describing the distribution of this typically coastal strand species (*W. biflora*) on Pohnpei as "common and becoming locally abundant in a number of secondary growth areas up to 700 ft or more in altitude."

Uses in Pohnpei: In Pohnpeian, this plant is known as **Ngkahu** or **Moaresed**. Its reported therapeutic uses include steam from the heated small fruits of *W. biflora* being used as a traditional medicine for illness of the groin in Pohnpei. Moreover, according to Balick et al. (2009: 338–339), this beach sunflower (which they list as *Melanthera biflora*) has many medicinal uses on Pohnpei:

"To treat *soumwahu en kau* (sickness sent by sorcerers)... To treat burns... To treat conjunctivitis... To treat fungal skin infection in children along with the coconut meat... To use for toothache... [and] To treat arthritis..."

According to Lee et al. (2010: 59), the dense, mat-forming perennial herb or shrub is known in Pohnpeian as 'Ingkahu' (= **Ngkahu**, which Lee et al. listed as *Melanthera biflora*, here recognized as *Wollastonia biflora*), and they reported that it has been used on Pohnpei to heal burns on people's skin:

"One mature leaf of Ingkahu [**Ngkahu**] that has fallen to the ground is collected to treat skin burn. The 'meat' (*upw*) from one coconut fruit (*Cocos nucifera* L.) is scraped out. The 'Ingkahu' leaf and the coconut meat are wrapped together in a piece of cloth. The juice is squeezed out and rubbed onto the rash. This treatment is applied twice daily until the burn has healed."

Uses in Mwoakilloa: According to Glassman (1953: 296), the beach sunflower species, which he identified botanically as *Wedelia biflora* (= *Wollastonia biflora*), was uncommon along the strand of Mwoakilloa when he visited the island in the middle of 20th century. He noted its name as 'Morishish', in which we can recognize the currently used spelling, **Moarjij**.

▸ **Sawahn Awai** (P), **Jawahn Jeipɛn** (Mw), **Sawa Seipaen** (Pi), **Dalo Balaau** (Nu), **Dala Taibana** (Ka) *Xanthosoma sagittifolium*

Description: *Xanthosoma sagittifolium* is a large herb, originally from the New World tropics, with an erect, above-ground stem approximately 1–1.5 m (~2–6 ft) tall. Its large, light green, arrowhead-shaped (sagittate) leaves are about 20 cm (8 in) long and 15 cm (6 in) wide. The stalks that join with leaves (petioles) are up to 3–3.6 m (10–12 ft) long and are grooved on the underside. Note: An individual leaf stalk of *X. sagittifolium* is joined to the edge of the leaf, not to the underside of the leaf and inside of the leaf margin, as it is in "true taro," *Colocasia esculenta*. *X. sagittifolium* is sometimes also confused with *Alocasia macrorrhizos*, which is known in English as the "elephant's ear" or "giant taro" (**Ohd**, **Sepwikin** or **Wehd** in Pohnpeian). Manner (2011a/b) provided a guide to differentiating New World taro (*X. sagittifolium*) from giant taro (*A. macrorrhizos*):

> "... a simple way to distinguish *Xanthosoma* from *Alocasia* is to compare the angle of the leaf to the petiole (stem). In *Alocasia*, the leaf blade and the petiole are in the same plane. For *Xanthosoma*, mature leaves are angled about 30 degrees off the petiole. Furthermore, the leaf surfaces of *Alocasia* are usually shiny, while those of *Xanthosoma* are less so."

Distribution: There are about 40 species in the genus *Xanthosoma*. Three species in this aroid genus are important food crops in their native regions in the New World tropics, including *X. sagittifolium*, a New World "taro" known in much of its tropical Central and northern South American range as **Yautia**, or more widely as **Tannia** or **Cocoyam** (e.g., see Ramanatha et al. 2010). Not only is *X. sagittifolium* a staple food reportedly consumed by hundreds of million people worldwide, it is also one of the world's oldest "root" crops and is believed to have been brought to the different Micronesian islands from a variety of areas. Some scholars believe that *X. sagittifolium* originally came from South America to Pohnpei during the Japanese administration; in Chuuk, at least some people believe that *X. sagittifolium* was also first introduced there during the Japanese administration—its Chuukese name, **Woten Sapaan**, means it is perceived as a Japanese variety of sweet taro, *Colocasia esculenta* (which it is not). In Pohnpei, *X. sagittifolium* is also classified by the local people as a type of **Sawa** (*Colocasia esculenta*, true taro), and its Pohnpeian, Kapingamarangi, and Nukuoro names suggest respectively that the plant came directly from Hawai'i, Saipan, and Palau.

It should also be noted here that *X. sagittifolium* can become an invasive species in wet tropical environments such as those found on many Pacific islands, including some in Micronesia (e.g., on Puluwat Atoll). Indeed, beyond its native geographical range in the New World tropics, *X. sagittifolium* has become naturalized along some stream banks and in some humid shady places. *X. sagittifolium* can be grown where the "annual rainfall is between 1,000–2,000 mm" (about 40–80 in/year), but "planted just before the start of the wet season." However, under minimum amounts of 2,000 mm (80 in) and maximum amounts of 5,000 mm (200 in) per year, *X. sagittifolium* "can be planted at any time" (see Manner

2011b; PIER 2006; Weightman and Moros 1982). "It grows well in a wide range or soil types, especially in rich alluvial soils, but does not thrive in the sandy, nitrogen, phosphorous, manganese and iron deficient soils of the atolls" (Sproat 1968: 28–29).

General uses: The lateral tubers, or cormels (which are small corms growing at the side of a mature corm), of some *Xanthosoma* species, such as *X. sagittifolium*, are consumed as a starchy food. Note: The central or primary tuber or cormel is acrid and not used for human consumption. Today, the New World taro, *X. sagittifolium*, is very widely used in the tropical islands of the Pacific region; as a crop plant, it matures within about 9–12 months after planting (maturity is indicated by yellowing of its older leaves). However, it has been reported that the edible cormels of *X. sagittifolium* "can be harvested as needed after 5 months of growth" (Manner 2011b: 5):

> The corms of Yautia are usually washed and boiled then peeled in preparation for eating. These may also be roasted on a hot charcoal base. Cooked corms may be mashed and mixed with banana or coconut milk. Its tubers contain quantities of protein, fats, potash, calcium, phosphorous, iron and carbohydrate and small quantities of vitamin A, B1, C, niacin and riboflavin (Sproat 1968: 28–29).

Although some consider *X. sagittifolium* to be inferior to *Colocasia* taro (e.g., Smith 1985; Cambie and Ash 1994), this New World taro "is often grown because of the higher yield per unit labor, larger tuber size, taste and other factors." Two of these advantageous other factors are the ease with which the cormels of *X. sagittifolium* can be dried and processed into a starch powder, and the use of its leaves, like those of sweet taro, *Colocasia esculenta*, as a form of green forage fodder for livestock.

Uses in Pohnpei: The introduction of *X. sagittifolium* into Pohnpei occurred after European contact. This relatively new crop plant in Pohnpei superficially resembles the other taro crop plants long cultivated on Pohnpei. In fact, *X. sagittifolium* belongs to Araceae, the same family as the three other food-producing aroids found in Pohnpei: "giant swamp taro" (*Cyrtosperma merkusii*, **Mwahng** in Pohnpeian), "true taro" (*Colocasia esculenta*, **Sawa** in Pohnpeian), and "giant taro" (*Alocasia macrorrhizos*, **Ohd**, **Wehd**, or **Sepwikin** in Pohnpeian). In the late 1960s, only small quantities of *X. sagittifolium* were being grown in Yap and Chuuk, "and hardly at all" in Pohnpei, even though "ten named varieties of *X. sagittifolium* were introduced from the West Indies" to Pohnpei during this time (Sproat 1968: 28–29).

On the other hand, although *X. sagittifolium*, known in Pohnpeian as **Sawahn Awai**, has only been cultivated in Pohnpei in modern times, it was an important crop when food was scarce during World War II. For example, according to Bascom (1965), Pohnpeian Traditional Chief Iso Nahnken of Nett "used each section in his district to plant an estimated total of 200,000 plants of *Xanthosoma* [which] were set out between December 1944 and February 1945. Unfortunately, as Bascum (1965) noted "by June 1945 the entire crop [of *Xanthosoma*] had been destroyed by the leafhopper." In this case, the devasting insect was the taro leaf hopper, known scientifically as *Tarophagus proserpina* (see Englberger et al. 2009a: 154). In spite of these setbacks, at least two varieties of **Sawahn Awai** are still found and cultivated on Pohnpei (Englberger et al. 2009b: 119), and because of the abundant rainfall on Pohnpei, there is no need to dig a taro patch depression in which to cultivate certain taro crops, such as **Sawahn Awai**:

> To plant *Cyrtosperma* [**Mwahng**] and *Xanthosoma* [**Sawahn Awai**] on Pohnpei Island, a small hole about a foot in depth is dug in the ground and planting material is put in place. Usually the small plant of sucker (*ili*) growing beside the mother plant is used. The corm crown (*paki*), the lower part of the petiole with the top part of the corm, may be used when planting material is limited, but it grows less well and takes a longer time to reach maturity compared to the *ili*." (Englberger et al. 2009a: 152; Sproat 1968)

When natural hazards such as parasitic outbreaks, severe storms, or debilitating drought occur, it is essential that both food availability and ecological stability are supported by crop diversity (Declerck et al. 2011), and this can be realized by maintaining a diverse assemblage of plants that provide similar nutritional and ecological functions. For example, **Sawa** (sweet taro, *Colocasia esculenta*) and **Sawahn Awai** can both be used as a source of starch (from the corms or cormels), as well as a source of dark leafy greens (adequately cooked taro leaves). Furthermore, they both have relatively large leaf areas, which contributes to biomass accumulation and so may indirectly improve water quality and reduce runoff. Cultivating both taro crops helps ensure that if one is lost, their function in the system is still performed by another similar kind of crop (e.g., see Dignan et al. 2004; Hodbod and Eakin 2015).

Uses in Mwoakilloa: *X. sagittifolium* on Mwoakilloa Atoll includes one or more of the following varieties ***Jawahn Jeipɛn***, ***Jawahn Palau***, and ***Jawahn Ruk***. These names derive from the term for sweet taro (*Colocasia esculenta*, known as **Jawa** on Mwoakilloa), modified by geographic provenance of the specific varieties when they were brought into Mwoakilloa.

Uses in Pingelap: According to Englberger et al. (2009b: 119), *X. sagittifolium* is known on the atoll as ***Sawah Seipaen*** (meaning "**Sawa** (*Colocasia esculenta*) from Saipan"). However, Yamamoto et al. (2015: 163) found three Pingelapese names used for varieties of *X. sagittifolium* and/or *Colocasia esculenta*: ***Sawah Pingelap***, ***Sawah Seipaen***, and ***Sawah wai*** (meaning "from abroad", nowadays referring mainly to USA). Yamamoto et al. also commented on the results of their comparative study of the home garden cultivation of many edible plants on Pingelap, including *X. sagittifolium* taro:

> "More than 25 species of edible plants were observed in home gardens... *Cocos nucifera* was present at the highest frequency (82.8%), followed by *Musa* spp. (79.0%), *Pandanus* spp. (66.9%), *Artocarpus* spp. (52.9%), and *Carica papaya* (44.6%). These crops were consumed, or used in other ways on a daily basis, as starch staples, fruit, drink, food wrapping material, handicrafts, canoes, or as construction material. The frequencies of root and tuber crops were low. They included *C. merkusii* (8.3%), *Dioscorea* spp. (7.0%), *Xanthosoma sagittifolium* (6.4%), *Colocasia esculenta* (1.9%), and *Ipomoea batatas* (1.3%). This is partly because there is a big patch of *C. merkusii* on the island and it is difficult to cultivate some crops, e.g., *Dioscorea* spp. which are one of the most important crops in Pohnpei Island, due to alkali soil derived from coral [Kawai et al. 2010]. *Alocasia macrorrhizos* (12.1%) and *Tacca leontopetaloides* (1.9%), which were regarded by villagers as famine food, were found in home gardens as both cultivated and spontaneous forms." (Yamamoto et al. 2015: 163)

▸ **Pwulok** (P) (Mw), **Pwuloak** (Pi)
Xylocarpus granatum

Description: *Xylocarpus granatum* is a native mangrove tree that grows to heights of 5–20 m (25–65 ft). It has smooth white bark and reddish, colorful wood. It has an unusual, elaborate curtain or ribbon-like system of buttressed roots that help stabilize the tree and allow it to breathe in its natural habitat in (or near) the landward edge of mangrove swamps. The tree has pinnate leaves with 2–6 elliptic to oblong leaflets, each about 8–20 cm (3–8 in) long and 4–10 cm (1.5–3.5 in) wide. The flowers are cream colored, and the fruits are large, round, green to brown capsules that are about 10–15 cm (4–6 in) in diameter at maturity. Because of the shape of its fruit, one of the English names for *X. granatum* is the "cannonball tree," and its oddly angled seeds are the source of another of its English names, the "puzzle-nut tree."

Distribution: *X. granatum* is native to various areas in Southeast Asia and in various tropical islands in the Western Pacific region. As a tree species adapted to wetlands, *X. granatum* is usually found at the landward edge of the mangrove, frequently in dense stands.

Uses in Pohnpei: *X. granatum* is called **Pwulok** in the Pohnpeian language. Its reddish wood is used for making canoes, house construction, and carvings. Parts of this large tree are also used medicinally; for example, Balick et al. (2009: 441) referred to the treatment of muscle, joint, or back pain:

> "... the scraped bark is placed on hot stones; then the person, sitting over it and covered with a sheet, breathes in the vapors from the heated bark. This *umwulap* treatment is done daily until [the patient is healed]."

▸ **Oangen pele** (P), **Laagau hagamaluu libok** (Ka)
Zingiber zerumbet

Description: *Zingiber zerumbet* is known in English as the "shampoo ginger." This herbaceous ginger species is extremely variable in form. Generally, *Z. zerumbet* produces numerous leafy shoots 0.75–2 m (2–6 ft) long. It is a deciduous plant (i.e., it loses its leaves during part of the year). Its small white to yellow tubular flowers are surrounded by many relatively large and eventually colorful bracts that are green when young but become pale and finally bright red. As the flower heads (inflorescences—bracts and flowers) mature, they gradually fill with an aromatic, slimy liquid and turn a brighter red color; the flower heads usually remain hidden beneath the leaf stalks.

Distribution: *Zingiber* is a moderately large ginger genus of herbs that belong to the family Zingiberaceae. According to botanists, *Zingiber* presently includes about 141 species, mainly distributed in the tropics of mainland and insular Asia and many Pacific islands. *Z. zerumbet* is assumed by biogeographers to be native to tropical Asia, perhaps originally to India or Ceylon, and was probably introduced into Micronesia by early Pacific Ocean voyagers. It now grows wild in some forested areas of the high islands in Micronesia, and as a perennial, tuberous root species, it is adapted to damp, shaded parts of lowland area or hill slopes, where it can be found as scattered plants or in thickets.

Uses in Pohnpei: *Z. zerumbet*, called **Oangen pele** in Pohnpeian, was probably introduced to Pohnpei and other islands in Micronesia as a "canoe plant" by early Pacific Islanders who found and settled on these islands, including Pohnpei. Now, *Z. zerumbet* is naturalized widely in some forested areas of this island. In other words, besides cultivation of this introduced fragrant, decorative, and medicinal plant species, shampoo ginger is now also growing on its own on Pohnpei and on some other Micronesian islands. *Z. zerumbet* has long been a useful plant over a wide geographical range. The entire shampoo ginger plant is aromatic, and when its mature reddish inflorescence is squeezed, it produces a fragrant, frothy fluid. This squeezed-out liquid has been traditionally used as a shampoo and hair conditioner in many tropical islands in the Pacific region, including Pohnpei. Long ago, Tuyama (1941) noted that the juices squeezed "from the inflorescences of this plant are used for [a] kind of pomade on the hair." Moreover, this "shampoo ginger" is also used medicinally in Pohnpei. For example, the root serves as a laxative (e.g., see Merlin et al. 1992; Merlin and Juvik 1996). According to Balick et al. (2009: 317), shampoo ginger plants have a variety of medicinal and other uses on Pohnpei:

> "The stem is eaten for food by children... To remove bad taste in the mouth after drinking *sakau*... To use to stimulate hair growth in women... To give body to hair... To use as a shampoo... To treat itchy skin... To treat muscle pain... To treat headache... To treat fever in children... For pregnant women during labor, four pieces of the rhizome are pounded, squeezed in a glass of water, and sipped by the woman throughout the day to ease the process. Up to four cups are drunk in a day if the labor is particularly difficult...

Lee et al. (2010: 152) referred to the use of the rhizome of this ginger "to treat painful urination," and described a drink prepared using the juice squeezed out of a shampoo ginger rhizome after it is ground, cleaned, and pounded, which is then swallowed to relieve the symptoms of this urination ailment. Shampoo ginger leaves are also reported as a wrap for food put in the *uhmw* [Pohnpeian earth oven] when they are used as a layer between the outer banana leaves and the food to be cooked (see www.comfsm.fm/~dleeling/ethnobotany/text/flora.html).

In the following chapters, we focus on some aspects of ecological science and its relationship to the general study of living things, and more specifically to an understanding of the environments and biological diversity of Pohnpei. We also consider the question of "How is the natural resource wealth of Pohnpei and the outer islands intimately associated the traditional knowledge of the people who have lived in these islands for millennia?"

4

ECOSYSTEM OF POHNPEI

CHAPTER 4

ECOSYSTEM OF POHNPEI

This chapter introduces the fundamentals of ecology, starting with key ecological terms and concepts, and then briefly describes how nutrients cycle and energy flows through Pohnpei's ecosystems. Building on this foundation, we examine the development of biological communities and how they change over time through ecological succession. We then focus on topics relevant to Pohnpei and other high islands in Micronesia, particularly the importance of the upland forest—its rich biodiversity, ecological value, and the need for its conservation. Forest conservation has long been a traditional Pohnpeian value, and we hope this cultural commitment will continue to support the island's fragile and threatened ecosystems, along with its remarkable plant (flora) and animal (fauna) life. At the same time, we acknowledge the growing impact of human activity on Pohnpei's biodiversity, leading to the endangerment of many species and the risk of their disappearance or extinction. Finally, while this book's primary focus is on plants and people, we also highlight some of Pohnpei's threatened animals, including marine and mangrove-adapted species and several vulnerable native forest birds.

4.1. ECOLOGICAL DEFINITIONS

Ecology is the scientific study of the relationships between living organisms and their environment, including non-living elements like climate, soil, and water. One of the key concepts studied by ecologists is the **ecosystem**, a place or environment where living (organic) things interact with one another and their physical (inorganic or abiotic) surroundings. Three examples of ecosystems in Pohnpei include the following: coral reefs (where countless marine organisms such as coral, algae, sea snails, and fish live among layers of sediment atop the bedrock of an ancient volcano); mangrove swamps (where large trees and many smaller plants and animals live on the substrate of mud and fine sand piled up by streams, rivers, and tides); and native upland forests (where many plants, insects, snails, reptiles, and birds live above a base of rocks, soil, and leaf litter). Note that every ecosystem is made up of both living (biotic) and non-living (abiotic) components.

An ecosystem can also be much smaller or much larger than the places we see around us. For example, a whole human body may be viewed as an ecosystem. Indeed, a human body with all its interacting bacteria, fungi, and parasites, as well as the food we eat and clothes we wear, can be considered as a kind of ecosystem cycling nutrients using an ongoing flow of energy. Another conceptual ecosystem that is much larger would be the entire Earth with all of its living and non-living things; this massive ecosystem is also referred to as the earth's **biosphere**. Just like reefs, swamps, and forests, all ecosystems, including the relatively small body of a human or the planet Earth in its vast entirety, cannot be considered independent of inputs of energy. For the vast majority of ecosystems, the input is solar energy from the Sun (which plants convert to food that people and other animals consume). We should note here that we now know that some recently discovered marine, super-heated, hot vent ecosystems are not powered by solar energy but function via geothermal energy (e.g., see https://deepoceaneducation.org/topic/hydrothermal-vents/).

All the living things within an ecosystem are called a **biological community**. A biological community is thus made up of all the organisms in an ecosystem. They are interdependent. That means that they depend on one another in various ways to survive.

There are many biological communities in Pohnpei. Some examples include:
(a) all the plants, animals, and microorganisms living on the fringing coral reef surrounding Temwen Island; (b) all the geckos, insects, birds, plants, and other living things found in Kolonia town; and (c) the plants, animals, and other living things adapted to life in the highest reaches of Pohnpei in the dwarf, mossy forest of Nahnalaud and adjacent high mountain environments.

Within a biological community, we find different groups of organisms belonging to different species. Any group of living things of the same species that lives in the same community is called a **biological population**. It includes all individuals of a particular species in a particular place. Examples of biological populations are: (a) all the **Roamoak** (*Scaevola taccada*) shrubs growing along the coast of Mwoakilloa Atoll; (b) all the **Sara** squirrelfish in Sapwuahfik lagoon; and (c) all the **Kotop** or **Koatop** (*Clinostigma ponapense*) palm trees in the rainforests of Pohnpei.

Finally, each biological population comprises many individuals. Each of these individuals is referred to as an **organism**. It is either a single plant, such as one **Sadak** tree (*Elaeocarpus carolinensis*), a single animal, such as one **Serehd** (*Trichoglossus rubiginosus*), or a single fungus or microorganism.

Figure 4.1. Levels of organization in nature, from an individual organism to an ecosystem.

4.2. ECOLOGICAL FUNCTIONS

Based on its ecological function in the energy flow and nutrient cycling of an ecosystem, every organism can be classified as being a producer, a consumer, or a decomposer.

- **Producers** in an ecosystem include plants, such as the large native fig trees, **Aiau** (*Ficus prolixa*), and the common breadfruit trees, **Mahi** (*Artocarpus altilis*). These organisms use inorganic (non-living) chemicals, water, and sunlight to produce organic molecules such as carbohydrates, proteins, and fats. This process is called **photosynthesis**. Most plants use photosynthesis to sustain themselves and grow.

Figure 4.2. Examples of producers.

- **Consumers** include animals and many microorganisms. Some consumers in an ecosystem eat the producers (plants). These are called **primary consumers** or **herbivores**. For example, some crabs in the mangrove swamps are primary consumers because they feed on the fallen leaves of mangrove trees. Insects of many kinds that feed on plant material are also primary consumers. Similarly, purple-capped fruit dove, **Kiniwed** (*Ptilinopus ponapensis*), is a primary consumer because it feeds on fleshy fruits. The endangered Pohnpei fruit bat, **Pwehk** (*Pteropus molossinus*), is also a primary consumer since it feeds on fruits and seeds. Some consumers eat other consumers. These animals are called **secondary consumers** or **carnivores**. Examples of secondary consumers on Pohnpei include **Menin kahnlipw**, the large dragonfly, which feeds on flying insects; **Limenimeninseri**, the blue-tailed lizard that eats insects on the ground; **Likoht**, the short-eared owl (*Asio flammeus*), which feeds on a variety of small animals; and **Kutoahr**, the Pohnpei kingfisher (*Todiramphus reichenbachii*), which eats insects, lizards, small fish, shrimp, and crabs. Finally, consumers that eat both plants and animals are called **omnivores**. One such animal is the **Pwiliet**, the Micronesian honeyeater (*Myzomela rubratra*); this bird feeds mostly on nectar from flowers but also eats snails and insects. Human beings (*Homo sapiens*) and pigs (*Sus scrofa*) are also omnivores because they too eat both plant and animal material.

Figure 4.3. Examples of consumers.

- **Decomposers** recycle organic matter and release nutrients in an ecosystem. They feed on dead producers (plants) and consumers (animals). As they feed, they break down the bodies of the dead organisms into basic inorganic chemical components, making these "building blocks" available to the new producers again in an endless cycling of nutrients. Most decomposers are bacteria and fungi, such as mushrooms; but they also include many small animals, such as earthworms and soil insects.

Figure 4.4. Examples of decomposers.

4.3. CYCLING OF NUTRIENTS

We will now look at how nutrients cycle through an ecosystem, as they move from producers to consumers and finally to decomposers. Fig. 4.5 below shows an example from Pohnpei.

Producers (green plants) use nutrients and sunlight to grow and develop. They are, in turn, eaten by the consumers (animals), passing the nutrients along the chain. Decomposers (fungi, bacteria, etc.) then feed on dead producers and consumers, breaking down their organic matter into simpler inorganic compounds that are released back into the ecosystem where they can be reused by producers. This movement of nutrients through an ecosystem from a producer to one or more consumers and then to decomposers can be thought of as a **food chain**. Often, this nutrient-cycling process becomes quite complex, with several consumers feeding on a single type of producer and being eaten themselves by several other types of consumers. Thus, the term **food web** can often be more descriptive than food chain.

The decomposers play an important role in the ecosystem. They break down the dead organisms and provide the essential nutrients to new generations of producers so that the cycling of nutrients and life itself can continue. Fig. 4.6 on the following page shows the cycling of nutrients from the soil into organic compounds in a tree which, in turn, drops leaves and other parts of the plant on to the soil or leaf litter; these decaying or dead plant parts, in turn, are decomposed into inorganic compounds that are recycled back into the tree.

Figure 4.5. When a plant or animal dies, decomposers feed on it, breaking down its organic material into simpler inorganic compounds. These nutrients return to the soil, where they can be taken up and reused by plants—an example of how nutrient cycling works.

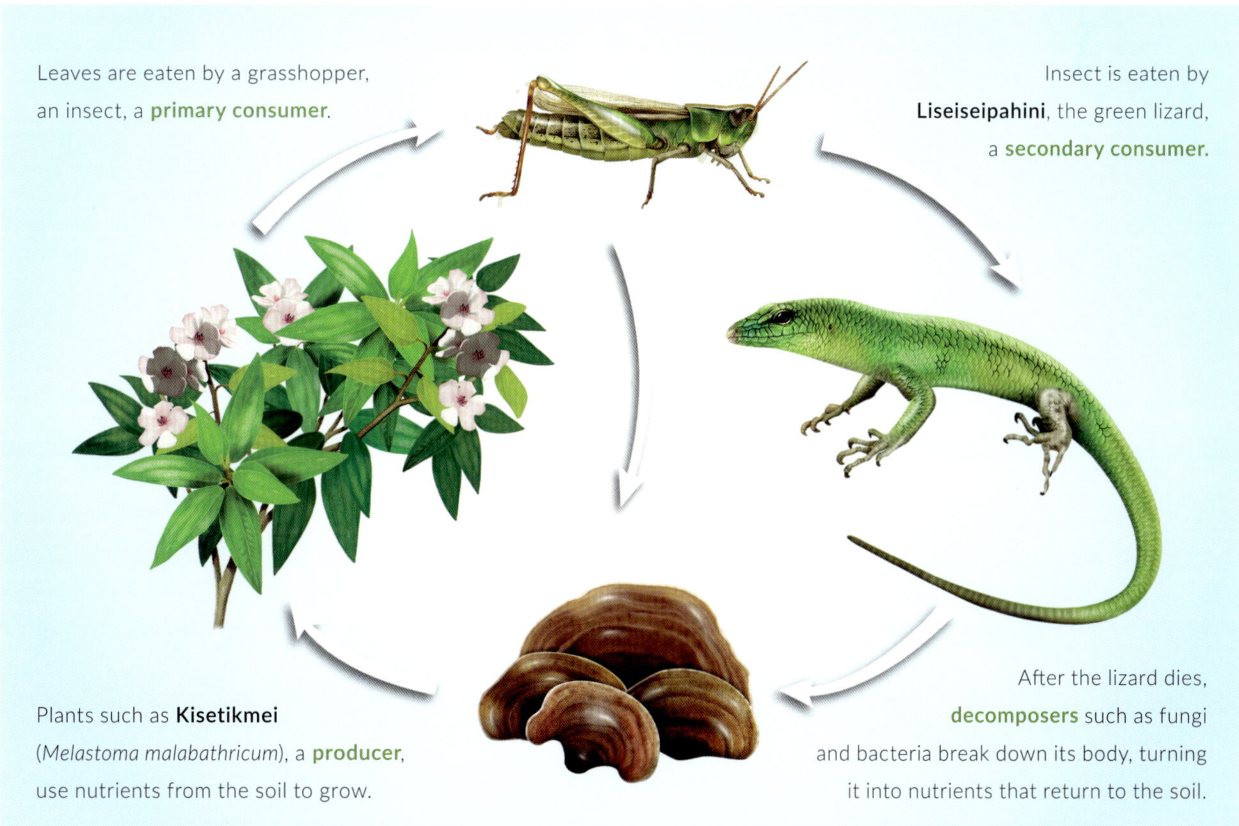

Figure 4.6. This diagram shows the basic cycling of nutrients between a producer (the tree) and decomposers (in the soil). Root uptake or absorption of nutrients (*alahldan mwenge*) from the soil (*pwel*) allows the tree to grow (*keirdahr tuhke*). Eventually, leaves and branches drop (*mweredi en teh, wah, oh rah ape*) from the tree canopy (*kadok en tuhke*) and fall into the litter layer (*mwasoahngoat en tuhke*) below. In this layer, breakdown (*mwasoahngoat*) of the fallen plant material takes place. Once organic compounds are converted into inorganic compounds by decomposers, uptake of the nutrients by the roots (*pwel*) can happen once again and so the cycle repeats, powered by a flow of energy from the Sun.

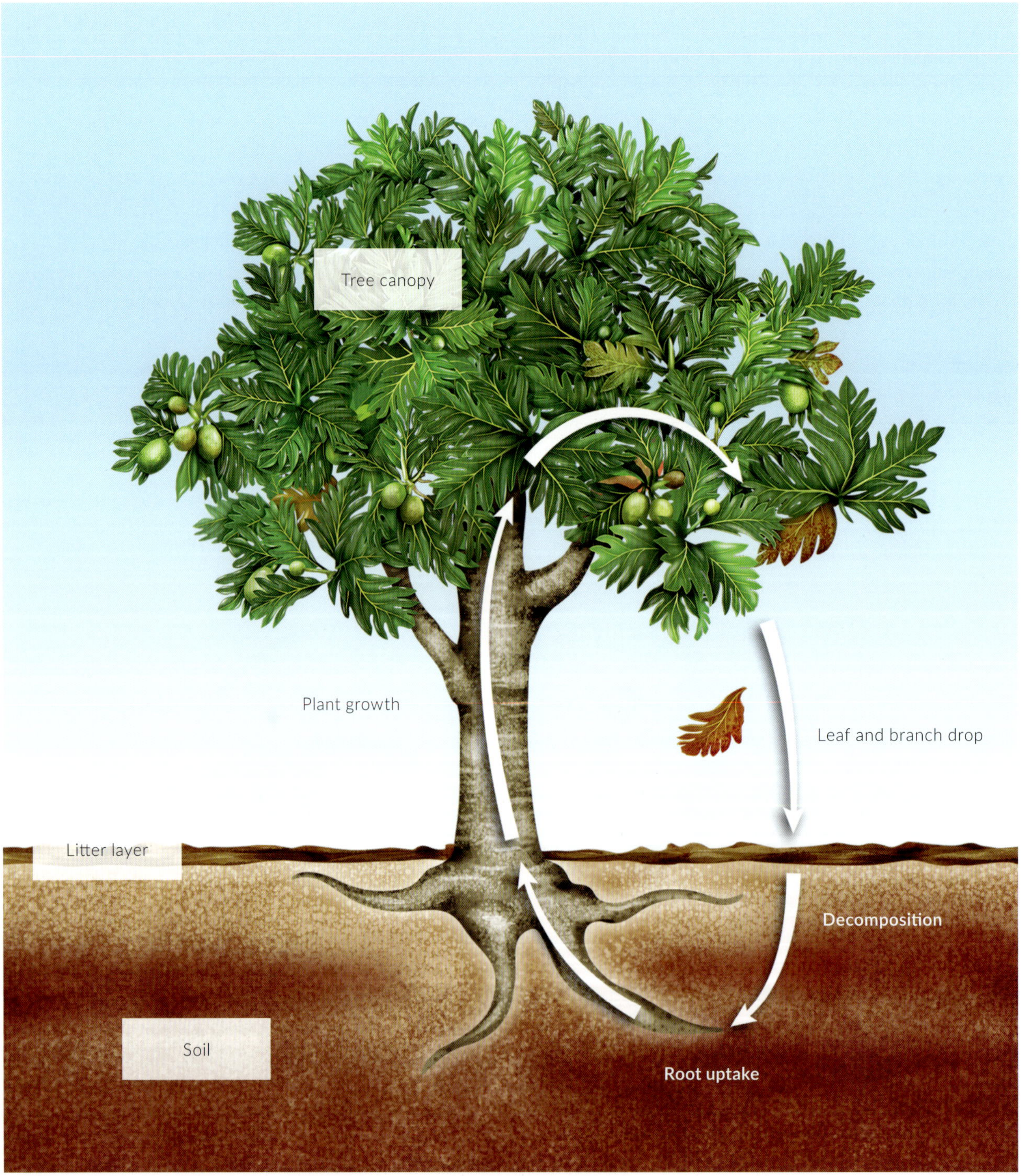

4.4. ENERGY FLOW

We will now look at how energy flows through an ecosystem, using Pohnpei as an example. As we noted, energy typically moves within an ecosystem from producers to consumers and finally to decomposers.

Nutrients cycle through ecosystems with minimal or no net loss of **biomass** (living matter). In contrast, energy flows through the system with considerable loss at each step in the food chain because heat energy is released by living organisms through their own **metabolism** (living processes). Energy lost as heat or in other ways is unavailable for growth or to produce biomass. Therefore, the total weight of living organisms (biomass) at each **trophic level** (position in the food chain or food web) is less than the weight of organisms in the next feeding level below. This relationship is often depicted by scientists as a biological "pyramid" of biomass, as shown in Fig. 4.7 and Fig. 4.8.

Biological pyramids illustrate that a large percentage of solar energy is converted into heat or other forms of energy as it flows through the ecosystems, and is thus "lost" from the food chain. Although energy is never "created" or "destroyed," it can be transformed into forms that are unavailable to organisms; therefore, a continuous input of energy into an ecosystem is necessary.

Figure 4.7. Biological pyramid for a mangrove forest in Pohnpei.

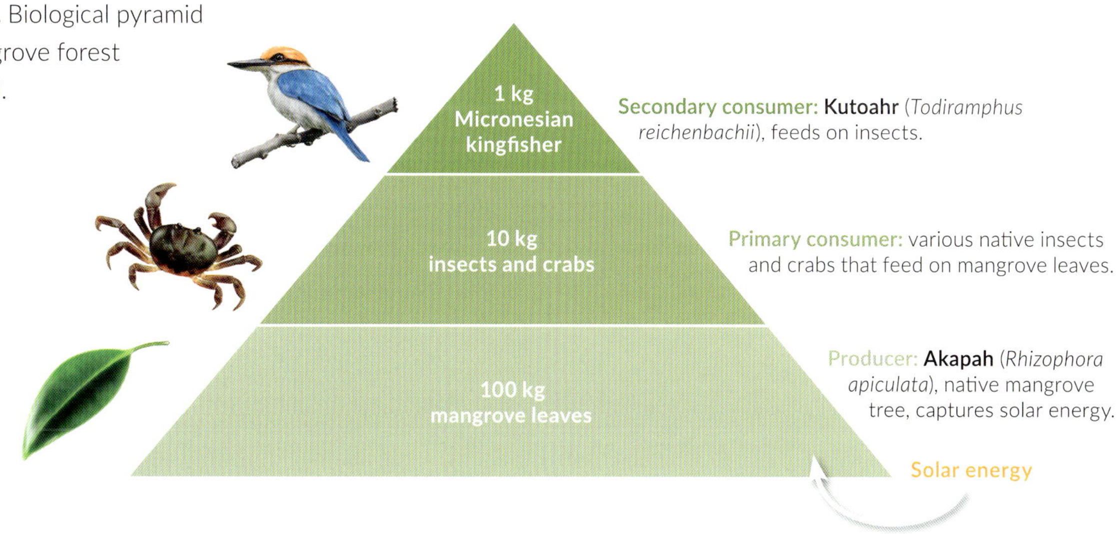

Figure 4.8. Biological pyramid for an agricultural area in Pohnpei.

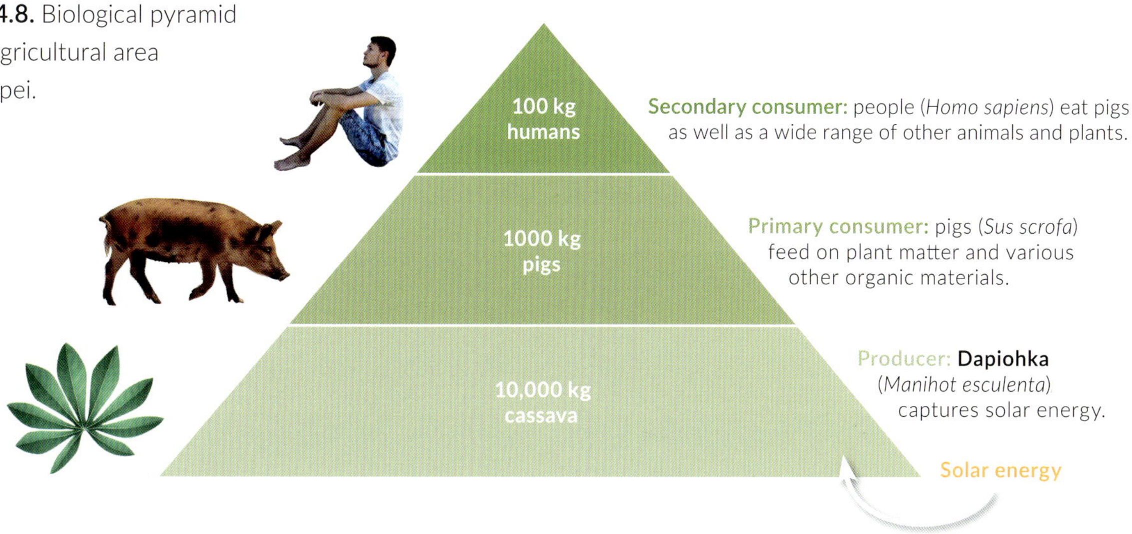

4.5. ECOLOGICAL SUCCESSION

An ecosystem is made up of living organisms (like plants, animals, and microbes) and non-living components (such as rocks, chemicals, and sunlight), all interacting and influencing one another. Overall, an ecosystem functions like a living thing, with the cycling of nutrients and the movement of energy working together to sustain its balance, health, and development over time. As nutrients cycle and energy is transferred within an ecosystem, different living things reproduce, feed, grow, interact, and die. Some organisms may leave, while others arrive, gradually reshaping the ecosystem through a process that scientists call **ecological succession**.

There are two basic types of ecological succession:

1. **Primary succession** occurs in places where no biological community existed before. For example, (1) a bare rock surface newly exposed by a landslide that carried away all the soil and everything that was there before, (2) a recently cooled field of lava from an active volcano, or (3) a new empty stretch of dry sand piled up by winds and currents on a shallow reef are all places that will experience primary succession as soon as the first organisms begin living there.

2. **Secondary succession** occurs where a biological community exists but was nearly destroyed or severely disturbed, leaving some soil, together with any surviving buried seeds and microbes. For example, secondary succession can occur after a taro patch is devastated by saltwater intrusion and abandoned, when a forest is burned by wildfire or clear cut by humans, or when an area is heavily impacted by flooding.

A wall of stacked basalt rocks in Kolonia offers a new surface to be colonized by pioneer species: moss, ferns, and other small plants.

The first organisms to colonize a newly disturbed available site are called **pioneer species**. The most common pioneer species are lichens, grasses, and other producers that can live with very little soil, water, or nutrients. In humid places, such as Pohnpei, mosses and ferns are important pioneers (see opposite page for the photo of these plants colonizing basalt rock surfaces of the Spanish Wall in Kolonia). Shortly after these plants take root and grow, they attract consumers, usually insects that feed on their leaves or other parts. As they die, these pioneer plant and animal species leave their remains, which decomposers, such as bacteria, fungi, worms, and insects begin to turn into new soil. This soil will enable new colonizing species to take root and survive in the area, eventually changing the biological community through the process of ecological succession.

A **pioneer community** can typically be seen in an area that has recently been burned or cleared. Weedy herbs and grasses may emerge from any seeds previously buried in the remaining soil or become dispersed into the area by birds or wind. In Pohnpei, typical early plant colonizers of such areas include **Dipwinmal** or **Mwedilinmal** (false staghorn fern, *Dicranopteris linearis*) and **Limpar** or **Kidi en mal** (clubmoss, *Lycopodiella cernua*). In the outer islands, the most common place to see pioneer communities forming is where high waves pile up sediment and create a new sandy area above the reach of high tide. Typical plant colonizers of barren sand are **Titin** (*Heliotropium arboreum*) and **Remek** (*Scaevola taccada*).

As soil and nutrient conditions improve, the plants and other organisms that form the pioneer community are replaced by taller shrubs, trees, and different animals in the process of succession. Eventually, a **mature community** is formed. These mature communities are more stable and change little in appearance or species composition over time, unless a major disturbance occurs, such as a typhoon, fire, serious earthquake, flooding, or human activity.

A row of sprouted coconuts taking root on a coral rubble bar, newly formed by waves on Ahnd Atoll.

4.6. EVOLUTION AND BIODIVERSITY

Building upon our understanding of ecological succession, we now turn to the concept of biological diversity, which is fundamental to the stability and resilience of ecosystems. Biodiversity encompasses not only the variety of species but also the genetic diversity within them and the variety of ecosystems, all of which evolve over time through natural processes. Given widespread concerns about declining global biodiversity (e.g., Kolbert 2024; Pimm et al. 2014), we offer a discussion on the origins and importance of biological **evolution** and **biodiversity**.

Members of the same species can mate and produce offspring but cannot mate with different species due to biological or geographical barriers. Scientists believe that the variety or diversity of species on Earth came about over a very long period of time, through a process known as evolution through **natural selection**. This process is also called "survival of the fittest," and includes adapting to the environment, survival, and reproduction over time.

Some individuals in a population may be better equipped or suited than others to survive and reproduce in their environment (the ecosystem or place where they live); these individual plants or animals are more likely to live to produce offspring. The individuals who are not well suited to their environment are less likely to produce offspring. Over many generations, populations adapt and become better suited, or "fit" to their environment; this change is called **adaptation** and may be so significant that it eventually leads to the evolution of new species.

4.6.1. Extinction

When the environment changes, or when a species moves into a new habitat, some members of the population must adapt to the new conditions for the species to survive. If the change in the environment is too great, the population may die out. If that is the only population of that species, then the species is gone forever from the face of the Earth. This is called **extinction** and represents a natural component of evolution.

One way humans drive species to extinction is through overhunting or over-exploitation. Humans as predators hunt and fish for different animals and harvest different plants. Driving a bird species like the Micronesian pigeon or **Mwuroi** (*Ducula oceanica*) to extinction through excessive hunting would be a tragic example of over-exploitation.

Similarly, over-harvesting a particular type of plant could lead to it disappearing altogether. This is especially dangerous in the case of species that are already threatened because they exist in low numbers or are endemic—and are therefore often particularly vulnerable to endangerment and even extinction. For example, the tree known as *Lepinia ponapensis*, found only on Pohnpei, has not been seen for decades despite intensive searches by botanists. Could it still be hiding in the upland forest, or has it already gone extinct?

Lepinia ponapensis

4.6.2. Endemic species

An **endemic species** is one that is native to a particular place and found nowhere else in the world. If a species is endemic to Pohnpei and disappears from the island, it means that species—which may have taken thousands or even millions of years to evolve—becomes extinct, lost forever not only from Pohnpei but from the entire world.

Endemic plants are very special aspects of our natural heritage and cannot be found growing naturally anywhere else on Earth. Some examples of trees endemic to Pohnpei are:

- *Astronidium ponapense*
- *Cinnamomum sessilifolium*
- *Dendrobium ponapense*
- *Discocalyx ponapensis*
- *Garcinia ponapensis*
- *Ilex volkensiana*
- *Lepinia ponapensis*
- *Melicope ponapensis*
- *Myrsine carolinensis*
- *Parkia korom*
- *Phyllanthus senyavinianus*
- *Timonius ponapensis*

Timonius ponapensis
Kehn (P)

Ilex volkensiana
Oahr en wel (P)

Herrera et al. (2010: 8) provided the counts of native and alien vascular plant species in Pohnpei, summarized in Table 4.1. By adding up numbers in each column of the table, we can see that Pohnpei has at least 942 species of plants, of which 50 are **endemic** (unique to Pohnpei) and an additional 345 **indigenous** (also native but not unique to Pohnpei). The remaining species are considered **introduced**, having been brought by people from elsewhere. These include 180 **naturalized** species (not native but growing wild) and 367 **cultivated** species.

According to another source, Pohnpei has 47 endemic species of vascular plants (Costion and Lorence 2012: 93). Whatever the exact number of endemic plant species, it is important to remember that the endemic species are special to Pohnpei. In fact, the number of endemic plant species in Pohnpei may not seem that impressive when compared with the level of 90% (or more) endemism estimated for plant species native to the Hawaiian Islands. However, if we consider the fact that Pohnpei has just 2% of the total land area of the Hawaiian Islands, then perhaps we can appreciate the importance of the endemic flora of Pohnpei and the need to protect it for this and future generations.

Table 4.1. Plant species by taxonomic category and their status in Pohnpei.

Category	Indigenous	Endemic	Naturalized	Cultivated	Total in flora
Pteridophyta	101	4	4	2	**111**
Gymnospermae	0	0	0	6	**6**
Monocotyledonae	94	21	52	98	**265**
Dicotyledonae	150	25	124	261	**560**

Note: *Pteridophyta* = ferns, *Gymnospermae* = cone bearing (open seeded) plants, *Monocotyledonae* = flowering plants containing a single cotyledon (seed leaf) in their embryos, and *Dicotyledonae* = flowering plants containing two cotyledons (seed leaves) in their embryos.

4.6.3. Invasive species

While endemic species are unique to specific regions and have evolved in isolation, **invasive species** are those that came from somewhere else, can reproduce on their own, and therefore may pose a significant threat to the local ecosystems, perhaps nowhere more so than on islands. These non-native species can disrupt the balance, outcompete native organisms, and cause irreversible changes to local biodiversity. Therefore, for people living on islands, it is of paramount importance to prevent introductions of species that do not belong there. Many alien species turn out to be highly aggressive and spread quickly and extensively, harming native plants and animals in the process (see Donnegan et al. 2011: 46; Merlin and Raynor 2004, 2005; and Oleiro 2014):

> "The unique characteristics of islands provide habitat and resources for avian species, including resident, migrants and seabirds. For many species, islands and archipelagos [are] extremely important, [as] they represent full habitat ranges, and for others may provide key resources, including migration stopover sites and suitable habitat for reproduction and nesting. Insular organisms are often naïve to foreign forces, and island species are susceptible to introduced competitors and predators from continental origins and the effect of anthropogenic processes. Almost without exception, island habitats and species have been impacted by invasive species and other anthropogenic processes—island systems exhibit the highest rate of avian extinctions. Habitat alteration and introduced species often have detrimental effects in insular biota. The effects range from native habitat disturbance to the annihilation of native species, including endemics (e.g., Tern Island and Guam)." (Oleiro 2014)

In the Hawaiian Islands, for example, many species of plants and animals have certainly become extinct—they have disappeared from the Earth forever. One explanation for a large number of these extinctions involves the intentional or accidental activities of humans. Many of these extinctions in Hawai'i occurred between about 1800 and the present. During this time, humans introduced and successfully established numerous, aggressive alien plants and animals. The environmental impact of many of these organisms, such as the destructive effects of wild goats, cattle, pigs, and others, has been very severe. This impact drastically changed many Hawaiian ecosystems so that today they are very different from the original, relatively stable ecosystems in which the now-extinct biota (plants and animals) evolved over millions of years (e.g., see Merlin and Juvik 1992; Weller et al. 2018).

In French Polynesia, *Miconia calvescens*, an alien tree in the *Melastome* family, was introduced to a botanical garden on Tahiti in 1937. Unfortunately, this aggressive woody pest species quickly escaped from the garden as birds carried seeds far and wide throughout much of the island. *Miconia* quickly became a major problem on Tahiti, with over 60% of the island now heavily invaded with the dark groves of this dangerous alien pest (Meyer and Florence 1996). Scientists in French Polynesia estimate that about 25% of the native Tahitian plants and animals are now threatened with extinction due to loss of habitat to the *Miconia* tree.

Miconia calvescens

In addition, this tree has very shallow roots that allow excessive soil erosion. Despite its attractive leaves, *Miconia* should never be introduced to Pohnpei! It was introduced into Hawai'i in the 1960s, and many efforts have been undertaken to control its rapid and pervasive spread there. Indeed, the highly invasive *Miconia* tree has spread to a number of islands in the Hawaiian Islands and French Polynesia in the past century or less. Meyer et al. (2011: 91) provided this update of the geographical and management problems associated with this dangerously invasive "Green Cancer" plant species:

> "More than 80,000 ha are currently invaded in French Polynesia, 10,000 ha in the Hawaiian Islands and 140 ha in New Caledonia. Control [programs] have been under way in the Hawaiian Islands (O'ahu, Maui, Hawai'i, Kaua'i) and French Polynesia (Raiatea, Tahaa, Nuku Hiva, Fatu Hiva) since the early 1990s, and in New Caledonia (Province Sud) since 2006. Despite more than 15 years of intensive control efforts and millions of plants destroyed, eradication has not been achieved in any of these islands, mainly because the species has multiple features that thwart its elimination (e.g., prolific seed production, active dispersal by alien and native frugivorous birds, large and persistent soil seed bank, shade-tolerance), combined with the difficulty of detecting and destroying plants on rough terrain and steep slopes, insufficient control frequency, and limited financial and human resources."

Boiga irregularis

Another potential threat to native animals of Pohnpei is found on Guam. On this island, nearly all the existing native birds have been eliminated by the feeding habits of an alien reptile, the brown tree snake (*Boiga irregularis*). This very aggressive, mildly poisonous snake was accidentally introduced into Guam from Melanesia. As a fast-spreading alien species, the brown tree snake can rapidly increase in numbers, preying on or outcompeting native species for habitat. Although it has not been established in Pohnpei, this noxious snake species remains one of the main threats to the birds of Pohnpei and other Pacific islands. In fact, there is still a serious possibility of the brown tree snake spreading throughout Micronesia, as well as into Hawai'i. The people of the FSM should remain on guard to report any sighting of this dangerous and destructive snake. Even if nobody is foolish enough to bring this threatening snake to the FSM, this species is capable of spreading by hiding itself and "hitchhiking" inside aircraft (especially within landing gear) and cargo arriving by plane or ship.

4.7. HUMAN-INDUCED HABITAT ALTERATION

While invasive species represent a major threat, they are just one of many environmental challenges caused by human activities. When people destroy or severely transform the natural habitat of a species—the place where it lives, grows, and reproduces—that species may become extinct if it cannot migrate to a similar habitat or adapt quickly enough to survive in a different environment. These adaptations are often unsuccessful, and as humans continue to significantly alter or even completely eliminate natural habitats, an extraordinary number of species are disappearing forever. As a result, human-driven habitat destruction is accelerating biodiversity loss and destabilizing ecosystems around the world. Widespread human activities are driving the irreversible loss of countless plant and animal species. In many areas of the world, ample evidence suggests that we are in the early stages of an extinction spasm, marked by a sudden and pronounced global decline in the abundance and diversity of several ecological groups of organisms (e.g., see Kolbert 2024).

> "Humankind has converted forest land to agricultural use for thousands of years as part of the process of economic development. **Deforestation** was most prevalent in the temperate climatic domain until the late 19th century and is now greatest in the tropical climatic domain." (FAO 2016, underlined for emphasis).

Tree stumps and scattered branches mark a recently cleared section of forest in Pohnpei.

Although tropical forests are "life's greatest species storehouse," many of these forests are now under severe threat in several tropical areas because of human disturbance. This disturbance includes logging, cattle ranching, burning, and various forms of agriculture. Scientists predict that a significant percentage of plant and animal species could vanish by the second half of this century.

> "The most common pressures causing deforestation and severe forest degradation are agriculture, unsustainable forest management, mining, infrastructure projects and increased fire incidence and intensity... Some infrastructure activities, such as road building, have a large indirect effect through opening up forests to settlers and agriculture. Poor forest management and unsustainable fuelwood collection degrade forests and often instigate a 'death by a thousand cuts' form of deforestation." (World Wide Fund for Nature 2020)

Since the end of the last great ice age, about 11,700 years ago, the Earth has lost one-third of its forests (e.g., see Kump et al. 2004). This is a large amount of deforestation, approximately two billion hectares (almost five billion acres) of the world's forests, which is about twice the size of the United States. Most of this massive forest loss has occurred over the past 100 years, driven by population growth and the expansion of agriculture, livestock, and firewood collection. This widespread loss of the world's forests accelerated in the second half of the 20th century, particularly in tropical regions.

Although the rate of worldwide deforestation has declined over the past quarter of a century, humans have destroyed millions of hectares of natural forest during the last few decades (e.g., see Ritchie et al. 2021; Lewis et al. 2015; Williams et al. 2020). For both ethical and practical considerations, many scientists and world leaders tell us that large tracts of undisturbed rainforest must be set aside as preserves to maintain biological diversity. On Pohnpei, that certainly includes the upland forest, which must be a priority for protecting the island's unique biodiversity.

4.8. SIGNIFICANCE OF THE UPLAND FOREST IN POHNPEI

Pohnpei's diverse ecosystems include marine environments, coastal areas, mangrove swamps, agricultural lands, and upland native forest. Such ecological variety makes Pohnpei home to many different species of plants and animals. Indeed, its biodiversity—including native and useful introduced species—has led to Pohnpei being described as the "emerald of the Pacific" (Oleiro 2014). Yet, there are threats to the long-term preservation of the ecological communities, especially the unique and essential upland forest. Recognition of the conservation value and wise use of this forest is vital for the long-term sustainability of Pohnpei and its people. This has been true for centuries, rooted in traditional Pohnpeian values:

Preservation of forest is a traditional Pohnpeian value.

Traditional conservation and management of resources, as well as the cultural activities and knowledge associated with it, are vital on Pohnpei and in so many other places around the world. Researchers and authorities involved in conservation worldwide have long recognized that appreciation of local livelihoods and cultural practices is essential for successful integration of modern scientific principles into the conservation plans. The people of Pohnpei have customarily collected plant and animal species from both the *nansapw* (settled and cultivated areas) and *nanwel* (wild forest regions). Here, we draw attention to the latter, focusing on the sustainability of water and biological resources (i.e., the resources managers commonly refer to as **ecosystem services**). We will highlight some of the key reasons for protecting the upland forest of Pohnpei, which supports diverse ecosystems where numerous native species have thrived for thousands of years. These native plants and animals, and their local Pohnpeian uses, have contributed to long-term preservation of traditional knowledge, guided more recently by scientific principles (see Chapter 3 for a detailed discussion of individual plant species and their uses in Pohnpei State).

4.8.1. The Pohnpei watershed: a crucial natural resource

The interior of Pohnpei is mountainous, with its highest elevations over 760 m (~2,500 ft) and with 11 peaks exceeding 600 m (~1950 ft). From the rugged summit region, thickly forested ridges and valleys descend outward and downward. The upland forest, or rainforest, is home to a large percentage of the island's native terrestrial plant and animal species. This rainforest is dominated by native trees, such as **Dohng** (*Campnosperma brevipetiolata*), **Sadak** (*Elaeocarpus carolinensis*), **Seir** (*Fagraea berteroana*), and **Ais** (*Atuna racemosa*), which cover much of the area between 200 m and 600 m (~650–2,000 ft) elevation. The forest floor is rich in mosses and ferns, and dense patches of native sedges, such as **Pwoaki** (*Thoracostachyum pandanophyllum*), which grow about 1 m high. Native palm forest is a subtype within this habitat, dominated by the endemic **Kotop** palm (*Clinostigma ponapensis*). Nearly pure stands of **Piht** (*Pandanus patina*), the woody, endemic pandanus tree are also scattered throughout this wet high-elevation area. The flatter and better-drained summit regions have dense thickets that include several high-elevation endemic shrubs and trees. Tree ferns, **Katar** (*Sphaeropteris* spp.) are common throughout. At the highest elevations, the rainforest transitions to another special subtype called tropical montane cloud forest, also known locally as dwarf forest or moss forest (*nannahna*). This is some of the lowest tropical montane cloud forest in the world.

These dense upland rainforest areas are extremely important because of their relationships to the water resources of the island. The native vegetation is essential for protecting and maintaining the Pohnpeian **watershed**. If that vegetation is removed, water resources on the island will be significantly altered with consequences for all the organisms, including people. This is because plant cover controls the collection and purification of drinking water. Vegetation helps protect the soil from damaging erosion and allows the rainwater time to seep into the ground, rather than accumulate and run off over the ground surface. The dwarf forest on mountaintops may also capture water as fog drip from clouds passing through the vegetation at or near ground level. As rain and fog drip through the vegetation, they fall to the ground and are naturally purified while seeping slowly through the island's volcanic rock. This creates a body of **groundwater** stored inside the island and eventually feeds springs and streams, or is pumped out of wells.

4.8.2. Major threats to the upland watershed of Pohnpei

The traditional concept of *nahk* as conservation of "storage areas" in the forest and other natural environments is ecologically vital for the protection of many natural resources. However, in the 21st century, the human population is increasing and patterns of resource use have been changing. New technologies enable people to have a much greater impact on the natural environment, and commercial markets encourage greater exploitation of natural resources. In the past several decades, infrastructure developments, such as roads, and unsustainable agricultural practices have led to deforestation.

Significant damage to native upland forest areas in recent decades resulted from expanded clearing for the cultivation of **Sakau** plants. Trustrum (1996) reported a decline in the native mountain forest areas from 42% of the total vegetation cover in 1975 to 15% in 1995. By the end of the first decade of the 21st century, the forests had been significantly altered:

> "The majority of the island's forests have been to various degrees converted or at least degraded to mixed forest (native species mixed with lowland secondary species), largely attributable to the cultivation of **Sakau** (*Piper methysticum*) as a major cash-crop (B. Raynor in litt. 2012). The fragmentation of such forest by **Sakau** clearings also introduces and encourages the spread of invasive species in isolated areas throughout the forest." (BirdLife International 2021)

Although we have referred to the "**Sakau** problem" in previous chapters, we return to it here to emphasize our cautionary warnings. The critical need to preserve Pohnpei's upland native rainforest for biodiversity and to maintain adequate water resources makes upland cultivation particularly concerning for this **keystone cultural species**.

It is widely recognized that Pohnpeian people consider **Sakau** to be the most important plant in the island's culture:

> "It is used to ask forgiveness when a family member harms a person in another family. It is used to settle disputes between parties, always mediated by a chief or titled clan member. It is used to ask a father for his daughter's hand in marriage, and during all of the required feasts, such as the yam festival or the beginning of the bread fruit harvest. It is the plant that symbolizes respect for oneself, one's family, and one's culture. It holds the culture together through the most difficult of times. If a visitor comes to your house, an offer of *sakau* shows respect for the visitor even more than food or television." (Balick and Lee 2002: 97)

Even though **Sakau** is a crucial and deeply respected plant in Pohnpeian culture, it is essential that people acknowledge that native biodiversity and the vital watershed services of their upland forest regions are potentially threatened by unregulated farming, now or in the future. It is also important to understand that people responded to this rapidly expanding conversion of native upland forest on Pohnpei with the successful "Grow Low" campaign. This was sponsored by the Conservation Society of Pohnpei and The Nature Conservancy and was initiated to conserve intact upland forests by promoting the cultivation of agricultural crops at lower elevations: "Efforts to involve the community in natural resource management include the participatory rural appraisal component of the watershed program [and encouraging] **Sakau** producers to grow lowland **Sakau** and thus spare forested uplands..." (FSM 2001: 16).

A study of the watershed management plan and its implementation (Ogura 2003) provided encouraging support for "enhanced collaboration" (i.e., people of various kinds working together for a common cause) in resource conservation efforts: "Perhaps the most telling impact of the watershed management program is that all interviewed (non-supporters included) responded that if not for the watershed management program, watershed areas on the island would be either completely destroyed or damaged more severely."

The following attributes of the watershed management program for forests were considered important in helping protect and preserve the island watershed for present and future generations:

- The "Grow Low" campaign showed local communities that **Sakau** could be grown successfully in the lowland areas. This resulted in movement of **Sakau** cultivation from the upland forests to lowland areas, as well as the clearing of pig pens and latrines near streams, and associated clean-up efforts in villages.
- Success of the watershed management program can be attributed to key leaders within communities adopting the practice as well as efforts by The Nature Conservancy to provide nursery-grown plantings for communities, allowing them to experiment with the new planting style. The special involvement of the late William (Bill) Raynor, former director of The Nature Conservancy on Pohnpei, was cited most frequently as a critical part of the program's success due to his unique expertise and commitment.
- Growing **Sakau** in the lowlands allows the elderly, women, and children to participate, whereas its production in the upland forests is typically reserved for young men due to the long, strenuous hikes required to reach and harvest the plants.
- Awareness of forest conservation and biocultural diversity issues has become more prevalent on the island due to education efforts focusing on the importance of watershed areas and threats.

Cultivation in upland native forest

Forests have been valuable to the people of Pohnpei for centuries, thanks to the ecological services they provide and also for the cultural uses of many plants and animals living there. However, some observers believe that current management practices may not be sustainable into the future:

Upland forest on Pohnpei may not easily recover from clearing because many native trees cannot compete with fast-growing ferns or invasive herbs. Those smaller plants quickly colonize open areas and prevent trees from growing.

> "Pohnpei's vegetation and wildlife are some of the most diverse in Micronesia. The upland forests serve as habitat for at least 269 different species of plants, of which 110 are endemic to Pohnpei. The upland forest also provides a home for the island's wildlife, and the people of Pohnpei depend on many birds and animals for subsistence or income. There are over 24 different species of birds, which nest in the upland forest, of which five species and eight subspecies are found only in Pohnpei." (Scrimgeour and Gallen 2003; also see Raynor 1994; Merlin and Raynor 2004, 2005; Donnegan et al. 2011)

Rapid human population growth often results in high dependence on agriculture and consequently puts pressure on upland forest environments. This causes many native upland forests to be converted into traditional agricultural areas. This is not an environmental concern unique to Pohnpei; other Pacific islands have been experiencing the same issue with farmers depending on the native forests for survival. Many of these areas have been, and may continue to be, converted for the cultivation of different crops; in Pohnpei, the cultivation of **Sakau** as a **cash crop** has been "perceived to be the dominant activity causing deforestation, where the farmers' main intention is to earn fast cash to support family needs" (Scrimgeour and Gallen 2003).

If **Sakau** plantations or other farmed areas continue to replace upland forest on Pohnpei, it will cause an additional human environmental impact on biodiversity and the mountain watershed. The expansion of upland farming can influence biodiversity in two ways: (1) growth of current agroforest patches, which occurs when undisturbed vegetation surrounding these agricultural areas is removed to enlarge existing plots; and (2) opening up new agricultural plots, which can be established by clearing sections of undisturbed vegetation. New agriculture plots in recent times have frequently been located in areas with intact native forest vegetation (Merlin and Raynor 2005; Balick 2009).

These two patterns of forest use both have the potential to affect the native plants, birds, and other animals in Pohnpei, but in different ways. An expansion of existing agricultural plantations removes some undisturbed vegetation and expands forest edge, although usually not by much; the expansion of existing plots where staple (food) crops are cultivated often follows this process and is unlikely to impact undisturbed vegetation in other areas currently characterized by little **habitat fragmentation**.

On the other hand, new **Sakau** plantations have often been established in previously undisturbed regions of intact native vegetation, and these new plots have a greater potential to affect what are currently large and unbroken patches of native forest habitat. When deforestation occurs in undisturbed forest to establish new farming plots, fragmentation of forest area increases in these undeveloped landscapes—these new patches can introduce the forest edge to locations where it may not yet exist, thus allowing alien weedy plants and animals easier access to undisturbed forest.

Expansion of secondary vegetation

Another serious concern related to potential loss of native biodiversity and watershed protection is the expansion of secondary vegetation that typically colonizes recently cleared forest areas. Several factors can lead to the conversion of mature (climax) native forest to secondary vegetation, a process some predict will accelerate in the future:

> "... secondary vegetation would likely expand under the range of future conditions Pohnpei might experience. Secondary vegetation is common on the island in locations where anthropogenic activities, such as farming or urban development, [have] disturbed native habitats and then the areas were abandoned. Secondary vegetation has also developed in areas where wildfires previously occurred, where landslides occurred, or where there was tree mortality. We thus predicted that under future scenarios that include substantial anthropogenic development and altered climates, with increased likelihoods for drought and with altered precipitation, secondary vegetation could expand in areas where it already exists." (Oleiro and Kesler 2015: 22)

4.9. IMPACTS OF GLOBAL CLIMATE CHANGE

Future environments will likely differ from those currently established on Pohnpei because of anthropogenic change, including global warming. For example, changes in rainfall patterns and increased storm frequency associated with global warming have the potential to significantly impact forested habitats. Rising temperatures on high islands like Pohnpei are expected to alter the vertical distribution of different island vegetation zones. Temperature warming will involve upward shifts and displacement of native plant and animal species that may challenge their ultimate survival.

Currently, Pohnpei remains largely unaffected by the mid-Pacific typhoons that have drastically changed habitats on other islands (e.g., Chowdhury et al. 2010; Kauffman and Cole 2010). Under climate-altered scenarios, Pohnpei may experience an increased number of intense storms as well as drought frequency. Thus, global warming poses a significant threat to both lowland and upland native forests of Pohnpei. The low-lying atolls in the Outer Islands are even more vulnerable to climate change. Many scientists predict that by 2050, most atolls in the tropical Pacific and other oceans will become uninhabitable due to sea level rise and intensified wave-driven flooding.

In their key publication, Storlazzi et al. (2018) argued that people living on the majority of atolls will not be able to survive on their traditional homelands by the middle of this century because of the intensifying effects of sea level rise on increasingly damaging and dangerous storm surges and flooding:

> "Sea levels are rising, with the highest rates in the tropics, where thousands of low-lying coral atoll islands are located. Most studies on the resilience of these islands to sea-level rise have projected that they will experience minimal inundation impacts until at least the end of the 21st century. However, these have not taken into account the additional hazard of wave-driven overwash or its impact on freshwater availability. We project the impact of sea-level rise and wave-driven flooding on atoll infrastructure and freshwater availability under a variety of climate change scenarios. We show that, on the basis of current greenhouse gas emission rates, the nonlinear interactions between sea-level rise and wave dynamics over reefs will lead to the annual wave-driven overwash of most atoll islands by the mid-21st century. This annual flooding will result in the islands becoming uninhabitable because of frequent damage to infrastructure and the inability of their freshwater aquifers to recover between overwash events. This study provides critical information for understanding the timing and magnitude of climate change impacts on atoll islands that will result in significant, unavoidable geopolitical issues if it becomes necessary to abandon and relocate low-lying island states."
> (Storlazzi et al. 2018, underlined for emphasis).

To prepare for severe climate shifts and the possible intensification of tropical storms, it is crucial to understand the specific threats facing each island or atoll, as well as the perceptions of local communities regarding their predicament (Kei and Sueo 2013).

4.10. CONSERVATION OF ENDANGERED AND THREATENED ANIMAL SPECIES

While this book primarily focuses on the plants of Pohnpei and their uses by people, it also touches on aspects of animal life and their significance to humans. Many native animals play crucial roles in nutrient cycling, pollination, seed dispersal, and other ecological functions, supporting the health of native vegetation and island ecosystems. In this context, we will briefly highlight several iconic species—ranging from turtles and mangrove crabs to forest birds—whose conservation is especially important. Protecting Pohnpei's native animals, alongside its plant species, is vital, and the effective management of natural resources—both in the ocean, on reefs, and on land—is a critical issue not just for the people of Pohnpei, but for global communities as well. This is something the people of Pohnpei have understood for centuries. As previously mentioned, Pohnpei has a rich cultural tradition of resource management, encapsulated in the local word *nahk*, meaning "conservation," which reflects their enduring commitment to preserving natural resources from loss, waste, and harm.

4.10.1. Marine animals

There are many examples of traditional marine resource conservation methods that were once strictly observed within local communities on Pohnpei. For instance, turtles could only be hunted traditionally during the *isol* period (i.e., "yam season, time of scarcity"), and the first turtle caught had to be offered to the *Nahnmwarki* as a sign of respect. However, in today's cash economy, with the introduction of more efficient hunting methods and devices, some individuals now hunt turtles for sale, bypassing these traditional rules.

Similarly, a *soun sed* ("ocean expert") traditionally adhered to the conservation practices set by the *Nahnmwarki*, ensuring they did not take more fish than needed. Yet today, the lure of cash incentives has led some fishermen to harvest far more than would have been acceptable in the past, eroding these time-honored methods of resource management. These practices, once vital for ensuring sustainability, are increasingly difficult to uphold in a cash-driven economy.

A prime example of a species deeply intertwined with the health of marine ecosystems around Pohnpei is the mud or mangrove crab, **Elimoang** (*Scylla serrata*). Found widely across the Indian Ocean and the tropical Pacific, this crustacean thrives in close association with mangrove habitats, where it plays key ecological roles. **Elimoang** was a traditional food of the people of Pohnpei for centuries. It remains highly prized as a delicacy, with its sweet, tantalizing flesh popular among locals and visitors alike, who are willing to pay a premium for it. However, its abundance has been diminishing over time due to increased hunting and habitat loss. In recent years, many of these crabs have been caught, sold, and even exported to locations such as Guam and Saipan. Without proper conservation measures, this species is at high risk of overexploitation, particularly due to its reproductive strategy:

> The larvae of *S. serrata* [**Elimoang**] can drift at sea as part of the plankton for up to 75 days, enabling them to [colonize] habitats far from the home of their parents... The life cycle of these crabs is complex, involving an extended period in the planktonic phase where they go through five developmental stages before settling into habitat close to mangroves... Mud crabs typically only live for a maximum of three to four years, such that constant recruitment is necessary to keep the population replenished and stable." (Halford and Bosserelle 2019)

Scylla serrata
Elimoang (P)

More than 40 years after the 1977–1978 survey of **Elimoang** in Pohnpei and other areas of the FSM, the population size and distribution of this mangrove crab were reassessed in 2018. The earlier study emphasized "strong spatial differences in the size structure of mud crab populations between locations where crab trapping occurred," noting that these differences correlated with the density of nearby human populations and the extent of local mangrove forests. While the 1970s survey concluded that the crab populations "were not being overexploited and that fishing restrictions were unnecessary," the more recent study revealed that:

> FSM's human population has risen by ~70% and the pressure on natural resources is

high. Responding to increased concerns from constituents about an apparent long-term decline in the abundance and size of mud crabs, a national senator requested help from the Pacific Community to undertake an assessment of the health of Pohnpei's mud crab stocks, and to provide advice on the most effective management strategies." (Halford and Bosserelle 2019)

The first step in conserving this valuable resource is recognizing that there is a problem. Equally important is the second step—"collecting data to inform decision-making," and this process is already underway. The challenge lies in implementing management plans that effectively ensure the long-term sustainability of this traditional Pohnpeian food resource (Halford and Bosserelle 2019).

4.10.2. Birds

Just as traditional practices once guided the sustainable use of marine resources like turtles, fish, and crabs, they also played a crucial role in the conservation of Pohnpei's land animals—most notably birds, whose ability to fly and reach distant locales makes them especially significant on oceanic islands:

> "Islands exhibit the planet's most unique flora and fauna, but the diversity on islands is also vulnerable to impending forces of global change. Scattered across the Pacific Ocean >20,000 islands range from sand spits to large islands and support some of the most unique biotic assemblages. Shaped by island size, topography and degree of isolation from other landmasses, including continents and other islands, insular systems support the largest number of endemic flora and avifauna [birds]. Because of the mobility associated with flight, birds have colonized the most remote islands, and are one of the most important taxa in insular systems." (Oleiro 2014, underlined for emphasis).

Pohnpei still contains substantial areas of native upland forest, including some of the world's lowest-elevation tropical montane cloud forests. It also supports the largest mangrove forest areas in the FSM and extensive areas of gallery forest, which form as corridors along streams or wetlands. These forests offer habitats to many resident bird species, including six species that are endemic to Pohnpei. Unfortunately, during the past several decades, changes in how local forests are used have caused significant changes to the island's landscape and bird populations.

The first systematic avian (bird) survey was carried out in 1983, with its results and recommendations published some years later (see Engbring et al. 1990). A second survey was carried out in 1994 by Buden (2000), who reported a sizeable decline in the number of birds observed on Pohnpei since the previous survey:

> "During the same period of time large tracts of native forest were transformed to anthropogenic habitats, mainly through new cultivation of local staple crops and **Sakau** (*Piper methysticum*)... Pohnpei does not have introduced snakes and introduced species (e.g., rats, lizards, and birds) are present in low densities, suggesting that habitat alterations are the major drivers of Pohnpei avifauna populations." (Buden 2000)

More recently, Oleiro and Kesler carried out an extensive bird survey of Pohnpei in 2012. It assessed the status of Pohnpeian avifauna by recording the presence and relative numbers of individual bird species across different ecosystems. A brief description of the objectives and results of this research is presented here:

> "How will Pohnpei's birds deal with future forest changes? This project models bird species' responses to long-term and large-scale landscape change on Pohnpei Islands, Federated States of Micronesia. Data from past bird surveys spanning back to 1983, combined with 30 years of vegetation mapping, informed modeling for the islands' endemic bird population. The results are a spatially explicit predictive model of species change to guide reforestation and forest bird management over the long term. This is the first effort of its kind in the Pacific Islands outside of Hawai'i." (Kesler et al. 2013, underlined for emphasis; also see Oleiro 2014; Oleiro and Kesler 2015).

There are believed to be 32 resident bird species in Pohnpei. All but two—the chicken or jungle fowl, introduced by early Micronesian settlers, and the rock dove, a more recent introduction—are native to Pohnpei. There are also 26 migrant bird species, which are regular visitors to the island but normally do not breed in Pohnpei. The migratory species include shorebirds, some water birds, and many seabirds. These migrants generally fly through or to Pohnpei on a seasonal basis. Finally, a number of vagrant bird species also visit Pohnpei occasionally, as the island lies well outside their normal migratory range.

The majority of the resident birds of Pohnpei live in the forests and can be used as forest-health indicators. Among the forest birds, some have specific habitat and food preferences. For example, **Kiniwed** (*Ptilinopus ponapensis*), the purple-capped fruit dove, nests in the forest and eats fruits of trees such as **Kehnpap** (*Syzygium carolinense*), **Sadak** (*Elaeocarpus carolinensis*), **Nihn** (*Ficus tinctoria*), **Aiau** (*Ficus prolixa*), and **Topwuk** or **Oahr** (*Premna obtusifolia*).

In contrast, **Sihk** (*Phaethon lepturus*), the white-tailed tropicbird, eats small fish, rather than fruits of the forest, and may nest directly on stony surfaces or rocky cliffs like many seabirds do, but it also frequently nests in forests, especially in large trees like **Pwulok** (*Xylocarpus granatum*), **Sadak** (*E. carolinensis*), and **Dohng** (*Campnosperma brevipetiolata*), or sometimes in large **Tehnlik** ferns (*Asplenium nidus*).

Some native birds, such as **Sioahk** (*Aplonis opaca*), the Micronesian starling, **Pwiliet** (*Myzomela rubratra*), the Micronesian honeyeater, and **Kutoahr** (*Todiramphus reichenbachii*), the Pohnpei kingfisher, are commonly sighted near settlements as well as in the forest.

In fact, most of the forest birds on Pohnpei are found in a variety of habitats, including mature and secondary forests, coastal lowlands, and uplands. In addition, some native birds of Pohnpei are adapted to freshwater wetland habitats. These marshes and swamps are currently, and apparently always have been, small in size on Pohnpei. Therefore, concentrated hunting or habitat alteration (such as draining or filling in a marsh or swamp) can result in the death of many, or even all, of the birds adapted to these environments. To ensure the survival of the island's native birds, it is crucial to protect Pohnpei's diverse ecosystems—especially upland forests, mangrove forests, and remaining freshwater wetland habitats.

Threats to the native land birds: past and present

High islands, such as those in the tropical Pacific region, exhibit some of the planet's most unique flora and fauna, but this biodiversity is very vulnerable to the impending forces of global change. Pohnpei is highly representative of the diversity of these oceanic islands and has numerous endemic plants and animals, including species of land birds found nowhere else in the world.

Aplonis opaca
Sioahk (P)

Myzomela rubratra
Pwiliet (P)

Ptilinopus ponapensis
Kiniwed (P)

In the past, under traditional chiefly authority, bird hunting and egg collection were regulated for sustainability. Now, however, with population growth and changing social dynamics, the island's long tradition of bird hunting has intensified. This is largely due to the use of guns and the decline of chiefly authority over resource management, with some forest hunters not following either traditional or government rules about legal limits to killing and consuming forest birds. Consequently, some bird species face the risk of dramatic population decline and even possible extinction. Similar threats face seabirds in some parts of Micronesia, which, although they nest on uninhabited remote islets, are now commonly hunted by people using motor-powered boats for quick and frequent trips to seabird colonies.

In a survey of forest use at the turn of this century by Scrimgeour and Gallen (2003), most Pohnpeian households interviewed stated concern regarding the wildlife that they depend on for their subsistence. Most people interviewed believed that bird populations had declined, likely due to forest cover loss, particularly from upland **Sakau** cultivation and the increased frequency and effectiveness of bird hunting with guns. Today, there are three major ways of hunting birds and other wild animals in Pohnpei: shooting, the traditional use of very sticky breadfruit sap, and the setting of traps.

> "The majority of the interviewees confirmed that they hunt birds by shooting... The village members use slingshots and rifles. Breadfruit sap is used in hunting birds. They generally use the sap during fruit bearing season. Traps are designed for wild pigs, deer and chickens... Dogs are also used to hunt wild pigs, deer and chicken." (Scrimgeour and Gallen 2003: 5)

Excessive predation, such as that associated with commercialized bird hunting, has caused the extinction of many species around the world, especially in recent years, and hunting remains one of the potential threats to a number of native Pohnpeian birds, which are discussed further below. However, an even more serious threat to biodiversity loss in Pohnpei and worldwide is **habitat alteration**—or even **habitat destruction**—caused directly by human activity or in combination with other environmental pressures.

Two legally protected native birds in Pohnpei

Micronesian pigeon, Ducula oceanica, is a native species found in forests across Micronesia, from Palau to Nauru. However, its population size is declining due to hunting and deforestation; indeed, the International Union for the Conservation of Nature (IUCN) has assessed it as a near-threatened species. On Pohnpei, it is called **Mwuroi** and is now rarely seen around settlements. As a shy species that avoids human interaction, it is extremely dependent on its intact forest habitat. Recent studies have shown that loss of native forest and expansion of anthropogenic vegetation poses a serious threat to **Mwuroi** and "minimizing mangrove alteration and setting aside reserve areas could aid in maintaining a stable population" (Oleiro and Kesler 2015).

Ducula oceanica
Mwuroi (P)

Serehd (*Trichoglossus rubiginosus*), the Pohnpei lorikeet, is a parrot species endemic to Pohnpei and nearby Ahnd Atoll. The "Designation of State Bird Act" established the **Serehd** as the state bird of Pohnpei, and also protected the bird from being harmed or killed. Hunting this species is completely prohibited and violators of this law face fines or imprisonment (Boer 1996: 97).

Trichoglossus rubiginosus
Serehd (P)

Three of the rarest native birds in Pohnpei

Three endemic and highly endangered species are very uncommon forest-dwelling birds that deserve special attention and protection if they are to survive.

Sie (*Aplonis pelzelni*), the Pohnpei mountain starling, may be extinct, though it is currently recognized as a critically endangered species by the IUCN (IUCN 2013; see Oleiro and Kesler 2015: 29). It has only been observed and recorded with certainty once in recent times (in 1995; see Buden 2000). In fact, this unique bird had declined considerably since 1930, probably through one or a combination of habitat loss, hunting, and predation by alien invasive rats.

Aplonis penzelni
Sie (P)

> "However, it may remain extant since there have also been several unconfirmed reports, and further intensive surveys are required. Any remaining population is likely to be tiny and continuing to decline, and for these reasons it is treated as Critically Endangered." (BirdLife International 2021)

Tiht (*Rukia longirostra*), the Pohnpei greater long-billed white-eye, is an endemic bird whose continued survival depends on conservation in Pohnpei. Specifically, it could benefit from ongoing enforcement of upland habitat preservation, and it could be helped by efforts to restrict the expansion and establishment of agroforest or areas where the "edge effect" (at or near the boundary of natural forest and a cultivated area) becomes detrimental to the population. The observed natural habitat of Tiht is "associated with undisturbed climax forest [which is] extremely important for their populations" (Oleiro and Kesler 2015: 17, 56).

Rukia longirostra
Tiht (P)

Likoht (*Asio flammeus*), the short-eared, native owl was not detected during the 2012 extensive surveys, but its continued presence in Pohnpei was confirmed by researchers' observations of owl pellets (regurgitated remains of prey) and detection "outside of the survey areas when traveling between sites" (Oleiro and Kesler 2015: 26). This native predator is very rare and should be given special attention by resource managers.

Asio flammeus
Likoht (P)

4.10.3. Reasons for protecting the native species of Pohnpei

Although the native land biota (plants and animals) of Pohnpei has not been severely damaged or diminished to the extent seen in many other Pacific islands, we need to act now to understand, appreciate, and protect the native species of Pohnpei. Why is that?

1. Native plants and animals have their own **right to exist**. They have been living here for a very long time, long before any humans arrived on Pohnpei thousands of years ago. Local people can be proud of these native plants and animals as special parts of the unique natural history of Pohnpei.

2. Native plants and animals may play important, even crucial, **ecological roles** in the environment. As vital components of the island ecosystem, they may help provide food, control erosion, protect shorelines, offer essential habitats, and exert predation pressure on other species within the natural food webs of Pohnpei.

3. Native plants and animals are integral to the unique **natural beauty** and character of Pohnpei's island environment, with many species being uniquely Pohnpeian and found nowhere else in the world. They hold deep cultural significance, serve traditional purposes, and bring joy and inspiration through their sheer presence. Preserving them ensures that future generations of Pohnpeians can continue to appreciate and benefit from these irreplaceable natural treasures.

4. These native plants and animals may provide economically **valuable resources**: through using parts of them, cultivating them, studying their medicinal potential, generating interest and funding for scientific studies; or even by attracting a special kind of tourism. For example, these species and the interesting natural history of Pohnpei may be a resource for nature-based, educational tourism development, which can bring economic gain to the country. In fact, this kind of tourism has increased phenomenally in the past decades in various other areas of the world.

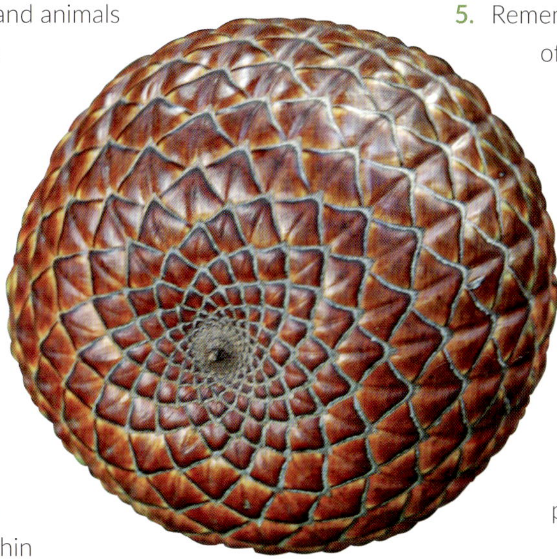

5. Remember all the ways the people of Pohnpei have used the plants and animals of these islands. Many of these **traditional uses** are listed and discussed in Chapter 3 of this book. Perhaps you have thought of many more uses for food, famine food, green mulch, medicine, ceremonial goods, tools, spices, perfume, and cosmetics.

At one time, the people of Pohnpei and its outlying atolls were completely self-sufficient; in other words, they obtained everything they needed from the plants and animals of these islands and the surrounding seas. Again and again, the proverbs and legends show how much the people of old knew about the plants and animals of Pohnpei and how much they valued them.

"Pirakih me se pwahpwa, me mwangih nek rong oh kainenehla."

"Take what you need, and save the rest."

5

NAMES OF PLANTS

CHAPTER 5

NAMES OF PLANTS

This chapter explores the rich and complex world of plant names, both local and scientific, and the important role they play in preserving cultural heritage and ensuring clear communication. It looks at how traditional names reflect deep connections between people and nature, how these names evolve over time and across communities, and how scientific names provide a universal system for identifying species worldwide. By understanding both naming systems, we gain insight into the diversity of life in Pohnpei State and the stories, knowledge, and relationships associated with these names.

5.1. COMMON NAMES

Every language has its own set of common names for different living organisms. In Micronesia, with its many languages and dialects, the same type of tree or bird might have many different names on different islands. For example, In Pohnpei State, the banana is called **Uht** in Pohnpei, **Wus** and **Wis** on the atolls of Mwoakilloa and Pingelap, and **Gahudi** and **Hudi** on the Polynesian outliers of Nukuoro and Kapingamarangi.

The local names of plants are very important and are part of our culture and heritage. They reflect the deep, centuries-old connections between people and the natural world. They remind us that, even if we forget certain plants or become unfamiliar with them, these plants were once significant and indispensable to our ancestors. In this book, we aimed to establish and use these names to remind readers of their existence and preserve them for future generations. Of course, this is not to suggest that the names presented here are the only correct ones. Our goal has been to determine at least one reliable and correctly spelled name for each species in every language where such a name is available. We assume that there are many more names that we did not yet record, which we hope students and others in this and future generations will document and appreciate as part of their heritage.

Haplopteris elongata
Mwosou (P)

The naming of plants can often be surprisingly complex. A single plant may have multiple names within the same language. One village may use a different name for the same plant than another village. For example, let's look at the Pohnpeian names for several fern species. These observations come from Dana Lee Ling, a professor at the College of Micronesia, whose help answering our questions throughout this project has been invaluable. One species of fern is called **Mwosou** by some people, while others call it **Alis en kewelik**, which translates to "beard of the heron." Dialectical variations add another layer of complexity to this, so that the aforementioned **Alis en kewelik**, as pronounced in the Northern dialect of the Pohnpeian language, becomes **Alis en koawoalik** in the Kitti dialect. In addition, when using *meing*—Pohnpei's honorific "high language"—the name changes to **Alis en nahnsou sed**, as the word for "heron" shifts from **Kewelik** in everyday speech to **Nahnsou sed** in honorific speech. While personal preference, dialects, and honorific speech often result in a single plant having multiple names, the reverse is also true: a single name can refer to several species. For instance, **Peipei eni** is a name used for several distinct fern species in Pohnpei.

Confusion about plant names is common. Dana Lee Ling has observed many of his students confusing the names of two ferns, **Kideu** and **Ulung en kieil**. Some students get them right, but about half mistakenly swap the names, each insisting their version is correct. Why do these names get reversed? The two ferns may seem similar and often grow together on the same trees. In some cases, the names may even be reversed deliberately for specific purposes. To keep a medicinal recipe secret, a healer might swap the plants' names when sending someone to collect them, confusing anyone who might overhear while the collector understands that the true name is being deliberately concealed.

Plant names also change over time. Many early documents indicate that **Mwosou** is the primary name of one of the ferns previously mentioned. However, for years now, many of Dana Lee Ling's students have been familiar only with the name **Alis en kewelik** and its variations. The name **Mwosou** is unfamiliar to them, perhaps because **Alis en kewelik** became more widely used over time. Unfortunately, in recent years, students are not aware of the newer name either and have a hard time naming any specific fern, even with help from family members at home.

Taeniophyllum marianense
Patehn kewelik (P)

Interestingly, while hanging feathers from the "beard" of the *kewelik* bird inspired the name of this fern, the bird's "foot" lent its name to another unusual plant on Pohnpei: the leafless orchid *Taeniophyllum marianense*. Despite lacking leaves, it still performs photosynthesis through its strange, wiry, green roots. Clinging to surfaces and grasping wood like bird toes, these roots earned the plant its local name, **Patehn kewelik**, which translates to "foot of the heron."

Since local languages evolve constantly, how do we decide which names to use for different plants? People continuously reshape their language, and with so much flexibility in usage between individuals, families, and communities, how do we ensure clarity in a book like this and assign a consistent name to each plant? This was a challenging process for us and is explained in detail on pp. 16–19.

5.2. SCIENTIFIC NAMES

In contrast to the rich tapestry of local plant names, which reflect traditional heritage, cultural diversity, and evolving language, scientific names provide a consistent universal system for naming species worldwide. By assigning a unique scientific name to every species, scientists ensure that everyone refers to the same species by the same name, regardless of their location. This prevents confusion and establishes standardized, globally recognized names.

5.2.1. Binomial nomenclature

According to accepted biological classification, the names scientists assign to every known kind of living organism follow the system called **binomial nomenclature**. It was invented by the Swedish botanist Carl Linnaeus in the 18th century. He introduced it as a standardized way to name and classify organisms using two Latin names: the genus and the species, which helped bring order and consistency to biological classification. For instance, the seedless breadfruit tree, called **Mahi** in Pohnpeian, has the official scientific name *Artocarpus altilis*. The first word (*Artocarpus*) refers to the genus, which is a broader group of related species, while the second word (*altilis*) specifies the individual species within that genus. In this book, we use scientific names (binomials) as well as Pohnpeian, Mokilese, Pingelapese, Nukuoro, and Kapingamarangi names for plants (and occasionally animals). In this way, readers both from Pohnpei State and beyond will know which plant or animal species we are referring to or describing. Let's use Micronesia's most useful plant, the coconut palm, as another example. Scientists refer to it as *Cocos nucifera*. If a scientist from Kosrae comes to Pohnpei to study coconut palms, she may not know its local name, **Nih**, just as Pohnpeians may not be familiar with its Kosraean name, **Koacnu**. To avoid confusion, she will informally use the English name "coconut palm" for easier communication, but if she writes about it in a scientific paper that researchers from all over the world may read, she will definitely use the scientific name, *Cocos nucifera*, to ensure clarity for everyone.

5.2.2. Synonyms

It's important to note that a fundamental principle of the scientific method is that scientific knowledge is always evolving. As new information is discovered or old data is reinterpreted, scientific explanations and theories may change. This approach encourages scientists to stay open-minded and ready to revise their thinking based on the best available data. Although binomial nomenclature ensures clarity and uniformity, scientific names can and do change due to new scientific discoveries, improved understanding of evolutionary relationships, or corrections to earlier classifications that may have been based on incomplete or outdated data. For example, herbaceous yellow-flowered coastal plant in the sunflower family (Asteraceae), known as **Moaresed** or **Ngkahu** in Pohnpei (**Moarjij** on Mwoakilloa, **Kisiwael** on Pingelap, **Aigi mea** on Nukuoro, and **Dilidai** on Kapingamarangi), has the scientific name *Wollastonia biflora*. However, some older books, including dictionaries, may mention it under other scientific binomials, such as *Wedelia biflora* or *Melanthera biflora*. These older names are considered synonyms and there can be many of them.

In other words, for each plant species, the scientific community recognizes only one valid name—typically because it was the first to be published in accordance with the rules of botanical nomenclature or because subsequent research supports its use. All other names applied to the same species are treated as synonyms—historical alternatives that may appear in older literature but no longer reflect the most accurate taxonomic understanding. Comprehensive online databases, such as Plants of the World Online (see powo.science.kew.org) and World Flora Online (see www.worldfloraonline.org), cross-reference any synonyms with currently accepted names.

5.2.3. Author citations

To help scientists better understand and keep track of changes in nomenclature, professional botanical literature usually includes author citations alongside scientific binomials. For example, the coastal woody species known as **Titin** in Pohnpei (**Sisin** in Mwoakilloa, **Sesen** in Pingelap, **Manu gili maau** in Nukuoro, and **Dogodogongo** in Kapingamarangi) has undergone multiple taxonomic revisions as scientific understanding improved over time. It began its nomenclatural history as *Tournefortia argentea* L.f. (1782)—where "L.f." denotes Carl Linnaeus the Younger (*filius*), who first described the species, building upon the pioneering work of his father, Carl Linnaeus. The species was later reclassified as *Messerschmidia argentea* (L.f.) I.M.Johnst. (1928), then moved to *Argusia argentea* (L.f.) Heine (1976). After a brief restoration to *Tournefortia argentea*, it was then transferred to the genus *Heliotropium* as *Heliotropium arboreum* (L.f.) Fosberg (1987). Research then revealed that it was actually Francisco Manuel Blanco who first described this species under the *Heliotropium* genus in 1825, leading to the current accepted name *Heliotropium arboreum* (Blanco) Mabb. (2003)—where "Mabb." stands for David Mabberley, a leading botanist in tropical taxonomy who formalized the transfer while honoring Blanco's earlier work. In this book, for simplicity sake, we use only binomials (e.g., *Heliotropium arboreum*) in the main text, but full author citations are provided in the index of plant names (pp. 461–465).

5.3. NAMES OF VARIETIES

Now that we've discussed how different plant species get their local and scientific names, let's consider the fact that a single plant species may contain several different types, usually called varieties. How do people distinguish and name them? If the distinction is important enough to communities, they may simply give the special variety a separate name. For example, in the Marshall Islands, the extremely common coastal plant *Scaevola taccada* is known as **Kōṇṇat**. However, a rare type that has pink instead of white flowers is referred to as **Kōḷaime**. In contrast, scientists use the same binomial name for all varieties of a single species, and append the abbreviation 'var.' followed by the variety name. For example, scientists have described a unique variety of *Scaevola taccada* with small hairs on the leaves, giving them a distinctly silky texture. They named it *Scaevola taccada* var. *sericea* to distinguish it from the typical variety with smooth leaves, which then became known as *Scaevola taccada* var. *taccada*. While such small differences in wild plants may not concern most people, identifying and naming distinct varieties becomes much more important for cultivated plants, as these variations often reflect differences in taste, yield, resilience, or specific uses valued by the communities that grow them.

5.4. NAMES OF CULTIVARS

While some varieties emerge naturally, human cultivation has played a key role in developing many distinct plant types. Such varieties, known as cultivars, often result from long-standing cultivation and emerge through deliberate selection for desirable traits and careful management by humans. When a cultivar becomes widely known, scientists assign it a name by appending the cultivar name in single quotations to the species' scientific name. For example, globally recognized coconut palm cultivars are *Cocos nucifera* 'Green Dwarf', *Cocos nucifera* 'King Coconut', and *Cocos nucifera* 'Golden Malay'. Locally, there are many more cultivars, each with unique names and characteristics. For instance, on the atoll of Mwoakilloa, one local cultivar, called ***Adohl***, is notable for its sweet, edible husk. Another cultivar, known as ***Nihn pwinjo***, has trees so short that coconuts can be harvested directly from the ground, without the need to climb. Though the names of these and many other cultivated varieties are well-known locally, they often lack standardized English or scientific names. Despite sharing the scientific name *Cocos nucifera* or the general Pohnpeian name **Nih**, they represent a range of distinct cultivars with varying characteristics and uses.

Chapter 6 of this book provides more information on the diversity of cultivars. It lists well over 500 named cultivars in Pohnpei State, some widely known and some rare, known only to the people who cultivate them. These growers use time-honored techniques to preserve and enhance desirable characteristics. To propagate a desirable cultivar and ensure it remains identical to the mother plant, the growers must clone (vegetatively reproduce) the plant using a cutting of material from the mother plant. For example, to multiply a **Mwahng** (*Cyrtosperma merkusii*) cultivar known as ***Sounpwong weneu***—which reportedly can produce a "good-sized corm [underground swollen stem] in a year's time under good growing conditions"—a farmer cuts and plants part of the corm from the mother plant and cultivates it in a suitable farming area.

Sometimes, individual plants from different **Mwahng** cultivars cross-pollinate without human help. When one cultivar fertilizes another, the resulting seeds often produce a new type of **Mwahng** that differs from both its parent plants and is, more often than not, of little use. While most hybrids are undesirable, occasionally a new plant shows a novel and desirable trait, such as unique color, form, smell, or taste, that the farmer values. In these cases, the farmer may decide to perpetuate this new type by using cuttings (clones) from it to grow new individual plants identical to the mother plant. Scientists call this process vegetative reproduction, and it helps explain how so many different cultivars exist today. The names of some cultivars in Pohnpei and other parts of Micronesia often reflect this process, honoring the people who created, discovered, or introduced them. Similarly, some cultivars are named after their place of origin. For example, some cultivars of a major yam species, **Kehp** (*Dioscorea alata*) in Pohnpei have been named after the assumed regions from which they were introduced to this island: ***Kehpin seini*** (China), ***Kehpin sapahn*** (Japan), ***Kehpin palau pwetepwet*** (Palau), ***Nuhkaledohnia*** (New Caledonia), and ***Kehpin awai sohkatepe*** (Hawai'i). As seen in the next chapter, Pohnpeians use many other creative and meaningful methods to name their plant cultivars.

5.5. NAMES OF HYBRIDS

In rare instances, the above-mentioned process of cross-fertilization can involve two different, but related species. The anthropologist Alkire (1974: 3), studying the cultivation of breadfruit tree types in Micronesia, reported that some breadfruit varieties are possible hybrids between two distinct species, the native *Artocarpus mariannensis* and the introduced *Artocarpus altilis*. For example, a popular local variety on Pohnpei known as ***Mein padahk***, is precisely such a hybrid. It results from the cross-fertilization of the native, seeded breadfruit known as **Meipa** or **Meikole** and the introduced, "seedless" breadfruit known as **Mahi**. Similarly, many types of bananas originated as hybrids between two different species within the genus *Musa*. In fact, the scientific name of the most widespread banana types globally is *Musa* × *paradisiaca*, with the symbol × indicating that it is a vegetatively reproduced hybrid, sterile and unable to produce viable pollen or seeds (see p. 294). In contrast, a lime, *Citrus* × *aurantifolia*, is also a hybrid of two *Citrus* species, but it is a fertile hybrid, producing viable seeds and reproducing naturally, which is part of the reason why the × symbol is usually omitted from its name.

Hybridization is possible only between closely related species. Species that are distinct enough not to be classified in the same genus normally cannot fertilize each other and will not produce offspring. For example, while one *Citrus* species can fertilize another and bring forth a new type of tree from the seeds, a *Citrus* species cannot fertilize a *Cocos* species to produce something else.

We hope you enjoy learning the names of different plants and animals, not only in your own language but also in scientific nomenclature and other languages. Knowing these names is the first step toward gaining deeper knowledge about these organisms. It can offer clues to their characteristics, uses, origins, and even the history of when and from where they were brought to the islands of Pohnpei State. You can also gain insights into your own language and its connections to other languages spoken in Pohnpei State and Micronesia. Comparing the names of the same plant or animal across different languages can be both insightful and rewarding. If the names are similar, scientists call them **cognates**. This can reveal shared histories, ancient or modern contacts between people, exchanges of plants, animals, vocabulary, knowledge, and, ultimately, the bonds that connect us all across the Pacific and the world.

Karat bananas are all scientifically classified as *Musa troglodytarum*, yet Pohnpeian farmers identify a variety of distinct cultivars.

6

LISTS OF CULTIVARS

CHAPTER 6

LISTS OF CULTIVARS

This chapter brings together a comprehensive overview of named cultivars of major food plants across Pohnpei and its outlying atolls, including Mwoakilloa, Pingelap, Nukuoro, Kapingamarangi, and Sapwuahfik. Drawing on dictionaries, ethnobotanical studies, historical accounts, and recent linguistic fieldwork, it documents the remarkable diversity of breadfruit, banana, taro, pandanus, coconut, yam, sugarcane, kava, and other cultivated plants. Beyond listing names, the chapter touches on the local systems of classification, patterns of introduction and exchange, and the cultural meanings embedded in cultivar nomenclature. Considerable effort has been made to reconcile variant spellings, identify probable synonyms, and flag uncertain or poorly documented entries, while prioritizing standardized orthographies used in contemporary dictionaries. The result is not a fixed or definitive taxonomy, but a carefully curated snapshot of living agricultural knowledge—one that recognizes historical change, regional variation, and the central role of local expertise in understanding the biodiversity of plants used in Micronesian agriculture and food systems.

6.1. BREADFRUIT

6.1.1. Breadfruit cultivars on Pohnpei

In line with Pohnpeian cultural classification, breadfruit cultivars are organized into two categories: smooth-skinned (**Meiniwe**) and rough-skinned (**Meinsahrek**). We compiled their list by comparing Ragone and Raynor (2009: Table 3.2), the Pohnpeian-English dictionary by Rehg and Sohl (1979), and its up-to-date online edition (www.trussel2.com/pnp), and additional sources, such as Lawrence et al. (1964: 60–65).

To reconcile differences among sources, we adopted the Pohnpeian-English dictionary spellings. For example, ***Lukuwal***, ***Meinpeimwas***, ***Meinpwuht***, ***Meinpwuht in lihmwahni***, ***Meiserihseng***, and ***Meipwilipwil***, instead of the variants '*Luhkual*', '*Meipeimwas*', '*Meipwuht*', '*Meipwuht in lhmwahni*', '*Meinseriseng*', and '*Meipuilipuil*' used by Ragone and Raynor (2009). Some dictionary entries were merged as synonyms (e.g., **Pwumpwumw** and **Meipwumpwupw**). Some dictionary entries were excluded: **Meiapaup** and **Opwudenleng** because they appear in no other sources whatsoever, and **Peleu**, which is a banana cultivar despite being labeled breadfruit in the dictionary. Englberger et al. (2005c) list additional breadfruit names not found in other sources: '*Mei sengou*', '*Meiuhpw in Hawaii*', '*Mahi lohlo*', '*Mahi Philippine*', '*Mei ice*', '*Loki*', '*Mein bie*', '*Mei pali*', '*Mei pihs*', '*Mei tam*' (possibly **Meitamwoarok**?), '*Mei tehlik*', '*Meiuhpw en wai*', '*Meiuhpw en Kosrae*', and '*Mei wusol*'. According to the same source, **Mei pwidikdik** is a unique cultivar found only on Sapwuahfik Atoll. Raynor (1991) also provides a list that includes some names we could not corroborate elsewhere. These may be valid or be synonyms of cultivars already listed.

Many cultivars have distinct traits that shape their cultural or practical importance. For instance, **Meiuhpw**, a **Meiniwe**-type cultivar, fruits early but is excluded from tribute and other customary uses on Pohnpei because it is not considered indigenous. **Meikalak**, also of the **Meiniwe** type, is notable for initiating the *rahk* (breadfruit) season. **Meiserihseng**, a **Meinsahrek** type, means "crying child breadfruit" because it cooks quickly, and therefore is useful for calming a hungry child.

Breadfruit cultivars on Pohnpei			
Meiniwe (smooth-skinned)			
Kirimwot	*Meikalak en meikuet*	*Meikuwet*	*Meisei*
Meiais	*Meikalak en meiniwe*	*Meipwiliet*	*Meitehid*
Meiarepe	*Meikalak en meisei*	*Meipwuhleng*	*Meitehid en meinsahrek*
Meikalak	*Meikidi*	*Meisaip*	*Meiuhpw*
Meikalak en meiarepe	*Meikipal*	*Meisapwahpw*	*Meiuhpw en Samoa*
Meinsahrek (rough-skinned)			
Amarepe	*Meilong*	*Meinpeimwas*	*Meipahnie*
Animwure	*Meiluhr*	*Meinpeimwas*	*Meipahte* (= *Meinpahten*)
Lehtemp	*Meimarahra*	*Meinpeipei*	*Meipikos*
Lipet	*Meimoang*	*Meinpeisahle*	*Meipohntakai*
Lukuwal	*Meimuhle*	*Meinpek*	*Meipwet*
Lukuwal lukiamwas	*Meimwed*	*Meinpoakod*	*Meipwetepwet*
Lukuwal lukielel	*Meinahdor*	*Meinpohnkeweneie*	*Meipwilipwil*
Lukuwal opohp	*Meinanihmwoll*	*Meinpohnkoatoa*	*Meise*
Meiakohnd	*Meinanuhsar*	*Meinpohnle*	*Meiserihseng*
Meiauleng	*Meinenguhp*	*Meinpohnoulal*	*Meitakai*
Meideipw	*Meinenimwure*	*Meinpohnpahnwi*	*Meitamwoarok*
Meiepil	*Meinihmwed*	*Meinpohnsakar*	*Meitek*
Meikapas	*Meinihwor*	*Meinpwahr*	*Meitempap*
Meikarat	*Meinimpei*	*Meinpwuht*	*Meiti*
Meikatakot	*Meinintoal*	*Meinpwuht en lihmwahni*	*Meitikapau*
Meikatik	*Meinipwidi*	*Meinpwuht en sokehs*	*Meitoal*
Meikeke (= *Meiwoke*)	*Meinkohrari*	*Meinsakaresei*	*Meitoal en lapar*
Meiketieu	*Meinlitokpwuri*	*Meinsanipwur*	*Meitoal en oarou*
Meikewelik	*Meinmall*	*Meinsapwerehk*	*Meitoantoal*
Meikihmwer	*Meinmesehl*	*Meinseniak*	*Meiumpwei*
Meikiol	*Meinmweli*	*Meintihnpoap*	*Meiupenserek*
Meikitik	*Meinmwelihtik*	*Meinuhsar*	*Meiutuhnpei*
Meikole	*Meinmwoakot*	*Meinutuhn*	*Meiwehwe*
Meikole meikolemwot	*Meinmwohnei*	*Meinwet*	*Meiweke* (= *Meiwoke*)
Meikole meipa (= *Kuru kiniti*)	*Meinohwa*	*Meinwol*	*Nahnmwal*
Meikole meiwid	*Meinpadahk*	*Meioang* (= *Meipoang*?)	*Pwumpwumw* (= *Meipwumpwupw*)
Meili	*Meinpadahk en meiniwe*	*Meiolodung*	*Sangkamwahu*
Meilingkarahk	*Meinpahnwi*	*Meiomp*	*Soulik en lukuwal*
Meiloangon			

6.1.2. Breadfruit cultivars on Mwoakilloa

Our list of breadfruit cultivars from Mwoakilloa derives from Harrison and Albert (1977) and its online edition (www.trussel2.com/mok). It includes both seeded and seedless varieties. Older records can be difficult to match due to variant spellings; for example, Glassman's (1953) '*Moshaborok*' and Englberger et al.'s (2005c) '*Mein sapwerehk*' both refer to ***Moai joapwoahroak***. As expected, several Mwoakilloa cultivars closely resemble those on Pingelap; for instance, the nearly identical names of all three seeded types, or ***Moai upw*** of Mwoakilloa and ***Maei uhpw*** of Pingelap (a cultivar known to produce small fruit, but bearing almost continually).

Breadfruit cultivars on Mwoakilloa			
Seeded			
Moai pa	*Moai si*	*Moai soal*	
Seedless			
Moai id	*Moai in uhrɛk*	*Moai kalak*	*Moai soahid*
Moai in pahdak	*Moai joapwoahroak*	*Moai ngɛljoau*	*Moai upw*

6.1.3. Breadfruit cultivars on Pingelap

The names of breadfruit cultivars on Pingelap come from the Pingelapese-English Dictionary (Hattori 2012), which includes updated spellings of names found in Damas (1994: 235–236) and one additional cultivar (***Maei soal***) from Englberger et al. (2005c). Seeded and seedless varieties are listed separately, though some sources disagree on classification. For example, Damas (1994: 235) notes that St. John (1948) mistakenly identified '*Mei sabarek*' (***Maei sapwaeraek***) and '*Mei si*' (***Maei si***) as seedless, when in fact they contain seeds. Hattori (2012) also includes the term *saeipil*, defined as a "watery type of breadfruit," though it is unclear whether it refers to a distinct variety. The dictionary also offers insights into specific cultivars: ***Maei sapwaeraek*** produces sweet, elongated fruit; its mature tree trunk, like those of ***Maei pa*** and ***Maei si***, are "big and strong [...] ideal for canoe-making." As expected, Pingelap shares many cultivars with Mwoakilloa, often with similar names. For example, ***Maei sapwaeraek*** matches ***Moai joapwoahroak*** in Mwoakilloa.

Breadfruit cultivars on Pingelap			
Seeded			
Moai pa	*Moai si*	*Moai soal*	
Seedless			
Maei sapwaeraek	*Maei uhpw*	*Maein mwoakilloa*	*Maein wae*

6.1.4. Breadfruit cultivars on Nukuoro

Nukuoro has at least four breadfruit cultivars. The native seedless type is called *de henua* ("of the island"), and the seeded one *golee* ("seed"). Others are known as *malali* ("smooth," in reference to the fruit's texture) and *Saamoa*, indicating where it may have been introduced from. This information was provided by Emily Drummond, a linguist researching the Nukuoro language.

Breadfruit cultivars on Nukuoro			
Gulu de henua	*Gulu golee*	*Gulu malali*	*Gulu saamoa*

6.1.4. Breadfruit cultivars on Kapingamarangi

Englberger et al. (2005c) noted three breadfruit cultivars from Kapingamarangi: ***Gulu golee***, from *golee* ("seed"), and '*Gulu Kirinit*' and '*Gulu Pohnpei*', named after geographic locations. The two places are rendered in standard Kapingamarangi spelling as *Gilinid* (= "Greenwich" [Island], a colonial name for Kapingamarangi) and *Ponopei* (= Pohnpei).

Breadfruit cultivars on Kapingamarangi		
Gulu gilinid	*Gulu golee*	*Gulu ponopei*

6.2. BANANA

6.2.1. Banana cultivars on Pohnpei

The information in the table on the next page is based on well-researched and comprehensive studies of banana cultivars from Pohnpei: Daniells (2004), Daniells et al. (2004), and Englberger and Lorens (2004). We also consulted the online Pohnpeian-English dictionary (www.trussel2.com/pnp) and standardized all cultivar name spellings according to official orthography. During this process, we identified possible duplicate names in the dictionary—such as ***Akadahn***, ***Lakadahn***, and ***Kaladahn***—which may represent errors or synonyms. The resultant list includes all cultivars from Daniells et al. (2004) and Daniells (2004), with updated spellings and additional cultivars and synonyms from Englberger and Lorens (marked [1]) and from the dictionary (marked [2]).

Any other published names we have encountered, such as '*Utin mwohnki*', '*Utin nukuoro*' (Englberger et al. 2005c), '*Karat kohdo*' (Riesenberg 1948), '*Kalapwa*' and several others (Raynor 1991) were excluded as they may be synonyms or names that have fallen in disuse. Cultivars recently introduced by the Secretariat of the Pacific Community (e.g., '*Grande naine*', '*Yangambi km5*', and the FHIA hybrids 01, 02, 03, 17, 18 and 23) are also excluded, as they are not local types, but experimental plants introduced for testing resistance to black leaf streak disease (Nelson and Javier 2007). The list encompasses all *Musa* species found on Pohnpei. Most cultivars are *M.* × *paradisiaca*, but some are *M. troglodytarum* (see p. 301) and *M. textilis* (see p. 299). We acknowledge that not all names listed necessarily represent distinct cultivars; some may be duplicates not yet recognized as synonyms. Better understanding requires more research, along with flexibility, as cultivar names in Pohnpei—like elsewhere in the region—evolve over time and may differ among informants. When in doubt, refer to original sources to examine any descriptions and name origins of cultivars or, better yet, consult with Pohnpeian farmers.

Banana cultivars on Pohnpei			
Musa × paradisiaca			
Aiko[2]	*Inahsio mweimwei*	*Mangat en angaur*	*Utiak en angaur* (= *Utiak en palau*)
Akadahn (= *Lakadahn*[2] = *Kaladahn*[2])	*Inahsio pehsehs*	*Mangat en kariki*	*Utiak en pohnpei* (= *Utiak*)
Akadahn weitahta (= *Akadahn awai*[2])	*Inahsio poh rotorot*	*Mangat en pohnpei*	*Utihdol* (= *Utindol*[2] = *Utidol*[2])
Akihna[2] (= *Poupoulap* = *Utin ruk*)	*Inen pahniepur*[1]	*Mangat en ruk*	*Utimwas*
Daiwang (= *Taiwan*)	*Kaimana* (= some *Kundihna*[1])	*Mangat en seipahn*	*Utimwot*[2]
Dukerehda[2]	*Keren lasi*[1]	*Mangat kingit*	*Utin kerenis* (= *Kirou rohi*)
Dukuru	*Kirihm* (= *Uht rais* = *Utin rais* = *Kudud*)	*Preisihl* (= *Brazil*)	*Utin kouruhr* (= *Peleu*)
Epohn[2]	*Kudud en Palau*[2]	*Pwopwsalahd*[2]	*Utin kuam* (= *Ahlap* = *Uht laud*)
Iemwahn	*Kundihna*[1] (which is not *Kaimana*)	*Sapwtehreng*[1]	*Utin lihli* (= *Ilario*)
Iepwtehreng[2]	*Lipwopwnahna*[2]	*Uht kapakap* (= *Kapakap*)	*Utin pihsi* (= *Fiji*)
Ihpalihn palau	*Makao* (= *Maka*[2]?)	*Uht mwot*	*Utin seipahn2* (= *Inahsio*?)
Ihpalihn pohnpei (= *Ihpali*[2])	*Mangat en alohkapw*	*Uht tikitik* (= *Aktikitik* = *Utin menihle*)	*Utin wai* (= *Utin wai mwoatomoat*)
Ihpalihn seipahn			
Musa troglodytarum			
Ihpel[2]	*Karat en pahlil*[2]	*Karat pako* / *Karat poako*	*Utin iap*
Ihpwen pohnial[2]	*Karat ihsilu*[1]	*Karat pwehu*	*Uht mwas*
Karat en iap	*Karat kole* (= *Karat pwonopwon*[1])		
Musa textilis			
Tikahp (= *Utisel*[2] = *Utinwel*[2])			

6.2.2. Banana cultivars on Mwoakilloa

The cultivars listed below are from Harrison and Albert (1977), online edition (www.trussel2.com/mok), with the following modifications: the erroneous entry '*Kalpohnia*' has been merged with ***Kaldohnia***. One additional cultivar not listed may be '*Mash*' (Englberger et al. 2005c). Note also that ***Amɛrkoair*** corresponds to the Pohnpei cultivar ***Utin lihli*** / ***Ilario*** (Englberger et al. 2005c), and that many Mwoakilloa cultivar names closely resemble those from Pingelap and the Marshall Islands with only minor spelling variations.

Banana cultivars on Mwoakilloa			
Amɛrkoair	*Johrumw*	*Luhda*	*Wus in iap*
Daiwang	*Kaldohnia*	*Pihji*	*Wus in kuamw*
Inahjio	*Lakdahn*	*Prɛjil*	*Wus karas*

6.2.3. Banana cultivars on Pingelap

The cultivars listed below are from Hattori (2012), which incorporates earlier key sources and standardizes cultivar name spellings. The original information comes from Solis, a Pingelapese boatman and informant of St. John (1948: 108), who identified 10 locally named, generally "vigorous and productive" banana varieties. Glassman (1953: 301) later added two more. Damas (1994: 234) also reported observing 12 cultivars on the atoll, though the names he recorded do not fully match those cited earlier. This likely explains why Hattori (2012) lists 15 names. We cross-checked all previous sources and reconciled name variations whenever possible, despite highly inconsistent spellings. Even so, some names in Hattori (2012) could not be matched to earlier references. Additional possible cultivars not listed include '*Wihs in loko*' (Englberger et al. 2005c, = ***Wis lokohe***?), '*Panilo*' (from St. John, confirmed by Damas), '*Kutkut*' (from St. John, changed to '*Kudul*' by Damas, possibly same as Pohnpeian ***Kudud***?), and '*Sapari*' and '*Tinian*' (only in Glassman 1953). Note the following: ***Wisinwai*** and ***Wiskaeras*** cultivars may be one and the same or closely related (Englberger et al. 2005c); ***Wis inek*** corresponds to the Pohnpei cultivar ***Utin ruk*** / ***Poupoulap***; and ***Wis aro*** is Pohnpeian ***Inahsio***. Many Pingelap banana cultivar names also resemble those from Mwoakilloa and the Marshall Islands, with minor spelling differences.

Banana cultivars on Pingelap			
Wis amirikae	*Wis esera*	*Wis lakadahn*	*Wisiak*
Wis aro	*Wis iehmwahn*	*Wis lokohe*	*Wisinwai*
Wis daiwang	*Wis inek*	*Wis manihla*	*Wiskaeras*
Wis dukuru	*Wis ladin*	*Wis pihsi*	

6.2.4. Banana cultivars on Nukuoro

We found no published sources on banana cultivars of Nukuoro, but were provided very useful linguistic field data (Drummond and Rudolph 2019) for 10 cultivars. Their cultivar names reflect either individuals credited with their introduction (e.g., *Soosue*, *Naasio*) or the purported place of origin: *Fiisii* (Fiji), *Taivanga* (Taiwan), *Sainaa* (China), *Belesiili* (Brazil), *Luugu* (Chuuk), *Modolago* (Mortlock Islands). The type designated *gainamaa* is said to have arrived from Samoa via Pohnpei. The banana cultivar ***Galadi*** is the local rendition of Pohnpeian ***Karat***, the only one in the list below that is *Musa troglodytarum* (all others being *Musa* × *paradisiaca*).

Banana cultivars on Nukuoro			
Gahudi belesiili	*Gahudi luugu*	*Gahudi sainaa*	*Gahudi taivanga*
Gahudi fiisii	*Gahudi modolago*	*Gahudi soosue*	*Galadi*
Gahudi gainamaa	*Gahudi naasio*		

6.2.5. Banana cultivars on Kapingamarangi

Three cultivars were documented by Englberger et al. (2005c).

Banana cultivars on Kapingamarangi		
Daamoo dun	*Daamoo hagadane*	*Daamoo hagadii*

6.3. TARO

6.3.1. Taro cultivars on Pohnpei

The table below presents Pohnpei cultivars of all species collectively and commonly referred to as taro: **Mwahng** (*Cyrtosperma merkusii*), **Ohd** / **Wehd** (*Alocasia macrorrhizos*), and **Sawa** (both *Colocasia taro* and *Xanthosoma sagittifolium*). It is based primarily on the Pohnpeian-English Dictionary by Rehg and Sohl (1979), including its online edition (www.trussel2.com/pnp). Additional *Cyrtosperma* cultivars were added from Balick et al. (2009: table 5.2; marked [1]) and Englberger et al. (2005a; marked [2]), with all spellings adjusted to reflect standard orthography. Despite best efforts, many cultivars remain undetermined. Several listed by Englberger et al. (2005c) could not be matched with entries in the dictionary or Balick et al. (2009), including: '*Mwahng sawahn*' (= ***Mwahng suwain*** [cultivar with lacy-edged leaves]?), '*Mwahng saleng*' (= ***Saleng walek***?), '*Mwahng loale*', '*Sehnpaki*', '*Mwahng weitahta*' (= **Mwahng tekatek weitahta**), '*Deiterehi*' (= ***Deikoaroahi***?), '*Mwahng lohsapw*', '*Mwahng saidip*', and '*Mwahng keneng*' and its apparent opposite '*Mwahng sohte keneng*'. These names, provided by COM students, may include misnomers or undocumented valid cultivars. One additional cultivar, '*Arrowhead*' (Balick et al. 2009), is omitted for lack of a local name. Some names listed by from Raynor (1991) were also omitted because we found them in no other source (for example, *Alocasia* varieties '*Ohdeneihr*' and '*Ohdepwou*'). Finally, note that **Mwahng tekatek** comprises at least three types distinguished by stem color—greenish, purplish, and reddish (the last called ***Mwahng tekatek weitahta***).

Species identification remains uncertain for some taro cultivars. For example, ***Saleng walek*** is classified as *Cyrtosperma* by Balick et al. (2009: 143), but the dictionary assigns it to *Alocasia*. Likewise, possible *Cyrtosperma* cultivar ***Mwahng tekemuhr*** from Balick et al. (2009) may be the same as *Alocasia* cultivar ***Tekemwer*** from the dictionary. For more on cultivar history and recommendations, see Balick et al. (2009: 139–145); for a discussion of cultivar loss and traditional knowledge, see Englberger et al. (2009).

Taro cultivars on Pohnpei			
Mhwang			
Deikoaroahi	*Mwahng tekatek*[1]	*Mwahngin namwanamw*	*Mwahngin seidip*
Inen mwakot	*Mwahng tekemuhr*[1]	*Mwahngin ngetik*	*Mwahngin wel*
Lahsekir	*Mwahngeida* (= *Aida*[1]?)	*Mwahngin nukuwer*	*Mwatang*
Mwahng kieil	*Mwahngin kialap*	*Mwahngin palau*	*Nein aleks* (= "Alex"[2])
Mwahng ngeli	*Mwahngin lamerer*	*Mwahngin pehleng*	*Nein pwohpw* (= "Bob")
Mwahng pwiliet	*Mwahngin medel*[2]	*Mwahngin pohnpei*[1]	*Pohnengles*[2]
Mwahng sehm[2] (= *Nein sehm* = "Sam")	*Mwahngin meir*	*Mwahngin pohnusamah*	*Simihden*
Mwahng sewi	*Mwahngin nah*	*Mwahngin pwetepwet*	*Sounpwong weneu*
Mwahng suwain			
Sawa			
Analuhlu (= Honolulu)	*Keiwetik*	*Pasdohra*	*Sawahn india*
Derwesu	*Kuwet*	*Sarawihpe*	*Sawahn kusai*
Dinlahwe	*Lienkuweteleng*	*Sawa weitahta*	*Sawahn mwahng*
Enisoupedo	*Pahrotorot* (= *Pahtantal*)		

Taro cultivars on Pohnpei			
Ohd / Wehd			
Oden dawahk	*Oden mwekil*	*Pahnta*	*Taiang*
Oden katau	*Odenei*	*Saleng walek*	*Tekemwer*
Oden kepinne	*Odepe*	*Sepwikin*	

6.3.2. Taro cultivars on Mwoakilloa

The names of taro cultivars in Mwoakilloa in the table below derive from the dictionary by Harrison and Albert (1977) and its online version (www.trussel2.com/mok). The dictionary does not distinguish between different taro species, and thus we have consulted other sources for proper classification and identification of additional cultivars. Despite highly disparate spellings across sources, most **Mwɛhng** (*Cyrtosperma*) cultivars could be reconciled with dictionary entries. For example, Glassman's (1952: 297) forms such as '*Shumbung-unu*', '*Chaleng welek*', and '*Shirieh*' correspond to Bentzen's (1949: 32) '*Sonbóngwenu*', '*Shálingwálik*', and '*Seri**', and match ***Johnpwong wonow***, ***Jalɛngwalɛk***, ***Jɛria***, respectively, in standard orthography. However, several cultivars could not be corroborated in the dictionary or other sources, including the Mwoakilloa cultivars '*Pákilámon*' and '*Ts*nts*l*', and alleged introductions from Ngatik, '*Tongivim*', '*Paylingayling*', and '*Mwengshónang*' (Bentzen 1949), and '*Dohleidok*' (Englberger et al. 2005c). For *Colocasia* [and/or *Xanthosoma*] (**Sawa**), Glassman (1952: 297) lists five varieties: '*Chawasa*', '*Peeter*', '*Tawah*', '*Pehmeru*', and '*Chawa-n-jaban*', but we could only match the final one, to ***Jawahn jeipɛn***, one one of just three clearly identified *Colocasia* cultivars in the dictionary. Another unconfirmed *Colocasia* is '*Kimweng*', noted by Bentzen (1949) as a wetland variety introduced from Pohnpei around 1942. For *Alocasia* (**Wod**), Glassman (1952: 297) mentions the poisonous variety '*Wut-en-mokil*' (= ***Wod en Mwoakilloa***), which he later (1953: 301) also refers to as '*Sehbuken*' (= ***Sepwikin*** in Pohnpei).

Taro cultivars on Mwoakilloa			
Mwɛhng (*Cyrtosperma*)			
Aikɛm	*Johnpwong wonow*	*Nihn jaimon*	*Pikar* 'red'
Dawa	*Jɛria*	*Nihn jɛhm*	*Pikar* 'white'
Jalɛngwalɛk	*Meiuhpw en samoa*	*Nihn ɛnri*	*Wahrau*
Jikohki	*Mwɛhngin ngoasik*	*Pamdɛhl*	*Wiklalɛ*
Jimihdin	*Nihn dijoan*	*Pijɛp*	*Wɛhnmwɛhng*
Jawa (*Colocasia* and/or *Xanthosoma*)			
Jawahn jeipɛn	*Jawahn palau*	*Jawahn ruk*	
Wod (*Alocasia*)			
Wod en Mwoakilloa	*Wodpa*		

6.3.3. Taro cultivars on Pingelap

The list of taro cultivars from Pingelap presented in the table on the next page is based on entries from the Pingelapese-English Dictionary (Hattori 2012), with minor modifications (e.g., merging the separate entries ***Mwaeiang naein silingidaen*** and ***Silingidaen***; ***Mwaeiang naein pisep*** and ***Pisep***, etc.) and cross-referenced with St. John (1948), Glassman (1953), and Damas (1994).

For *Cyrtosperma* (**Mwaeiang**), St. John and Glassman list five cultivars: '*Muiăng an ngatĭk*', '*Sĭmidĭn*', '*Nane pakĕleman*', '*Muiăn sdŏntŏl*', and '*Sĕrisĕng*'. Damas (1994: 233) repeats these, with improved spellings. However, only the first three could be reliably matched to dictionary entries: ***Mwaeiang aen ngatik***, ***Simihden***, and ***Mwaeiang naein pwaekilaemaen***. Notably, the last one—'*Sĕrisĕng*'—resembles Pohnpeian [**Mei**]***serihseng***, which refers to a breadfruit cultivar, not a swamp taro.

Damas adds seven more cultivars: '*Sounpwung walnu*', '*Nein aikem*', '*Pisop*', '*Ne silington*', '*Nesira*', '*Epawn*', and '*Nein seria*', all of which were properly rendered in the dictionary as ***Sounpwong waenau***, ***Naein aikem***, ***Pisep***, ***Naein silingidaen***, ***Naein esera***, ***Aepoahn***, and ***Naein seria***. Damas also notes distinguishing traits—for example, ***Sounpwong waenau*** means "six months," but is typically harvested after one to one-and-a-half years; ***Pisep*** is used in annual competitions and does not require fertilizer; and ***Naein aikem***, ***Naein seria***, and ***Naein silingidaen*** are named after a grandfather, son, and grandson who each independently discovered the varieties. These three are propagated from the seed rather than from stalk and cap (i.e., asexual vegetative reproduction, using a plant cutting that includes an upper section of the corm and part of the main stem). Furthermore, Damas (1994) reports that these varieties grow exceptionally large: "... the largest sizes of any of the aroids with a record corm of [***Naein silingidaen***] variety being marked on the wall of a municipal building said to be 10 feet above ground. The largest roots of these types extend well above the level of the mud and coconut husks and other fertilizer is piled upon them." For *Colocasia* (**Sawa**), not all cultivars cited by earlier sources could be reconciled with dictionary entries. Of the seven mentioned by St. John, Glassman, and Damas—'*Bokor*', '*Tawăng*', '*Mesawsŏl*', '*Sawa Pingelap*', '*Koso*', '*Sauk*', and '*Pemeru*'—only the first four were matched: ***Poakoahr***, ***Dawang***, ***Sawah soal***, and ***Sawah Pingelap***. An additional variety, '*Sawah wai*', is mentioned by Yamamoto (2019). For *Alocasia* (**Woad**), Glassman (1952: 301) states that in addition to the general type, there is a poisonous type called '*Seh-buken*' (clearly a cognate of Pohnpeian **Sepwikin**). Damas (1994: 234) lists **Woad** and two of its varieties, which are also found in the dictionary, though one, ***Wisili***, is labeled as "soft taro" rather than *Alocasia*. Lastly, note that many Pingelap cultivars have direct equivalents in Mwoakilloa and Pohnpei, with minor orthographic variations reflecting different languages. For example: ***Mwaeiang aen ngatik*** / ***Mwɛhngin ngoasik*** / ***Mwahngin ngetik***; ***Simihden*** / ***Jimihdin*** / ***Simihden***; ***Seria*** / ***Jeria***; ***Salaengwalaek*** / ***Jalɛngwalɛk*** / ***Saleng walek***, and so on.

Taro cultivars on Pingelap			
Mwaeiang			
Mwaeiang aen ngatik	*Mwaeiang naein pisep*	*Mwaeiang naein silingidaen*	*Salaengwalaek*
Mwaeiang aepoahn	*Mwaeiang naein pwaekilaemaen*	*Mwaeiang saekaesaek*	*Simihden*
Mwaeiang naein aikem	*Mwaeiang naein sehm*	*Mwaeiang soal*	*Sounpwong waenaeu*
Mwaeiang naein esera	*Mwaeiang naein seria*		
Sawa			
Dawang	*Sawah alahe*	*Sawah seipaen*	*Sawah weisahsa*
Poakoahr	*Sawah pingelap*	*Sawah soal*	
Wod (*Alocasia*)			
Wisili	*Woadihruk*		

6.3.4. Taro cultivars on Nukuoro

In Nukuoro, the names for taro species are as follows: **Ngaungau** for *Alocasia*, **Bulaga** and **Daagoli** (archaic forms: **Daogoli** and **Daogili**) for *Cyrtosperma*, **Dalo** for *Colocasia*, and **Dalo balaau** for *Xanthosoma*. The Nukuoro Lexicon (Carroll and Soulik 1973) lists only three specific varieties: **Mangila** and **Manu-gaugau** which belong to *Alocasia*, and **Manu-alioo**, which is unspecified. The dictionary also includes the term ***Manu de henua***, which likely refers collectively to all *Cyrtosperma* varieties (Emily Drummond, pers. comm.).

Taro cultivars on Nukuoro			
Mangila	*Manu de henua*	*Manu-alioo*	*Manu-gaugau*

6.3.5. Taro cultivars on Kapingamarangi

The Kapingamarangi Lexicon (Lieber and Dikepa 1974) lists one *Cyrtosperma* (**Bulaga**) cultivar, ***Bulaga-dimihoi***; and one *Alocasia* (**Ngaungau**) cultivar, ***Manu-gaugau-dalinga***. It also records five *Colocasia* (**Dala**) cultivars: ***Dala-duumalidu***, ***Dala-hagadili***, ***Dala-mada-lau-hala***, ***Dala-maliinaa***, and ***Dala-muuia***. Many other *Colocasia* cultivars appear to have been lost. In the mid-20th century, Hiroa (1950) worked with an informant—"an old man" named Tukohi—who could remember the following: '*Mihai*', '*Hamea*', '*Tara kapuranaihuku*', '*Hapurepure*', '*Tara ruri*', and '*Tara hakatiri*'. Of these, only the last can be matched to a name in the Kapingamarangi Lexicon, where it appears as ***Dala-hagadili***. In more recent research involving COM students, Englberger et al. (2005c) recorded references to *Cyrtosperma* cultivars '*Bulaga ngatik*' and '*Bulaga dumalidu*', but note the similarity of the latter with ***Dala-duumalidu***, a *Colocasia* cultivar.

Taro cultivars on Kapingamarangi				
Bulaga				
Bulaga-dimihoi				
Dala				
Dala-duumalidu	*Dala-hagadili*	*Dala-mada-lau-hala*	*Dala-maliinaa*	*Dala-muuia*
Ngaugau				
Manu-gaugau-dalinga				

6.4. PANDANUS

6.4.1. Pandanus cultivars on Sapwuahfik

Four pandanus cultivars from Sapwuahfik are listed by Englberger et al. (2005c). They are shown below, with the spelling **Kipar** changed to **Kiparh** to reflect the distinct pronunciation in the Sapwuahfik dialect, which uses a sound closer to the English 'r' than the Pohnpeian pronunciation, and represents it with 'rh'.

Pandanus cultivars on Sapwuahfik			
Kiparh en kapinga	*Kiparh en mwokil*	*Kiparh en nukuor*	*Wedak*

6.4.2. Pandanus cultivars on Mwoakilloa

The cultivars listed in the table on the next page include 23 entries from Harrison and Albert (1977), online edition (www.trussel2.com/mok), with four additional cultivars from Englberger et al. (2005c; marked [1]) and Englberger et al. (2009c; marked [2]). We also cross-checked the names with an earlier list of 19 cultivars recorded by Glassman (1953), though his spellings can be difficult to recognize today. For example, the cultivar ***Juaipwɛhpwɛ*** appears as '*Schwai-pue-ep*'. Similarly, [***Kipar in***] ***Jeliwij*** is barely identifiable as '*Shee-lahweh*', though it is more recognizable in Englberger et al.'s (2005c) form '*Jeliwig*'. Some of Glassman's names could not be reconciled with contemporary spellings and were not recognized by our informants. These include '*Arawan-en-mehluh*', '*Shaleh-show-wushuh*', '*Ruboh*', and '*Uhnbesch*'—they may no longer be found on Mwoakilloa Atoll or may still exist under different or obscure names. The cultivar ***Kipar in majal*** (Englberger et al. 2009c) may or may not be the same as ***Kipar in Jeliwij*** (considering that *Majal* = Marshall Islands, and *Jeliwij* = Jaluit, the Marshallese atoll where the cultivar presumably originated). When matching historically recorded names with the present-day situation it ismportant to consider any known descriptive clues—such as **Doapwoahdin** having very small keys (individual fruits) or **Uhnmahng** being only for weaving (evidently equivalent of weaving-only ***Wūnmaañ*** variety in the Marshall Islands).

Pandanus cultivars on Mwoakilloa			
Aida	*Joahmwinjoang*	*Kipar in pid*	*Mwajak*[2]
Deipw	*Johnmonoia*	*Luarmwɛ*	*Nenakaisak*[1]
Doapwoahdin	*Jorum*[2]	*Lɛhkmwahn*	*Nɛhnkɛhdak*
Dohpw	*Juaipwɛhpwɛ*	*Mahngoron*	*Oarwɛhn*
En kehlen[2]	*Jukpej*	*Mehkilkil*	*Oarwɛhnin pɛhn pajjo*
Insohl	*Kipar in jeliwij*	*Mehusik*	*Pwɛnpwɛl*
Jipwehrik	*Kipar in majal*[2]	*Moakoskos* (= *Makasokosok*[1]?)	*Uhnmahng*

6.4.3. Pandanus cultivars on Pingelap

The information presented in the table below was compiled from multiple sources. The earliest records of Pingelapese pandanus cultivars come from St. John (1948), who listed 14 names using highly subjective spellings. At the time, the Pingelapese language was still exclusively oral and lacked a standardized writing system. Glassman (1953: 302) also struggled with this when he added three more names, which appear to be alternate spellings of cultivars previously documented by St. John. Damas (1994: 231) also repeated St. John's list but used yet another set of spellings. Most recently, Hattori (2012), in the Pingelapese-English Dictionary, rendered the names of 14 cultivars using modern Pingelapese orthography. We compared the names reported by St. John (1948), Glassman (1953), Damas (1994), and Englberger et al. (2013), and reconciled them with the dictionary entries, which take precedence in spelling. These matches were not always straightforward. For example, St. John's '*Asibuirek*', '*Suioibueibuei*', and '*Tobodin*' correspond to **Asipwihrek**, **Soahipwehpw**, and **Saepwoahdin**, respectively. Additionally, Damas' '*Nanagaisal*' and '*Nanagasak*' were tentatively combined under **Naenaekaehsak**. One cultivar, **Kipar in mwahsael**, was included based solely on the dictionary, as it does not appear in earlier sources. Another cultivar was added from St. John (1948) and Damas (1994), where it was given as '*Meikilikil*' and we spelled it **Mehkilkil** based on a match with nearby Mwoakilloa Atoll. Many Pingelap cultivars have close analogues in Mwoakilloa—for instance, **Soahipwehpw** and **Juaipwɛhpwɛ**, or **Luwahramwa** and **Luarmwɛ**, respectively.

Pandanus cultivars on Pingelap			
Aeiseseul	*Kipar in mwahsael*	*Mwisamwis*	*Saepwoahdin* (= *Soapwoahdin*)
Aeraewahn (= *Arawahn*)	*Luwahramwa*	*Mwisikel*	*Soahipwehpw*
Asipwihrek	*Makaesoakaesoak*	*Naenaekaehsak*	*Soahnmeneia*
Esies	*Mehkilkil*		

6.4.4. Pandanus cultivars on Nukuoro

Stone (1966b), in his study of pandanus cultivars on Kapingamarangi Atoll, refers to six edible cultivars from Nukuoro: '*Hara kinij*', '*Hara masong*', '*Hara motu motu*', '*Hara mokil*', '*Hara aria*', and '*Hara kapinga*', along with one wild, non-clonal form: '*Hara vau*'. We matched these early records with cultivar names collected by Drummond and Rudolph (2019), whose spellings reflect modern Nukuoro orthography. The only exception is '*Hara kapinga*', which was omitted, as it could not be confirmed and is likely a synonym of '*Hara kinij*'—a Kapingamarangi import resembling the **Binu-modu** cultivar from there. Both 'Kapinga' and 'Kinij' refer to the same toponym: the former a shortened version of the English name, the latter a shortened form of *Kilinisi*, the Nukuoro name for Kapingamarangi. Drummond and Rudolph (2019) noted three additional cultivars: **Hala benu daalinga** (another import from Kapingamarangi, still with its original name: **Binu-daalinga** in Kapingamarangi), **Hala seu ebuebu** (a particularly juicy cultivar, with *ebuebu* meaning "release bubbles"), and **Hala kadi** (known for the ease of biting into the flesh to remove the seed; *kadi* means "bite"). Several more cultivars were recorded as '*Hara taki*' (Fosberg, cited in Stone 1966), and '*Hala maimai*', '*Hala masele*' (likely originating in the Marshall Islands and perhaps matching similarly named pandanus cultivars such as **Binu-maatele** on Kapingamarangi, **Kipar in mwahsael** on Pingelap, and **Kipar in majal** on Mwoakilloa), '*Hala modu*' (distinct from **Hala modumodu**), '*Hala benuvaidolo*' (possibly another import from Kapingamarangi, = **Binu-waidolo**), and '*Hala moobu*' (Englberger et al. 2003). These are also included in the table below, following orthographic corrections provided by Emily Drummond (pers. comm.).

Pandanus cultivars on Nukuoro			
Hala aaleo	*Hala kilinisi*	*Hala masele*	*Hala seu ebuebu*
Hala benu daalinga	*Hala maasonga*	*Hala modu*	*Hala tagi*
Hala benu vaidolo	*Hala magili*	*Hala modumodu*	*Hala vao*
Hala kadi	*Hala maimai*	*Hala moobu*	

6.4.5. Pandanus cultivars on Kapingamarangi

This list of *Pandanus* cultivars from Kapingamarangi is based on entries from Lieber and Dikepa (1974). It includes all 16 cultivars originally recorded by Stone (1966b), with spellings updated according to the Kapingamarangi-English Dictionary. For example, Stone's '*Binu-hakamhou*' is now written **Binu-hagammuu**, '*Binu-hononghei*' as **Binu-hana-ngeia**, and '*Binu-taring*' as **Binu-daalinga**. The only cultivar from Stone not found in the dictionary is '*Binu-pukute*', which corresponds to **Binu-bugudee** as confirmed by Englberger et al. (2003), and has been appended to the list. The current list appears comprehensive, and names encountered in other sources are likely the result of misspellings. For instance, names such as '*Binu maasala*', '*Binu tagabogia*', '*Binu talaadahi*', and '*Binu tadumaangala*' reported by Englberger et al. (2003) correspond to **Binu-maatele**, **Binu-hagaboogia**, **Binu-halaadahi**, and **Binu-hadu-maangala**, respectively. Only '*Binu hetat*' from the same source remains unresolved. Finally, it should be noted that Kapingamarangi has the highest pandanus cultivar diversity in Pohnpei State. While the highest total number of cultivars estimated for the entire state is 55 (Stone 1963, cited in Englberger et al. 2002), Kapingamarangi alone boasts well over 40.

Pandanus cultivars on Kapingamarangi			
Baehoo	*Binu-dua-weelua*	*Binu-hana-ngeia*	*Binu-magiaa*
Binu-balagahi	*Binu-gauhia*	*Binu-hingala*	*Binu-manga-lua*
Binu-binu	*Binu-gonoihe*	*Binu-hogihogi*	*Binu-modu*
Binu-bugudee	*Binu-hadu-maangala*	*Binu-hoo*	*Binu-nuguolo*
Binu-daalinga	*Binu-hadukene*	*Binu-huguniu*	*Binu-nunagida*
Binu-delei	*Binu-hagaboogia*	*Binu-kadi*	*Binu-tala-di-wawe*
Binu-di-adili	*Binu-hagadili*	*Binu-laudia*	*Binu-tala-duuli*
Binu-di-maduu	*Binu-hagammuu*	*Binu-luaaua*	*Binu-talaaduhu*
Binu-di-ngolu	*Binu-hala-duuli*	*Binu-maatele*	*Binu-togongo*
Binu-di-ulungi	*Binu-halaadahi*	*Binu-madiiloo*	*Binu-waidolo*
Binu-dolongohai	*Binu-halaamobu*	*Binu-magana*	*Hedebinu*
Binu-dua-ohigu			

6.4. COCONUT

6.5.1. Coconut cultivars on Pohnpei

Based on the Pohnpeian-English Dictionary by Rehg and Sohl (1979), online edition (www.trussel2.com/pnp), there are about a dozen named coconut cultivars in Pohnpei, each associated with specific characteristics. The cultivar ***Adohl*** is known for its sweet, edible husk. ***Nih mwotomwot*** is a dwarf cultivar. ***Nih pwetepwet*** produces medium-sized pink nuts and light green fronds. Cultivars ***Nihtoal*** and ***Nihweita*** are similar, distinguished by fruit color: suffix "*-toal*" indicates dark green nuts while "*-weita*" denotes reddish-orange nuts. This same contrast appears in three related cultivar groups: ***Nihkarat***, known for its round nuts, comprises ***Nihkarattoal*** and ***Nihkaratweita***; ***Nihrir***, recognized for its large bunches of medium-sized nuts, comprises ***Nihrirtoal*** and ***Nihrirweita***; and ***Nihtik***, noted for its small nuts, comprises ***Nihtiktoal*** and ***Nihtikweita***. In addition to cultivars listed in the dictionary, we have come across '*Nih lap*' and '*Nih rem*' in Raynor (1991).

Coconut cultivars on Pohnpei		
Adohl	*Nih pwetepwet*	*Nihtoal*
Nih mwotomwot	*Nihkengk*	*Nihweita*
	Nihkarat	
Nihkarattoal	*Nihkaratweita*	
	Nihrir	
Nihrirtoal	*Nihrirweita*	
	Nihtik	
Nihtiktoal	*Nihtikweita*	

6.5.2. Coconut cultivars on Mwoakilloa

The table below outlines the coconut cultivar names on Mwoakilloa, as listed in the dictionary by Harrison and Albert (1977), online edition (www.trussel2.com/mok). ***Nihn pwinjo*** is a short tree variety, reaching only about 3 m (10 ft) in height; ***Ni pwɛspwɛs*** is described as having a white nut; ***Adohl*** produces nuts with a sweet husk and may derive its name from the English word "atoll," as recorded by Glassman (1953). Additional cultivars are reported to be or to have been present in Mwoakilloa. Among those documented by Murphy (1949) and Glassman (1953), three remain unconfirmed by the dictionary: '*Nikarus*', '*Sukabish*', and '*Ni sikisik*'. The last, meaning "small coconut," may refer to ***Nihn pwinjo***.

Coconut cultivars on Mwoakilloa			
Adohl	*Ni rei*	*Ni soal*	*Nihn pwinjo*
Ni pwɛspwɛs	*Ni roam*	*Ni wahssa*	

6.5.3. Coconut cultivars on Pingelap

The cultivars listed below are drawn from Hattori (2012) and include varieties previously documented by Glassman (1953) under earlier spellings. ***Nihkusa*** refers to a short-tree variety (possibly equivalent to ***Nihn pwinjo*** in Mwoakilloa); ***Adoahl*** denotes a sweet-husked variety (known under the same name in both Mwoakilloa and Pohnpei); and ***Nihweisahsa*** is distinguished by its orange nuts (like ***Ni wahssa*** in Mwoakilloa).

Coconut cultivars on Pingelap		
Adoahl	*Nihmaho*	*Nihweisahsa*
Nihkusa	*Nihsoal*	

6.5.4. Coconut cultivars on Nukuoro

The cultivars listed here are drawn from the Nukuoro Lexicon (Carroll and Soulik 1973). The cultivar ***Daogave*** bears nuts seasonally; ***Dulivai*** has reddish nuts (compare with ***Ni wahssa*** in Mwoakilloa and ***Nihweisahsa*** in Pingelap); and ***Gadoi*** (more accurately spelled ***Gaadoi***; Emily Drummond, pers. comm.) has an edible husk, similar to sweet-husked cultivars in Pohnpei, Mwoakilloa, and Pingelap.

Coconut cultivars on Mwoakilloa			
Daogave	*Dulivai*	*Gaadoi*	*Huaalanga*

6.5.5. Coconut cultivars on Kapingamarangi

Kapingamarangi Lexicon (Lieber and Dikepa 1974) notes only two atypical types of coconut trees. ***Niu modumodu*** bears fruit seasonally and likely corresponds to ***Daogave*** in Nukuoro. ***Niu maangala*** produces nuts with sweet, edible husk and likely corresponds to ***Gaadoi*** in Nukuoro.

Coconut cultivars on Kapingamarangi	
Niu maangala	*Niu modumodu*

6. 6. YAMS

6.6.1. Yam cultivars on Pohnpei

The list of cultivars provided here was created by comparing those listed by Balick et al. (2009: table 2.1) with entries in the Pohnpeian–English Dictionary by Rehg and Sohl (1979), online edition (www.trussel2.com/pnp). In cross-checking the two sources, we found that most cultivars appear in both, although some are found only in the dictionary (marked [1]) and others only in Balick et al. (marked [2]). Certain entries were not treated as unique cultivars for the following reasons: (a) they are synonyms of other cultivars and are identified as such in parentheses (e.g., ***Elies*** and ***Soulik en mesihsou*** are the same as ***Kupwuneik***); (b) some appear to be duplicate names resulting from typographical errors and were therefore omitted (e.g., '*Kehpin namwe kepsu*' for ***Kehp en namwu kepeu***); and (c) some entries refer to broad types or groups of related cultivars rather than to specific named cultivars (e.g., ***Kehp en dol***, ***Kehpeneir***, ***Koakonehp***, ***Nahlik***, ***Ohpwet***, etc). The spellings used in the two sources are generally comparable. The principal difference is that the dictionary contracts nouns in the possessive case, so that, for example, ***Kehp en dol en Pohnpei*** becomes **Kehpin dolen Pohnpei**. Here we retain the former, uncontracted spellings for clarity, while also adhering to standard Pohnpeian orthography (for example, in the rendering of place names such as *Pwanuatu* for Vanuatu and *Awai* for Hawai'i). Finally, it is important to note that the most general term **Kehp** (yams) in Pohnpei can refer to several species, including *Dioscorea alata*, *D. esculenta*, *D. nummularia*, *D. bulbifera*, *D. hispida*, *D. pentaphylla*, and perhaps even *D. flabellifolia*. The correspondence between individual cultivars in Pohnpei and particular yam species has not been comprehensively studied; accordingly, the list below includes cultivars associated with all of these species. Only the set of names specifically related to *D. bulbifera*—**Kehmpalahi**, **Palahi**, **Palahirot**, **Ohdek**, and **Weisoahn**—is well defined, though it is unclear which individual terms stand for distinct types and which are synonyms referring to the species as a whole.

Yam cultivars on Pohnpei			
Ariarsin	Kehp en penieu weitahta	Kehpeneir kehpmwas	Koakonehp weitahta
Deulimau	Kehp en pihsi	Kehpeneir kehpuwalap	Koasiemen
Dolen langkair	Kehp en pwair	Kehpeneir kehpuwetik[2]	Kunukunda (= Kiewek en tamworohi)
Ehperem	Kehp en pwanuatu (= Vanuatu)	Kehpeneir kumwuten en kidi	Kupwuneik (= Elies = soulik en mesihsou)
Ineupile	Kehp en pweisaper	Kehpeneir mpo	Lepenmoar
Iuhsin dauk	Kehp en rota	Kehpwallal	Lepwak en kehp pwediwel
Kait en kehp en dol	Kehp en sapahn	Kehpwetik en kehmpwalap[1]	Limweilamalam
Kait en pohnpei	Kehp en sapwerehk	Kehpwetik[1]	Luhte
Kaniki en sapuatik	Kehp en seini	Kiewek en dehpehk[1]	Lukenaisais
Kehm mwuterek	Kehp en seipahn	Kiewek en kehp en kipar	Lukenaisais mwotomwot
Kehmmarepehn (= Kehmmarepe[1]?)	Kehp en sounet (= Sounet)	Kiewek en kehp en namwu pwetepwet	Manrahleng
Kehmmwas[1]	Kehp en souwiso	Kiewek en kehp en namwu weitahta	Masaloal (= Kehp pwengkin)
Kehmmwot en pohnpei (= Kehmmwot)	Kehp en tamwara	Kiewek en kehp en nihpwung[2]	Mpel en kehp en namwu
Kehmpalahi (= Palahi[1])	Kehp en tamwara pwomwahu	Kiewek en koakonehp	Nahlik en dolen katau
Kehmpekehi[1]	Kehp en wou	Kiewek en mahnd	Nahlik en sokole[2]
Kehmpuwetik[1]	Kehp eret	Kiewek en pasahn pwehk	Nahlik en souwiso
Kehmpwedawel[1]	Kehp kos	Kiewek en tamworohi[1]	Nahlik lapalap
Kehp akitik	Kehp lapwed	Kilimenip en ariarsin	Nahlikie en sokole
Kehp arar	Kehp meir kelik mwotomwot	Kilimenip en inarongemahl	Nahlikie pwetepwet
Kehp en and (= Akitik)	Kehp meir kelik pwetepwet	Kilimenip en inousohdi	Nahlikie weitahta
Kehp en awai sohkatepe	Kehp meir kelik weitahta	Kilimenip en kehp en and[2]	Nein soukihn (= Soukihn)
Kehp en depehk	Kehp mwotelel	Kilimenip en kehp en depehk	Neinpwadil
Kehp en dol en pohnpei	Kehp mwoten namwanamw	Kilimenip en kehp en kipar pwetepwet	Neinsahle
Kehp en dol en wai ohreirei	Kehp mwuten pahsed	Kilimenip en kehp en kipar weitahta	Ngisingisdo
Kehp en dol en wai pwetepwet	Kehp mwuterek en enipein	Kilimenip en kehp en lupwu	Nuh kaledonia
Kehp en dol en wai weitahta[1]	Kehp pahini	Kilimenip en kehp en marau	Nuh pertin
Kehp en eikesu[1]	Kehp palai	Kilimenip en kehp en mweli	Ohdek[1]

Yam cultivars on Pohnpei			
Kehp en iap pwetepwet	*Kehp palai en meitipw*	*Kilimenip en kehp en namwu pwetepwet*	*Ohpwet en pohnpei*
Kehp en iap weitahta	*Kehp pwediwel pwetepwet*	*Kilimenip en kehp en namwu weitahta*	*Ohpwet en salapwuk*
Kehp en kipar	*Kehp pwediwel weitahta*	*Kilimenip en kehp en nan peilam*	*Ohpwet en wai*
Kehp en kipar en rabaul	*Kehp pwet*	*Kilimenip en kehp en nihpwung*	*Oun enipoas*
Kehp en kipar weitahta	*Kehp pwetepwet en takaieu*	*Kilimenip en kehp meir kelik*	*Oundolen liwi*
Kehp en loangtakai ohnta	*Kehp pwetepwet*[2]	*Kilimenip en kehp mwuterek* (= *Sandohlina*)	*Oundolen nahpik*
Kehp en lohd	*Kehp pwomwahu pwetepwet*	*Kilimenip en kehp pwediwel pwetepwet*	*Ounsouna* (= *Kiewek en koakonehp*)
Kehp en lupwu	*Kehp pwomwahu weitahta*	*Kilimenip en kehp weitahta*	*Ounwel*
Kehp en mehlel	*Kehp sahle*	*Kilimenip en kehpen pwanuatu*	*Palahirot*[1]
Kehp en mweli	*Kehp sel*	*Kilimenip en kehpenei kopou*	*Pasahn pwehk*
Kehp en mweli pwohmwahu weitahta	*Kehp send*	*Kilimenip en kiewek en mwahnd*	*Pohnpeiros*
Kehp en namwu eremia	*Kehp silik*	*Kilimenip en kiewek en pasahn pwehk*	*Sau ne*
Kehp en namwu kepeu	*Kehp tipaker*	*Kilimenip en koakonehp pwetepwet*	*Saud awak*
Kehp en namwu pwetepwet	*Kehpenei en sapahn*	*Kilimenip en koakonehp weitahta*	*Sisenando*[1]
Kehp en namwu weitahta	*Kehpenei inarongemahl*	*Kilimenip en lukenaisais*	*Sokoahtikitik*
Kehp en nan peilam	*Kehpenei inosipehd* (= *Mpel*)	*Kilimenip en ounsouna*	*Sokolamwowe*
Kehp en nihpwung	*Kehpenei inousohdi*	*Kilimenip en pasahn pwehk*	*Sosmikieu*
Kehp en pahsed	*Kehpenei kopou*	*Kilimenip en pohnpei*	*Soulik en palikir*
Kehp en palau pwetepwet	*Kehpeneir ahmarepe*	*Kiped* (= *Kehp en pwanuatu*)	*Soulik en pohrasapw*
Kehp en palau weitahta	*Kehpeneir amarepen kehpuwetik*	*Koakonehp en denian*	*Soumwahnd* (= *Keseumwahnd*)
Kehp en pehleng	*Kehpeneir amarepen kehpwulap*	*Koakonehp en iap*	*Tohnkihpim*
Kehp en penieu pwetepwet	*Kehpeneir en wai*	*Koakonehp pwetepwet*	*Weisoahn*[1]

It should be noted that the Pohnpeian–English Dictionary also includes brief descriptions of some cultivars. For example, the dictionary indicates that ***Kehmmwas***, ***Kehmpuwetik***, and ***Kehmmarepe*** all grow wild; ***Kehpwallal*** and ***Kehpwetik*** are not associated with a particular season; ***Kehmmwot*** is seed-bearing; and ***Kehp en kipar*** is fruit-bearing.

6. 7. SUGARCANE

6.7.1. Sugarcane cultivars on Pohnpei

The table below lists sugarcane cultivar names included in the Pohnpeian–English Dictionary by Rehg and Sohl (1979) and its regularly updated online edition (www.trussel2.com/pnp). Four of those cultivars—***Seun nta***, ***Seuneir***, ***Seupwet***, and ***Rewenpwok***—are widely considered to be of Pohnpeian origin (Raynor 1991).

Sugarcane cultivars on Pohnpei			
Misikeni	*ehu kala* (= *Kalahltoal*)	*Seualahl*	*Seun wai*
Olomwot	*Sehu kala toahl*	*Seun nta* (= *Seuweita*)	*Seuneir*
Rahmwot	*Sehu permwohsa*	*Seun nukini* (= *Seutoal*)	*Seupwet*
Rewenpwok	*Seuahlek*	*Seun palau*	*Tipwatipw*

6.7.2. Sugarcane cultivars on Mwoakilloa

Harrison and Albert (1977), online edition (www.trussel2.com/mok), includes the names of six sugarcane cultivars on Mwoakilloa.

Sugarcane cultivars on Mwoakilloa		
Dauwas	*Doa kalai*	*Doahn wai*
Doa alahl	*Doahn kuas*	*Johkala*

6.7.3. Sugarcane cultivars on Pingelap

In her linguistic work in Pingelap, Hattori (2012) identified eight varieties of sugarcane on this atoll and provided contemporary Pingelapese spellings for cultivars previously recorded by St. John (1948: 107) and Damas (1994: 232). Hattori (2012) only excludes the variety '*Teimos*' reported by St. John (1948) but adds two previously unrecorded varieties: ***Seu soal*** and ***Kaepina***. According to tradition, sugarcane types on Pingelap "were introduced from far to the west" (Damas 1994: 232). The varieties named after Palau and Taiwan (*Parmosa* = Formosa) are named for the islands from which they were reportedly introduced into Pingelap during the Japanese period.

Sugarcane cultivars on Pingelap			
Kaelai	*Seni*	*Seu palau*	*Seu soal*
Kaepina	*Seu iehsehng*	*Seu parmosa*	*Seuweisahsa*

6. 8. KAVA

6.8.1. Kava cultivars on Pohnpei

Sakau grown in Pohnpei comprises two principal types. ***Rahmedel*** (or ***Rahmoadoal***) is distinguished by having smooth, light-colored stems. In contrast, ***Rahmwanger*** has series of purplish or blackish spots on the stem. Each of these types has additional names or subtype designations by which it may be known. When the two are grown together, they are referred to as ***Malahd***.

Kava cultivars on Pohnpei	
Rahmedel (= *Rahmoadoal*)	
Kalaidong	*Kohre*
Rahmwanger	
Kohkore	*Naniepw*

6. 9. OTHER PLANTS

The table below summarize the diversity of several other trees and herbs cultivated on Pohnpei, based on cultivar names included in the Pohnpeian–English Dictionary by Rehg and Sohl (1979) and its regularly updated online edition (www.trussel2.com/pnp). The dictionary excludes any names for cassava (*Manihot esculenta*) because it is a non-traditional crop whose varieties on the island lack well-established local names. Its cultivars result from more recent introductions and experimentation, with names drawn from Pretrick and Tuivavalagi (2017).

Varieties of other trees and herbs cultivated on Pohnpei			
Dihng, Ti plant (*Cordyline fruticosa*)			
Dingmwahn		*Dingpein*	
Sinter, White ginger lily (*Hedychium coronarium*)			
Sinter oangen waii		*Sinter pwetepwet*	
Oahng, Turmeric (*Curcuma longa*)			
Oangalap	*Oangen pahsed*	*Oangen palau*	*Oangitik* (= *Kisinioang*?)
Keleu, Sea hibiscus (*Hibiscus tiliaceus*)			
Pahtakai			
Kehngid, Mango tree (*Mangifera indica*)			
Inen mwoakoatoa	*Kehngid en pohnpei*	*Kiewek en kehngid en pohnpei*	*Sallong*
Kehngid en meinuwe	*Kehngid pweinaper*	*Kiewek en sallong*	
Kehp tuhke, Cassava (*Manihot esculenta*)			
'Benga'	*'Navavalu'*	*'Palikir'*	
'Japan'	*'Palau a'*	*'Tolonier'*	
'Marsi'	*'Palau b'*	*'Yap'*	

GLOSSARY

Abiotic ▸ Pertaining to factors or things that are separate and independent from living things; nonliving.

Acute ▸ Pointed, with an angle less than 90 degrees.

Adaptation ▸ A change in a species that results in an increased ability to survive and reproduce in a particular environment.

Alga, pl. **algae** ▸ Any of numerous kinds of photosynthetic plants that live and reproduce entirely immersed in water. Many species, the planktonic forms, exist as single or small groups of cells that float freely in the water. Other species, the "seaweeds," may be large and attached.

Alien species ▸ Species found in but not native to a particular area; also known as introduced or exotic.

Alternate ▸ Plant parts such as leaves not opposite each other along a stem, for eample, one leaf arising alone at a node.

Annual ▸ A plant that grows from seed, sets seed, and dies within a year.

Apex ▸ The tip or terminal point of a plant structure such as a leaf, stem, root, or fruit.

Archipelago ▸ A group or chain of related islands.

Aroid ▸ A member of the Araceae, the *Taro* family.

Artificial selection ▸ Plant and animal breeders' practice of selecting individuals with the greatest expression of desired traits to be the parents of the next generation.

Axillary ▸ Arising from leaf or flower axils, an axil being the angle made by a petiole or peduncle with the stem.

Bacteria ▸ Any of numerous kinds of microscopic organisms which exist as simple, single cells that multiply by simple division. Along with fungi, they comprise the decomposer component of ecosystems.

Base ▸ Bottom, near the attachment of an organ.

Biodegradable ▸ Able to be consumed and broken down to natural substances such as carbon dioxide and water by biological organisms, particularly decomposers.

Biological (pest) control ▸ Control of a pest population by introduction of predatory, parasitic, or disease-causing organisms.

Biological diversity ▸ Range or number of species or subspecies found in a particular area.

Biomass ▸ Total amount of living or formerly living matter in a given area, measured as dry weight.

Biosphere ▸ The worldwide system within which all life functions; composed of smaller systems, including the atmosphere, hydrosphere, and lithosphere. The overall ecosystem of the earth. It is the sum total of all the biomes and smaller ecosystems, which are ultimately all interconnected and interdependent through global processes such as water and atmospheric cycles.

Biota ▸ refers to any and all living organisms and the ecosystems in which they exist.

Biotic ▸ Living or derived from living things.

Blade ▸ The flat part of a leaf.

Bract ▸ Structure produced below a flower, sometimes leaf-like but often colored.

Bristle ▸ Stiff hair.

Calyx ▸ The outermost whorl of floral parts, frequently green as opposed to the often colored petals.

Canopy tree ▸ The tallest plants in a vegetation type that would be seen from aerial view.

Capsule ▸ A dry fruit that splits open (dehisces) at maturity.

Caroline Islands ▸ A cluster of five main island groups and numerous atolls south of Guam and north of the Equator. The main Caroline islands are Chuuk, Pohnpei and Kosrae (Eastern Caroline Islands) and Yap and Belau (Western Caroline Islands).

Cell (biological) ▸ The basic unit of life, the smallest unit that still maintains all the attributes of life. Many microscopic organisms consist of a single cell. Large organisms consist of trillions of specialized cells functioning together.

Climate ▸ A general description of the average temperature and rainfall conditions of a region over the course of a year.

Coastal strand ▸ The area near the ocean where the vegetation is affected by strand influences such as sandy substrate and salt spray.

Community ▸ A collection of organisms occupying a specific geographic area.

Compound (plants) ▸ With more than one part, as in a leaf with several leaflets or an inflorescence with many branches.

Conical ▸ Cone shaped.

Conservation ▸ The management of a resource in such a way as to assure that it will continue to provide maximum benefit to humans over the long run. Conservation may include various degrees of use or protection, depending on what is necessary to maintain the resource over the long run.

Consumers ▸ In an ecosystem, those organisms that derive their energy from feeding on other organisms or their products.

Corm ▸ An underground stem; for example, the taro "root" used as a source of food.

Corolla ▸ All the petals spoken of as a unit.

Creeping ▸ Habit of plant that roots at its nodes and retains a generally low stature.

Cultivar ▸ A cultivated variety of a plant species.

Cylindrical ▸ Long slender rod-shaped structure round in cross section.

Decomposers ▸ Organisms the feeding action of which results in decay, rotting, or breakdown of complex to simpler components. The primary decomposers are fungi and bacteria.

Deciduous ▸ Losing all leaves at once.

Detritus ▸ The dead organic matter, such as fallen leaves, twigs, and other plant and animal wastes, that exist in any ecosystem.

Detritus feeders ▸ Organisms such as termites, fungi, and bacteria that obtain their nutrients and energy mainly by feeding on dead organic matter. Also known as decomposers.

Dioecious species ▸ With separate male and female plants.

Domesticated species ▸ Species that have been bred for specific characteristics that humans value, sometimes resulting in species that are dependent on humans for continued survival.

Dominant ▸ The most conspicuous species in a vegetation type, often with the most biomass.

Dynamic ▸ In reference to a system, exhibiting the tendency to change in response to changing conditions.

Ecologists ▸ Scientists who study ecology, i.e., the ways in which organisms interact with each other biomeand their environment.

Ecology ▸ The study of the interrelationships between living organisms and the living and nonliving components and processes that make up the environment.

Ecosystem ▸ A grouping of plants, animals, and other organisms interacting with each other and their environment in such a way as to perpetuate the grouping more or less indefinitely. Ecosystems have characteristic forms such as deserts, grasslands, tundra, deciduous forests, and tropical rainforests (also see ▸ Biome).

Ellipsoid ▸ Oval in outline (3-dimensional).

Elliptic ▸ Oval in outline (2-dimensional).

Endangered species ▸ A species whose total population is declining to relatively low levels, a trend which if continued will result in extinction.

Endemic ▸ Native to a given area and occurring nowhere else.

Endemic Species of Pohnpei ▸ Native plants or animals whose ancestors arrived on Yap without human help, later evolved into unique species only found on Yap.

Energy ▸ The ability to do work. Common forms of energy are light, heat, electricity, motion, and chemical bond energy inherent in compounds such as sugar, gasoline, and other fuels.

Entire ▸ Having leaf margins (edges) without teeth or lobes, continuous, and undivided.

Environment ▸ The combination of all things and factors external to the individual or population of organisms in question.

Environmental science ▸ The branch of science concerned with environmental issues.

Epiphyte ▸ A plant growing on the trunk, branches, or stems of another plant, but not obtaining nutrients from its host plant as parasites do.

Erect ▸ Upright position.

Erosion ▸ Removal of soil by running water or wind. The process of soil particles being carried away by wind or water. Erosion generally moves the smaller soil particles first, thus reducing the soil to a more coarse, sandy, stony texture over time.

Estuary ▸ The area of tidal mixing with river waters.

Evaporation ▸ Molecules leaving the liquid state and entering the vapor or gaseous state as, for example, water evaporates to form water vapor.

Evolution ▸ The theory that all species now on earth are descended from ancestral species through a process of gradual change brought about by natural selection.

Extinction ▸ The process by which a species ceases to exist.

Family ▸ A grouping of related genera, the family name usually ends with the "-aceae" suffix.

Fertile ▸ Bearing reproductive material i.e., pollen, spores, fruit.

Food chain ▸ A linear path that food energy takes in passing from producer to consumers to decomposers in an ecosystem.

Food web ▸ A complex, interlocking set of paths that food energy takes in passing from producer to consumers to decomposers in an ecosystem. The combination of all the feeding relationships that exist in an ecosystem.

Fronds ▸ The leaves of ferns, palms, and some other plants are often referred to as fronds.

Fungus, pl. fungi ▸ Any of numerous species of molds, mushrooms, brackets, and other forms of nonphotosynthetic plants. They derive energy and nutrients by consuming other organic materials. Along with bacteria they form the decomposer component of ecosystems.

Genera ▸ Plural of ▸ genus, or groups of closely related species, in classification of plants or animals.

Genetics ▸ The study of heredity and the processes by which inherited characteristics are passed from one generation to the next.

Genus, pl. **genera** ▸ A grouping of related species; the first of two words in a plant name, e.g., *Colocasia* is the genus of taro, *Colocasia esculenta*.

Germinate ▸ To begin to grow from seed.

Glabrous ▸ Without hairs, smooth.

Globose ▸ 3-dimensional globe shaped.

Grass ▸ A member of the grass family Poaceae, or a plant resembling a grass.

Groundwater ▸ Water that is accumulated in the ground, completely filling and saturating all pores and spaces in rock and/or soil. Groundwater is free to move more or less readily. It is the reservoir for springs and wells. It is replenished by infiltration of surface water.

Habitat ▸ Biologically, it refers to the type of environmental conditions in which an organism lives, or to which it is adapted. The specific environment (e.g., woods, desert, swamp) in which an organism lives.

Hawaiian Islands ▸ The northern most island group of Polynesia forming an archipelago with eight main islands (Hawai'i, Maui, Moloka'i, Kaho'olawe, O'ahu, Lana'i, Kaua'i, and Ni'ihau) and a leeward chain of islands including Midway.

Herbivore ▸ An organism such an insect or a bird that feeds primarily on green plants, or plant products such as seeds or nuts.

Humidity ▸ The amount of water vapor in the air (also see ▸ **relative humidity**).

Humus ▸ A dark brown or black, soft, spongy residue of organic matter that remains after the bulk of dead leaves, wood, or other organic matter has decomposed. Humus does oxidize, but relatively slowly. It is extremely valuable in enhancing physical and chemical properties of soil.

Hybrid ▸ A plant or animal resulting from a cross between two closely related species which do not normally cross.

Hybridization ▸ Cross-mating between two more or less closely related species.

Hypocotyl ▸ Part of the elongating stem of a germinating plant.

Indigenous ▸ Native to a particular area but also found elsewhere growing without human help.

Indigenous species of Pohnpei ▸ Native species of plants or animals which arrived on Pohnpei without human help but is not native only to Pohnpei.

Infiltration ▸ The process of water soaking into soil as opposed to its running off the surface.

Infiltration capacity ▸ The maximum rate at which soil can absorb water.

Inflorescence ▸ The arrangement of a group of flowers.

Inorganic ▸ Classical definition: All things such as air, water, minerals, and metals that are neither living organisms nor products uniquely produced by living things. Chemical definition: All chemical compounds that do not contain carbon atoms as an integral part of their molecular structure.

Insecticide ▸ Any chemical used to kill insects.

Interface ▸ The boundary between two zones.

Lanceolate ▸ Leaf type that is much longer than wide, with the widest part near the base.

Larva, pl. **larvae**, adj. **larval** ▸ A free-living immature form that occurs in the life cycle of many organisms and that is structurally distinct from the adult. For example, caterpillars are the larval stage of moths and butterflies.

Leaf ▸ An often green flat structure found below a bud on a stem and generally consisting of a petiole (stalk) and a blade.

Leaflet ▸ One part of a compound leaf.

Legume ▸ A member of the Pea family, Fabaceae. The group of land plants that is virtually alone in its ability to fix nitrogen (nitrogen fixation with the sympiotic assistance of certain microscopic organisms).

Life cycle ▸ The various stages of life, progressing from the adult of one generation to the adult of the next.

Limiting factor ▸ A factor primarily responsible for limiting the growth and/or reproduction of an organism or a population. The limiting factor may be a physical factor such as temperature or light, a chemical factor such as shortage of a particular nutrient, or a biological factor such as a competing species.

Linear ▸ A long narrow shape.

Litter ▸ In an ecosystem, the natural cover of dead leaves, twigs, and other dead plant material. This natural litter is subject to rapid decomposition and recycling in the ecosystem, whereas much of human litter, such as bottles, cans and plastics, is not.

Lobes ▸ Rounded projections on a leaf or fruit.

Mangrove ▸ A member of the Mangrove family, the Rhizophoraceae, which occurs in saltwater-influenced swamp vegetation dominated by members of the Rhizophoraceae; many other species also occur in such vegetation.

Mariana Islands ▸ The archipelago of islands north of the Caroline Islands including Guam, Tinian, Rota, Saipan and several smaller islands to the north.

Marsh ▸ A wetland vegetation type dominated by herbaceous rather than woody species.

Microorganism ▸ Any microscopic organism particularly bacteria, viruses and protozoa.

Micronesia ▸ One of the three major divisions of the Pacific comprising the small islands north of the Equator, east of the Philippines and west of Polynesia.

Midrib ▸ The middle vein in a leaf.

Monocropping ▸ The practice of growing the same crop year after year on the same land.

Monoculture ▸ An agricultural system in which a sole crop is grown repeatedly over a large area.

Monotypic ▸ With only one species.

Multiple use ▸ The use of land for as many different purposes as possible to gain maximum benefit from it.

Mutation ▸ A random change in one or more genes of an organism. Mutations may occur spontaneously in nature, but their number and degree are vastly increased by exposure to radiation and/or certain chemicals. Mutations generally result in physical deformity and/or metabolic malfunction.

Native ▸ Naturally occurring in an area without the aid of man.

Natural ▸ An adjective used to describe a substance or factor occurring or produced as a normal part of nature apart from any activity or intervention of humans. Opposite of artificial, synthetic, human-made, or caused by humans.

Natural selection ▸ The process whereby the natural factors of environmental resistance tend to eliminate those members of a population that are least well adapted to cope and thus, in effect, select those best adapted for survival and reproduction.

Naturalized ▸ Introduced to an area by people on purpose or by accident and now growing wild without the aid of man.

Niche ▸ The ecological role of an organism. Essentially, its feeding and reproductive behavior.

Node ▸ The joint where a leaf is attached.

Nut ▸ A dry, hard, one-seeded fruit.

Nutrient ▸ For a plant, a nutrient is an essential element in a particular ion or molecule that can be absorbed and used by the plant. For example, carbon, hydrogen, nitrogen, and phosphorous are essential elements; carbon dioxide, water, nitrate, and phosphate are respective nutrients. For an animal, a nutrient is a material such as protein, vitamins, and minerals that are required for growth, maintenance, and repair of the body and also materials such as carbohydrates that are required for energy.

Nutrient cycles ▸ The repeated pathway of particular nutrients or elements from the environment through one or more organisms back to the environment. Nutrient cycles include the carbon cycle, the nitrogen cycle, the phosphorous cycle, and so on.

Oblanceolate ▸ Much longer than broad with the apical portion wider than the base.

Oblong ▸ Longer than broad with almost parallel sides.

Obovate ▸ The shape of a hen's egg in outline, with the narrow end toward the base (2-dimensional).

Obovoid ▸ Obovate in outline (3-dimensional), larger above.

Omnivore ▸ An animal that feeds more or less equally on both plant material and other animals.

Opposite ▸ Two leaves borne on either side of a branch at a single node.

Orbicular ▸ Circular, (2-dimensional).

Organic ▸ Substance derived from living matter. Classical definition: All living things and products that are uniquely produced by living things such as wood, leather, and sugar. Chemical definition: All chemical compounds, natural or synthetic, that contain carbon atoms as an integral part of their structure.

Organic compounds or molecules ▸ Chemical compounds, the structure of which is based on bonded carbon atoms with hydrogen atoms attached. Natural organic compounds are made by living organisms and comprise their bodies, the most common examples being cellulose making up wood, proteins, fats, starches and sugars.

Ovary ▸ Part of flower that matures to fruit and bears seeds.

Ovate ▸ The shape of a hen's egg in outline, with the large end toward the base (2-dimensional).

Ovoid ▸ Ovate in outline (3-dimensional), larger below.

Panicle ▸ A compound, branched inflorescence.

Pantropical ▸ Occurring throughout the tropics.

Parasites ▸ Organisms (plant, animal, or microbial) that attach themselves to another organism, the host, and feed on it over a period of time without killing it immediately but usually doing harm to it. Commonly divided into ectoparasites, those that attach to the outside of their host organism, and endoparasites, those that live inside their host.

Parent material ▸ The mineral matter from which soil is formed.

Peat ▸ An accumulation of partially decomposed organic material sometimes of considerable thickness.

Peduncle ▸ The stalk of an inflorescence.

Peltate ▸ Aa leaf or other structure with the stem attached to the lower surface rather than at the margin as leaves of a water lily or an umbrella.

Perennial ▸ A plant that lives longer than one year.

Pesticide ▸ A chemical used to kill pests. Pesticides are further categorized according to the pests they are designed to kill—for example, herbicides to kill plants, insecticides to kill insects, fungicides to kill fungi, and so on.

Petiole ▸ The stalk of a leaf.

Photosynthesis ▸ The chemical process carried on by green plants through which light energy is used to produce glucose from carbon dioxide and water, and oxygen is released as a byproduct.

Pinnate leaf ▸ A compound leaf bearing leaflets along a single axis.

Plant community ▸ The diversity of plant species, including numbers, ages, distribution, etc., that occupies a given area.

Pneumatophores ▸ "Breathing roots" (see discussion of mangroves in Chapter 2).

Pollutant ▸ A substance the presence of which is causing pollution.

Pollution ▸ Human additions of undesirable substances to the environment. Contamination of air, water, or soil with undesirable amounts of material or heat. The material may be a natural substance, such as phosphate, in excessive quantities, or it may be very small quantities of a synthetic compound such as dioxin which is exceedingly toxic.

Population ▸ A group group of individuals of the same species living and interbreeding within a specific geographic area at a given time.

Potential resource ▸ A portion of the natural or human environment that is not today considered of value but that one day may gain value as a result of technological, cognitive (perception by society), or economic developments.

Predator-prey relationship ▸ A feeding relationship existing between two kinds of animals. The predator is the animal feeding upon the prey. Such relationships are frequently instrumental in controlling population size of herbivores.

Preservation ▸ The non-use of resources. Limited resource development for the purpose of saving resources for the future.

Primary consumer ▸ An organism such as a rabbit or deer that feeds more or less exclusively on green plants or their products such as seeds and nuts. Synonym ▸ **herbivore**.

Principle of ecosystem stability ▸ Ecosystem diversity provides ecosystem stability. Wide population swings, "booms and busts," are common in simple ecosystems, rare in diverse ecosystems.

Principle of limiting-factors ▸ The principle that whatever factor (nutrient, water, sunlight, etc.) is in short supply limits the growth and development of an organism or a community.

Principle of population change ▸ The population of a species is the result of a dynamic balance between its biotic potential and the environmental resistance it faces.

Producer ▸ A green plant that produces organic matter through photosynthesis.

Prop roots ▸ Stilt-like roots of some mangrove species and pandanus trees.

Prostrate ▸ Lying on the ground.

Protein ▸ The class of organic macromolecules that is the major structural component of all animal tissues and that function as enzymes in both plants and animals.

Pubescence ▸ Hairiness.

Raceme ▸ Unbranched inflorescence with flowers produced on short stalks.

Recycling ▸ The practice of processing wastes and using them as raw material for new products as, for example, scrap iron is remelted and made into new iron products.

Relative humidity ▸ The measure in percent of how much moisture is in the air compared to how much the air can hold at the given temperature.

Renewable resource ▸ A resource that can be depleted but can be replenished by natural processes i.e., forests and fisheries. Biological resources such as trees that may be renewed by reproduction and regrowth. However, conservation to prevent overcutting and protection of the environment is still required.

Rhizome ▸ Horizontal underground stem bearing shoots, leaves and roots.

Rosette ▸ A radiating cluster of leaves as in a dandelion.

Ruderal ▸ Growing in disturbed, open places.

Run-off ▸ That portion of precipitation which runs off the surface as opposed to soaking in.

Saltwater intrusion ▸ The phenomenon of seawater moving back into aquifers or estuaries. It occurs when the normal outflow of fresh water is diverted or removed for use.

Savannah ▸ A grassland with scattered trees; tropical or subtropical semiarid grassland with scattered trees.

Scales ▸ Wide, flat hairs; or, in grasses and sedges the flat, often straw-colored structures of the inflorescence.

Scientific fact ▸ An observation regarding some object or phenomenon that has been repeated and confirmed by the scientific community. An objectively verified observation.

Scientific method ▸ The methodology by which scientific information is generated. Involves observations, formulating specific questions and hypotheses regarding the question's answer, then testing the hypothesis through experimentation.

Secondary consumer ▸ An organism such as a fox or coyote that feeds more or less exclusively on other animals that feed on plants.

Sedge ▸ A member of the Sedge family, Cyperaceae.

Sepals ▸ The individual segments of the calyx, the often green structures produced outside the petals.

Serrate ▸ Having saw-toothed leaf edges or margins with pointed teeth.

Sessile ▸ Attached directly by the base, as a leaf without a petiole, or a flower without a pedicel.

Sheath ▸ In grasses and sedges, part of the leaf surrounding the stem.

Soil erosion ▸ The loss of soil due to particles being carried away by wind and/or water.

Soil fertility ▸ Ability of soil to supply essential nutrients to plants.

Soil structure ▸ The phenomenon of solid particles (sand, silt, and clay) being loosely stuck together to form larger clumps and aggregates, generally with considerable air spaces in between. Structure enhances infiltration and aeration. It develops as a result of organisms feeding on organic matter in and on the soil.

Soil texture ▸ The relative size of the mineral particles that make up the soil. Generally defined in terms of the sand, silt, and clay content.

Solar energy ▸ Energy derived from the sun. Includes direct solar energy (the use of sunlight directly for heating and/or production of electricity) and indirect solar energy (the use of wind, which results from the solar heating of the atmosphere, and biological materials such as wood, which result from photosynthesis).

Sori ▸ The cluster of reproductive material in ferns. The dusty round or linear structures on the underside of a fern frond.

Spadix ▸ A spike on a succulent axis; in the Aroid family, Araceae, the inflorescence consists of the spadix surrounded by a sheathing bract, the spathe.

Spathe ▸ The bract borne below and often enclosing the spadix of plant species in Araceae family; or a similar structure in other families.

Speciation ▸ The evolutionary process whereby populations of a single species separate and, through being exposed to differing forces of natural selection, gradually develop into distinct species.

Species ▸ A naturally occurring population of individuals which are reproductively isolated from similar species; in a plant name it is the second word, but a plant species is always referred to by its binomial combination—the genus and species together such as *Colocasia esculenta*; All the organisms (plant, animal, or microbe) of a single kind. The "single kind" is determined by similarity of genetics and appearance and/or by the fact that members do or potentially can mate together and produce fertile offspring. Physical, chemical, or behavioral differences block breeding between species.

Spikelet ▸ In grasses and sedges a spikelet is the basic unit of an inflorescence consisting of a group of overlapping scales; each scale represents a flower in most sedges, while in grasses a flower usually is contained within two scales, with usually two scales borne at the base of a spikelet not representing flowers.

Spore ▸ The reproductive structures of ferns.

Stamen ▸ The male part of the flower which includes the pollen bearing anther and the anther stalk, or filament.

Stigma ▸ The sticky receptive part of the flower's pistil (female organ) that can capture pollen.

Stipe ▸ The stem of a fern frond.

Stipule ▸ A scaly, leaf-like appendage at the base of the petiole (stalk) of the leaf of some plants.

Submerged aquatic vegetation ▸ Aquatic plants rooted in bottom sediments growing under water. Depend on light penetrating through the water for photosynthesis, such as seagrasses.

Succession ▸ The order in which one vegetation type or ecological community replaces another following some change or disturbance. The gradual, or sometimes rapid, change in species that occupy a given area, with some species invading and becoming more numerous while others decline in population and disappear. Succession is caused by a change in one or more abiotic or biotic factors benefiting some species but at the expense of others. Primary succession: The gradual establishment, through a series of stages, of a climax or mature ecosystem in an area that has not been occupied before, e.g., a rock face. Secondary succession: The reestablishment, through a series of stages, of a climax or mature ecosystem in an area from which it was previously cleared.

Succulent ▸ Fleshy, thick, juicy.

Surface water ▸ Water and ice found in rivers, lakes, swamps, and other above-ground water bodies.

Survival of the fittest ▸ The concept that individuals best adapted to cope with both biotic and abiotic factors in their environment are the "fittest" and most likely to survive and reproduce.

Sustainable agriculture ▸ Agriculture that maintains the integrity of soil and water resources such that it can be continued indefinitely. Much of modern agriculture is depleting these resources and is, hence, not sustainable.

Sustainable development ▸ Development that provides people with a better life without sacrificing/ depleting resources or causing environmental impacts that will undercut future generations.

Sustained yield ▸ Management of renewable resources in such a way as to allow a constant rate of harvest indefinitely.

Swamp ▸ A wetland vegetation type dominated by woody species.

Symbiosis ▸ The intimate living together or association of two kinds of organisms, especially in a way that provides a mutual benefit to both organisms.

Synthetic ▸ Human-made as opposed to being derived from a natural source. For example, synthetic organic compounds are those produced in chemical laboratories, whereas natural organic molecules are those produced by organisms.

System ▸ An entity consisting of a set of parts that work together, forming a whole. The human body, transportation networks, and the earth are all systems.

Taxon, pl. **taxa** ▸ A category of classification such as family, genera, species, variety and form of plants and animals.

Taxonomy ▸ The science of identification and classification of organisms according to evolutionary relationships.

Technology ▸ The application of scientific information to solve practical problems or achieve desired goals.

Terracing ▸ A soil and water conservation technique in which ridges are built on the contour, or level areas constructed on a slope. The practice of grading sloping farmland into a series of steps and cultivating only the level portions in order to reduce erosion.

Theory ▸ A conceptual formulation which provides a rational explanation or framework for numerous related observations.

Threatened species ▸ A species that is not endangered but has a rapidly declining population. A species the population of which is declining quickly because of direct or indirect human impacts.

Topsoil ▸ The surface layer of solid, which is rich in humus and other organic material, both living and dead. As a result of the activity of organisms living in the topsoil, it generally has a loose, crumbly structure as opposed to being a compact mass. In many cases, because of erosion, development, or mining activity, the topsoil layer may be absent.

Trophic level ▸ One of the steps in a food chain where all organisms are the same number of steps away from primary producers. Feeding level with respect to the primary source of energy. Green plants are at the first trophic level, primary consumers at the second, secondary consumers at the third, and so on.

Tuber ▸ A swollen underground stem, as a potato (*Solanum tuberosum*) family Solanaceae.

Tufted ▸ With the stems arising from one place, rather than arising from a creeping rhizome.

Understory ▸ The underlying vegetation in a forest.

Venation ▸ The arrangement of veins.

Viviparous ▸ Bearing young plants on the parent plant.

Water cycle ▸ The movement of water from points of evaporation through the atmosphere, through precipitation, and through or over the ground, returning to points of evaporation.

Water vapor ▸ Water molecules in the gaseous state.

Weathering ▸ The gradual breakdown of rock into smaller and smaller particles, caused by natural, chemical, physical, and biological factors.

Wetlands ▸ Areas that are constantly wet and are flooded at more or less regular intervals. Especially marshy areas along coasts that are regularly flooded by tide.

Whorled ▸ With three or more leaves at a node.

BIBLIOGRAPHY

Abraham, D., Cash, H.L., Durand, A.M., Denholm, J., Moadsiri, A., Gopalani, S.V. and E. J. Abraham. 2012. From Yam to Spam: The Evolution of Pacific Islander Food Culture. In: (eds.) *First Marianas History Conference: One Archipelago, Many Stories.* Guam Protection Trust, June 12-16 in Saipan, pp. 151–164.

Abraham, D., Cash, H.L., Durand, A.M., Denholm, J., Moadsiri, A., Gopalani, S.V. and E. Johnson. 2018. High Prevalence of Non-Communicable Disease Risk Factors among Adolescents in Pohnpei, Micronesia. *Hawai'i Journal of Medicine and Public Health* 77: 283–286.

Ahuja, S.C., Ahuja, S. and U. Ahuja. 2014. Coconut – History, Uses, and Folklore. *Asian Agri-History* 18(3): 221–248.

Allison, S., Nielson, C. and R. Hughes, R. 2006. Elevated enzyme activities under the invasive nitrogen fixing tree *Falcataria moluccana. Soil Biology and Biochemistry* 1537–1544.

Anderson, G.D.S., Heinrich, H. and K.D. Harrison. 2013. *Kapingamarangi Talking Dictionary.* Living Tongues Institute for Endangered Languages. www.talkingdictionary.org/kapingamarangi

Anson, H. and W. Raynor. 1993. Traditional resource management and conservation of biological diversity on Pohnpei Island, Federated States of Micronesia. In: L.S. Hamilton (ed.) *Ethics, Religion and Biodiversity.* Cambridge: White Horse Press, pp. 133–146.

Aporosa, S.A. 2017. Kava (*Piper methysticum*): Demythifying the Pacific's cultural keystone species. *Tālanga Seminar Series.* University of the South Pacific, Tonga, 12 October (see: researchcommons.waikato.ac.nz/handle/10289/12255).

Aporosa, S. A. 2019a. De-mythologizing and re-branding of kava as the new 'world drug' of choice. *Drug Science, Policy and Law* Volume 5, January–December. https://doi.org/10.1177/2050324519876131

Aporosa, S.A. 2019b. Kava and ethno-cultural identity in Oceania. In: Ratuva S (ed.) The Palgrave Handbook of Ethnicity. Chapter 134-1. London: Springer-Nature, pp. 1–15.

Aporosa, S.A., Atkins, M. and R. Brunton. 2020. Kava drinking in traditional settings: Towards understanding effects on cognitive function. *Human Pharmacology: Clinical and Experimental* 35(2) March e2725; First published: 16 February. doi.org/10.1002/hup.2725

Applegate, W.L. and V. Lebot. 2006. Validation of *Piper methysticum* var. *wichmannii* (Piperaceae). *Novon: A Journal for Botanical Nomenclature* 16(1): 3–4.

Arnau, G., Bhattacharjee, R., Sheela, M., Chair, H., Malapa, R., Lebot, V., et al. 2017. Understanding the genetic diversity and population structure of yam (*Dioscorea alata* L.) using microsatellite markers. *PLoS ONE* 12(3): e0174150. https://doi.org/10.1371/journal.pone.0174150

Ashby, G. 1993. *A Guide to Pohnpei: An Island Argosy.* Eugene, Oregon and Kolonia, Pohnpei: Rainy Day Press. Revised Edition.

Assi, R.A., Darwis, Y., Abdulbaqi, I.M., Khan, A.A., Vuanghao, L. and M.H. Laghari. 2017. *Morinda citrifolia* (Noni): A comprehensive review on its industrial uses, pharmacological activities, and clinical trials. *Arabian Journal of Chemistry* 10: 691–707.

Athens, J.S. 1997. Hawaiian Native Lowland Vegetation in Prehistory. In: P.V. Kirch and T.L. Hunt (eds.) *Historical Ecology in the Pacific Islands: Prehistoric Environmental and Landscape Change.* New Haven, Connecticut: Yale University Press, pp. 248–270.

Athens, S. J. and J. Stevenson. 2012. Pohnpei coring records: The natural distribution of *Cyrtoperma chamissonis. Journal of Pacific Archaeology* 3(1): 35–48.

Athens, J.S., Ward, J. and G.M. Murakami. 1996. Development of an agroforest on a Micronesian high island: prehistoric Kosraean agriculture. *Antiquity* 70: 834–846.

Austin, D. 1991. *Ipomoea littoralis* (Convolvulaceae): Taxonomy, Distrbution, and Ethnobotany. *Economic Botany* 45(2): 251–256.

Ayres, W.S. 1990. Pohnpei's position in eastern Micronesian Prehistory. *Micronesica Supplement* 2: 187–212.

Ayres, W.S. and A.E. Haun. 1985. Archaeological Perspectives on Food Production in Eastern Micronesia. In: I.S. Farrington (ed.) *Prehistoric Intensive Agriculture in the Tropics*, pp. 455–473. Oxford: B.A.R.

Ayres, W.S. and A.E. Haun. 1990. Prehistoric Food Production in Micronesia. In *Pacific Production Systems: Approaches to Economic Prehistory. Papers from a Symposium at the XVth Pacific Science Congress, New Zealand* (D. Yen and M.J.M. Mummery, eds.): 211–227. Canberra: Research School of Pacific Studies, Australian National University.

Ayres, W.S., Haun, A.E. and C. Severance. 1981. Subsistence and Settlement on Ponape, Micronesia. Micronesian Archaeological Survey Reports No. 4 (Ponape Archaeological Survey Report 78-2). Historic Preservation Office, Saipan.

Bachman, S. and H. Chadburn. 2015. Caroline Ivory Nut: *Metroxylon amicarum. The IUCN Red List of Threatened Species 2015*: e.T38610A44533123. http://dx.doi.org/10.2305/IUCN.UK.2015-2.RLTS.T38610A44533123.en.

Balick, M.J. et al. 2009 (ed.). *Ethnobotany of Pohnpei: Plants, people, and island culture*. Honolulu: University of Hawai'i Press and New York Botanical Garden [Note: this large volume has many local and other contributors].

Balick, M.J. and R.S. Lee. 2002. Traditional use of sakau (kava) in Pohnpei: lessons for integrative medicine. *Alternative Therapies in Health Medicine* 8(4): 96–98.

Balick, M.J. and R.S. Lee. 2009. The Sacred Root: sakau en pohpei. In: Balick, M.J., collaborators (eds.). *Ethnobotany of Pohnpei: Plants, people, and island culture*. Honolulu: University of Hawai'i Press and New York Botanical Garden.

Bally, I.S.E. 2006. *Mangifera Indica* (Mango). ver. 3.1. In: Elevitch, C.R. (ed.). 2006a. *Traditional Trees of Pacific Islands: their Culture, environment, and Use*. Holualoa, Hawai'i: Permanent Agriculture Resources <http://www.traditionaltree.org>.

Baltrushes, N. 2006. *Medical Ethnobotany, Phytochemistry, and Bioactivity of the Ferns of Mo'orea, French Polynesia*. Senior Honors Thesis, Department of Integrative Biology, University of California, Berkeley.

Barker, G., Lloyd-Smith, L., Barton, H., Cole, F., Hunt, C., Piper, P., Rabett, R., Paz V. and K. Szabó. 2011. Foraging-farming transitions at the Niah Caves, Sarawak, Borneo. Antiquity 85: 1–18.

Barrau, J. 1960. Plant Introduction in the Tropical Pacific: Its Role in Economic Development. *Pacific Viewpoint* 1(1): 1–9.

Barrau, J. 1961. Subsistence Agriculture in Polynesia and Micronesia. *B.P. Bishop Museum Bulletin* 223. Honolulu: Bishop Museum Press.

Bascom, W.R. 1946. *A Pacific economy in transition*. U.S. Commercial Company Economic Survey of Micronesia, Honoulu.

Bascom, W.R. 1948. Ponape prestige economy. *Southwestern Journal of Anthropology* 4: 211–221.

Bascom, W.R. 1949. Subsistence farming on Ponape. *New Zealand Geographer* 5: 115–129.

Bascom, W.R. 1965. Ponape: A Pacific Economy in Transition. *Anthropological Records* Vol. 22. Berkeley and Los Angeles: University of California Press (originally published as a government report in 1947).

Bascom, W.R. 1970. Ponapean Prestige Economy. In: T.G. Harding and B.J. Wallace (eds.), *Cultures of the Pacific Selected Readings*. New York: Free Press, pp. 85–93.

Bauman, H. 2016. Food as Medicine: Mango (*Mangifera indica*, Anacardiaceae). *HerbalGram* 13(6) June. https://www.herbalgram.org/

Bayliss-Smith, T. 2012. Taro, Turmeric, and Gender. In: Feinberg, R. and R. Scaglion (eds.). *Polynesian Outliers: the State of the Art*. Ethnology Monographs No. 21, Department of Anthropology, University of Pittsburgh, Chapter 7, pp. 109–138.

Bayliss-Smith, T. and E. Hviding. 2015. Landesque capital as an alternative to food storage in Melanesia: irrigated taro terraces in New Georgia, Solomon Islands. *Environmental Archaeology* 20: 425–436.

Benson, W. 2012. *The kingdom of plants: a journey through their evolution*. London: Harper Collins Publishers.

Benstead, J.P., March, J.G., Pringle, C.M., Ewel, K.C. and J.W. 2009. Biodiversity and ecosystem function in species-poor communities: community structure and leaf litter breakdown in a Pacific island stream. *Journal of the North American Benthological Society* 28(2): 454–465.

Bentzen, C. 1949. *Land and Livelihood on Mokil: An Atoll in the Eastern Carolines*. Issue 25 of the Coordinated Investigation of Micronesian Anthropology (CIMA) Final Report, part 2. Number 11. National Research Council, USA, Pacific Science Board, Washington, D.C.

Bernart, L. 1977. The Book of Luelen. Translated and edited by J. Fischer, S. Riesenberg and M. Whiting. *Pacific History Series* 8: 1–193. Honolulu: University Press of Hawai'i.

Bissa, S., Songara, D. and A. Bohra. 2007. Traditions in oral hygiene: Chewing of betel (*Piper betle* L.) leaves. *Current Science* 92(1): 26–28.

Bodner, C.C. and R.E. Gereau. 1988. A Contribution to Bontoc Ethnobotany. *Economic Botany* 42(3): 305–369.

BirdLife International. 2021. Species factsheet: *Aplonis pelzelni*. Downloaded from http://www.birdlife.org on 28/02/2021.

Boer, B. (ed.). 1996. *Environmental Law in the South Pacific*. Gland, Switzerland: International Union for the Conservation of Nature and Natural Resources.

Bose, L.V., Varghese, G.K. and S. Habtemariam. 2013. Identification of acteoside as the active antioxidant principle of *Premna serratifolia* root wood tissues. *Phytopharmacology* 4: 228–236.

Bosland, P.W. and E.J. Votava. 2000. *Peppers: Vegetable and spice capsicums*. CABI Publishing, Oxon, UK and New York. 2nd Edition.

Boucher, B.J. and N. Mannan. 2002. Metabolic effects of the consumption of Areca catechu. *Addiction Biology* 7: 103–110.

Bourke, R.M. 2013. History of agriculture in Papua New Guinea. In: Bourke, R.M. and T. Harwood (eds.). *Food and Agriculture in Papua New Guinea*. Canberra: Australia National University Press, pp. 10–26.

Bradbury, J.H. and W.D. Holloway. 1988. *Chemistry of Tropical Root Crops: Significance for Nutrition and Agriculture in the Pacific* Monographs, Australian Centre for International Agricultural Research, Canberra.

Brosi, B.J., M.J. Balick, R. Wolkow, R. Lee, M. Kostka, W. Raynor, R. Gallen, A. Raynor, P. Raynor and D. L. Ling. 2007. Cultural erosion and biodiversity: canoe-making knowledge in Pohnpei, Micronesia. *Conservation Biology* 21: 875–879.

Brown, F.B.H. 1935. *Flora of Southeastern Polynesia.* Bernice P. Bishop Museum Bulletin 130: 1–386.

Bruce, L. 1976. Preliminary Study of Three Polynesian Sources for Celestial Navigation. In: K. Peacock and C.J. Daeufer (eds.). *Micronesian and Polynesian Voyaging, Three Readings*, Pacific Islands Program, University of Hawai'i, Honolulu, Hawai'i, Reading One: pp. 1–25.

Buck, Sir Peter. 1950 (see under Hiroa, Te Rangi).

Buden, D.W. 1998. The Reptiles of Kapingamarangi Atoll, Micronesia. *Atoll Research Bulletin* 453: 1–8.

Buden, D.W. 2000. A comparison of 1983 and 1994 Bird Surveys of Pohnpei Island, Federated States of Micronesia. *Wilson Bulletin* 112(3): 403–410.

Buden, D.W. and D. Taboroši. 2016. *Reptiles of the Federated States of Micronesia*. Pohnpei: Island Research & Education Initiative.

Buente, W., Dalisay, F., Pokhrel, P., Kramer, H.K. and I. Pagano. 2020. An Instagram-Based Study to Understand Betel Nut Use Culture in Micronesia: Exploratory Content Analysis. *Journal of Medical Internet Research* 22(7): www.jmir.org/2020/7/e13954.

Burkill, I.H. 1951. Dioscoreaceae. In: C.G.G.J. van Steenis (ed.). *Flora Malesiana*, ser. I, 4 Djakarta: Noordhoff-Kolff, pp. 293–335.

Burkill, I.H. 1966. *A Dictionary of the Economic Products of the Malay Peninsula* (2 Vols.). Kuala Lumpur: Ministry of Agriculture of Malaysia (2nd edition; original edition 1935).

Carroll, V. and T. Soulik. 1973. *Nukuoro Lexicon.* Honolulu: University Press of Hawai'i.

Catala, R. L.A. 1957. Report on the Gilbert Islands: Some aspects of human ecology. *Atoll Research Bulletin* No. 59. Washington, D.C.: Smithsonian Institution.

Chamisso, A. von. 1836. *Reise um die Welt mit der Romanzoffischen Entdeckungs-Expedition in den fahren 1815–1818 auf der Brigg 'Rurik', Capt. Otto von Kotzebue.* Tagebuch, Bemerkungen, und Ansichten (in 2 vols).

Chimner, R.A. 2004. Soil respiration rates of tropical peatlands in Micronesia and Hawai'i. *Wetlands* 24: 51–56.

Chimner, R.A. and K.C. Ewel. 2004. Differences in carbon fluxes between forested and cultivated Micronesian tropical peatlands. *Wetlands Ecology Management* 12: 419–427.

Chowdhury, M. R., Chu, P. S., Zhao, X., Schroeder, T. A. and J. J. Marra. 2010. Sea level extremes in the US-Affiliated Pacific Islands – a coastal hazard scenario to aid in decision analyses. *Journal of Coastal Conservation* 14: 53–62.

Christian, F. 1899. *The Caroline Islands, travel in the sea of the little islands*. London: Methuen and Company; London: Cass, 1967. Reprint.

Christman, S. 2018. *"Citrus meyerii." Floridata.* https://floridata.com/plant/1067.

Clarke, W.C. 2008. Agroforestry in Pacific Islands. In: Selin, H. (ed.) *Encyclopaedia of the History of Science, Technology, and Medicine in Non-Western Cultures*, Springer Science & Business Media 2nd edition, pp. 87–89.

Clarke, W.C. and R.R. Thaman (editors). 1993. *Agroforestry in the Pacific Islands: Systems for Sustainability.* Tokyo: United Nations University Press.

Comer, D.C., Comer, J.A., Dumitru, I.A., Ayres, W.S., Levin, M.J., Seikel, K.A., White, D.A. and M.J. Harrower. 2019. Airborne LiDAR Reveals a Vast Archaeological Landscape at the Nan Madol World Heritage Site. *Remote Sensing* 11(18), 2152. https://doi.org/10.3390/rs11182152.

Conservation Society of Pohnpei. 2006. *A Rapid Ecological Assessment of the Coral, Fish and Seagrasses of Pohnpei, And and Pakin, Federated States of Micronesia – Findings and Recommendations*. Pohnpei, FSM.

Conservation Society of Pohnpei. 2013. *Conservation society of Pohnpei & Pohnpei iStop report Mirconesia Regional Invasive Species Council (RISC).*

Corner, E.J.H. 1997. *Wayside Trees of Malaya: in two volumes.* Fourth edition, Malayan Nature Society, Kuala Lumpur.

Costion, C.M. and D.H. Lorence. 2012. The Endemic Plants of Micronesia: A Geographical Checklist and Commentary. *Micronesica* 43: 51–100.

Coulter, J.W. 1941. Land Utilization in American Samoa. *Bernice Bishop Museum, Bulletin* 170. Honolulu, Hawai'i.

Coursey, D.G. 1967. *Yams: an account of the nature, origins, cultivation and utilisation of the useful members of the* Dioscoraceae. Longmans and Green, London, UK.

Coursey, D.G. 1969. The future of the yam in world agriculture: an appraisal of some problems. *Tropical Root Tuber Crops News* 12: 4–8.

Coursey, D.G. 1972. The civilizations of the yam: interrelationships of man and yams in Africa and the Indo-Pacific region. *Archeology and Physical Anthropology in Oceania* 7: 215–33.

Coursey, D.G. and R.W. Martin. 1970. The past and future of yams as crop plants. In: *Proceedings of 2nd Sympsoium on Tropical Root Crops*. Honolulu, Hawai'i, Volume I, pp. 87–90; 99–101.

Cox, P.A. 1982. Vertebrate Pollination and the Maintenance of Dioecism in *Freycinetia. The American Naturalist* 120(1): 65–80.

Cox, P.A. 1990. Pollination and the evolution of breeding systems in Pandanaceae. *Annals of the Missouri Botanical Garden* 77: 816–840.

Crowley, T. 1994. Who Drank Kava? In: Andrew Pawley and Malcolm Ross (eds.), *Austronesian Terminologies: Continuity and Change. Pacific Linguistics*, C127. Canberra: Australian National University, pp. 87–100.

Dahl, C. 1993. Tourism development on the island of Pohnpei (Federated States of Micronesia): Sacredness, control and autonomy. *Ocean & Coastal Management* 20(3): 241–265.

Damas, D. 1985. Pingelap Politics and American-Micronesian Relations. *Ethnology* 24(1): 43–55.

Damas, D. 1994. *Bountiful Island: A Study of Land Tenure on a Micronesian Atoll*. Waterloo, Ont.: Wilfrid Laurier University Press. Online 2006: Project MUSE muse.jhu.edu/book/40268.

Dampier, W. 1697. *A New Voyage round the World*. London. Reprinted by Adam & Charles Black, London, 1937.

Daniells, J.W. 2004. Consultancy report on training in the identification and characterization of banana varieties and development of the banana industry in Pohnpei, Federated States of Micronesia. www.musalit.org/seeMore.php?id=10797.

Daniells, J., Englberger, L. and A. Lorens. 2004. *Pohnpei banana varieties* – a work in progress. www.musalit.org/seeMore.php?id=15460.

DaSilva, E.J., Murukesan, E. Nandwani, V.K., Taylor, M. and P.C. Josekutty. 2004. The Pacific Islands: a biotechnology resource bank of medicinal plants and traditional intellectual property. *World Journal of Microbiology & Biotechnology* 20: 903–934.

De Langhe, E. 1995. *Banana and Plantain: the Earliest Fruit Crops?* INIBAP Annual Report. Montpellier, France: International Network for the Improvement of Banana and Plantain.

De Langhe, E., Vrydaghs, L., de Maret, X., Perrier, P. and T. Denham. 2009. Why bananas matter: An introduction to the history of banana domestication. *Ethnobotany Research and Applications* 7: 165–177.

Declerck, F.A.J., Fanzo, J., Palm, C. and R. Remans. 2011. Ecological Approaches to Human Nutrition. *Food and Nutrition Bulletin* 32(1):S41–50.

Del Rosario, A.G. and N.M. Esguerra. 2003. *Medicinal plants in Palau, Volume 1*. PCC–CRE Publication 28/03. Palau Community College, Koror, 51 pp.

Dela Cruz et al. 2022. Exploring Foods of the Pacific: Cultural Food Identity in the US Affiliated Pacific Region. *Hawai'i Journal of Health and Social Welfare* 81(9): 247–252.

Des Rochers, K. 1992. Women's Fishing on Kosrae: A Description of Past and Present Methods. *Micronesica* 25(1): 1–22.

Devall, M.S. 1992. The biological flora of coastal dunes and wetlands. 2. *Ipomoea pes-caprae* (L.) Roth. *Journal of Coastal Research* 8(2): 442–456.

Dignan, C.A., Burlingame, B.A., Arthur, J.M., Quigley, R.J. and G.C. Milligan. 1994 (1st ed.) / 2004 (2nd ed.). *The Pacific Islands Food Composition Tables*. Noumea, New Caledonia: South Pacific Commission; Rome: FAO.

Dongre, R.S. 2014. Bio-chemical compositional analysis of *Annona muricata*: a miracle fruit's review. *International Journal of Universal Pharmacy and Bio Sciences* 3(2): 82–104.

Donnegan, J.A., Butler, S.L., Kuegler, O. and B.A. Hiserote. 2011. *Federated States of Micronesia's Forest Resources, 2006*. Resource Bulletin PNW-RB-262. U.S. Department of Agriculture, Forest Service, Pacific Northwest Research Station, Portland, OR.

Donohue, M. and T. Denham. 2009. Banana (*Musa* spp.) Domestication in the Asia-Pacific Region: Linguistic and archaeobotanical perspectives. *Ethnobotany Research & Applications* 7: 293–33.

Drummond, E., Anderson, G., Rudolph, J., Henry, M., Erwin, K. and K.D. Harrison. 2016. *Nukuoro Talking Dictionary*. Living Tongues Institute for Endangered Languages. http://www.talkingdictionary.org/nukuoro

Duke, N.C. 2006. *Rhizophora apiculata, R. mucronata, R. stylosa, R. × annamalai, R. × lamarckii* (Indo-West Pacific stilt mangroves), ver. 2.1. In: Elevitch, C.R. (ed.). *Species Profiles for Pacific Island Agroforestry*. Permanent Agriculture Resources (PAR), Hōlualoa, Hawai'i, pp. 641–660. http://www.traditionaltree.org.

Duke, N.C. 2010. Overlap of eastern and western mangroves in the SW Pacific: hybridization of all three *Rhizophora* combinations in New Caledonia. *Blumea* 55 (2): 171–188.

Duke, N.C. 2011. Mangrove Islands. In: D. Hopley (ed.). *Encyclopedia of Modern Coral Reefs. Structure, Form and Process*. Dordrecht, The Netherlands, Springer, pp. 653–655.

Duke, N.C. 2016. Oil spill impacts on mangroves: recommendations for operational planning and action based on a global review. *Marine Pollution Bulletin* 109(2): 700–715.

Duke, N.C., Ball, M.C. and J.C. Ellison. 1998. Factors influencing biodiversity and distributional gradients in mangroves. *Global Ecology and Biogeography Letters* 7: 27–47.

Ehrlich, C. 1989. Special Problems in an Ethnobotanical Literature Search: *Cordyline terminalis* (L.) Kunth, the "Hawaiian Ti Plant" *Journal of Ethnobiology* 9(1): 51–63.

Ehrlich, C. 1999. The ethnobotany of *Cordyline fruticosa* (L.) A. Chev.: the "Hawaiian ti plant". Ph. D. Dissertation, University of Buffalo.

Ehrlich, C. 2000a. The Efficacy of Ti. *Newsletter of the Native Hawaiian Center of Excellence John A. Burns School of Medicine*, University of Hawai'i at Mānoa, vol. 3, April 2000.

Ehrlich, C. 2000b. Inedible to Edible: Firewalking and the Ti Plant [*Cordyline fruticosa* (L.) A. Chev.]. *Journal of the Polynesian Society* 109(4): 371–400.

Eilers, A. 1934. *Inseln um Ponape, Ergebnisse der Sudsee-Expedition*: 1908–1910. II (B) 8: 1–464. Friederichsen, de Gruyter & Co., Hamburg.

Elbert, S.H. 1948. *Grammar and Comparative Study of the Language of Kapingamarangi*. Report to Pacific Science Board, National Research Council, CIMA (Co-ordinated Investigations in Micronesian Anthropology), Washington, D.C. (Mimeograph copy).

Eldredge, L.G. 1987. Poisons for fishing on coral reefs. In: B. Salvat (ed.) *Human Impacts on Coral Reefs: Facts and Recommendations*. Antenne Museum E.P.H.E., French Polynesia, pp. 61–66.

Elevitch, C.R. (ed.). 2006. *Traditional Trees of Pacific Islands: their Culture, Environment, and Use*. Holualoa, Hawai'i: Permanent Agriculture Resources. tradionaltree.org

Elevitch, C.R. (ed.). 2006b. *Species Profiles for Pacific Island Agroforestry: Ecological, Economic and Cultural Renewal*. Holualoa, Hawai'i: Permanent Agriculture Resources. https://agroforestry.org/free-publications/traditional-tree-profiles.

Elevitch, C.R. and L.A.J. Thomson. 2006. *Hibiscus tiliaceus* (beach hibiscus). In: Elevitch, C.R. (ed.). 2006. *Traditional Trees of Pacific Islands: their Culture, Environment, and Use*. Holualoa, Hawai'i: Permanent Agriculture Resources, pp. 391-404: tradionaltree.org

Ellison, J., Koedam, N.E., Wang, Y., Primavera, J., Jin Eong, O., Wan-Hong Yong, J. and V. Ngoc Nam. 2010a. *Lumnitzera littorea. The IUCN Red List of Threatened Species 2010*: e.T178854A7628170.

Ellison, J., Koedam, N.E., Wang, Y., Primavera, J., Jin Eong, O., Wan-Hong Yong, J. and V. Ngoc Nam. 2010c. *Nypa fruticans. The IUCN Red List of Threatened Species 2010*: e.T178800A7610085.

Emory, K.P. 1965. Kapingamarangi, social and religious life of a Polynesian Atoll. *B.P. Bishop Museum Bulletin* 228: 1–357.

Endress, B.A. 2002. The Importance of Endemic Species to Forest Succession in Palau. *Micronesica* 34 (2): 141–153.

Engbring, J., Ramsey, F.L., and Wildman, V.J. 1990. *Micronesian Forest Bird Surveys, the Federated States: Pohnpei, Kosrae, Chuuk, and Yap*. Honolulu: U.S. Fish and Wildlife Service.

Englberger, K. 2009. *Invasive weeds of Pohnpei: A guide for identification and public awareness*. Kolonia, Federated States of Micronesia: Conservation Society of Pohnpei.

Englberger, L. 2003. Carotenoid-rich bananas in Micronesia. Info*Musa* 12(2): 2–5.

Englberger, L. Darton-Hill, I., Coyne, T., Fitzgerald, M.H. and G.C. Marks. 2003a. Carotenoid-rich bananas: A potential food source for alleviating vitamin A deficiency. *Food and Nutrition Bulletin* 24: 303–318.

Englberger, L., Schierle, J., Marks, G.C., and M.H. Fitzgerald. 2003b. Micronesia banana, taro, and other foods: Newly recognized sources of provitamin A and other carotenoids. *Journal of Food Consumption and Analysis* 16: 237–247.

Englberger, L., Marks, G.C. and M.H. Fitzgerald. 2003c. Insights on food and nutrition in the Federated States of Micronesia: a review of the literature. *Public Health Nutrition* 6(1): 5–17.

Englberger, L. and A. Lorens. 2004. Pohnpei Bananas: A photo collection: Carotenoid-rich varieties. Island Food Community of Pohnpei and Secretariat of the Pacific Community, Suva, Fiji.

Englberger, L., Aalbersberg, W., Dolodolotawake, U., Schierle, J., Humphries, J., Iuta, T., Marks, G.C., Fitzgerald, M.H., Rimon, B., and M. Kaiririete. 2005a. Carotenoid content of pandanus fruit cultivars and other foods of the Republic of Kiribati. *Public Health Nutrition* 9(5): 631–643.

Englberger, L., Aalbersberg, W., Schierle, J., Marks, G.C., Fitzgerald, M.H., Muller, F., Jekkein, A., Alfred, J., and N. Vander Velde. 2005b. Carotenoid content of different edible pandanus fruit cultivars of the Republic of the Marshall Islands. *Journal of Food Composition and Analysis* 19: 484–494.

Englberger, L., Fitzgerald, M.H. and G.C. Marks. 2005c. Cultivar recognition in Micronesia: banana, breadfruit, giant swamp taro and pandanus. *PGR Newsletter FAO Biodiversity* 142: 1–9.

Englberger, L., Schierle, J., Kraemer, K., Aalbersberg, W., Dolodolotawake, U., Humphries, J., Graham, R., Reid, A.P., Lorens, A., Albert, K., Levendusky, A., Johnson, E., Paul, Y., and F. Sengebau. 2008. Carotenoid and mineral content of Micronesian giant swamp taro (*Cyrtosperma*) cultivars. *Journal of Food Composition and Analysis* 21: 93–106.

Englberger, L., Albert, K., Lorens, A. and A. Levendusky. 2009a. Taro: An Important Pohnpei Staple Food. In: Balick, M.J. (ed.). *Ethnobotany of Pohnpei: Plants, people, and island culture*. Honolulu: University of Hawai'i Press and New York Botanical Garden, pp. 132–164.

Englberger, L., Lorens, A., Levendusky, A., Pedrus, P., Albert, K., Hagilmai, W., Paul, Y., Nelber, D., Moses, P., Shaeffer, S. and M. Gallen. 2009b. Documentation of the traditional food system of Pohnpei. In: *Indigenous Peoples' Food Systems: the many dimensions of culture, diversity and environment for nutrition and health.* Rome: Food and Agriculture Organization of the United Nations Centre for Indigenous Peoples' Nutrition and Environment, pp. 109–138.

Englberger, L., Schierle, J., Hofmann, P., Lorens, A., Albert, K. Levendusky, A., Paul, Y., Kickaneth, E. Elymore, A., Maddison, M., deBrum, I., Nemra, J., Alfred, J., Vander Velde N. and K. Kraemer. 2009c. Carotenoid and mineral content of Micronesian atoll foods: Pandanus: (*Pandanus tectorius*) and garlic pear (*Crataeva speciosa*) fruit. *Journal of Food Composition and Analysis* 22: 1–8.

Englberger, L., Lorens, A., Pretrick, M., Raynor, B., Currie, J., Corsi, L., Kaufer, L., Naik, R.A., Spegal, R. and H.V. Kuhnlein. 2011. Approaches and Lessons Learned for Promoting Dietary Improvements, Pohnpei, Micronesia. In: Thompson, B. and L. Amoroso (eds.) *Combating Micronutrient Deficiencies: Food-based Approaches.* CABI & FEO of the United Nations, pp. 224–253.

Englberger, L., Lorens, A., Pedrus, P., Albert, K., Levendusky, A., Hagilmai, W., Paul, Y., Moses, P. Jim, R., Jose, S., Nelber, D., Santos, G., Kaufer, L. Larsen, K., Pretrick, M.E., and H.V. Kuhnlein. 2013. Let's Go Local! Pohnpei promotes local food production and nutrition for health. In: Kuhnlein, H.V., Erasmus, B., Spigelski, D. and B. Burlingame (eds.). *Indigenous Peoples' food systems and well-being.* Rome: Food and Agriculture Organization of the United Nations, pp. 191–220.

Eshbaugh, W.H. 1993. Peppers: history and exploitation of a serendipitous new crop discovery. In: Janick, J. and J.E. Simon (eds.) *New Crops.* New York: John Wiley and Sons, pp. 132–139.

Eshbaugh, W.H. 2012. The Taxonomy of the Genus *Capsicum*. In: Russo, V.M. (ed.) *Peppers: Botany, Production and Uses.* Wallingford, UK: CABI, pp. 14–28.

Etoh, T. 2001. Edible plants and poisonous plants in Micronesia (1) Poisonous plants in Yap. *The Progress report of the 1999 Survey of the Research Project "Social Homeostasis of small Islands in an Island-zone."* Kagoshima University Research Center for the Pacific Islands. Occasional Papers. No. 34, 141-144. Part 2, Section 2, Report 1.

Evans, D. 1975. *Musical Instruments of the Eastern Caroline Islands.* Honolulu: East-West Center, Culture Learning Institute.

Falanruw, M. 1992. Seagrass Nets. *Atoll Research Bulletin* 364: 1–12.

Falanruw, M.C. 2002. *Terrestrial Biodiversity of the Federated States of Micronesia.* FSM National Biodiversity Strategy and Action Plan Project funded by the Global Environmental Facility, implemented by the United Nations Development Programme and nationally executed by the FSM Department of Economic Affairs.

Falanruw, M.V.C. 2015. *Trees of Yap: A Field Guide.* United States Department of Agriculture, Forest Service, Pacific Southwest Research Station. Institute of Pacific Island Forestry, Hilo, Hawai'i. General Technical Report, PSW-GTR-249.

Falanruw, M.C., Maka, J.E., Cole, T.G. and C.D. Whitesell. 1990. *Common and scientific names of trees and shrubs of Mariana, Caroline, and Marshall Islands.* Resource Bulletin, PSW-26. Berkeley, CA: Pacific Southwest Research Station, Forest Service, U.S. Department of Agriculture.

Fall, P.L. 2010. Pollen evidence for plant introductions in a Polynesian tropical island ecosystem, Kingdom of Tonga. In: S. Haberle, J. Stevenson, and M. Prebble (eds.) *Altered ecologies: fire, climate and human influence on terrestrial landscapes.* Terra Australis 32. Canberra: ANU E Press, pp. 253–271.

FAO. 2016. *State of the World's Forests 2016.* Forests and agriculture: land-use challenges and opportunities. Food and Agriculture Organization of the United Nations, Rome.

FAO and WHO. 2016. *Kava: a review of the safety of traditional and recreational beverage consumption.* Food and Agriculture Organization of the United Nations and the World Health Organization, Rome.

Federated States of Micronesia. 2014. *Fifth National Report to the Convention on Biological Diversity.* Palikir, Pohnpei.

Fischer, J.L. and A.M. Fischer. 1957. *The Eastern Carolines.* HRAF Press, New Haven, Conn.

Fischer, J.L., Riesenberg, S.H. and M.G.Whiting. 1977. *The Book of Luelen.* Australian National University Press, Canberra.

Fitzpatrick, S.M., Nelson, G.C. and R. Reeves. 2003. The Prehistoric Chewing of Betel Nut (*Areca catechu*) in Western Micronesia. People and Culture in Oceania 19: 55–65.

Fletcher, C. and B.M. Richmond. 2010. *Climate change in the Federated States of Micronesia: Food and water security, climate risk management, and adaptive strategies.* University of Hawai'i Sea Grant College Program, Honolulu.

Flood, B., Strong, B.E. and W. Flood. 2002. *Micronesian Legends.* Bess Press, Honolulu, Hawai'i.

Florentine, S.K. and M.E. Westbrooks. 2004. Restoration on abandoned tropical pasturelands – do we know enough? *Journal for Nature Conservation* 12: 85–94.

Fosberg, F.R. 1960. The vegetation of Micronesia. 1. General descriptions, the vegetation of the Marianas Islands, and a detailed consideration of the vegetation of Guam. *Bulletin of the American Museum of Natural History* 119(I): 1–75 and plates 1–40.

Fosberg, F.R. and M.-H. Sachet. 1975b. Flora of Micronesia, 2: Casuarinaceae, Piperaceae, and Myricaceae. *Smithsonian Contributions to Botany* 24: 1–28.

Fosberg, F.R. and M.-H. Sachet. 1977. Convolulaceae. In: *Flora of Micronesia, 3. Smithsonian Contributions to Botany* 36: 1–33.

Fosberg, F.R. and M.-H. Sachet. 1987b. Noteworthy Micronesian Plants 5. *Micronesica* 20: 131–156.

Fosberg, F.R., Sachet, M.H, and R. Oliver. 1979. A geographical checklist of the Micronesian Dicotyledonae. *Micronesica* 15(12): 41295.

Fosberg, F.R., Sachet, M.H., and R. Oliver. 1987. A Geographical Checklist of the Micronesian Monocotyledonae. Micronesica. 20: (1-2): 19–129.

Fosberg. F.R., Sachet, M.H. and R. Oliver. 1993. Flora of Micronesia, 5: Bignoniaceae-Rubiaceae. *Smithsonian Contributions to Botany* 81: 1–135.

Francis, J.K. (ed.). 2004. *Wildland shrubs of the United States and its Territories: thamnic descriptions: volume 1*. Gen. Tech. Rep. IITF-GTR-26. San Juan, PR: U.S. Department of Agriculture, Forest Service, International Institute of Tropical Forestry, and Fort Collins, CO: U.S. Department of Agriculture, Forest Service, Rocky Mountain Research Station.

Franklin, J. and M. Merlin. 1992. Species-environment patterns of forest vegetation on the uplifted reef limestone of Atiu, Mangaia, Ma'uke, and Miti'aro, Cook Islands. *Journal of Vegetation Science* 3(1): 3–14.

Friday, J.B., and R. Ogoshi. 2011 (revised). Farm and Forestry Production and Marketing Profile for Tamanu (*Calophyllum inophyllum*). In: Elevitch, C.R. (ed.). Specialty Crops for Pacific Island Agroforestry. Permanent Agriculture Resources (PAR), Holualoa, Hawai'i. http://agroforestry.net/scps.

Friday, J.P. and D. Okano 2006a. *Calophyllum inophyllum* (kamani). In: C.R. Elevitch, (ed.). *Traditional Trees of Pacific Islands: their Culture, environment, and Use*. Holualoa, Hawai'i: Permanent Agriculture Resources, pp. 183–198. www.tradionaltree.org.

Friday, J.P. and D. Okano 2006b. *Cordia subcordata* (kou). In: C.R. Elevitch, (ed.). *Traditional Trees of Pacific Islands: their Culture, environment, and Use*. Holualoa, Hawai'i: Permanent Agriculture Resources, pp. 303–316. www.tradionaltree.org.

Friday, J.P. and D. Okano 2006c. *Thespesia populnea* (milo). In: C.R. Elevitch, (ed.). *Traditional Trees of Pacific Islands: their Culture, environment, and Use*. Holualoa, Hawai'i: Permanent Agriculture Resources, pp. 357–374. www.tradionaltree.org.

FSM 2001. Grow low campaign on Pohnpei (see: www.equatorinitiative.org/wp-content/uploads 2017/05/case_1348152141).

Funk, E. 1987. Spontaneous spread of *Angiopteris evecta* (Marattiales) in the central Ko'olau Mountains, O'ahu, Hawai'i. Hawaiian Botanical Society Newsletter 26: 58–59.

Furey, L. 2017. Archaeological Evidence for Tattooing in Polynesia and Micronesia. In Krital. L. and A. Deter-Wolf (Eds.). *Ancient Ink: The Archaeology of Tattooing* Seattle; London: University of Washington Press, pp. 159–184.

Gallaher, T.J. 2014. The Past and Future of Hala (*Pandanus tectorius*) in Hawai'i. In: Keawe, L.M.A., MacDowell, M. and C.K. Dewhurst. *'Ike Ulana Lau Hala: The vitality and vibrancy of lau hala weaving traditions in Hawai'i. Hawai'inuiākea* 3: 94–112. Honolulu: University of Hawai'i Press.

Gallaher, T.J., Callmander, M.S., Buerki, S. and S.C. Keeley. 2015 A long distance dispersal hypothesis for the Pandanaceae and the origins of the *Pandanus tectorius* complex. *Molecular and Phylogenetic Evolution* 83: 20–32.

Garg, A., Chaturvedi. P. and P.C. Gupta. 2014. A review of the systemic adverse effects of areca nut or betel nut. *Indian Journal of Medical and Paediatric Oncology* 35(1): 3–9.

Glassman, S.F. 1952. The Flora of Ponape. *Bulletin of the Bernice P. Bishop Museum* 209. Honolulu: Bernice P. Bishop Museum.

Glassman, S.F. 1953. New Plant Records from the Eastern Caroline Islands, with a Comparative Study of the Native Plant Names. *Pacific Science* 7(3): 291–311.

Gohil, K.J., Patel, J.A. and A.K. Gajjar. 2010. Pharmacological review on *Centella asiatica*: a potential herbal cure-all. *Indian Journal of Pharmaceutical Sciences* 75(5): 546–556.

Good, E.M., and W. Welley. 1989. A preliminary grammar sketch, text and vocabulary of Pingelapese. In: *Papers in Kosraean and Ponapeic*, 1–114. Series C-112. Canberra: Pacific Linguistics.

Gramling, C. 2020. Sea level rise could drown mangroves. *Science News* 198(1): 7.

Green, E.P. and F.T. Short (eds.). 2003. *World Atlas of Seagrasses*. University of California Press, Los Angeles.

Guppy, H.B. 1906. *Observations of a naturalist in the Pacific between 1896 and 1899*. London: Macmillan and Co.

Gupta, P.C. and S. Warnakulasuriya. 2002. Global epidemiology of areca quid usage. *Addiction Biology* 7: 77–83.

Habtemariam, S. and G.K. Varghese. 2015. A Novel Diterpene Skeleton: Identification of a highly aromatic, cytotoxic and antioxidant 5-methyl-10-demethyl-abietane-typediterpene from *Premna serratifolia*. *Phytotherapy Research* 29(1): 80–85.

Hadfield, M. 2010. *Pagan Island Tree-Snail Surveys: a report to the U.S. Fish & Wildlife Service*. Kewalo Marine Laboratory, University of Hawai'i at Mānoa, Honolulu, Hawai'i.

Halford, A.R. and P. Bosserelle. 2019. The mangrove crabs of Pohnpei Island, Federated States of Micronesia: A timely intervention to ensure sustainability of a favoured resource. *SPC Fisheries Newsletter* 158: 42–52.

Hall. I. 2018. *The Federated States of Micronesia National Biodiversity Strategy and Action Plan 2018–2023*. https://www.cbd.int/doc/world/fm/fm-nbsap-v2-en.pdf.

Hanlon, D. 1988. *Upon a Stone Altar*. Honolulu: University of Hawai'i Press.

Harris, S. 2020. Plant 119, *Musa × paradisiaca* L. (Musaceae) Banana.https://herbaria.plants.ox.ac.uk/bol/plants400/Profiles/MN/Musa

Harrison, S.P. and S. Albert. 1977. *Mokilese-English Dictionary*. Honolulu: University Press of Hawai'i.

Hather, J.G. and M.I. Weisler. 2000. Prehistoric Giant Swamp Taro (*Cyrtosperma chamissonis*) from Henderson Island, Southeast Polynesia. *Pacific Science* 54(2): 149–156.

Hattori, R. 2012. *Preverbal particles in Pingelapese: A language of Micronesia*. Unpublished Ph.D. dissertation. University of Hawai'i at Mānoa, Honolulu, Hawai'i.

Haun, A. 1984. *Prehistoric Subsistence, Population, and Sociopolitical Evolution on Ponape, Micronesia*. Unpublished Ph.D. dissertation. University of Oregon, Eugene, Oregon.

Haws, M. (ed.) 2006. *Natural resources management needs for coastal and littoral marine ecosystems of the U.S. Affiliated Pacific Islands: American Samoa, Guam, Commonwealth of the Northern Marianas Islands, Republic of the Marshall Islands, Federated States of Micronesia and the Republic of Palau*. Hawai'i Cooperative Studies Unit, Technical Report HCSU-002. University of Hawai'i at Hilo.

Hay, A. 1988. *Cyrtosperma* (Araceae) and its Old World allies. *Blumea* 33: 427–469.

Herrera, K., Lorence, D.H., Flynn, T. and M.J. Balick. 2010. Checklist of the vascular plants of Pohnpei, Federated States of Micronesia with local names and uses. *Allertonia* 10: 1–204.

Hezel, F.X. 1984. *A Brief Economic History of Micronesia. Past Achievements and Future Possibilities*. Majuro: Micronesian Seminar. 1984: 11–62.

Hezel, F.X. 1989. *From Conquest to Conquest*. Saipan: CNMI Divison of Historical Preservation.

Hezel, F.X. 2001. *The New Shape of Old Island Cultures: A Half Century of Social Change in Micronesia*. Honolulu: University of Hawai'i Press.

Hezel, F.X. 2013. *Strangers in Their Own Land: A Century of Colonial Rule in the Caroline and Marshall Islands*. Honolulu: University of Hawai'i Press.

Hiroa, Te Rangi (Peter H. Buck). 1950. *Material Culture of Kapingamarangi*. Honolulu: Bernice P. Bishop Museum, Bulletin 200.

Hocart, C.H., Fankhauser, B. and D.W. Buckle. 1993. Chemical archaeology of kava, a potent brew. *Rapid Communications in Mass Spectrometry* 7(3): 219–24.

Hoch, P.C., Wagner, W.L. and P.H. Raven. 2015. The correct name for a section of *Ludwigia* L. (Onagraceae). *PhytoKeys* 50: 31–34.

Hodbod, J. and H. Eakin. 2015. Adapting a Social-Ecological Resilience Framework for Food Systems. *Journal of Environmental Studies and Sciences* 5(3): 474–84.

Holm, L., Doll, J., Holm, E., Pancho, J. and J. Herberger. 1997. *The World's Worst Weeds: Natural Histories and Distribution*. New York: John Wiley & Sons, Inc. (revised edition).

Homhual, S. Bunyapraphatsara, N., Kondratyuk, T., Herunsalee, A., Chaukul, W., Pezzuto, J., Fong, H. and H-J Zhang. 2006. Bioactive dammarane triterpenes from the mangrove plant *Bruguiera gymnorrhiza*. *Journal of Natural Products* 69: 421–424.

Horrocks, M. and S. Bedford. 2010. Introduced *Dioscorea* spp. starch in Lapita and later deposits, Vao Island, Vanuatu. *New Zealand Journal of Botany* 48(3–4): 179–183.

Horrocks, M. and P. Nunn. 2007. Evidence for Introduced Taro (*Colocasia esculenta*) and Lesser Yam (*Dioscorea esculenta*) in Lapita era (c. 3050-2500 cal. BP) deposits from Bourewa, Southwest Vitu Levu Island, Fiji. *Journal of Archaeological Science* 34(5): 739–748.

Horrocks, M. and M. Weisler 2006. Analysis of plant microfossils in archaeological deposits from two remote archipelagos: The Marshall Islands, Eastern Micronesia, and the Pitcairn Group, Southeast Polynesia. *Pacific Science* 60(2): 261–280.

Hughes, R.F. and J.S. Denslow. 2005. Invasion by an N-fixing tree, *Falcataria moluccana* alters function and structure of wet lowland forests of Hawai'i. *Ecological Applications* 15: 1615–1628.

Hughes, R.F., Johnson, M.T. and A. Uowolo. 2011. The Invasive Alien Tree *Falcataria moluccana*: Its Impacts and Management. *Proceedings of the XIII International Symposium on Biological Control of Weeds*. Waikoloa, Hawai'i, pp. 218–223.

Hunter-Anderson, R.L. 1991. A Review of Traditional Micronesian High Island Horticulture in Belau, Yap, Chuuk, Pohnpei, and Kosrae. Micronesia 24(1): 1–56.

Hunter-Anderson, R. L. 2005. *Archaeological monitoring and data recovery atTN-1-0857 (S4) and Vicinity for the IBB Low Band Antenna Project, Northwest Tinian Island, Commonwealth of the Northern Mariana Islands IBB, Saipan*. Micronesian Archaeological Research Services, Guam.

Hunter-Anderson, R.L. 2009. Savannah anthropogenesis in the Mariana Islands, Micronesia: Re-interpreting the paleoenvironmental data. *Archaeology in Oceania* 44: 125–141.

Iese.V. 2005. *Characterisation of Giant swamp taro (Cyrtosperma chamissonis (Merr.) Schott) using morphological features, genetic finger printing and tarditional knowlegde from farmers.* M.S. Thesis, University of the South Pacific, Suva, Fiji.

International Potato Institute CIP. 2018. cipotato.org/crops/sweetpotato/sweetpotato-facts-and-figures.

Iqbal, M. and C. Gnanaraj 2012. *Eleusine indica* L. possesses antioxidant activity and precludes carbon tetrachloride (CCl4)-mediated oxidative hepatic damage in rats. *Environmental Health and Preventative Medicine* 17(4): 307–315.

IUCN. 2010. *IUCN Red List of Threatened Species* (ver. 2010.2). www.iucnredlist.org.

IUCN. 2013. *IUCN Red List of Threatened Species* (ver. 2010.2). www.iucnredlist.org. [reference from Oleiro and Kesler 2015].

Iwasa, S. 1975. *Useful Plants in the Tropics.* Norin-tokei-kyokai, Tokyo [in Japanese].

Janick, J. and R.E. Paull. 2008. *The encyclopedia of fruit & nuts.* CABI digital books. http://www.cabi.org/cabebooks/ebook/20113366221.

Joffry, S.M., Yob, N.J., Rofiee, M.S., Affandi, M.M.R.M.M., Suhaili, Z., Othman, F. and Z.A. Zakaria. 2012. *Melastoma malabathricum* (L.) Smith ethnomedicinal uses, chemical constituents, and pharmacological properties: A review. *Evidence-based Complementary and Alternative Medicine* [258434]. DOI: 10.1155/2012/258434

Jolly, M. 2002. From Darkness to Light? Epidemiologies and Ethnographies. In: V. Lukere and M. Jolly, eds. *Birthing in the Pacific: Beyond tradition and Modernity?* University of Hawai'i Press, pp. 148–177.

Jones, H., Nickel, B., Srebotnjak, T., Turner, W., Gonzalez-Roglich, M., et al. 2020. Global hotspots for coastal ecosystem-based adaptation. *PLoS ONE* 15(5): e0233005. https://doi.org/10.1371/journal.pone.0233005.

Jones, W.P. and A.D. Kinghorn. 2008. Biologically active natural products of the genus *Callicarpa. Current Bioactive Compounds* 4(1): 15–32.

Joo, Y.J., Newcombe, D. Nosa, V. and C. Bullen. 2020. Investigating Betel Nut Use, Antecedents and Consequences: A Review of Literature. *Substance Use & Misuse* 55(9): 1422–1442.

Josekutty, P.C., Waguk, E.E. and M.J. Joseph. 2002. Invasive/weedy angiosperms in Kosrae, Federated States of Micronesia. In: *Invasive species and their management. Micronesica Supplement* 6: 61–65.

Kanehira, R. 1915 and 1916. [Forests and plants of the South Seas Occupied Islands]. *Nai Nihon Sanrin Kaihō* [Japanese Forestry Association] various pages; see Sachet and Fosberg 1971.

Kanehira, R. 1931. An enumeration of the woody plants collected in Micronesia, Japanese Mandate (in 1929 and 1930). *The Botanical Magazine* (also known as *Shokubutsugaku Zasshi*) 45: 271–296.

Kanehira, R. 1933. *Flora of Micronesia.* South Sea Bureau under the Japanese Mandate: Tokyo.

Kanehira, R. 1935. An Enumeration of Micronesian Plants. *Journal of the Department of Agriculture Kyushu Imperial University* 4(6): 237–435.

Kauffman, J.B., and T.G. Cole. 2010. Micronesian mangrove forest structure and tree responses to a severe typhoon. *Wetlands* 30: 1077–1084.

Kay, D.E. 1987. *Root Crops*, 2nd edition, revised by E.G.B. Gooding, *Crop and Product Digest*, No. 2. London: Tropical Development and Research Institute.

Kayser, A. 1936. Die Fischerei auf Nauru. *Mitteilungen der Anthropologishen Gesellschaft in Wien* 66: 92–149.

Kawai, K., Kuwahara, S., Onjo, M., Noda, S., Nishimua, A., Tominaga, S. and S. Nagashima. 2010. The Influence of Environmental Changes on the Micronesian Area: a Case Study of Islands in Pohnpei State, Federated States of Micronesia. *South Pacific Studies* 30(2): 23–43.

Kei, K. and K. Sueo (eds.). 2013. *Research on the Influence of Globalization and Global Warming on Life in Chuuk Atoll, Federated States of Micronesia.* Kagoshima University Occasional Research Paper 53, Research Center for the Pacific, Kagoshima, Japan.

Kennedy, J. 2008. Pacific bananas: Complex origins, multiple dispersals? *Asian Perspectives* 47: 75–94.

Kennedy, J. 2009. Bananas and People in the Homeland of Genus *Musa*: Not just pretty fruit. *Ethnobotany Research & Applications* 7: 179–197.

Kennedy, J. and W. Clarke. 2004. *Cultivated Landscapes of the Southwest Pacific.* Resource Management in Asia-Pacific Working Paper No. 50. Management in Asia-Pacific Program. Research School of Pacific and Asian Studies. Canberra: Austrailian National University.

Kepler, A. and F.G. Rust. 2011. *The World of Bananas in Hawai'i: Then and Now – Traditional Pacific & global varieties, cultures, ornamentals, health & recipes.* Pali-O-Waipi'o Press, Haiku, Hawai'i.

Kesler, D.C., Oleiro, P.C., Joseph, E. and D.B. Lynch. 2013. *Predicting Future Bird Distributions in Micronesia Final Report to PICCC*: University of Missouri, Conservation Society of Pohnpei, and College of Micronesia.

Kihleng, E. 2015. *Menginpehn Lien Pohnpei: A Poetic Ethnography of Urohs (Pohnpeian Skirts).* Ph.D. dissertation, Victoria University at Wellington, New Zealand.

Kihleng, K. 1996. *Women in Exchange: Negotiated Relations, Practice and the Constitution of Female Power in Cultural Reproduction and Change in Pohnpei, Micronesia*. Ph.D. dissertation, University of Hawai'i.

Kitalong, A.H., Balick, M.J., Rehuher, F. Besebes, M., Hanser, S. Soaladaob, K., Ngirchobong, G., Wasisang, F., Law, W. Lee, R. Tadeo, V.R., Kitalong, C. and C. Kitalong. 2011. Plants, people and culture in the villages of Oikull and Ibobang, Republic of Palau. In: J. Liston, G. R. Clark and D. Alexander (eds.), *Pacific Island Heritage: Archaeology, Identity and Community*. Canberra: Australian National University Press, pp. 63–84.

Kitalong, A.H., DeMeo, R.A. and T. Holm. 2008. *Native Trees of Palau: A Field Guide*. Koror, Palau: The Environment Inc.

Kitalong, A.H., DeMeo, R.A. and T. Holm. 2014. Native Trees of Palau: *A Field Guide* 2nd Edition. Koror, Palau: The Environment, Inc.

Kiviat, E. 2013, Ecosystem services of *Phragmites* in North America with emphasis on habitat functions. *AoB PLANTS* 5: plt008, doi: 10.1093/aobpla/plt008.

Kock, R. L. and R. T. Tsuda. 1978. Seagrass assemblages of Yap, Micronesia. *Aquatic Botany* 5: 245–249.

Kolbert, E. 2024. *The Sixth Extinction: An Unnatural History* [10th anniversary edition]. New York: Henry Holt and Company.

Krishnapillai, M.V. 2005a. Assessment of yam cultivar diversity in Yap State, Federated States of Micronesia. Paper presented during United Nations International Meeting on Sustainable Development of Small Island Developing States, Port-Louis, Mauritius. January 10–11.

Krishnapillai, M.V. 2005b. Women's role in the conservation of yam genetic resources in Yap Island. Paper presented during United Nations International Meeting on Sustainable Development of Small Island Developing States, Port-Louis, Mauritius.

Krishnapillai, M.V. 2010. Traditional Therapeutic Uses of Noni (*Morinda citrifolia* L. var. *citrifolia* L.) in Micronesia. In: Rethinam, P. and T. Marimuthu (eds.). *Proceedings of Fifth National Symposium "Noni for Global Wellness."* First World Noni Congress, Chennai, India. October 1–3, pp. 185–202.

Kulip J. 1997: A preliminary survey of traditional medicinal plants in the west coast and interior of Sabah. *Journal of Tropical Forest Science* (Malaysia) 10(2): 271–274.

Kump, L. R., Kasting, J. F. and R.G. Crane. 2004. *The Earth System* (Vol. 432). Upper Saddle River, NJ: Pearson Prentice Hall.

Kwok, J. 2008. Cyanide Poisoning and Cassava. Food Safety Focus. Issue 19. www.cfs.gov.hk/english/multimedia/multimedia_pub/multimedia_pub_fsf_19_01.html.

Lacap, C.D.A., Verymaat, J.E., Rollan, R.N. and H.M. Nacorda. 2002. Propagule dispersal of the SE Asian seagrasses *Enhalus acoroides* and *Thalassia hemprichii*. *Marine Ecology Progress Series* 235: 75–80.

Lai, J.C., Lai, H.Y., Rao, N.K. and S.F. Ng. 2016. Treatment for diabetic ulcer wounds properties. *Journal of Ethnopharmacology* 289: 277–289.

Lai, H.Y., Lim, Y.Y. and K.H. Kim. 2010. *Blechnum orientale* Linn – a fern with potential as antioxidant, anticancer and antibacterial agent. *BMC Complementary and Alternative Medicine*. doi: 10.1186/1472-6882-10-15.

Laird, W.E. 1983. *Soil Survey of Island of Kosrae, Federated States of Micronesia*. Washington: U.S.D.A. Soil Conservation Service.

Lambeth, L. 2001. *An assessment of the role of women in fisheries in Pohnpei, Federated States of Micronesia*. Secretariat of the Pacific Community (SPC), Noumea, New Caledonia.

Landers, M.A. and S. Khosrowpanah. 2004. *Rainfall climatology for Pohnpei Islands, Federal States of Micronesia*. http://www.weriguam.org/docs/reports/100.pdf.

Langeland, K.A. and K.C. Burks. 1998. *Identification and Biology of Non-Native Plants in Florida's Natural Areas*. IFAS Publication SP 257. University of Florida, Gainesville.

Lansdown, R.V. 2013. *Phragmites karka*. The IUCN Red List of Threatened Species 2013. http://dx.doi.org/10.2305/IUCN.UK.2013-1.RLTS.T168955A1256797.en

Lawrence, P., Hadley M. and T. McKnight. 1964. Breadfruit cultivation practices and beliefs in Ponape. In: *Breadfruit Cultivation practices and beliefs in the Trust Territory of the Pacific Islands*, Anthropological Working Papers. Saipan: Trust Territory of the Pacific Islands. No. 7-8: 44–64.

Lebot, V. 1991. Kava (*Piper methysticum* Forst. f.): The Polynesian distribution of an Oceanian plant. In: Cox, P.A. and S.A. Bannack (eds.) *Islands, Plants, and Polynesians*. Portland, Oregon: Dioscorides Press, pp. 169–201.

Lebot, V., Arhadya M. and R. Manshardt. 1991. Geographical survey of genetic variation in kava (*Piper methysticum*). *Pacific Science* 45(2): 169–185.

Lebot, V., Merlin, M. and M. Lindstrom. 1997. *Kava: The Pacific Elixir*. Inner Traditions Press, Rochester, VT., [original edition: Lebot, V., Merlin, M. and M. Lindstrom. 1992. Kava: The Pacific Drug. Yale University Press, New Haven and London].

Lebot, V. and P. Simeoni. 2004. Is the Quality of Kava (*Piper methysticum* Forst. f.) Responsible for Different Geographical Patterns? *Ethnobotanical Research and Applications* 2: 19–28. https://ethnobotanyjournal.org/index.php/era/article/view/34/33

Lee, R.A., Balick, M.J., Sohl, F., Roberts, A.S., Herrera, K., Dahmer, S., Lieskovsky, M., Raynor, W., Raynor. P., Albert. E., Trauernicht, C., Offringa, L., Adam, I., Law, W., Hunt, M. and A. Dores. 2010. *Pohnpei Primary Health Care Manual: Health*

Care in Pohnpei, Micronesia: Traditional Uses of Plants for Health and Healing. CreateSpace Independent Publishing Platform.

Lee, R.A. and M.J. Balick. 2005. Sweet Wood – Cinnamon and Its Importance as a Spice and Medicine. *EXPLORE: The Journal of Science and Healing* 1(1): 61–64.

Lee Ling, D. No Date. *Micronesian Ethnobotany: A Course Companion*. A compilation of handouts used in SC/SS 115 Ethnobotany. College of Micronesia–FSM Pohnpei, Federated States of Micronesia. Edition 4.1. and 4.2. http://www.comfsm.fm/~dleeling/ethnobotany/text/ethnobotany

Lee Ling, D. 2007 *Micronesian Healing Plants Presentations [2005–2007] in SC/SS 115 Ethnobotany*. College of Micronesia, Pohnpei, FSM. www.comfsm.fm/~dleeling/ethnobotany/healing_plants_presentations.html

Lee Ling, D. 2014. Healing Plants: Healing plants presentations in SC/SS 115 Ethnobotany. September 29. College of Micronesia–FSM Pohnpei, Federated States of Micronesia.

Lemmens, R.H.M.J. 2013. *Ipomoea pes-caprae* (L.) R.Br. In: Schmelzer, G.H. and A. Gurib-Fakim (eds.). *Prota 11(2): Medicinal plants/Plantes médicinales 2*. PROTA, Wageningen, Netherlands. lso see: https://plantuse.plantnet.org/en/Ipomoea_pes-caprae_(PROTA)

Lessa, W. 1977. Traditional Uses of the Vascular Plants of Ulithi Atoll, with Comparative Notes. *Micronesia* 13(2): 129–190.

Levin, M.J. 2015a. *Food Production, Environment, and Culture in the Tropical Pacific: Evidence for Prehistoric and Historic Plant Cultivation in Pohnpei, Federated States of Micronesia*. Ph.D. Dissertation, Department of Anthropology and the Graduate School of the University of Oregon, Eugene, Oregon.

Levin, M.J. 2015b. Roasting Breadfruit in the Pacific: A Combined Plant Macroremains. Oceania 51: 70–76.

Levin, M.J. 2016. *Pohnpeian and Pingelapese Food Production in the Archaeological Record: A Pilot Project*. Report Submitted to Historic Preservation Office. Pohnpei, Federated States of Micronesia.

Levin, M.J. 2018. Breadfruit Fermentation in Pohnpei, Micronesia: Site Formation, Archaeological Visibility, and Interpretive Strategies. *The Journal of Island and Coastal Archaeology* 13(1): 109–131.

Levin, M.J. 2019. Prehistoric Agricultural Transformations in the Pacific Islands: Towards a Comprehensive Phytolith Analysis. *Quaternary International*. DOI: 10.1016/j.quaint.2019.03.002.

Levin, M.J., M. Shelton, and W.S. Ayres. 2017. Swiddening Practices in the Micronesian Past: Macrocharcoal Quantification from a Gardening Site in Pohnpei. *The Journal of Island and Coastal Archaeology* 12(3): 451–457.

Levin, M.J., Seikel, K. and A. Miles. 2018. *Subsistence and Settlement on Pingelap Atoll: Results from the 2017 Field Season*. Report submitted to the Pohnpei State Historic Preservation Office.

Levin, M.J., Seikel, K. and A. Miles. 2019. A Partial Chronological Sequence of Human Habitation for Pingelap Atoll (Pohnpei State, Federated States of Micronesia). *Radiocarbon* 61(1): 765–776.

Lewis, S.L., Edwards, D.P. and D. Galbraith. 2015. Increasing human dominance of tropical forests. *Science 349* (6250): 827–832.

Lichvar, R.W., D.L. Banks, W.N. Kirchner, and N.C. Melvin. 2016. *The National Wetland Plant List: wetland ratings. Phytoneuron* 2016-30: 1–17.

Lieber, M.D. and K. H. Dikepa. 1974. *Kapingamarangi Lexicon*. PALI Language Texts-Polynesia. Honolulu: University Press of Hawai'i.

Lim, T.K. 2012. *Edible Medicinal and Non-Medicinal Plants: Volume 1, Fruits*. Dordrecht, The Netherlands: Springer.

Lincoln, N.K., Ragone, D., Zerega, N., Roberts-Nkrumah, L.B., Merlin, M. et al. 2018 Grow us our daily bread: A review of breadfruit cultivation in traditional and contemporary systems. *Horticulture Review* 46: 299–384.

Lindstrom, L. 2007. Brunton, Ron, The Abandoned Narcotic: Kava and Cultural Instability in Melanesia [book review]. *Social History of Alcohol and Drugs* 22: 90–92.

Little, E.L. Jr. and R.G. Skolmen. 1989. *Agriculture Handbook* (no. 679). Forest Service, U.S. Department of Agriculture (reprint version published by the College of Tropical Agriculture and Human Resources, University of Hawai'i at Mānoa, 2003).

Liwy, T. 2010. *Indigenous Mathematics of the Nahsen Pohnpei (Pohnpeian Feast House)*. Term Paper submitted to PREL (Pacific Resources for Education and Learning).

Lobban, C.S., and R.T. Tsuda. 2003. Revised checklist of benthic marine macroalgae and seagrasses of Guam and Micronesia. *Micronesica* 35-36: 54–99.

Long-Solis, J. 1986. *Capsicum y Cultura: La Historia del Chilli*. Fondo de Cultural Económica S. A. de C. V. Av. De la Universidad 975, 31000 México, DF.

Lovelock, C.E. Blue carbon from the past forecasts the future. 2020. *Science* 368: 1050. June 5. doi: 10.1126/science.abc3735.

Lucas, M.P. and D. Ragone. 2012. Will breadfruit solve the world hunger crisis? New developments in an innovative food crop. *ArcNews* Summer: 6–7.

MacCaughey, V. 1917. The genus Artocarpus in the Hawaiian Islands. Torreya 17(2): 33–49.

MacLean, D., Cole, T., Whitesell, C., Falanruw, M. and A. Ambacher. 1986. Vegetation of Pohnpei, Federated States of Micronesia. USDA Forest Service. *Resource Bulletin*

PSW18. Pacific Southwest Forest and Range Experiment Station. Forest Service, U.S. Department of Agriculture.

Mahony, F.J. 1960. Taro Cultivation in Chuuk. *Anthropological Working Papers* No. 6. Taro Cultivation practices and Beliefs Part II. The Eastern Carolines and the Marshall Islands. A series issued from the Office of the Staff Anthropologist Trust Territory of the Pacific Islands. Guam, M. I.

Mahony F. and T. Bikajle. 1960. Taro cultivation practices and beliefs. II. The Eastern Carolines and Marshall Islands. *Anthropological Working Papers* 6: 69–140. Trust Territory of the Pacific Islands.

Mali, P.Y. 2015. *Premna integrifolia* L.: A review of its biodiversity, traditional uses and phytochemistry. *Ancient Science of Life* 35(1): 4–11.

Mali, P.Y. 2016. Pharmacological potentials of *Premna integrifolia* L. *Ancient Science of Life* 35(3): 132–142.

Mani, S. 2013. *Saccharum spontaneum*. The IUCN Red List of Threatened Species 2013. https://www.iucnredlist.org/species/164377/19233469

Manner, H.I. 1990. A review of atoll agriculture and ecology with respect to climate change and sea level rise in the Pacific basin. In: P.J. Hughes & G. McGregor (eds.), *Global warming-related effects on agriculture and human health and comfort in the South Pacific*. A report to SPREP and UNEP. University of Papua New Guinea, Port Moresby, pp. 41–64.

Manner H.I. 1993a. Taro (*Colocasia esculenta* (L.) Schott) in the atolls and low islands of Micronesia. In: Ferentinos, L. (ed.). *Proceedings of the Sustainable Taro Culture for the Pacific Conference*. Sustainable Taro Culture for the Pacific Conference; 1992 Sept 24–25; Honolulu: University of Hawai'i, pp. 88–100.

Manner, H.I. 1993b. A Review of Traditional Agroforestry in Micronesia. *USDA Forest Service General Technical Report*. PSW-GTR-140: 32–36.

Manner, H.I. 2008. Directions for long term research in traditional agricultural systems of Micronesia and the Pacific Islands. *Micronesica* 40: 63–86.

Manner, H.I. 2011a (revised). Farm and Forestry Production and Marketing Profile for Giant Taro (*Alocasia macrorrhiza*). In: Elevitch, C.R. (ed.). *Specialty Crops for Pacific Island Agroforestry*. Permanent Agriculture Resources (PAR), Holualoa, Hawai'i.

Manner, H.I. 2011b. Farm and Forestry Production and Marketing Profile for Tannia (*Xanthosoma sagittifolium*). In: Elevitch, C.R. (ed.). *Specialty Crops for Pacific Island Agroforestry*. Permanent Agriculture Resources (PAR), Holualoa, Hawai'i.

Manner, H.I. 2011c (revised). Farm and Forestry Production and Marketing Profile for Giant Swamp Taro (*Cyrtosperma chamissonis*). In: Elevitch, C.R. (ed.). Specialty Crops for Pacific Island Agroforestry. Permanent Agriculture Resources (PAR), Holualoa, Hawai'i.

Manner, H.I. and C.R. Elevitch. 2006a. *Cananga odorata* (ylang-ylang). In: C.R. Elevitch (ed.). *Traditional Trees of Pacific Islands: their Culture, Environment, and Use*. Holualoa, Hawai'i: Permanent Agriculture Resources, pp. 199–209. http://www.traditionaltree.org

Manner, H.I. and C.R. Elevitch. 2006b. *Tournefortia argentea* (tree heliotrope). In: C.R. Elevitch (ed.). *Traditional Trees of Pacific Islands: their Culture, Environment, and Use*. Holualoa, Hawai'i: Permanent Agriculture Resources, 12 pp. http://www.traditionaltree.org

Manner, H.I. and E. Mallon. 1989. An Annotated list of the vascular plants of Pulawat Atoll. *Micronesica* 22: 23–63.

Manner, H.I. and M. Taylor. 2011. Farm and Forestry Production and Marketing Profile for Taro (*Colocasia esculenta*). In: C.R. Elevitch (ed.). 2011. *Specialty Crops for Pacific Islands*. Holualoa, Hawai'i: Permanent Agriculture Resources, pp. 431–464.

Marler, T.E. and J.H. Lawrence. 1994. Ifit: *Intsia bijuga*: The Territorial Tree of Guam. Department of Agriculture, Forestry and Soil Resources Division, USDA Forest Service and the U.S. Department of Defense. https://www.uog.edu/publications/ceo/IfitIntsiabijuga157

Marshall, J.T. 1957. Atolls visited during the first year of the Pacific Islands Rat Ecology Project. *Atoll Research Bulletin* 56: 1–11.

Marshall, M. 2013. *Drinking Smoke: The Tobacco Syndemic in Oceania*. Honolulu, University of Hawai'i Press.

Marshall, M. and L.B. Marshall. 2002. Holy and unholy spirits: The effects of missionization on alcohol use in Eastern Micronesia. In: N.J. Goetzfridt and K.M. Peacock (eds.). *Micronesian histories: An analytical bibliography and guide to interpretations*. Connecticut: Greenwood Press, pp. 130–141.

Martin, F.W. 1974. *Tropical yams and their potential. Part I, Discorea esculenta*. Agricultural Research Service, Agriculture Handbook No. 457. U.S. Dept. of Agriculture in cooperation with U.S. Agency for International Development.

Massal, E. and J. Barrau. 1956. *Food plants of the South Sea Islands*. Technical paper No. 94. Noumea: South Pacific Commission.

Masterson, J. 2007. *Dioscorea bulbifera*. Smithsonian Marine Station at Fort Pierce, Florida.

Matanane, L.B.R., Gibson, W.J., Yiu, E., Hollyer, J. and M.K. Fialkowski. 2014. *The United States Affiliated Pacific Food Guide: A Children's Healthy Living Program Resource for Nutrition*.

USDA. National Institute of Food and Agriculture. Agriculture and Food Research Initiative No. 2011-68001-30335.

Mather, J. 2012. *Study for the Management of Invasive Plants within a Proposed Protection Area, Olum Watershed, Kosrae, Federated States of Micronesia.* Pacific Invasives Initiative, Auckland, New Zealand.

Maxwell, B. 1982. Floristic Description of Native Upland Forests on Kosrae, Eastern Caroline Islands. Micronesica 18(2): 109–120.

McClatchey, W.C. 1996. The ethnopharmacopoeia of Rotuma. *Journal of Ethnopharmacology* 50: 147–156.

McClatchey, W., Manner, H.I. and C.R. Elevitch. 2006. *Metroxylon amicarum, M. paulcoxii, M. sagu, M. salomonense, M. vitiense,* and *M. warburgii* (sago palm), ver. 2.1. In: Elevitch, C.R. (ed.). *Species Profiles for Pacific Island Agroforestry.* Permanent Agriculture Resources (PAR), Hōlualoa, Hawai'i.

McConkey, K.R., Meehan, H.J. and D.R. Drake. 2004. Seed dispersal by Pacific Pigeons (*Ducula pacifica*) in Tonga, Western Polynesia. Emu 104: 369–376.

McDermid, K.J, and A. Edward. 1999. Seagrass Community Composition and Biomass at Nahpali Island, Pohnpei. *Micronesica* 31(2): 255–262.

McGrath, T. B. 1973. Sakau in town, Sarawi in town. *Oceania* 44: 64–67.

McLean, M. 2014. *Music, Lapita, and the Problem of Polynesian Origins.* ebook ISBN 978-0-473-28873-0. Auckland: self-published by the author.

Melteras, M., Lebot, V., Asher, C. and J.N. O'Sullivan. 2008. Crop development and root distribution in lesser yam (*Dioscorea esculenta*): Implications for fertilization. *Experimental Agriculture* 44: 209–221.

Melville, H. 1847. *Omoo: Narrative of Adventures in the South Seas.* J.M. Dent, London and New York, reprint 1921.

Merlin, M. 1985. Woody Vegetation in the Upland Region of Rarotonga, Cook Islands. *Pacific Science* 39(1): 81–99.

Merlin, M. 1989. The Traditional Geographical Range and Ethnobotanical Diversity of *Cordyline fruticosa* (L.) Chevalier. Ethnobotany 1: 25–39.

Merlin, M. 1991. Woody vegetation on the raised coral limestone of Mangaia, southern Cook Islands. *Pacific Science* 45(2): 131–151.

Merlin, M. 2002. Traditional uses of plants for fishing in Micronesia. *South Pacific Commission Women in Fisheries Information Bulletin* 11: 27–31.

Merlin, M. 2005. Pacific Ocean Islands, Coastal Ecology. In: Schwartz, M. (ed.). *Encyclopedia of Coastal Science.* Encyclopedia of the Earth Sciences Series. Springer, Dordrecht, The Netherlands, pp. 746–754.

Merlin, M. 2011. Seagrasses. In: Hopley, D. (ed.) *Enclyclopedia of Modern Coral Reefs.* Encyclopedia of Earth Sciences Series. Springer-Verlag, Germany, pp. 973–977.

Merlin, M., Jano, D., Raynor, W., Keene, T., Juvik, J. and B. Sebastian. 1992. *Tuhke en Pohnpei: Plants of Pohnpei.* Environment and Policy Institute, East-West Center, Honolulu, Hawai'i [Second printing, edition 1997].

Merlin, M. and J. Juvik. 1992. Relationships between Native and Alien Plants on Pacific Islands with and without Significant Human Disturbance and Feral Ungulates. In: C.P. Stone, C.W. Smith, and J.T. Tunison (eds.). *Alien Plant Invasions in Native Ecosystems of Hawai'i: Management and Research.* University of Hawai'i Cooperative National Park Resources Studies Unit Press, pp. 597–625.

Merlin, M. and J. Juvik. 1995. Montane Cloud Forest in the Tropical Pacific: Some Aspects of their Floristics, Biogeography, Ecology, and Conservation. In: Hamilton, L., Juvik, J. and F. Scatena (eds). *Tropical Montane Cloud Forests.* Springer-Verlag, New York, pp. 234–253

Merlin, M. and J. Juvik. 1996. *Irá me Neeniier nón Chuuk: Plants and their Environments in Chuuk.* Program on Environment, East-West Center, Honolulu, Hawai'i.

Merlin, M. and W. Raynor. 2004. Modern Use and Environmental Impact of the Kava Plant in Remote Oceania. In: M. Steinberg, J. Hobbs and K. Mathewson (Eds.). *Dangerous Harvest: Drug Plants and the Transformation of Indigenous Landscapes.* New York: Oxford University Press, pp. 274–293.

Merlin, M. and W. Raynor. 2005. Kava Cultivation and Integrated Watershed Resource Management on Pohnpei Island. *Pacific Science* 59(2): 41–260.

Merlin, M., Taulung, R. and J. Juvik. 1993. *Sahk Kap Ac Kain In Acn Kosrae: Plants and Environments of Kosrae.* Program on Environment, East-West Center, Honolulu, Hawai'i.

Meyer, J-Y. 2000. Preliminary review of the invasive plants in the Pacific Islands (SPREP Member Countries), *Invasive species in the Pacific: A technical review and draft regional strategy.* Apia, Samoa: South Pacific Regional Environment Programme.

Meyer, J-Y. 2010. The Miconia saga: 20 years of study and control in French Polynesia (1988–2008). In: Loope, L.L., Meyer, J.-Y., Hardesty, B.D. and Smith, C.W. (eds.). *Proceedings of the 2009 International Miconia Conference,* Keanae, Maui, Hawai'i, May 4–7, 2009. http://www.hear.org/conferences/miconia2009/proceedings/

Meyer, J-Y. and J. Florence. 1996. Tahiti's Native Flora Endangered by the Invasion of *Miconia calvescens* DC. (Melastomataceae). *Journal of Biogeography* 23(6)775–781.

Meyer, J-Y., Loope, L. and A-C. Goarant. 2011. Strategy to control the invasive alien tree *Miconia calvescens* in Pacific islands: eradication, containment or something else? In: Veitch, C. R.; Clout, M. N. and Towns, D. R. (eds.). Island invasives: eradication and management. IUCN, Gland, Switzerland, pp. 91–96.

Migvar, L. 1965. *The Coconut in Micronesia*. Agricultural Extension Circular Number 3. Division of Agriculture Department of Resources and Development Trust Territory of the Pacific Islands. Saipan, Mariana Islands.

Milgrom, P., Tut, O.K., Gilmatam, J., Gallen, M. and D.L. Chi. 2013. Areca use among adolescents in Yap and Pohnpei, the Federated States of Micronesia. *Harm Reduction Journal* 10: 26.

Milgrom, P., Tut, O.K., Gallen, M., Mancl, L., Spillane, N., Chi, D.L. and D.S. Ramsay. 2016. Symptoms with betel nut and betel nut with tobacco among Micronesian youth. *Addiction Behavior* 53: 120–124.

Miller, C.D., Murai, M. and F. Pen. 1956. The use of Pandanus fruit as food in Micronesia. *Pacific Science* 10: 3–16.

Miller, R.E. 1953. Health report of Kapingamarangi. *Atoll Research Bulletin*. Washington, DC: The Pacific Science Board, National Academy of Sciences–National Research Council, pp. 19–23.

Moghadamtousi, S.Z., Fadaeinasab, M., Nikzad, S., Mohan, G., Ali, H.M. and H.A. Kadir. 2015. *Annona muricata* (Annonaceae): A Review of its Traditional Uses, Isolated Acetogenins and Biological Activities. *International journal of molecular sciences* 16(7): 15625–15658.

Moore, D.R. 2005. Archaeological Evidence of a Prehistoric Farming Technique on Guam. *Micronesica* 38(1): 93–120.

Moore, D.R. 2012. What's New and What's Cooking in the Latte Period Pots. *Micronesica* 42: 121–147.

Moore, P.H. and P.D. McMakin. 1979. *Plants of Guam*. Mangilao: University of Guam, College of Agriculture and Life Sciences, Cooperative. Extension Service.

Moore. R. 1950. *Feast Day in Kapingamarangi – Polynesia*. National Geographic Society, Washington D.C.

Morton, J.F. 1987. *Fruits of Warm Climates*. Florida Flair Books, Miami, Fl.

Motley, T.J. 2004. The ethnobotany of *Fagraea* Thunb. (Gentianaceae): The Timber of Malesia and the scent of Polynesia. *Economic Botany* 58(3): 396–409.

Motooka, P., Ching, L. and G. Nagai, G. and L. Ching. 2002. *Herbicidal Weed Control Methods for Pasture and Natural Areas of Hawai'i*. Cooperative Extension Service, College of Tropical Agriculture and Human Resources, University of Hawai'i.

Mueller-Dombois, D. and F.R. Fosberg. 1998. *Vegetation of the Tropical Pacific Islands*. Ecological Studies vol. 132. Amsterdam: Springer-Verlag.

Muimba-Kankolongo, A. 2018. *Root and Tuber Crops*. In: *Food Crop Production by Smallholder Farmers in Southern Africa: Challenges and Opportunities for Improvement*. 2018. Elsevier Science Publishing Co. Inc., United States.

Müller, K. 1908. *Jahresbericht über die Tätigkeit der Kapuziner der rheinisch-westfälischen Kapuziner-Ordensprovinz auf den Karolinen, Marianen, und Palau-Inseln*. Jahresbericht 1908. Limburg: Limburger Vereinsdruckerei: 17.

Murai, M. 1954. Nutrition Study in Micronesia. *Atoll Research Bulletin* 27: 1–239.

Murai, M., Pen, F., and C.D. Miller. 1958. *Some tropical South Pacific Island foods. Description, history, use, composition, and nutritive value*. Hawai'i Agricultural Experiment Station Bulletin 110. Honolulu: University Press of Hawai'i.

Murphy, R.E. 1949. "High" and "low" islands in the Eastern Carolines. *Geographical Review* 39: 425–439.

Murphy, R.E. 1950. The Economic Geography of a Micronesian Atoll. *Annals of the Association of American Geographers* 40(1): 58–83.

Nakajima, Y., Matsuki, Y., Lian, C., Fortes, M.D., Uy, W.H., Campos, W.L., Nakaoka, M. and K. Nadaoka. 2014. The Kuroshio Current influences genetic diversity and population genetic structure of a tropical seagrass, *Enhalus acoroides*. *Molecular Ecology* 23: 6029–6044.

Nagaoka, T. 2014. Sailing Canoe Building on Mwoakilloa Atoll. In: Lee, S. (ed.) *Traditional Knowledge and Wisdom: Themes from the Pacific Islands*. Jeonju, Korea: ICHCAP. UNESCO publication.

Nayar, N.M. 2017. *The Coconut Palm: Phylogeny, Origins, and Spread*. Academic Press, London.

Neal, M.C. 1965. *In Gardens of Hawai'i*. Special Publication 50. Bishop Museum Press, Honolulu (New and Revised Edition).

Nelson, K.M, Dahlin, J.L., Bisson, J., Graham, J., Pauli, G.F. and M. Walters. 2017. The Essential Medicinal Chemistry of Curcumin. *Journal of Medicinal Chemistry* 60(5): 1620–1637.

Nelson, S.C. 2006. *Morinda citrifolia* (noni): Rubiaceae (coffee family). In: Elevitch, C.R. *(ed.)*. *Species Profiles for Pacific Island Agroforestry: Ecological, Economic and Cultural Renewal*. Holualoa, Hawai'i: Permanent Agriculture Resources. http://www.agroforestry.net/tti/Morinda-noni.pdf

Nelson, S.C. 2008. *Cassytha filiformis*. Plant Disease, July PD-42. Cooperative Extension Service, CTAHR, Honolulu: University of Hawai'i at Mānoa.

Nelson, S. C. and F. B. Javier. 2007. Trials of FHIA bananas for performance and for resistance to black leaf streak in Pohnpei, Federated States of Micronesia. *Proceedings of the 38th Hawai'i Banana Industry Association Annual Conference*, Kalaeloa, Hawai'i.

Nelson, S.C., Ploetz, R.C.and A.K. Kepler. 2006. *Musa* species (banana and plantain). Vers. 2.2 In: Elevitch, C.R. (ed.). *Species Profiles for Pacific Island Agroforestry: Ecological, Economic and Cultural Renewal*. Holualoa, Hawai'i: Permanent Agriculture Resources, http://www.agroforestry.net/images/pdfs/Musa-banana-plantain.pdf

Nick, A., Rali, T. and O. Sticher. 1995. Biological screening of traditional medicinal plants from Papua New Guinea. *Journal of Ethnopharmacology* 49(3): 119–185.

Niering, W.A. 1956. Bioecology of Kapingamarangi Atoll, Caroline Islands: terrestrial aspects. *Atoll Research Bulletin* 49: 1–32.

Niering, W.A. 1963. Terrestrial ecology of Kapingamarangi Atoll. *Ecological Monographs* 33: 131–160.

NRI (National Research Institute).1987. *Root crops. Crop and product digest*. 2nd ed. London: Tropical Development and Research Institute.

Ogura, C.S. 2003. Watershed Management on Pohnpei: Lessons for Enhanced Collaboration. MS in Natural Resources and the Environment. University of Michigan.

Okabe, M. 1941c. On the Exploitation of Forest Resources in Micronesia. *Sangyo no Nanyo* (Industry of the South Seas), Autumn Number, pages 1–7. Translated in 1952 by H. Takeda, Military Geology Branch, U.S. Geological Survey for Intelligence Division, Office of the Engineer Headquarters, Far East Command, Tokyo, Japan.

Olayemi, J.O. and E.O. Ajaiyeoba. 2007. Anti-inflammatory studies of yam (*Dioscorea esculenta*) extract on wistar rats. *African Journal of Biotechnology* 6(16): 1913–1915.

Oleiro, P.C. 2014. Avian Population Responses to Anthropogenic Landscape Changes in Pohnpei, Federated States of Micronesia. M.S. Thesis, University of Missouri.

Oleiro, P.C. and D. Kesler 2015. *Effect of landscape change and factors likely to influence future avifauna population change on Pohnpei Islands*. Department of Fisheries and Wildlife Sciences, University of Missouri. Final Report. Prepared in partnership with: Eugene Joseph, Conservation Society of Pohnpei, D. Brian Lynch, College of Micronesia. Submitted to Pacific Islands Climate Change Cooperative, Honolulu.

Ostertag, R., Cordell, S., Michaud, J., Cole, Colleen T.C., Schulten, J.R., Publico, K.M. and J.M. Enoka. 2009. Ecosystem and Restoration Consequences of Invasive Woody Species Removal in Hawaiian Lowland Wet Forest. *Ecosystems* 12(3): 503–515.

Partap, S., Kumar, A., Sharma, N.K. and K.K. Jha. 2004. *Luffa cylindrica*: An important medicinal plant. *Journal of Natural Products and Plant Resources* 2(1): 127–134.

Patel, A.D., Freyer, A.J., Reichwein, R. Hofmann, G. and R.K. Johnson. 1997. Piperpense, a tetrahydropyridine alkaloid from *Piper ponapense* stems and leaves. *Natural Product Letters* 9(3): 217–223.

Pattanayak, P., Behera, P., Das, D. and S.K. Panda. 2010. *Ocimum sanctum* Linn. A reservoir plant for therapeutic applications: An overview. *Pharmacognosy reviews* 4(7): 95–105.

Pauku, R.L. 2006. *Inocarpus fagifer* (Tahitian Chestnut). In: Elevitch, C.R. (ed.). 2006b. *Species Profiles for Pacific Island Agroforestry: Ecological, Economic and Cultural Renewal*. Holualoa, Hawai'i: Permanent Agriculture Resources, pp. 407–424.

Paulino, Y., Novotny R., Miller M.J. and S.P. Murphy. 2011. Areca (betel) nut chewing practices in Micronesian populations. *Hawai'i Journal of Public Health* 3(1): 19–29.

Paulson, D.R., and D.W. Buden. 2003. The Odonata of Pohnpei, eastern Caroline Islands. *International Journal of Odonatology* 6: 39–64.

Pelep, P.O. and M.G. Hadfield. 2011. The status of the endemic snails of the genus *Partula* (Gastropoda: Partulidae) on Pohnpei, Federated States of Micronesia. *Micronesica* 41(2): 253–262.

Perry, L., Dickau, R., Zarrillo, S., Holst, I., Pearsall, D.M., Piperno, D.R., Berman, M.J., Cooke, R.G., Rademaker, K., Ranere, A.J., Raymond, J.S., Sandweiss, H., Scaramelli, F., Tarble, K. and J.A. Zeidler. 2007. Starch fossils and the domestication and dispersal of chili peppers (*Capsicum* spp. L.) in the Americas. Science 5: 986–988.

Petersen, G. 1977. *Ponapean Agriculture and Economy: Politics, Prestige and Problems of Commercialization in the Eastern Caroline Islands*. Ph.D. Dissertation. New York: Columbia University.

Petersen, G. 1995. The Complexity of Power: The Subtlety of Kava. *Canberra Anthropology* 18(1-2): 34–60.

Petersen, G. 2006. Micronesia's Breadfruit Revolution and the Evolution of a Culture Area. *Archaeology in Oceania* 41(2): 82–92.

Petteys, E., Peter, S., Rugg, R. and T. Cole. 1986. *Timber volumes in the Mangrove Forests of Pohnpei, Federated States of Micronesia*. Pacific Southwest Research Station, Forest Service, U.S. Department of Agriculture, Resource Bulletin PSW19.

Pickersgill, B. 1969. The archaeological record of chili peppers (*Capsicum* spp. L.) and the sequence of plant domestication in Peru. *American Antiquity* 23: 54–61.

PIER. 2002 & 2006. *Pacific Island Ecosystems at Risk (PIER)* (3.3) database. Institute of Pacific Islands Forestry.

Pimm, S.L., Jenkins, C.N., Abell, R., Brooks, T.M., Gittleman, J.L., Joppa, L.N. and J.O. Sexton. 2014. The biodiversity of species and their rates of extinction, distribution, and protection. *Science* 344 (6187): 987+.

Piperno, D.R. 2006. *Phytoliths: A Comprehensive Guide for Archaeologists and Paleoecologists*. Lanham: AltaMira Press.

Ploetz, R.C.; Kepler, A.K.; Daniells, J. and S.C. Nelson. 2007. Banana and Plantain: An Overview with Emphasis on Pacific Island Cultivars. In: Elevitch, C.R (ed.). *Species Profiles for Pacific Island Agroforestry*. Hōlualoa, Hawai'i: Permanent Agriculture Resources (PAR), pp.1–27.

Pollock, N. 1992. *These roots remain: food habits in islands of the Central and Eastern Pacific since Western contact*. University of Hawai'i Press, Honolulu.

Pollock, N. 1995. Introduction. *Canberra Anthropology* 18(1-2): 1–19.

Pollock, N. 2006. Taro as the foundation of Pacific food security. L. Guarino, M. Taylor and T. Osborn (eds.) *The Third Taro Symposium*, Proceedings of an International Scientific Meeting jointly organized by the Secretariat of the Pacific Community and the International Plant Genetic Resources Institute, Nadi, Fiji, pp. 123–131.

Pollock, N. 2009. Sustainability of the Kava Trade. *The Contemporary Pacific* 21(2): 265–297.

Poteate, A.S., Fitzpatrick, S.M., Ayres, W.S. and A. Thompson. 2016. First Radiocarbon Chronology for Mwoakilloa (Mokil) Atoll, Eastern Caroline Islands, Micronesia. *Radiocarbon* 58(1): 169–178.

Prance. G.T. 2004. The Uses of *Atuna racemosa* Raf. (Chrysobalanaceae) in Samoa. *Economic Botany* 58(3): 470–475.

Preston, D. and M. Preston. 2004. *A Pirate of Exquisite Mind. Explorer, Naturalist, Buccaneer: the Life of William Dampier*. Walker & Company, New York.

Pretrick, M. and N. Tuivavalagi. 2017. The effect of an extended growing period on the yield and taste of cassava (*Manihot esculenta*) varieties grown in Pohnpei Island, Micronesia. Presentation given at the STEP-UP Conference, Washington DC, U.S.A.

Primo, A. 1992. *Colocasia* taro on Pohnpei Island. In: *Proceedings of the Sustainable Taro Culture for the Pacific Conference, Honolulu, Hawai'i, 24–25 September 1992*; Hawai'i Institute of Tropical Agriculture and Human Resources: Mānoa, HI, USA, 1993; pp. 6–8.

Pukui, M. 1983. *'Olelo No'eau: Hawaiian Proverbs and Poetical Sayings*. Honolulu: Bishop Museum Press.

Quattrocchi, U. 2012. *CRC World Dictionary of Medicinal and Poisonous Plants: Common Names, Scientific Names, Eponyms, Synonyms, and Etymology*. Boca Raton, Fl.: CRC Press. 5 vols.

Ragone, D. 1997. Breadfruit, *Artocarpus altilis* (Parkinson) Fosberg. In: *Promoting the conservation and use of underutilized and neglected crops series*, vol. 10. IPGRI, Rome, Italy.

Ragone, D. 2002. Breadfruit storage and preparation in the Pacific Islands. In: *Vegeculture in Eastern Asia and Oceania*, JCAS Symposium Series 16. S. Yoshida and P.J. Matthews (eds.) The Japan Center for Area Studies, National Museum of Ethnology, Osaka, Japan.

Ragone, D. 2006. *Artocarpus altilis* (breadfruit). In: Elevitch, C.R. (ed.). 2006. *Traditional Trees of Pacific Islands: their Culture, Environment, and Use*. Holualoa, Hawai'i: Permanent Agriculture Resources, pp. 85–100.

Ragone, D., Lorence, D.H., and T. Flynn. 2001. Plant Introductions to Pohnpei, Micronesia and the Role of the Pohnpei Agriculture Station. *Economic Botany* 55(2): 290–324.

Ragone, D. and H. Manner. 2006. *Artocarpus mariannensis* (dugdug). In: Elevitch, C.R. (ed.). 2006. *Traditional Trees of Pacific Islands: their Culture, environment, and Use*. Holualoa, Hawai'i: Permanent Agriculture Resources, pp. 127–138.

Ragone, D. and W. Raynor. 2009. Breadfruit and its traditional cultivation and use on Pohnpei. In: Michael J. Balick (ed.) *Ethnobotany of Pohnpei: Plants, People, and Island Culture*. University of Hawai'i Press and New York Botanical Garden, pp. 64–88.

Ramanatha, R.V., Matthews, P.J., Eyzaguirre, P.B. and D. Hunter. 2010. The global diversity of taro: Ethnobotany and conservation. Rome, Italy: Biodiversity International.

Rao, S., Taylor, M. and A. Jokhan. 2014. A descriptor list for Giant Swamp Taro (*Cyrtosperma merkusii*) and its cultivars in the Federated States of Micronesia *Telopea: Journal of Plant Systematics* 6: 95–117.

Raulerson, L. and A. Rinehart. 1991. *Trees and Shrubs of the Northern Mariana Islands*. Coastal Resources Management, Office of the Governor, CNMI. 1991. 2nd Edition 2018 published by University of Guam Press.

Raynor, W. 1989. *Structure, Production, and Seasonality in an Indigenous Pacific Island Agroforestry System: A Case Study on Pohnpei Island, F.S.M.* M.S. thesis. Honolulu, Hawai'i: University of Hawai'i at Mānoa.

Raynor, W. 1991. *Agroforestry systems in Pohnpei: Practices and strategies for development*. RAS/86/036 Field Document 4. Pohnpei: FAO/UNDP South Pacific Forestry Development Programme.

Raynor, W. 1994. Resource management in upland forests of Pohnpei: Past practices and future possibilities. *Journal of Micronesian Studies* 2: 47–66.

Raynor, W. 2006. The Federated States of Micronesia (FSM). In: Haws, M. (ed.) *Natural resources management needs for coastal and littoral marine ecosystems of the U.S. Affiliated Pacific Islands*. Hawai'i Cooperative Studies Unit, Technical Report HCSU-002. University of Hawai'i at Hilo, pp. 91–154.

Raynor, W.C. and J.H. Fownes. 1991. Indigenous Agroforestry of Pohnpei: 2. Spatial and successional vegetation patterns. *Agroforestry Systems* 16: 139–157.

Raynor, W.C. and M. Kostka. 2003. Back to the Future: Using Traditional Knowledge to Strengthen Biodiversity Conservation in Pohnpei, Federated States of Micronesia. *Ethnobotany Research & Applications: A Journal of Plants, People, and Applied Research* 1: 59–64.

Raynor, W.C., Lorens, A. and J. Phillip. 1992. Traditional Yam Cultivation on Pohnpei, Eastern Caroline Islands, Micronesia. *Economic Botany* 46(1): 25–33.

Raynor, W.C., Lorens, A. and J. Phillip. 2009. Yams and Their Traditional Cultivation on Pohnpei. In: Balick, M.J. (ed.). *Ethnobotany of Pohnpei: Plants, people, and island culture*. Honolulu: University of Hawai'i Press and New York Botanical Garden, pp. 40–63.

Razafimandimbison, S. G., McDowell, T. D., Halford, D. A. and B. Bremer. 2010. Origin of the pantropical and nutriceutical *Morinda citrifolia* L. (Rubiaceae): comments on its distribution range and circumscription. *Journal of Biogeography* 37(3): 520–529.

Rehg, K. and B. Bender. 1990. Lexical Transfer from Marshallese to Mokilese: A Case of Intra-Micronesian Borrowing. *Oceanic Linguistics* 29(1): 1–26.

Rehg, K. and D. Sohl. 1979, 1983. *Ponapean-English Dictionary*. Honolulu: University of Hawai'i Press [also see Sohl, Rehg, and Andreas, 2025].

Reynertson, K.A., Balick, M.J., Lee R., Raynor, W., Pelep, Y. and E.J. Kennelly. 2005. A traditional method of *Cinnamomum carolinense* preparation eliminates safrole from a therapeutic Pohnpean tea. *Journal of Ethnopharmacology* 102(2): 269–74.

Riesenberg, S.H. 1948. Magic and Medicine in Ponape. *Southwestern Journal of Anthropology* 4(4): 406–429.

Riesenberg, S.H. 1950. *The cultural position of Ponape in Oceania* Ph.D. Dissertation in Anthropology. University of California, Berkeley.

Riesenberg, S.H. 1968. *Native Polity of Ponape*. Smithsonian Contributions to Anthropology, vol. 10. Washington, D.C.: Smithsonian Institution Publication.

Riesenberg, S. H. and J. L. Fischer. 1955. Some Ponapean Proverbs. *The Journal of American Folklore* 68 (267): 9–18, https://doi.org/10.2307/537106.

Riesenberg, S.H. and A.H. Gayton. 1952. Caroline Island belt weaving, *Southwestern Journal of Anthropology* 8(3): 342–375.

Rigamoto, R.R. 2001. *A floristic survey of the coastal littoral vegetation of Rotuma*. M.S. Thesis. University of the South Pacific.

Ritchie, H., Spooner, F. and M. Roser. 2021. "Forests and Deforestation." Data adapted from Food and Agriculture Organization of the United Nations. https://archive.ourworldindata.org/20251210-143340/grapher/deforestation-share-forest-area.html

Roberts, C. 1997. *Kapingamarangi Canoe-Building Project*. Micronesia and South Pacific Program via the Historic Preservation and Cultural Affairs Division, Pohnpei State, FSM.

Romanchak, E.A., Criley, R.A. and N. Sugii. 2005. *The Propagation of Uluhe Fern (Dicranopteris linearis): Vegetative Versus Spores*. International Plant Propagators' Society, Combined Proceedings 5: 517–519.

Ross, M., Pawley, A. and M. Osmond (eds.). 2008. *The Lexicon of Proto Oceanic: The Culture and Environment of Ancestral Oceanic Society*. Vol. 3, Plants. Canberra: Pacific Linguistics, Research School of Pacific and Asian Studies, Australian National University.

Russell, A.E., Raich, J.W. and P.M. Vitousek. 1998. The ecology of the climbing fern *Dicranopteris linearis* on windward Mauna Loa, Hawai'i. *Journal of Ecology* 86: 765–779.

Sacks, O. 1998. *Island of the Color Blind*. Vintage Books, New York (original edition, 1996, Alfred Knopf publisher).

Safford, W. E. 1905. Useful Plants of the Island of Guam. *Contributions from the United States National Herbarium*, Vol. 9.

Saintilan, N., Khan, N.S., Ashe, E., Kelleway, J.J., Rogers, K., Woodroffe, C.D. and B. P. Horton. 2020. Thresholds of mangrove survival under rapid sea level rise. *Science* 368: 1118–1121.

Schattenberg, P. 1976. Food and Cultivar Preservation in Micronesian Voyaging. In: K. Peacock and C.J. Daeufer (eds.) *Micronesian and Polynesian Voyaging*. Pacific Islands Program, University of Hawai'i, Honolulu, Hawai'i, pp. 26–53.

Scott, S. and C. Thomas. 2000. *Poisonous plants of paradise: First Aid and Medicial Treatment of Injuries from Hawai'i's Plants*. Honolulu: University of Hawai'i Press.

Scrimgeour, F. and J. Gallen. 2003. *Use Values of Forests in Pohnpei*. Contributed Paper presented to the 47th Annual Conference of the Australian Agricultural and Resource Economics Society at Fremantle, Australia. February 12–14.

Secretariat of the Pacific Community. 2006. *Healthy Pacific Lifestyle Section of the Secretariat of the Pacific* (SPC). Noumea, New Caledonia.

Selvam V. 2007. *Trees and Shrubs of the Maldives*. Ministry of Fisheries, Agriculture and Marine Resources, Maldives.

Shah, K.A., Patel, M.B., Patel, R.J. and P.K. Parmar. 2010. *Mangifera indica* (Mango). *Pharmacognosy Reviews* 4(7): 42–48.

Sharief, M.U. and R.R. Rao 2007. Ethnobotanical studies of Shompens – A critically endangered and degenerating ethnic community in Great Nicobar Island. *Current Science* 93: 1623–1628.

Sharrock, S. 2001. Diversity in the genus *Musa*: focus on *Australimusa*. In: INIBAP (ed.), *Networking Banana and Plantain: INIBAP Annual Report 2000*. Montpellier, France: International Network for the Improvement of Banana and Plantain, pp. 14–19.

Sheely, D.L. and T.R. Meagher. 1996. Genetic diversity in Micronesian island populations of the tropical tree *Campnosperma brevipetiolata* (Anacardiaceae). *American Journal of Botany* 83(12): 1571–1579.

Shimizu, A. 1982. Chiefdom and the Spatial Classification of the LifeWorld: Everyday Life, Subsistence, and the Political System on Pohnpei. In: M. Aoyagi (ed.). *Islanders and Their Outside World: A Report of the Cultural Anthropological Research in the Caroline Islands of Micronesia in 1980–1981*. Committee for Micronesian Research, St. Paul's (Rikkyo) University, Tokyo, pp. 153–215.

Shresthra, P.M. 1990. *Twelve steps to growing healthy sweet taro*. F.S.M. Agriculture Division, Palikir, Pohnpei.

Singh, D., Jackson, G., Hunter, D., Fullerton, R., Lebot, V., Taylor, M., Iosefa, T., Okpul, T. and J. Tyson. 2012. Taro Leaf Blight – A Threat to Food Security. *Agriculture* 2: 182–203.

Sirichamorn, Y., Adema, F. A. C. B. and P. C. van Welzen. 2012. The Genera *Aganope*, *Derris*, and *Paraderris* (Fabaceae, Millettieae) in Thailand. *Systematic Botany* 37(2): 404–436.

Slocum, M.G. 2000. Logs and fern patches as recruitment sites in a tropical pasture. *Restoration Ecology* 8: 408–413.

Smith, A.C. 1985, 1979 *Flora Vitiensis nova: A new flora of Fiji*. Lawai, Kaua'i, Hawai'i. National Tropical Botanical Garden. 5 Volumes.

Smith, C.W. 1985. Impact of alien plants on Hawai'i's native biota. In: C.P. Stone and J.M. Scott (eds.). *Hawai'i's Terrestrial Ecosystems Preservation and Management*. University of Hawai'i Cooperative National Park Resource Studies Unit, Honolulu, pp. 180–250.

Smith, N.M. 1991. Ethnobotanical field notes from the Northern Territory, Australia. *Journal of the Adelaide Botanic Garden* 14(1): 1–65.

Sohl, D., Rehg, K. and R. Andreas. (eds.) 2025. *New Pohnpeian-English Online Dictionary: https://www.trussel2.com/pnp/*

Sopher, D.E. 1964. Indigenous uses of Turmeric (*Curcuma domestica*) in Asia and Oceania. *Anthropos* 59: 93–127.

Soucie, E.A. 1978. Types of Breadfruit in Ponape. In: *Tropical Horticulture for Secondary Schools*. Appendix II. Saipan: Department of Education. Trust Territory Pacific Islands.

Space, J.C. and M. Falanruw. 1999. *Observations on invasive plant species in Micronesia*. USDA Forest Service, Honolulu. Report to the Pacific Islands Committee, Council of Western State Foresters. USDA Forest Service, Honolulu.

Space, J.C. and T. Flynn. 2002. *Report to the Government of the Cook Islands on invasive plant species of environmental concern*. USDA Forest Service, Honolulu.

Space, J.C., Waterhouse, B., Denslow, J.S., Nelson, D. and E. Waguk. 2000. *Invasive plant species in Kosrae, Federated States of Micronesia*. U.S.D.A. Forest Service, Pacific Southwest Research Station, Institute of Pacific Islands Forestry, Honolulu, Hawai'i, USA.

SPREP. 2019. Federated State of Micronesia State of the Environment Report 2018. digital report, Apia, Samoa.

Sproat, M.N. 1968. *A guide to subsistence agriculture in Micronesia. Agricultural Extension Bulletin* 9. Division of Agriculture, Department of Re-sources and Development, Trust Territory of the Pacific Islands. Saipan: Publications Office.

Staples, G.W. and R.F. Bevacqua. 2006. *Areca catechu* (betel nut palm). In: Elevitch, C.R. (ed.) *Traditional Trees of Pacific Islands: their Culture, environment, and Use*. Holualoa, Hawai'i: Permanent Agriculture, pp. 69–84.

Staples, G.W. and D.R. Herbst. 2005. *A Tropical Garden Flora. Plants Cultivated in the Hawaiian Islands and Other Tropical Places*. Bishop Museum Press, Honolulu, Hawai'i.

Standal, B.R. 1979. Tropical Foods of the Pacific Region. In: G. Inglett and G. Charalambous. (eds.). *Tropical Food: Chemistry and Nutrition*. New York: Academic Press, Volume 1, pp. 1–24.

Starr, F., Starr, K. and L. Loope. 2003. New plant records from the Hawaiian Archipelago. *Bishop Museum Occasional Papers* 74: 23–34.

St. John, H. 1948. Report on the flora of Pingelap Atoll, Caroline Islands, Micronesia, and observations on the vocabulary of the native inhabitants. *Pacific Science* 2: 96–113.

St. John, H. 1951. Plant Records from Aur Atoll and Majuro Atoll, Marshall Islands, Micronesia. Pacific Plant Studies 9. *Pacific Science* 5: 279–286.

Stemmermann, L. 1981. *A Guide to Pacific Wetland Plants*. U.S. Army Corps of Engineers. Honolulu District.

Stone, B.C. 1959. The Flora of Namonuito and the Hall Islands. *Pacific Science* 13(1): 88–104.

Stone, B.C. 1960a. Corrections and Additions to the Flora of the Hall Islands and to the Flora of Ponape. *Pacific Science* 14(4): 408–410.

Stone, B.C. 1963. The Role of *Pandanus* in the Culture of the Marshall Islands: Marshallese Cultivar Index and Distribution of the Names by Atoll. In: J. Barrau (ed.). *Plants and the Migrations of Pacific Peoples: A Symposium*. Honolulu: Bishop Museum Press, pp. 61–82.

Stone, B.C. 1966a. Some Vernacular Names of Plants from Kapingamarangi and Nukuoro Atolls, Caroline Islands. *Micronesica* 2: 131–132.

Stone, B.C. 1966b. Cultivated *Pandanus* in Kapingamarangi Atoll. *Journal of the Polynesian Society* 75(4): 430–436.

Stone, B.C. 1967a. The Flora of Romonum Island, Truk lagoon, Caroline Islands. *Pacific Science* 29: 98–114.

Stone, B.C. 1967b. The Genus *Pandanus* in Micronesia I. The Marianas Species *Micronesica* 3: 105–128.

Stone B.C. 1970. The Flora of Guam. *Micronesica* 6: 1–659.

Storlazzi, C.D., Gingerich, S.B., van Dongeren, A., Cheriton, O.M., Swarzenski, P.W., Quataert, E., Vos, C.I., Field, D.W., Annamalai, H., Piniak, G.A. and R. McCall. 2018. Most atolls will be uninhabitable by the mid-21st century because of sea-level rise exacerbating wave-driven flooding. *Science Advances* (4) 4: eaap9741. DOI: 10.1126/sciadv.aap9741.

Swanholm, C.E. 1959. *A chemical study of the bitter principle of Pia (Tacca Leontopetaloides* (L.) O. Ktze.). Ph.D. Dissertation, Chemistry Department, Univesity of Hawai'i at Mānoa.

Swarbrick, J.T. 1997a. *Environmental weeds and exotic plants on Christmas Island, Indian Ocean: a report to Parks Australia*. Weed Science Consultancy.

Swarbrick, J.T. 1997b. *Weeds of the Pacific Islands*. Technical paper No. 209. Noumea, New Caledonia: South Pacific Commission.

Taboroši, D. and E. Allen. 2015. *Water for Life: A Pacific Island Handbook for Education, Health, Community Resilience*. Pohnpei: Island Research & Education Initiative.

Taboroši, D. and M. Sánchez Collazo. 2011. *Mwoakilloa Atoll: freshwater resources and their usage, state, and infrastructure*. Pohnpei: Island Research & Education Initiative.

Tan, P.Y., Corlett, R.T. and H.T.W. Tan (eds.). 2010. *A Field Guide to the Native Garden @HortPark: an Urban Oasis of the Native Flora and Fauna of Singapore*. Singapore: Centre for Urban Greenery and Ecology (National Parks Board) and National University of Singapore.

Teschke, R., Sarris, J., Glass, X. and J. Schulze. 2011. Kava, the anxiolytic herb: back to basics to prevent liver injury? *British Journal of Clinical Pharmacology* 71(3): 445–448.

Thaman, R.R. 1990. Kiribati Agroforestry: Trees, People and the Atoll Environment. *Atoll Research Bulletin* 333: 1–29.

Thaman, R.R. 1992a. Batiri Kei Baravi: The Ethnobotany of Pacific Island Coastal Plants *Atoll Research Bulletin* 361: 1–62.

Thaman, R. R. 1992b. Vegetation of Nauru and the Gilbert Islands: Case studies of poverty, degradation, disturbance, and displacement. *Pacific Science* 46(2): 128–158.

Thaman, R.R. 1994. Ethnobotany of Pacific Island Coastal Plants. In: J. Morrison, P. Geraghty and L. Crowl (eds.). *Science of Pacific Island Peoples: Fauna, Flora, Food and Medicine*, Vol. 3. Suva, Fiji: Institute of Pacific Studies, pp. 147–184.

Thaman, R.R., Elevitch, C.R. and J. Kennedy. 2006. Urban and homegarden agroforestry in the Pacific Islands: Current status and future prospects. In: Kumar, B.M. and P.K.R. Nair (eds.). *Tropical Homegardens. Advances in Agroforestry*, vol. 3. Springer, Dordrecht.

Thaman, R.R., Fosberg, F.R., Manner, H.I. and D.C. Hassall. 1994. Flora of Nauru. *Atoll Research Bulletin* 332: 1–223.

Thaman, R.R., Gregory, M. and S. Takeda. 2012 *Trees of Life: A Guide to Trees and Shrubs of the University of the South Pacific*. Suva, Fiji. University of the South Pacific.

Thaman, R.R. and P.M. Thomas. 1982. The *cassava* invasion: The cultural, nutritional and ecological impact of *cassava* on Pacific island food systems. In: R.M. Bourke and V. Kesavan (eds.). *Proceedings of the Second Papua New Cuinea Food Crops Conference: Part Two*. Port Moresby: Department of Primary Industries, pp. 330–350.

Thaman, R.R. and P.M. Thomas. 1985. *Cassava* and change in Pacific island food systems. In: D.J. Cattle, and K.H. Schwerin (eds.). *Food Energy in Tropical Ecosystems*. New York: Gordon and Breach, pp. 191–228.

Thaman, R.R., Thomson, L.A.J., DeMeo, R., Areki, F. and C.R. Elevitch. 2006. *Intsia bijuga* (vesi). In: Elevitch, C.R. (ed.). *Traditional Trees of Pacific Islands: their Culture, environment, and Use*. Holualoa, Hawai'i: Permanent Agriculture Resources, pp. 225–240.

Thaman R.R. and W.A. Whistler. 1996. *A Review of Uses and Status of Trees and Forests in Land-Use Systems in Samoa, Tonga, Kiribati and Tuvalu with Recommendation for Future Action*. South Pacific Forestry Development Programme, Suva, Fiji.

Thompson, A. 2011. *The Early Settlement of Mwoakilloa Atoll*. Historic Preservation Funds, National Park Service, U.S. Dept. of Interior, as well as the Government of the Federated States of Micronesia and the Government of Pohnpei. https://www.academia.edu/12305142/The_Early_Settlement_of_Mwoakilloa_Atoll.

Thompson, S. 1982. *Cyrtosperma chamissonis* (Araceae): Ecology, distribution, and economic importance in the South Pacific. *Journal d'agriculture traditionnelle et de botanique appliquée* 29(2): 185–203.

Thomson, L.A.J., Englberger, L., Guarino, L., Thaman, R.R. and C.R. Elevitch. 2006. Pandanus tectorius (pandanus). In: Elevitch, C.R. (ed.). *Traditional Trees of Pacific Islands: their Culture, environment, and Use*. Holualoa, Hawai'i: Permanent Agriculture Resources, pp. 563–590.

Thothathri, K. 1982. Leguminosae: Genus *Derris*. *Fascicles of Flora of India* 8: 1–33.

Tomlinson, P.B. 1986. *The Botany of Mangroves*. Cambridge, U.K.: Cambridge University Press.

Torres, M.A.O., Magalhães, I., Mondêgo-Oliveira, R., de Sá, J.C., Rocha, A.L. and A.L. Abreu-Silva. 2017. One Plant, Many Uses: A Review of the Pharmacological Applications of *Morinda citrifolia*. *Phytotherapy Research* 31(7): 971–979.

Trustrum, N.A. 1996. *Pohnpei's watershed spatial plan and management guidelines* (Consultant's report prepared for the Asian Development Bank, T.A. no. FSM-1925; Watershed Management and Environment). Landcare Research, New Zealand, Ltd.

Tsuda, R.T., Fosberg, F.R. and M.-H. Sachet. 1977. Distribution of seagrasses in Micronesia. *Micronesica* 13: 191–198.

Tsuda, R.T. and N. Sukhraj. 2016. Reassessment of Seagrass Species in the Marshall Islands. *Micronesica* 4: 1–10.

Tummons, P. 2013. All About Albizia (*Falcataria moluccana*). *Environment Hawai'i*. http://www.environment-hawaii.org/

Tuyama, T. 1941, *On the Names of Plants of the Palau Group [Nomina vernaula in insula Palau (Pelew), Caroline, Micronesia* [original title: *Kagaku Nanyo* (The Science of the South Sea), Part I, Vol. 3, No. 3, pp. 1–17; Part II, Vol. 4, No. 1, pp. 15–35; Part III, Vol. 4, No. 2, pp. 97–107. Translated by Ryuji Endo, Sept. 1948. "T" book of H. St. John miscellaneous papers].

Valenzuela, H. 2011 (revised). Farm and Forestry Production and Marketing Profile for Chili Pepper (*Capsicum annuum*). In: Elevitch, C.R. (ed.) *Specialty Crops for Pacific Island Agroforestry.* Permanent Agriculture Resources (PAR), Holualoa, Hawai'i.

Verma, V.M. 2014. *Soft Taro Cultivation Guide*. College of Micronesia, Kosrae Agricultural Experiment Station, Federated States of Micronesia. www.COM-FSM-Soft-Taro-Cultivation-Guide-Verma-2014.

Verma, V.M. 2014a. *Sweet Potato Cultivation Guide*. College of Micronesia, Kosrae Agricultural Experiment Station. Kosrae: Federated States of Micronesia. www.COM-FSM-Sweet-Potato-Cultivation-Guide-Verma-2014-1

Verma, V.M. 2018. Integrated research and outreach on in vitro multiplication, sustainable, organic and climate-smart commercial cultivation of black pepper. Micronesia Land Grants.

Victor, N. 2000. *Dioscorea bulbifera* [on Pohnpei]. http://www.comfsm.fm/~dleeling/botany/2000/vhp/dioscoreabulbifera.html

Wagner, W.L., Herbst, D.R. and S.H. Sohmer. 1999. *Manual of the flowering plants of Hawai'i*. Revised edition. Bernice P. Bishop Museum special publication. Honolulu: University of Hawai'i Press/Bishop Museum Press, 2 volumes, 1st Edition 1990.

Wagner, W.L., Hoch, P.C. and P.H. Raven. 2007. Revised classification of the Onagraceae. *Systematic Botany Monographs* 83: 1–240.

Wagner, W. L. and D.H. Lorence. 2011. A nomenclator of Pacific oceanic island *Phyllanthus* (Phyllanthaceae), including *Glochidion*. *PhytoKeys* 4: 674.

Waldien, D.L., Wiles, G.J., Buden D.W. and B. Obispo. 2020. *Pteropus molossinus*, Pohnpei flying fox. *IUCN Red List of Threatened Species*. e.T18741A22084572

Walter, A. and V. Lebot. 2007. *Gardens of Oceania*. ACIAR Monograph. No. 122 [Translated from the French by P. Ferrar]. Australian Centre for International Agricultural Research. First published in French by IRD Éditions and CIRAD as *Jardins d'Océanie*, 2003.

Warnakulasuriya, S., Trivedy, C. and T.J. Peters. 2002. Areca nut use: an independent risk factor for oral cancer: The health problem is under-recognised. *BMJ* (formerly the *British Journal of Medicine*) 324(7341): 799–800.

Weber, E. 2003. *Invasive plants of the World*. CABI Publishing, CAB International, Wallingford, UK.

Weightman, B.L. and I.M. Moros. 1982. The cultivation of taro *Xanthosama* sp. In: M. Lambert. Taro Cultivation in the South Pacific. SPC Handbook, No 22. South Pacific Commission, Noumea, New Caledonia, pp. 75–83.

Weisler, M.I. 1999a. The antiquity of aroid pit agriculture and significance of buried A Horizons on Pacific Atolls. *Geoarchaeology* 14: 621–654.

Weisler, M.I. 1999b. Atolls as settlement landscapes: Ujae, Marshall Islands. *Atoll Research Bulletin* 460: 1–51.

Weller, S., Sakai, A., Clark, M. Lorence, D.H., Flynn, T., Kishida. W., Tangalin, N. and K. Wood. 2018. The effects of introduced ungulates on native and alien plant species in an island ecosystem: Implications for change in a diverse mesic forest in the Hawaiian Islands. *Forest Ecology and Management* 409: 528–526.

Whistler, W.A. 1990. Ethnobotany of the Cook Islands: the plants, their Maori names and their uses. *Allertonia* 5(4): 347–424.

Whistler, W.A. 1991. Polynesian plant introductions. In: Cox, P.A and S.A. Banack. I*slands, Plants and Polynesians: an Introduction to Polynesian Ethnobotany.* Portland, OR: Dioscorides Press, pp. 41–46.

Whistler, W.A. 1992a. *Polynesian Herbal Medicine*. Lawai, Kaua'i: National Tropical Botanical Garden.

Whistler, W.A. 1992b. *Flowers of the Pacific Islands Seashore*. Honolulu: Isle Botanica.

Whistler, W. A. 2000. *Tropical ornamentals: a guide*. Portland, Oregon: Timber Press

Whistler, W.A. 2009. *Plants of the Canoe People: An Ethnobotanical Voyage through Polynesia*. Lawai, Kaua'i: National Tropical Botanical Garden.

Whistler, W.A. and C.R. Elevitch. 2006a. *Casuarina equisetifolia* (beach-oak) and *C. cunninghamiana* (river she-oak). In: Elevitch, C.R. (ed.). 2006. *Traditional Trees of Pacific Islands: their Culture, environment, and Use*. Holualoa, Hawai'i: Permanent Agriculture Resources, pp. 227–242. <http://www.traditionaltree.org>

Whistler, W.A. and C.R. Elevitch. 2006b. *Erythrina variagata* (Coral Tree). In: Elevitch, C.R. (ed.). 2006. *Traditional Trees of Pacific Islands: their Culture, environment, and*

Use. Holualoa, Hawai'i: Permanent Agriculture Resources, pp. 329–344. <http://www.traditionaltree.org>

Whistler, W.A. and C.R. Elevitch. 2006c. *Fagraea berteroana* (pua kenikeni). In: Elevitch, C.R. (ed.). 2006. *Traditional Trees of Pacific Islands: their Culture, environment, and Use*. Holualoa, Hawai'i: Permanent Agriculture Resources, pp. 345–354. <http://www.traditionaltree.org>

Whistler, W.A. and C.R. Elevitch. 2006d. *Syzygium malaccense* (Malay apple), ver. 2.1. In: Elevitch, C.R. (ed.). *Species Profiles for Pacific Island Agroforestry*. Permanent Agriculture Resources (PAR), Hōlualoa, Hawai'i. <http://www.traditionaltree.org>

Whittaker, R.J., Bush, M.B. and K. Richards. 1989. Plant recolonization and vegetation succession of the Krakatau Islands, Indonesia. *Ecological Monographs* 59(2): 59–123.

Wiens, H. J. 1956. The geography of Kapingamarangi Atoll in the Eastern Carolines. *Atoll Research Bulletin* 48: 1–86.

Wiens, H. J. 1957. Field notes on atolls visited in the Marshalls, 1956. *Atoll Research Bulletin* 54: 1–23.

Wiens, H.J. 1962. *Atoll environment and Ecology*. New Haven: Yale University Press.

Wiggins, I. L. and D. M. Porter. 1971. *Flora of the Galapagos Islands*. Palo Alto: Stanford University Press.

Wilder, G.P. 1928. The Breadfruit of Tahiti. *Bernice P. Bishop Museum Bulletin* 50: 1–83.

Wilkinson, K.M. and C.R. Elevitch. 1994. *Agroforestry Guides for Pacific Islands*. Permanent Agriculture Resources.

Williams, D.R., Clark, M., Buchanan, G.M., Ficetola, G.F., Rondinini, C. and D. Tilman. 2020. Proactive conservation to prevent habitat losses to agricultural expansion. *Nature Sustainability*, pp. 1–9.

Wilson, A.M and L.S. Hamilton. 1988. *A Practical Guide to Identifying Yams. The Main Species of Dioscorea in the Pacific Islands*. USP Institute for Research, Extension and Training in Agriculture (IRETA). Apia: Western Samoa.

Winter, W.P. de and V B. Amoroso (eds.). 2003. *Plant resources of South-East Asia No. 15(2). Cryptogams: Ferns and fern allies*. PROSEA Foundation, Borgor, Indonesia.

Wong, B.J. No Date. Effects of an invasive tree species (*Falcataria moluccana*) on understory diversity in Mo'orea, French Polynesia. MS submitted at University of California at Berkeley (also see Hughes et al. 2006 above).

World Health Organization. 1998. *Medicinal Plants of the South Pacific*. Manila, Philippines: Publications Office, World Health Organization, Regional Office for the Western Pacific, No. 19.

World Health Organization. 2009. *Medicinal Plants in Papua New Guinea: Information on 126 commonly used medicinal plants in Papua New Guinea*. Manila, Philippines: Publications Office, World Health Organization, Regional Office for the Western Pacific.

World Wide Fund for Nature. 2020. See: https://wwf.panda.org/discover/our_focus/forests_practice/deforestation_causes2/

Wortel, O.L. 2010. *Federated States of Micronesia. Federated States of Micronesia Fourth National Report: Implementation of Article 6 of the Convention on Biological Diversity*. http://www.cbd.int/doc/world/fm/fm-nr-04-en.pdf.

Yamamoto, S. 2011. Use of *Capsicum frutescens* on Pohnpei Island, Mokil Atoll, and Pingelap Atoll, Federated States of Micronesia. *People and Culture in Oceania* 27: 87–104.

Yamamoto, S. 2014. Food Security in Small Islands: Case Studies in the Federated States of Micronesia. Kagoshima University Research Center for the Pacific Islands. *Occasional Papers* 54: 27–33.

Yamamoto, S. 2018. Long-term food consumption survey in the Federated States of Micronesia: a case study in Pingelap Island, Pohnpei State. *South Pacific Newsletter* 29: 6.

Yamamoto, S. 2019. Long-Term Survey of Food Consumption on Pingelap Island, Pohnpei State, the Federated States of Micronesia. *The Journal of Island Studies* 20(2): 141–154.

Yamamoto, S., Kawanichi, M. and S. Nishimura. 2015. Dietary Patterns and Food Consumption Survey in the Federated States of Micronesia: A Case Study in Pingelap Island, Pohnpei State. *Tropical Agriculture and Development* 59(4): 161–169.

Yan, Y-B., Duke, N.C. and M. Sun. 2016. Comparative Analysis of the Pattern of Population Genetic Diversity in Three Indo-West Pacific *Rhizophora* Mangrove Species. *Frontiers of Plant Science* 7: 1434. doi: 10.3389/fpls.2016.01434.

Yen, D.E. 1974. The Sweet Potato and Oceania: An Essay in Ethnobotany. *Bernice P. Bishop Museum Bulletin* 236.

Yusuf, M., Begum, J., Hoque, M.N. and J.U. Chowdhury. 2009. *Medicinal Plants of Bangladesh* (Revised and Enlarged). Andarkilla, Chittagong, Bangladesh: Paramount Printing Service.

Zerega, N.J.C. and T.J. Motley. 2004. Complex origins of breadfruit (*Artocarpus altilis*, Moraceae): Implications for human migrations in Oceania. *American Journal of Botany* 91(5): 760–766.

Zerega, N.J.C., Ragone, D. and T.J. Motley. 2005. Systematics and Species Limits of Breadfruit (*Artocarpus*, Moraceae). *Systematic Botany* 30(3): 603–615.

Zerega, N.J.C., Ragone, D. and T.J. Motley. 2006. Breadfruit origins, diversity, and human-facilitated distribution. In T. J. H. Motley and N. Zerega (eds.) *Darwin's Harvest: Origins, Evolution, and Conservation of Crop Plants*. New York: Columbia University Press, pp. 213–238.

Zumbroich, T. J. 2007-2008. The origin and diffusion of betel chewing: A synthesis of evidence from South Asia, Southeast Asia and beyond. *Electronic Journal of Indian Medicine* 1: 87–140.

IMAGE CREDITS

All artwork, diagrams, maps, and illustrations featured in this volume were produced by the Island Research & Education Initiative (iREi). Among the individual contributors, we especially acknowledge Rosa Marín Ribas, who created the majority of the botanical illustrations, and Ewa Kmieciak, who designed the book and its layout. All photographs are by Danko Taboroši, except as noted here. We are deeply appreciative of the photographers whose work appears on the pages indicated below and remains their intellectual property and copyright.

pp. 1, 3 .... Aon168 / Shutterstock
p. 12 .... Mark Merlin
p. 17 .... Ho Thi Thanh Van
p. 30 .... "Kaselehlie Press"
p. 32 .... Patrick Nunn
p. 40 .... *Sakau en Nansapw* campaign
p. 54 .... Dawn Witherington
p. 76, *B. racemosa* .... Lienyuan Lee / Wikimedia Commons
p. 76, *B. aubertii* .... Coenobita / Wikimedia Commons
p. 76, *C. ramiflora* .... Lexazoeh / iNaturalist
p. 77, *R. corymbosa* .... Ong Jyh Seng / iNaturalist
p. 87, *M. indica* .... Shaylin Salas
p. 87, *M. esculenta* ..... Lotadegbemijowa / Wikimedia Commons
p. 89, *A. mariannensis* .... Halenhardy / Wikimedia Commons
p. 89, *C. odorata* .... B. Navez / Wikimedia Commons
p. 89, *F. tinctoria* .... Tau'olunga / Wikimedia Commons
p. 89, *G. ramiflorum* .... Michael J. Balick
p. 89, *M. hypargyraea* .. Hyongchan (David) / Wikimedia Commons
p. 93, *T. leontopetaloides* .... Anwar Mohamed / Shutterstock
p. 93, *T. catappa* .... A16898 / Wikimedia Commons
p. 103, *A. pavonina* .... Jessica / SNP
p. 103, *A. moluccana* .... Tony Rodd / Wikimedia Commons
p. 103, *C. bartramia* ... Wibowo Djatmiko / Wikimedia Commons
p. 103, *F. moluccana* .... Forest and Kim Starr
p. 103, *M. carolinensis* .... Kenneth R Wood / NTBG
p. 105, *S. spontaneum* .. Biswarup Ganguly / Wikimedia Commons
p. 109, *A. racemosa* .... Sofia's Farm
p. 109, *C. brevipetiolata* .... Conservation Society of Pohnpei
p. 109, *C. carolinense* .... Dana Lee Ling
p. 109, *E. carolinensis* .... Conservation Society of Pohnpei
p. 109, *F. berteroana* .... Dana Lee Ling
p. 109, *G. marianum* .... Conservation Society of Pohnpei
p. 109, *S. stelechantha* .... Conservation Society of Pohnpei
p. 113, *H. pallidum* .... P.B. Pelser and J.F. Barcelona
p. 113, *G. axillaris* .... P.B. Pelser and J.F. Barcelona
p. 113, *P. ponapensis* .... Conservation Society of Pohnpei
p. 113, *T. pandanophyllum* ..... Conservation Society of Pohnpei
p. 131, top .... bonhams.com
p. 135, bottom two .... British Museum
p. 136, top three .... Dana Lee Ling
p. 138 .... Sciencepics / Shutterstock
p. 161, bottom right .... Dana Lee Ling
p. 176, bottom right .... Stephen Fenech
p. 192 .... Alex Zuccarelli / Alamy
p. 199 .... Beate Münter / Alamy
p. 223 .... Bill Bachmann / Alamy
p. 303, top .... Greg Vaughn / Alamy
p. 303, karat closeup .... Travel Pix / Alamy
p. 303, bottom left .... Dana Lee Ling
p. 329 .... Boonchuay Iamsumang / Shutterstock
p. 355, top left .... Travel Pix / Alamy
p. 355, bottom .... ArteSub / Alamy
p. 391, *M. calvescens* .... Rachel Klyve

Black-and-white photograph on p. 29 is reproduced from Müller, Kilian. 1908. *Jahresbericht über die Tätigkeit der Kapuziner der rheinisch-westfälischen Kapuziner-Ordensprovinz auf den Karolinen, Marianen, und Palau-Inseln.* Limburg: Limburger Vereinsdruckerei.

Historical line engraving reproduced on p. 42 is from Hernsheim, Franz. 1883. *Südsee-Erinnerungen (1875–1880).* Berlin: Ernst Siegfried Mittler und Sohn, while those on p. 141 and p. 346 are from Hambruch, Paul, and Anneliese Eilers. 1932–1936. *Ponape.* 3 vols. In *Ergebnisse der Südsee-Expedition 1908–1910, II. Ethnographie: B. Mikronesien, Bd. 7.* Hamburg: Friederichsen, de Gruyter & Co.

INDEX OF PLANT SPECIES

These pages index all plant species mentioned in the book. Scientific names are listed alphabetically with plant families and page references; bold page numbers indicate main species accounts. Separate lists match local names from Pohnpei, Mwoakilloa, Pingelap, Nukuoro, and Kapingamarangi to corresponding scientific names.

SCIENTIFIC NAME INDEX

POHNPEI NAME INDEX

MWOAKILLOA NAME INDEX

PINGELAP NAME INDEX

NUKUORO NAME INDEX

KAPINGAMARANGI NAME INDEX

ABOUT THE AUTHORS

Mark D. Merlin

Mark Merlin is a distinguished ethnobotanist and Professor Emeritus at the University of Hawai'i at Mānoa. He studies how ancient and traditional cultures have utilized plants for many purposes, with a special focus on human alteration of native vegetation in tropical island ecosystems, particularly in Remote Oceania. He is the lead author of this book and the other volumes in this regional ethnobotanical and ecological series for several areas of Micronesia. His ardent hope is that these books will significantly help the Micronesian people preserve and continue to use their traditional ethnobiological knowledge effectively for generations to come.

Danko Taboroši

Danko Taboroši is the director of Island Research & Education Initiative (iREi), a Micronesia-based NGO. An environmental scientist and educator, he draws on experience from primary schools to universities to translate research findings into classroom-ready learning materials. He specializes in developing textbooks and educational resources that align with curriculum standards while remaining grounded in island cultures and local knowledge systems. He is especially passionate about reviving valuable educational materials from the past and making them accessible and relevant to new generations—an approach reflected in this book.

James O. Juvik

James Juvik is Professor Emeritus of Geography at the University of Hawai'i at Hilo. He has published extensively on the biogeography, climatology, and conservation of oceanic islands across the Indo-Pacific and Atlantic regions. He is co-author of the landmark Atlas of Hawai'i and other environmental education books used throughout the Pacific. His work includes several volumes in the Micronesian ethnobotany series published in the 1990s, and he is pleased to see these locally grounded educational resources renewed in fully revised and greatly expanded new editions, including this volume.